Geophysical Monograph Series

Including

IUGG Volumes
Maurice Ewing Volumes
Mineral Physics Volumes

Geophysical Monograph Series

77 **The Mesozoic Pacific: Geology, Tectonics, and Volcanism** *Malcolm S. Pringle, William W. Sager, William V. Sliter, and Seth Stein (Eds.)*

78 **Climate Change in Continental Isotopic Records** *P. K. Swart, K. C. Lohmann, J. McKenzie, and S. Savin (Eds.)*

79 **The Tornado: Its Structure, Dynamics, Prediction, and Hazards** *C. Church, D. Burgess, C. Doswell, R. Davies-Jones (Eds.)*

80 **Auroral Plasma Dynamics** *R. L. Lysak (Ed.)*

81 **Solar Wind Sources of Magnetospheric Ultra-Low Frequency Waves** *M. J. Engebretson, K. Takahashi, and M. Scholer (Eds.)*

82 **Gravimetry and Space Techniques Applied to Geodynamics and Ocean Dynamics (IUGG Volume 17)** *Bob E. Schutz, Allen Anderson, Claude Froidevaux, and Michael Parke (Eds.)*

83 **Nonlinear Dynamics and Predictability of Geophysical Phenomena (IUGG Volume 18)** *William I. Newman, Andrei Gabrielov, and Donald L. Turcotte (Eds.)*

84 **Solar System Plasmas in Space and Time** *J. Burch, J. H. Waite, Jr. (Eds.)*

85 **The Polar Oceans and Their Role in Shaping the Global Environment** *O. M. Johannessen, R. D. Muench, and J. E. Overland (Eds.)*

86 **Space Plasmas: Coupling Between Small and Medium Scale Processes** *Maha Ashour-Abdalla, Tom Chang, and Paul Dusenbery (Eds.)*

87 **The Upper Mesosphere and Lower Thermosphere: A Review of Experiment and Theory** *R. M. Johnson and T. L. Killeen (Eds.)*

88 **Active Margins and Marginal Basins of the Western Pacific** *Brian Taylor and James Natland (Eds.)*

89 **Natural and Anthropogenic Influences in Fluvial Geomorphology** *John E. Costa, Andrew J. Miller, Kenneth W. Potter, and Peter R. Wilcock (Eds.)*

90 **Physics of the Magnetopause** *Paul Song, B.U.Ö. Sonnerup, and M.F. Thomsen (Eds.)*

91 **Seafloor Hydrothermal Systems: Physical, Chemical, Biological, and Geological Interactions** *Susan E. Humphris, Robert A. Zierenberg, Lauren S. Mullineaux, and Richard E. Thomson (Eds.)*

92 **Mauna Loa Revealed: Structure, Composition, History, and Hazards** *J. M. Rhodes and John P. Lockwood (Eds.)*

93 **Cross-Scale Coupling in Space Plasmas** *James L. Horwitz, Nagendra Singh, and James L. Burch (Eds.)*

94 **Double-Diffusive Convection** *Alan Brandt and H. J. S. Fernando (Eds.)*

95 **Earth Processes: Reading the Isotopic Code** *Asish Basu and Stan Hart (Eds.)*

96 **Subduction Top to Bottom** *Gray E. Bebout, David Scholl, Stephen Kirby, and John Platt (Eds.)*

97 **Radiation Belts: Models and Standards** *J. F. Lemaire, D. Heynderickx, and D. N. Baker (Eds.)*

98 **Magnetic Storms** *Bruce T. Tsurutani, Walter D. Gonzalez, Yohsuke Kamide, and John K. Arballo (Eds.)*

99 **Coronal Mass Ejections** *Nancy Crooker, Jo Ann Joselyn, and Joan Feynman (Eds.)*

100 **Large Igneous Provinces** *John J. Mahoney and Millard F. Coffin (Eds.)*

101 **Properties of Earth and Planetary Materials at High Pressure and Temperature** *Murli Manghnani and Takehiki Yagi (Eds.)*

102 **Measurement Techniques in Space Plasmas: Particles** *Robert F. Pfaff, Joseph E. Borovsky, and David T. Young (Eds.)*

103 **Measurement Techniques in Space Plasmas: Fields** *Robert F. Pfaff, Joseph E. Borovsky, and David T. Young (Eds.)*

104 **Geospace Mass and Energy Flow: Results From the International Solar-Terrestrial Physics Program** *James L. Horwitz, Dennis L. Gallagher, and William K. Peterson (Eds.)*

105 **New Perspectives on the Earth's Magnetotail** *A. Nishida, D. N. Baker, and S. W. H. Cowley (Eds.)*

106 **Faulting and Magmatism at Mid-Ocean Ridges** *W. Roger Buck, Paul T. Delaney, Jeffrey A. Karson, and Yves Lagabrielle (Eds.)*

107 **Rivers Over Rock: Fluvial Processes in Bedrock Channels** *Keith J. Tinkler and Ellen E. Wohl (Eds.)*

108 **Assessment of Non-Point Source Pollution in the Vadose Zone** *Dennis L. Corwin, Keith Loague, and Timothy R. Ellsworth (Eds.)*

109 **Sun-Earth Plasma Connections** *James L. Burch, Robert L. Carovillano, and Spiro K. Antiochos (Eds.)*

110 **The Controlled Flood in Grand Canyon** *Robert H. Webb, John C. Schmidt, G. Richard Marzolf, and Richard A. Valdez (Eds.)*

Geophysical Monograph 111

Magnetic Helicity in Space and Laboratory Plasmas

Michael R. Brown
Richard C. Canfield
Alexei A. Pevtsov
Editors

American Geophysical Union
Washington, DC

Published under the aegis of the AGU Books Board

Library of Congress Cataloging-in-Publication Data

Magnetic helicity in space and laboratory plasmas / Michael R. Brown
   Richard C. Canfield, Alexei A. Pevtsov, editors.
      p. cm. -- (Geophysical monograph ; 111)
      Includes bibliographical references.
      ISBN 0-87590-094-1
      1. Magnetic reconnection. 2. Particles (Nuclear physics)-
   -Helicity. 3. Plasma (Ionized gases) 4. Plasma astrophysics.
   I. Brown, Michael Riley, 1959- . II. Canfield, Richard C.
   III. Pevtsov, Alexei A. IV. Series.
   QC809.P5M19 1999
   530.4'42--dc21   99-26374
                                                              CIP

ISBN 0-87590-094-1
ISSN 0065-8448

Copyright 1999 by the American Geophysical Union
2000 Florida Avenue, N.W.
Washington, DC 20009

Figures, tables, and short excerpts may be reprinted in scientific books and journals if the source is properly cited.

Authorization to photocopy items for internal or personal use, or the internal or personal use of specific clients, is granted by the American Geophysical Union for libraries and other users registered with the Copyright Clearance Center (CCC) Transactional Reporting Service, provided that the base fee of $1.50 per copy plus $0.35 per page is paid directly to CCC, 222 Rosewood Dr., Danvers, MA 01923. 0065-8448/99/$01.50+0.35.

This consent does not extend to other kinds of copying, such as copying for creating new collective works or for resale. The reproduction of multiple copies and the use of full articles or the use of extracts, including figures and tables, for commercial purposes requires permission from the American Geophysical Union.

Printed in the United States of America.

CONTENTS

Preface
Michael R. Brown, Richard C. Canfield, and Alexei A. Pevtsov ix

Utility of the Helicity Concept

Magnetic Helicity in Space Physics
Mitchell A. Berger 1

Implications of Magnetic Helicity Conservation
Allen H. Boozer 11

Influence of Geometry and Topology on Helicity
Jason Cantarella, Dennis DeTurck, Herman Gluck, Mikhail Teytel 17

Magnetic Energy and Helicity in Open Systems
B.C. Low 25

Helicity and Its Role in the Varieties of Magnetohydrodynamic Turbulence
David C. Montgomery and Jason W. Bates 33

Dynamos, Helicity, and the Solar Interior

Planetary Dynamos and Helicities
K.-H. Rädler 47

Helicity, Relaxation, and Dynamo in a Laboratory Plasma
S.C. Prager 55

Helicity in Large-Scale Dynamo Simulations
Axel Brandenburg 65

Creation of Twist at the Core-Convection Zone Interface
Peter A. Gilman and Paul Charbonneau 75

Flows in the Solar Convection Zone
A. G. Kosovichev 83

Twisted Flux Tubes and How They Get That Way
Dana Longcope, Mark Linton, Alexei Pevtsov, George Fisher, and Isaac Klapper 93

Helicity of the Photospheric Magnetic Field
Alexei A. Pevtsov and Richard C. Canfield 103

Balance and Solar-Cycle Variations of Magnetic Helicity
Alexander Ruzmaikin 111

CONTENTS

Plasma Relaxation and Helicity Conservation

Magnetic Helicity and Relaxation: Theory
Paul M. Bellan 119

Study of Magnetic Helicity and Relaxation Phenomena in Laboratory Plasmas
Masaaki Yamada 129

Magnetic Helicity and Relaxation Phenomena in the Solar Corona
E.R. Priest 141

Magnetic Helicity and Stability in Solar Corona
K. Kusano 149

The Role of Helicity in Magnetic Reconnection

The Evolution of Magnetic Helicity Under Reconnection
G. Hornig 157

Helicity, Reconnection, and Dynamo Effects
Hantao Ji 167

Measurements of Helicity and Reconnection in Electron MHD Plasmas
R.L. Stenzel, J.M. Urrutia, and M.C. Griskey 179

The Role of Helicity in Magnetic Reconnection: 3D Numerical Simulations
Spiro K. Antiochos and C. Richard DeVore 187

Helicity and Reconnection in the Solar Corona: Observations
Richard C. Canfield and Alexei A. Pevtsov 197

The Role of Reconnection in the Formation of Flux Ropes in the Solar Wind
J.T. Gosling 205

Solar Outer Atmosphere and Magnetosphere

Photospheric Motions as a Source of Twist in Coronal Magnetic Fields
A.A. van Ballegooijen 213

Magnetic Helicity in Solar Filaments and Coronal Mass Ejections
D.M. Rust 221

Solar Flares, Jets, and Helicity
Kazunari Shibata 229

Solar-Cycle, Radial and Latitudinal Variations of Magnetic Helicity: IMF Observations
Charles W. Smith 239

CONTENTS

Magnetic Helicity and Homogeneous Turbulence Models
William H. Matthaeus 247

Role of Magnetic Helicity in Cosmic Ray Scattering
John W. Bieber 257

The Role of Magnetic Helicity in Magnetospheric Physics
Andrew N. Wright 267

The Role of Coarse-Grained Helicity and Self-Organized Criticality in Magnetotail Dynamics
Tom Chang 277

Astrophysical Applications

Galactic and Accretion Disk Dynamos
Ethan T. Vishniac 285

Jets from Magnetized Accretion Disks
Ryoji Matsumoto 293

Conclusion

Magnetic Helicity in Space and Laboratory Plasmas: Editorial Summary
Michael Brown, Richard Canfield, George Field, Russell Kulsrud, Alexei Pevtsov, Robert Rosner, and Norbert Seehafer 301

PREFACE

Using the concept of magnetic helicity, physicists and mathematicians describe the topology of magnetic fields: twisting, writhing, and linkage. Mathematically, helicity is related to linking integrals, which Gauss introduced in the 19th century to describe the paths of asteroids in the sky. In the late 1970s the concept proved to be critical to understand laboratory plasma experiments on magnetic reconnection, dynamos, and magnetic field relaxation. In the late 1980s it proved equally important in understanding turbulence in the solar wind and the interplanetary magnetic field. During the last five years interest in magnetic helicity has grown dramatically in solar physics, and it will continue to grow as observations of vector magnetic fields become increasingly sophisticated.

The contributors to this interdisciplinary volume are leading solar and space physicists, laboratory experimentalists, astrophysicists, and pure mathematicians well versed in experimental and theoretical aspects of helicity. They share a knowledge of the concept that has developed in a variety of scientific fields that do not share a common scientific literature, and gather it in this single volume for graduate students and researchers in these fields. The authors introduce magnetic helicity, define its mathematical basis, and clarify its limitations for physical applications. They discuss the role of helicity in the generation of magnetic fields by plasma motions, i.e., dynamo action—a process that is observed in the laboratory and studied in stellar and planetary interiors, accretion disks, and galaxies. The authors describe magnetic helicity conservation—one of the attractive properties of the concept—and how it can be used to infer the topology of flows which cannot be observed directly. As well, they address another important issue—the role of helicity in magnetic reconnection—from the points of view of both laboratory and naturally occurring plasmas. They describe aspects of magnetic helicity that are relevant to the solar outer atmosphere and heliosphere, the generation of solar eruptions, and the interaction of magnetized solar plasmas with the Earth's magnetosphere. After a brief description of selected astrophysical applications they summarize current topics, from discussions at a recent Chapman conference, that will be important to solar and space physics, laboratory physics, and astrophysics in the coming millennium.

The Editors thank the following for their timely reviews: S.K. Antiochos, L. Bargatze, P.M. Bellan, M.A. Berger, J. Birn, A.H. Boozer, A. Brandenburg, P.K. Browning, C.R. DeVore, G.B. Field, T.G. Forbes, C.B. Forest, H.R. Gluck, S.E. Gibson, P.A. Gilman, J.T. Gosling, M.J. Hagyard, G. Hornig, H. Ji, R.M. Kiehn, I. Klapper, D. Kondrashov, A.G. Kosovichev, P. R. Kotiuga, R. Kulsrud, K. Kusano, P. Laurence, P. Liewer, A.W. Longbottom, D.W. Longcope, R. Matsumoto, W.H. Matthaeus, M.S. Miesch, D.C. Montgomery, A. Otto, C. Parnell, C.L. Rousculp, S.C. Prager, E.R. Priest, A.D. Roberts, D.M. Rust, A. Ruzmaikin, K. Schatten, N. Seehafer, K. Shibata, C.W. Smith, R.L. Stenzel, A.A. van Ballegooijen, B. Vasquez, E.T. Vishniac, P.G. Watson, J.G. Wissink, M. Yamada.

Michael Brown
Swarthmore College
Swarthmore, Pennsylvania

Richard Canfield
Montana State University
Bozeman, Montana

Alexei Pevtsov
Montana State University
Bozeman, Montana

Magnetic Helicity in Space Physics

Mitchell A. Berger

Mathematics, University College London, United Kingdom

The origins of magnetic helicity go back to Gauss in the early 19th century. This chapter traces the early history of magnetic helicity in the 1950s to the 1980s. We discuss the relation to field topology and to minimum energy Taylor states. The approximate conservation of helicity during reconnection is outlined. Also, we discuss how helicity is defined in open volumes and how helicity can be transferred across boundaries.

1. GAUSS AND ASTEROID TRACKING IN THE EARLY 19TH CENTURY

The origins of magnetic helicity go back to Karl Friedrich Gauss, the great 19th century mathematician. Gauss discovered a remarkable integral formula for the linking number of two curves. The motivation for this work derived from a space physics problem – the tracking of asteroids and comets [Epple 1998]. An astronomer who determines an asteroid's orbit about the sun must then map this orbit into a path across the sky, so that fellow astronomers know where to point their telescopes. Gauss attacked this mapping problem, and virtually founded a new branch of mathematics as a result. He showed that if the asteroid's orbit did not link the Earth's, then the path would be restricted to a range of latitudes. If the orbits were linked, on the other hand, then eventually all points on the celestial sphere would be visited by the asteroid.

One would expect solar system orbits to link at most once. However, the Gauss linking number has the power to measure multiple linking. Figure 1 shows two thin tubes with linking number $L_{12} = -3$. One of the tubes is circular, and the other passes through the circle three times. Because each tube has a direction (as shown by the arrows), the linking number has a sign. Curl the fingers of the right hand in the direction of the circular tube; if the thumb points in the direction of the second tube as it passes through the circle, then the linking is positive. Note that the linking number is symmetric, i.e. $L_{12} = L_{21}$.

The actual integral formula for L_{12} is useful for proving theorems, but people rarely directly calculate with it. For the record, it is

$$L_{12} = -\frac{1}{4\pi} \oint_1 \oint_2 \frac{d\mathbf{x}}{d\sigma} \cdot \frac{\mathbf{r}}{r^3} \times \frac{d\mathbf{y}}{d\tau} \, d\tau \, d\sigma \qquad (1)$$

where σ parametrises curve 1, τ parametrises curve 2, $\mathbf{x}(\sigma)$ and $\mathbf{y}(\tau)$ are points on curves 1 and 2, and $\mathbf{r} = \mathbf{x} - \mathbf{y}$ is the relative position vector. Gauss later employed his integral in studies of linked electrical circuits.

In the later part of the 19th century ideas in fluid mechanics led to the development of a general mathematical theory of knots and links. Maxwell, Tait, Lord Kelvin, and others sought a mechanical theory of the luminiferous aether, a medium thought to permeate space which carried light waves. Kelvin hoped that atomic theory could also be explained with the aether – atoms were visualized as knotted or linked vortex rings in the fluid aether, a different atom for each knot type ([Ricca & Berger 1996]). While these ideas did not succeed, they nevertheless stimulated early research into vortex dynamics.

Magnetic Helicity in Space and Laboratory Plasmas
Geophysical Monograph 111
Copyright 1999 by the American Geophysical Union

Figure 1. Two tubes with linking number -3.

2. HELICITY INTEGRALS

Elsasser (1956) first drew attention to the integral $H = \int \mathbf{A} \cdot \mathbf{B}\, d^3x$, where $\mathbf{B} = \nabla \times \mathbf{A}$, in a survey of magnetohydrodynamics, pointing out that H was conserved in ideal MHD. Kruskal and Kulsrud (1958) showed how this integral could be calculated for a field inside a torus, where the field lines live on nested flux surfaces. At the same time Woltjer (1958) brought the magnetic helicity integral H to the attention of the space physics and astrophysics community. Woltjer reasoned that static magnetic structures, such as quiescent loops and long-lived prominences in the solar corona, may be near equilibrium states for the field (of course observations now show quite a bit of dynamics and plasma flows inside these 'static' structures). An equilibrium state minimizes the energy given a set of physical constraints such as boundary conditions. Woltjer thought that the ideal invariant H should also serve as a constraint. Thus Woltjer tried to minimize

$$E = \frac{1}{8\pi}\int B^2\, d^3x - \frac{\alpha}{8\pi}\int \mathbf{A}\cdot\mathbf{B}\, d^3x \qquad (2)$$

where α is a Lagrange multiplier. This leads to the equation

$$\nabla \times \mathbf{B} = \alpha \mathbf{B}. \qquad (3)$$

Fields satisfying this equation are force-free, as the Lorentz force $\mathbf{J} \times \mathbf{B}$ vanishes. Also α is a constant, unlike more general force-free fields where $\nabla \times \mathbf{B} = \lambda(\mathbf{x})\mathbf{B}$. Sometimes constant α fields are called linear force-free fields because equation 3 is a linear PDE; thus two solutions with the same α can be summed to obtain a new solution. Woltjer's pioneering work demonstrated both the importance of helicity as a constraint, and the special place of the constant α field.

Moffatt (1969) elucidated the relation between helicity integrals and the Gauss linking number, and began a series of important papers advocating the use of topological concepts in fluid mechanics. The vector potential $\mathbf{A}$ appearing in the helicity integral is rather unpleasant, as it can be changed by a gauge transformation. If, however, we use the Biot-Savart formula to evaluate $\mathbf{A}$ (in Coulomb gauge), then

$$\mathbf{A}(\mathbf{x}) = -\frac{1}{4\pi}\int \frac{\mathbf{r}}{r^3} \times \mathbf{B}(\mathbf{x}')\, d^3x', \qquad (4)$$

where $\mathbf{r} = \mathbf{x} - \mathbf{x}'$. The helicity transforms into a double integral form similar to equation (1):

$$H = -\frac{1}{4\pi}\int\int \mathbf{B}(\mathbf{x}) \cdot \frac{\mathbf{r}}{r^3} \times \mathbf{B}(\mathbf{x}')\, d^3x\, d^3x'. \qquad (5)$$

The double integral form leads to the interpretation of magnetic helicity as the double sum of linking numbers *over all pairs of field lines*.

One may quibble, of course. A magnetic field contains an infinite number of field lines! It is preferable to think of dividing the field into a finite number N of small flux tubes. Each tube has flux $\Phi_i, i = 1\ldots N$, and tubes i and j have linking number L_{ij}. Then one can show

$$H = \sum_{i=1}^{N}\sum_{j=1}^{N} L_{ij}\Phi_i\Phi_j. \qquad (6)$$

Describing the field as N interlinked tubes may give only an approximation to the field, however large N

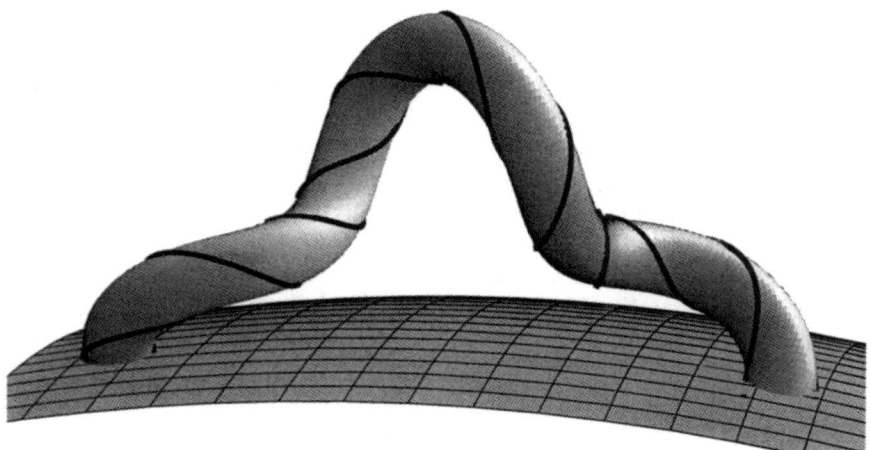

Figure 2. A kinked loop with writhe $Wr = 0.7\Phi^2$ and twist $Tw = 2.3\Phi^2$ for total helicity of $3\Phi^2$.

may be. Fortunately, the double sum converges nicely to the integral form in the limit $N \to \infty$ [Arnold & Khesin 1992].

If we do use the vector potential $\mathbf{A}$, then we must make sure that our results are gauge-invariant. First, the expression $H = \int \mathbf{A} \cdot \mathbf{B} \, d^3x$ is only valid when the volume of integration is bounded by a magnetic surface S, i.e. where $\mathbf{B} \cdot \hat{n}|_S = 0$. Inside such a surface a gauge variation of the vector potential of the form $\mathbf{A} \to \mathbf{A} + \nabla\psi$ changes H by

$$\delta H = \int \nabla\psi \cdot \mathbf{B} \, d^3x = \int \nabla \cdot \psi\mathbf{B} \, d^3x \quad (7)$$

$$= \oint \psi \mathbf{B} \cdot \hat{n}|_S \, d^2S = 0. \quad (8)$$

However, magnetic structures in the solar atmosphere have roots in the photosphere, where $\mathbf{B} \cdot \hat{n}|_S \neq 0$.

Other bad things happen if the volume of integration $\mathcal{V}$ has holes like a doughnut or torus. Then one can find vector fields $\mathbf{G}$ where $\nabla \times G = 0$ but $\mathbf{G} \neq \nabla\phi$ for any single valued function ϕ. (For example let ϕ measure angle the long way around inside the torus. Then ϕ jumps by 2π each time you go around.) Gauge transformations $\mathbf{A} \to \mathbf{A} + \mathbf{G}$ give $\delta H \neq 0$ even when $\mathbf{B} \cdot \hat{n}|_S = 0$. These gauge difficulties were not fully resolved until 1984.

3. TWIST AND WRITHE

The discovery of DNA structure in the 1950s provided a new application for the Gauss linking integral. The two strands of the double helix twist about a common central axis. This central axis coils and super coils in order to fit the centimetres-long molecule into a microscopic cell. The geometry of the twisting and coiling affects how different pieces of DNA reconnect with each other (biologists call this process 'recombination' rather than 'reconnection'). Consider for simplicity a DNA molecule which closes upon itself. We can calculate the linking number L of the two strands according to equation (1). Secondly, we can calculate the net angle Tw through which a strand twists about the molecular axis. Finally, we can also evaluate equation (1) for the axis alone, i.e. with both line integrals along the same axis. This last quantity is called the writhe Wr, and measures the coiling or helical structure of the axis. Unlike L, neither Tw or Wr are topological invariants for a closed DNA molecule. In 1961 the Romanian mathematician Călugăreanu discovered the simple formula $L = Tw + Wr$ to relate these three quantities.

The Călugăreanu formula can also be applied to magnetic helicity [Berger & Field 1984; Moffatt & Ricca 1992]: the magnetic helicity of a flux tube can be decomposed into a term measuring the twist of field lines about a central axis, plus a term measuring the writhe of the axis (see figure 2).

4. SELF HELICITY AND MUTUAL HELICITY

It is often said that helicity is a global quantity, as the helicity density $\mathbf{A} \cdot \mathbf{B}$ is not gauge-invariant. On a truly global scale, we may wish to calculate the total magnetic helicity of the universe. By symmetry, this might be expected to be zero; but symmetry breaking in the early universe could give a net handedness to fields on a cosmic scale.

Can helicity be sensibly used on smaller scales? In other words, can we measure the helicities of subregions of space? The definition of helicity in open volumes will be discussed later; for the moment we wish to see how the total helicity inside a volume relates to the helicities contained in subregions of the volume. Recall that equation (6) gives the helicity of a field consisting of N flux tubes. As a generalization of equation (6), we divide $\mathcal{V}$ into N regions, where each region is bounded by a magnetic surface except possibly at $\mathcal{S}$. For example, $\mathcal{V}$ could be the corona with $\mathcal{S}$ the photosphere, and the N regions would be N loops or arcades rooted in the photosphere. The sum

$$H_\mathcal{V} = \sum_i^N \sum_j^N H_{ij} \qquad (9)$$

has the following meaning: each term H_{ij} is calculated assuming magnetic flux exists only in regions i and j, with the field set to zero elsewhere. A similar situation occurs in the study of electrical circuits. The energy of N wires carrying currents $I_1, \ldots I_N$ is proportional to $\sum_{i=1}^{N} \sum_{j=1}^{N} M_{ij} I_i I_j$, where M_{ii} is a self inductance and M_{ij} is a mutual inductance.

For closed magnetic tubes with $i \neq j$, $H_{ij} = L_{ij}\Phi_i\Phi_j$ can be called the mutual helicity. For $j = i$, the diagonal terms H_{ii} can be called the self helicities. The self helicities measure linking of field lines within the same tube. If the field lines in tube i all twist about a central axis through the same angle Θ, then $H_{ii} = T\Phi^2$ where $T = \Theta/2\pi = Tw + Wr$.

5. RELAXATION AND DISSIPATION

The topology of a field at any instant of time refers to all the details of how field lines wind and braid about each other. In ideal conditions (which never quite exist in the real world) field lines never reconnect or pass through each other. Thus links and knots amongst the field lines never change. If a magnetic field in some arbitrary initial state loses energy in ideal MHD (perhaps by transferring energy to kinetic energy which is then drained by viscosity), then the topology will be preserved. But the final state may have the wrong topology to be a linear force-free field. In general resistivity is needed to change the topology.

In 1974 Taylor suggested that the dynamics of a Reversed Field Pinch could be approximately described as a relaxation process conserving the magnetic helicity. He made two conjectures: first that magnetic helicity would be roughly conserved, even in the presence of resistivity, and second that no other topological invariant would survive the turbulent relaxation phase of the RFP. He thus addressed the difficulty ignored by Woltjer: that the initial topology of the field may not be the right one for the minimum energy state. In Taylor's theory, the resistivity cuts through field lines, rapidly changing the topology. But helicity as a sum of linking numbers between flux tubes might survive (even though individual flux tubes and linking numbers might change). If the relaxation does not alter the overall field topology sufficiently, then the final state may not be a linear force-free field. For example, tokamak disruptions seem to leave unchanged other topological quantities [Bhattacharjee & Dewar 1982].

Studies of MHD turbulence lent support to these ideas. Pouquet, Frisch, and Léorat (1976) found in a numerical simulation that magnetic helicity would pile up at low wavenumbers as time progressed. As the energy cascade proceeds to higher wavenumbers, this was called an *inverse cascade*. Of course, at low wavenumbers there is little dissipation. They also pointed out that if magnetic helicity and energy were expressed in terms of power spectra $H(k)$ and $E(k)$, then the spectra must satisfy the relation $E(k) \geq kH(k)$. Thus for a given amount of helicity it takes far less energy to stuff the helicity into small k wavenumbers than into large wavenumbers.

In 1984 theorems were found which put strict upper bounds on magnetic helicity dissipation (Berger 1984). These bounds are independent of the details of the reconnection geometry. In the absence of helicity transfer across boundaries (to be discussed below), the time derivative of helicity is

$$\frac{dH}{dt} = -2\int \eta \mathbf{J} \cdot \mathbf{B}\, d^3x \qquad (10)$$

where we have expressed the electric field in terms of the electric current and the magnetic diffusivity, $\mathbf{E} = \eta\mathbf{J}$.

Let $W = \int B^2 dV$ measure the magnetic energy. The Ohmic dissipation rate is $|dW/dt| = 2\int \eta J^2 dV$. The integrals of $\mathbf{j}\cdot\mathbf{b}$, B^2, and j^2 are related by a Schwartz inequality, leading to

$$\left|\frac{dH}{dt}\right| \leq \sqrt{2\eta W \left|\frac{dW}{dt}\right|}. \qquad (11)$$

Consider an isolated volume, with no transfer of energy or helicity across the boundary (or at least small transfer on the timescale of reconnection). We define a length scale $L \equiv |H|/W$, which measures the effective size of helical field structure. For example, $L \approx 0.31R$

for a linear force free field inside a spherical magnetic surface of radius R. We also define a dissipation time $\tau_d = L^2/\eta$.

Consider an arbitrary reconnection or dissipation process occuring over a time Δt. Integrating (11) over time gives

$$\left|\frac{\Delta H}{H}\right| \leq \sqrt{\frac{\Delta t}{\tau_d}}. \qquad (12)$$

This inequality shows that ΔH is negligible for any fast reconnection event, where $\Delta t \ll \tau_d$. For example, consider a solar flare with $\Delta t \approx 1000s$, $L \approx 10^3$ Km, and $\eta \approx 10^{-6} \text{km}^2 s^{-1}$. With these values $\tau_d \approx 10^{12} s$, and $|\Delta H/H| < 3 \times 10^{-5}$.

The helicity decay inequalities allow for non-uniform resistivity, even for non-Ohmic resistivity. Simply replace the coefficient η by $\bar{\eta} = \int \eta_{\|} B^2 \, d^3x$, where $\eta_{\|} = E_{\|}/J_{\|}$ measures the effective resistivity parallel to $\mathbf{B}$. Also, we can sharpen these limits if we allow ourselves the luxury of making assumptions about the reconnection geometry. In most reconnection models the action of resistivity takes place in regions of high current density which occupy only a small fraction ϵ of the total volume. We can account for this by replacing η by $\bar{\eta} = \epsilon \eta$. The result is $|\Delta H/H| \leq \sqrt{\epsilon \Delta t/\tau_d}$. Finally we note that the scaling with $\eta^{1/2}$ (and hence magnetic Reynolds number $R_m^{-1/2}$) of the upper bounds may not reflect the scaling of ΔH itself (If the upper bound starts higher than the true value, and decreases more slowly, then there is room for different scalings). For example, Freedman & Berger (1993) give geometrical arguments for an R_m^{-2} scaling.

6. HELICITY OF OPEN FIELDS

Recall that the original definition of helicity in terms of a vector potential only works inside a simply connected volume (no holes) bounded by a magnetic surface (where $\mathbf{B} \cdot \hat{n}|_\mathcal{S} = 0$). A torus or doughnut, however, does have holes. Consider the line integral of the vector potential $\oint \mathbf{A} \cdot d\boldsymbol{\ell}$ along a circuit the long way around. By Stoke's theorem this equals the magnetic flux enclosed by the circuit. Some of this flux lies inside the torus, and can be obtained from $\mathbf{A}$ simply by calculating $\mathbf{B} = \nabla \times \mathbf{A}$. However, some of the flux may be *outside* the torus, passing through the central hole. The dynamics of the field inside the torus should be completely independent of this exterior flux (only with the quantum mechanical Aharonov-Bohm effect can the exterior flux be detected). Thus we would like the helicity to be independent of the exterior flux also. But if we are not careful, the linking of the exterior flux with the axial flux inside the torus will contribute a term to the helicity. The external linking can be readily eliminated by subtracting the product of the two fluxes (see Taylor 1986).

This leaves the problem of volumes not bounded by magnetic surfaces, the most prominent example being the solar corona, where flux passes through the photospherical boundary. In 1984 this problem was solved [Berger & Field 1984; Jensen & Chu 1984; Finn & Antonsen 1985]. Here we will give two geometrical descriptions of the general helicity integral. They both give the same answers, but can be pictured in different ways.

6.1. Helicity Integrals in General

First, helicity integrals in general can measure the mutual linking of any two divergence free vector fields $\mathbf{V}$ and $\mathbf{W}$ ($\nabla \cdot \mathbf{V} = \nabla \cdot \mathbf{W} = 0$). The volume of integration will be called $\mathcal{V}$, with boundary $\mathcal{S}$. For now, we will assume that $\mathcal{S}$ provides a closed surface for both fields ($\mathbf{V} \cdot \hat{n}|_\mathcal{S} = \mathbf{W} \cdot \hat{n}|_\mathcal{S} = 0$). The mutual linking of the two fields is measured by

$$H_\mathcal{V}(\mathbf{V}, \mathbf{W}) \equiv \int_\mathcal{V} \mathbf{A}_\mathbf{V} \cdot \mathbf{W} \, d^3 x \qquad (13)$$

where $\nabla \times \mathbf{A}_\mathbf{V} = \mathbf{V}$. For example the magnetic helicity in $\mathcal{V}$ is $H_\mathcal{V}(\mathbf{B}, \mathbf{B})$. Similarly, given a velocity field $\mathbf{v}$ and its vorticity ω, the kinetic helicity is $H_\mathcal{V}(\omega, \omega) = \int \mathbf{v} \cdot \omega \, d^3 x$.

The general helicity formula (13) has two simple algebraic properties. First, it is linear in each term, for example

$$H_\mathcal{V}(a\mathbf{V}_1 + \mathbf{V}_2, \mathbf{W}) = aH_\mathcal{V}(\mathbf{V}_1, \mathbf{W}) + H_\mathcal{V}(\mathbf{V}_2, \mathbf{W}). \qquad (14)$$

Secondly, if $\mathbf{V}$ and $\mathbf{W}$ are closed, then one can show that the helicity integral is symmetric, i.e.

$$H_\mathcal{V}(\mathbf{V}, \mathbf{W}) = H_\mathcal{V}(\mathbf{W}, \mathbf{V}). \qquad (15)$$

If both $\mathbf{V}$ and $\mathbf{W}$ are closed, then $H_\mathcal{V}(\mathbf{V}, \mathbf{W})$ is gauge-invariant, as in equation (8). What if only one is closed? Suppose in particular, that $\mathbf{V} = \mathbf{V}_{\text{open}}$ is open, and $\mathbf{W} = \mathbf{W}_{\text{closed}}$ is closed. If we start with equation (13), and let $\mathbf{A}_{\mathbf{V}_{\text{open}}} \rightarrow \mathbf{A}_{\mathbf{V}_{\text{open}}} + \nabla \psi$, then $H_\mathcal{V}(\mathbf{V}_{\text{open}}, \mathbf{W}_{\text{closed}})$ will still be gauge-invariant, by the same reasoning leading to equation (8). This shows that the linking of open lines by closed lines is well-defined, and will prove useful in the next section.

As a technical point, we must be careful about switching the positions of the vector fields in equation (15) when one is open and one is closed. They are no longer

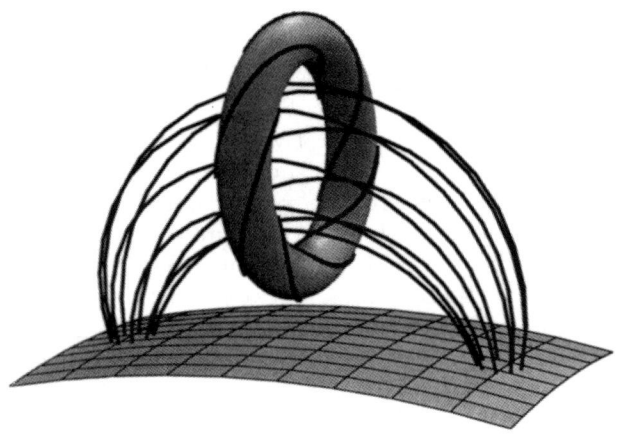

Figure 3. Any open field can be decomposed into a vacuum or potential field **P** plus a closed field $\mathbf{B}_{\mathrm{cl}}$. Here, for simplicity, the closed field is pictured as a circular ring.

very similar! In particular, the integral $\int_{\mathcal{V}} \mathbf{A}_{\mathbf{W}\text{closed}} \cdot \mathbf{V}_{\text{open}} \, d^3x$ is not gauge-invariant. Remember, however, that the linking is what we are after, and the formulas involving vector potentials are only calculational tools. This means that we must simply define $H_{\mathcal{V}}(\mathbf{W}_{\text{closed}}, \mathbf{V}_{\text{open}})$ to be the gauge-invariant integral no matter what. Thus we modify equation (13) to read

$$H_{\mathcal{V}}(\mathbf{W}_{\text{closed}}, \mathbf{V}_{\text{open}}) = H_{\mathcal{V}}(\mathbf{V}_{\text{open}}, \mathbf{W}_{\text{closed}}) \quad (16)$$
$$= \int_{\mathcal{V}} A_{\mathbf{V}\text{open}} \cdot \mathbf{W}_{\text{closed}} \, d^3x.$$

6.2. Magnetic Helicity in an Open Volume

There are several methods for deriving the form of magnetic helicity in an open volume (where flux passes through the boundary, $\mathbf{B} \cdot \hat{n}|_S \neq 0$). While there are conceptual differences between the various derivations, they all lead to equivalent expressions for the helicity. We wish to measure the helicity inside a volume $\mathcal{V}$ with boundary $\mathcal{S}$. The field **B** inside $\mathcal{V}$ (see figure 3) is generated in part by currents within $\mathcal{V}$, and in part by currents outside. More explicitly, we decompose **B** into the sum of two fields, as in [Kusano, Suzuki, & Nishikawa 1995]:

$$\mathbf{B} = \mathbf{B}_{\mathrm{cl}} + \mathbf{P} \quad (17)$$

where $\nabla \times \mathbf{P} = 0$ and $\mathbf{B} \cdot \hat{n}|_S = \mathbf{P} \cdot \hat{n}|_S$. The field **P** is called the vacuum field (or potential field as $\mathbf{P} = \nabla \psi$ in a simply connected volume). Using a simple variational principle, one can show that the vacuum field has the minimum magnetic energy given the boundary flux $\mathbf{B} \cdot \hat{n}|_S$. In a multiply connected volume like a torus, the vacuum field is also defined to have the same net fluxes through the holes as **B**. Meanwhile the closed field $\mathbf{B}_{\mathrm{cl}}$ has no flux through the boundaries.

Using equation (14),

$$H_{\mathcal{V}} = H_{\mathcal{V}}(\mathbf{B}, \mathbf{B})$$
$$= H_{\mathcal{V}}(\mathbf{B}_{\mathrm{cl}}, \mathbf{B}_{\mathrm{cl}}) + 2H_{\mathcal{V}}(\mathbf{P}, \mathbf{B}_{\mathrm{cl}}) + H_{\mathcal{V}}(\mathbf{P}, \mathbf{P}).$$

The first term on the right measure the self linking of a closed field. The next term is twice the mutual linking between an open and a closed field. Both of these are well defined (see previous section).

The last term measures the self-helicity of the vacuum field. However, we have not defined this helicity yet, since it is not the linking of closed with closed, or closed with open. We will simply define the vacuum field to have zero helicity. Similarly, in most of physics the vacuum can be assigned a zero energy density, as adding a constant to the energy does not affect dynamics (there is an exception to this rule in cosmology). As the vacuum field **P** occupies the minimum energy state of $\mathcal{V}$, the assignment of zero helicity seems natural.

We can recover an integral formula from these considerations. The self helicity of the closed field is

$$H_{\mathcal{V}}(\mathbf{B}_{\mathrm{cl}}, \mathbf{B}_{\mathrm{cl}}) = \int \mathbf{A}_{\mathrm{cl}} \cdot \mathbf{B}_{\mathrm{cl}} \, d^3x \quad (18)$$

while the mutual helicity between vacuum field and closed field is

$$2H_{\mathcal{V}}(\mathbf{P}, \mathbf{B}_{\mathrm{cl}}) = 2 \int \mathbf{A}_P \cdot \mathbf{B}_{\mathrm{cl}} \, d^3x. \quad (19)$$

These sum to give the Finn-Antonsen formula

$$H_{\mathcal{V}} = \int_{\mathcal{V}} (\mathbf{A} + \mathbf{A}_{\mathbf{P}}) \cdot (\mathbf{B} - \mathbf{P}) \, d^3x. \quad (20)$$

We can derive the helicity formula in a different way by considering the helicity of all space. The left diagram in figure 4 shows space divided into two regions by the boundary $\mathcal{S}$. We wish to calculate the helicity of the field in the upper region $\mathcal{V}$. The unknown field in $\mathcal{V}'$ below $\mathcal{S}$ is also pictured, but the structure of this field should not affect our calculation. If we integrate the helicity over both regions we obtain a gauge-invariant quantity H_{total}, which depends on fields both above and below.

Suppose, however, we replace the field above S by its corresponding vacuum field **P**, as in the right diagram. This gives us a reference field $\mathbf{B}_{\text{ref}}$, i.e. a vacuum field in $\mathcal{V}$ but the same field as **B** in $\mathcal{V}'$. The helicity of this field H_{ref} is also gauge-invariant.

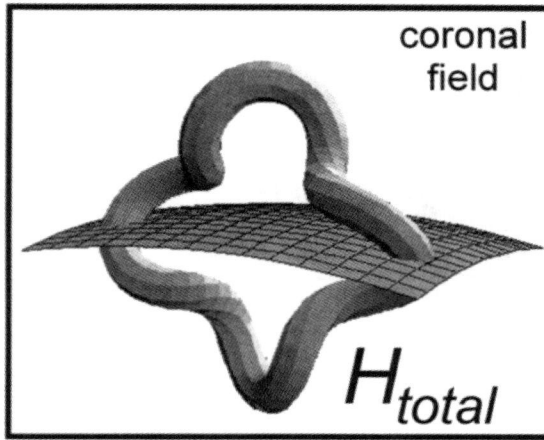 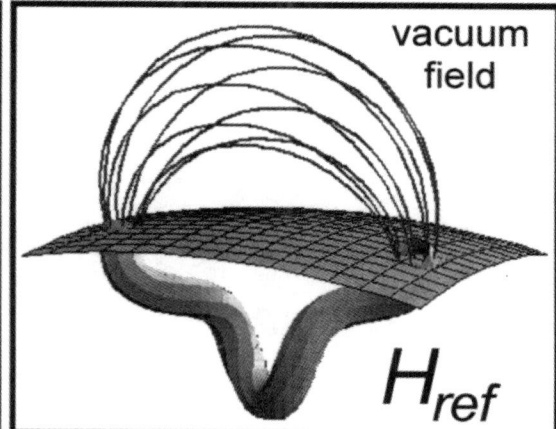

Figure 4. The left diagram shows a magnetic field on either side of a boundary. The total helicity of this field is H_{total}. In the right diagram, a vacuum field has replaced the true field in the upper volume. The helicity of this field is H_{ref}; $H_{\text{total}} - H_{\text{ref}}$ provides a well-defined helicity for the upper volume.

One can then show [Berger & Field 1984] that the difference between these two helicities $H_{\text{total}} - H_{\text{ref}}$ is independent of the field below $\mathcal{S}$ in $\mathcal{V}'$. Explicit calculation shows that this difference is given by equation (20) [Finn & Antonsen 1985].

7. HELICITY TRANSPORT

Once we have divided space into subvolumes, we can consider how helicity is transferred from one subvolume to another. The transfer can occur in two ways: first, a helical magnetic field may simply move across the boundary. Secondly, rotational motions at the boundary can twist the fields on either side. In both these processes, the total helicity of space is conserved, but the helicities of each subvolume may change.

Let $\mathbf{v}$ be the fluid velocity at the boundary. From the expressions for helicity in an open volume one can derive a Poynting theorem [Berger & Field 1984; Berger 1984]

$$\frac{dH}{dt} = 2 \oint_{\mathcal{S}} \left((\mathbf{A}_P \cdot \mathbf{v}) \mathbf{B} - (\mathbf{A}_P \cdot \mathbf{B}) \mathbf{v} \right) \cdot d^2 S, \quad (21)$$

where $\mathbf{A}_P$ is a unique vector field satisfying

$$\hat{\mathbf{n}} \cdot \nabla \times \mathbf{A}_P = B_n, \quad (22)$$
$$\nabla \cdot \mathbf{A}_P = 0, \quad (23)$$
$$\mathbf{A}_P \cdot \hat{\mathbf{n}} = 0. \quad (24)$$

The first term measures the effect of boundary motions, while the second measures transport of helical fields across the boundary.

The simplest example consists of a boundary $\mathcal{S}$ in the shape of a plane parallel to x and y. Given two points $\mathbf{x}$ and $\mathbf{x}'$, let $\mathbf{r} = \mathbf{x} - \mathbf{x}'$ be the relative position vector. Also define $\hat{\theta}_{\mathbf{xx}'} = \hat{\mathbf{z}} \times \hat{\mathbf{r}}$ to be the angular direction at $\mathbf{x}$ about the point $\mathbf{x}'$. Then

$$\mathbf{A}_P(\mathbf{x}) = \frac{1}{2\pi} \int_S \frac{B_z(\mathbf{x}')}{r} \hat{\theta}_{\mathbf{xx}'} \, d^2 S. \quad (25)$$

For example, consider flow of helicity into the solar corona through the photosphere. Let there be two flux tubes of flux Φ_1, Φ_2 centered at $\mathbf{x}_1$ and $\mathbf{x}_2$. The tubes rotate with angular velocities w_1 and w_2. Also, the tubes move about each other with angular velocity $d\theta_{12}/dt = (\mathbf{v}(\mathbf{x}_1) - \mathbf{v}(\mathbf{x}_2)) \cdot \hat{\theta}_{\mathbf{xx}'}$. Then

$$\frac{dH_\mathcal{V}}{dt} = -\frac{1}{\pi} \left(w_1 \Phi_1^2 + w_2 \Phi_2^2 + 2 \frac{d\theta_{12}}{dt} \Phi_1 \Phi_2 \right). \quad (26)$$

This equation describes the growth of the self helicities of the two tubes (first two terms), as well as their mutual helicity (last term).

The rotation of the sun as a whole leads to helicity transfers. Divide space up into four pieces: the north corona (everything outside the photosphere north of the equator), the south corona, the north solar interior, and the south solar interior. Then solar rotation sends negative helicity into the north corona from the north interior [Bieber et al 1987] and positive helicity into the south corona. Furthermore, differential rotation can lead to net build-ups of negative helicity in the north interior and net positive helicity in the southern interior [Babcock 1961; Rust & Kumar 1996; Berger

Figure 5. Two crossed loops have both mutual (linking) helicity and self (twist) helicity.

1998]. If this leaks into the corona, then it will affect the structure of coronal fields. The differential rotation effect corresponds correctly to the sign of helicity observed in coronal structures [Seehafer 1990; Martin et al 1992; Pevtsov et al 1994; Rust 1994].

8. HELICITY OF CORONAL LOOPS

Given a collection of coronal loops, how do we calculate their total magnetic helicity, as well as all the self and mutual helicities? Using vector potentials may be awkward. Instead, we can use the equations of helicity transport to simplify the calculation. To illustrate, we calculate the mutual helicity of two coronal loops (see figure 5). For simplicity, the photosphere will be taken as the plane ($z = 0$). We assume that the footpoints of the loops are small compared to the separations between them. They are located at the points $\mathbf{x}_1^+$, $\mathbf{x}_1^-$ for loop 1, and $\mathbf{x}_2^+$, and $\mathbf{x}_2^-$ for loop 2. Here $B_z > 0$ at $\mathbf{x}_1^+$ and $\mathbf{x}_2^+$. Also, if the loops cross over each other when viewed from above, then let the upper loop be loop 1. Finally, let θ_{1+2-} be the angle of the line segment from $\mathbf{x}_1^+$ to $\mathbf{x}_2^-$, measured with respect to the x axis.

Suppose loop 2 stays fixed, but loop 1 moves into place from far away. Initially the two loops have zero mutual helicity. As we move loop 1 into its final place, equation (26) keeps track of the helicity change (we will ignore the self helicities due to twist here). Because we are moving the upper loop, none of the angles change by more than π. The final mutual helicity is

$$H_\mathcal{V} = \frac{F_1 F_2}{\pi}(\theta_{1+2-} + \theta_{1-2+} - \theta_{1+2+} - \theta_{1-2-}). \quad (27)$$

If the two loops reconnect, this mutual helicity can be converted to self (twist) helicity.

REFERENCES

Arnold, V.I. & B.A. Khesin 1992 Topological methods in hydrodynamics, *Ann. Rev. Fluid Mech.* **24**, 145.
Babcock, H. W. 1961 The topology of the sun's magnetic field and the 22 year cycle, *Astrophys. J. 133*, 572.
Berger M. A. & G. B. Field 1984 The topological properties of magnetic helicity, *J. Fluid Mech. 147*, 133.
Berger, M. A. 1984 Rigorous new limits on magnetic helicity dissipation in the solar corona, *Geophys. & Astrophys. Fluid Dyn. 30*, 79.
Berger, M. A. 1998 Magnetic helicity and filaments, in *New Perspectives on Solar Prominences, IAU 167*, edited by Webb, D., Rust, D. & Schmieder, B. , Astronom. Soc. Pacific 150 102.
Bhattacharjee, A. & Dewar R. L. 1982 Energy principle with global invariants, *Phys. Fluids* **25**, 887.
Bieber, J. W., Evenson, P. A., & Matthaeus, W. H. 1987 Magnetic helicity of the Parker field, *Astrophys. J. 315*, 700.
Călŭgăreanŭ 1961 On isotopy classes of three-dimensional knots and their invariants, *Czechoslovak Math. J. T11*, 588.
Elsasser, W. M. 1956 Hydromagnetic dynamo theory, *Rev.Mod. Phys.* **28**, 135.
Epple, M. 1998 Orbits of asteroids, a braid, and the first link invariant, *Math. Intelligencer* **20**, 45.
Finn, J. & Antonsen 1985 Magnetic helicity: what is it, and what is it good for?, *Comments on Plasma Phys. and Contr. Fusion.* **9**, 111.
Freedman, M. & Berger, M. A. 1993 Combinatorial Relaxation, *Geophys. Astrophys. Fluid Dyn.* **73**, 91.
Jensen, T. & Chu, M. S. 1984 Current drive and helicity injection, *J. Plasma Physics* **25**, 459.
Kruskal, M. D. & Kulsrud, R. M. 1958 Equilibrium of a magnetically confined plasma in a toroid, *Phys. Fluids* **1**, 265.
Kusano, K., Suzuki, Y., & Nishikawa, K. 1995 A solar flare triggering mechanism based on the Woltjer-Taylor minimum energy principle, *Astrophys. J. 441*, 942.
Low, B.C. 1994 Magnetohydrodynamic processes in the solar corona - flares, coronal mass ejections, and magnetic helicity, *Physics Plasmas* **1**, 1684.

Martin, S. F., Marqette, & Bilimoria, R. 1992 The solar cycle pattern in the direction of the magnetic field along the long axes of polar filaments, in *The Solar Cycle*, edited by K.L. Harvey, ASP conf. series 27 53.

Moffatt, H. K. 1969 The degree of knottedness of tangled vortex lines, *J. Fluid Mech. 35*, 117.

Moffatt, H. K. 1978 , *Magnetic Field Generation in Electrically Conducting Fluids* , Cambridge Univ. Press, Cambridge.

Moffatt, H.K. & Ricca, R.L. 1992 Helicity and the Calŭgareanŭ invariant, *Proc. Royal Society London A 439*, 411.

Pevtsov, A.A., Canfield, R.C., & Metcalf, T.R. 1994 Latitudinal variation of helicity of photospheric fields, *Astrophys. J. 440*, L109.

Pouquet A., Frisch U., & Léorat, J. 1976 Strong MHD helical turbulence and the nonlinear dynamo effect, *J. Fluid Mech. 76*, 321.

Priest, E.R., van Ballegooijen, A. A., & Mackay, D. H. 1996 Model for dextral and sinistral prominences , *Astrophys. J. 460*, 530.

Ricca, R. L. & Berger, M. A. 1996 Topological ideas and fluid mechanics, *Phys. Today 49*, (12) 24.

Rust, D. 1994 Spawning and shedding helical magnetic fields in the solar atmosphere, *Geophysical Research Letters 21*, 241.

Rust, D.M. & Kumar, A. 1996 Evidence for helically kinked magnetic flux ropes in solar eruptions, *Astrophys. J. 464*, L199.

Seehafer N. 1990 Electric current helicity in the solar atmosphere, *Solar Physics 125*, 219.

Taylor J. B. 1974 Relaxation of toroidal plasma and generation of reverse magnetic fields, *Physical Review L. 33*, 1139.

Taylor J. B. 1986 Relaxation and magnetic reconnection in plasmas, *Rev. Mod. Phys. 58*, 741.

Woltjer, L. 1958 A theorem on force-free magnetic fields, *Proc.Natl. Acad. Sci. USA 44*, 480.

M. A. Berger, Mathematics, University College London, WC1E 6BT, United Kingdom.
(e-mail: m.berger@ucl.ac.uk)

Implications of Magnetic Helicity Conservation

Allen H. Boozer

Department of Applied Physics and Applied Mathematics, Columbia University, New York, NY

The conservation of magnetic helicity and the related conservation law for field line topology have important implications. Helicity conservation constrains the form of a magnetic dynamo driven by small scale fluid turbulence. In particular, helicity conservation is inconsistent with the well-known α-effect theory. Conservation laws for field line topology coupled with the physics of runaway electrons imply the existence of a corona about any star that has an outer convective zone with an embedded magnetic field. Magnetic helicity and field line topology are conserved if a magnetic field evolves while embedded in a perfectly conducting medium. An equivalent evolution is called ideal. For a generic magnetic field, the local evolution is always ideal. The global evolution is also ideal if the loop voltage, $\mathcal{V} \equiv \int (\mathbf{E} \cdot \mathbf{B}/B) d\ell$, on each closed magnetic field line is zero.

1. INTRODUCTION

The concept of magnetic helicity, $K = \int \mathbf{A} \cdot \mathbf{B} d^3x$, which was introduced by Elsässer [1956] and Woltjer [1958], became well-known after Brian Taylor [1974] used its conservation to explain properties of plasma confinement in reversed-field-pinch experiments. In this, and in many other applications, the important point is that helicity is better conserved than the magnetic energy, $U = \int (B^2/2\mu_o) d^3x$, when the electrical current $\mathbf{j}$ is spiky. This property is easily demonstrated. Helicity changes as $dK/dt = -2\int \eta \mathbf{j} \cdot \mathbf{B} d^3x$, see Section (2), while energy dissipation is $\int \eta j^2 d^3x$. Consequently, if the magnetic field $\mathbf{B}$ varies on a characteristic spatial scale $1/k$, the time derivative of helicity scales as k, but energy dissipation scales as k^2. In a highly turbulent plasma, in which k becomes large, the magnetic energy is rapidly dissipated while the helicity changes little.

Our emphasis will be on the magnetic field conservation properties that are associated with helicity conservation and their applications to physical problems.

Conservation properties will be discussed in Section (2). Applications to dynamo theory are discussed in Section (3) and to the formation of stellar coronas in Section (4).

2. CONSERVATION PROPERTIES

Magnetic helicity, $K = \int \mathbf{A} \cdot \mathbf{B} d^3x$, is created or destroyed according to the law $dK/dt = -2\int \mathbf{E} \cdot \mathbf{B} d^3x$. This conservation law follows directly from the equations $\mathbf{B} = \nabla \times \mathbf{A}$ and $\mathbf{E} = -\partial \mathbf{A}/\partial t - \nabla \Phi$. A surface integral, which arises in an integration by parts, has been ignored. This surface integral gives the flow of helicity from the outside world. Expressions for this flow have been given by Bevir and Gray [1981], Berger and Field [1984] and Boozer [1990].

The conservation law for helicity says there is something special about $\mathbf{E} \cdot \mathbf{B}$. What is special is that $\mathbf{E} \cdot \mathbf{B}$ controls the changes in topology of the magnetic field lines. A generic magnetic field has the important property of always obeying an ideal Ohm's law locally. The usual statement of the ideal Ohm's law is $\mathbf{E} + \mathbf{u} \times \mathbf{B} = 0$ with $\mathbf{u}(\mathbf{x},t)$ a velocity. Here, an apparently trivial generalization is made to the form

$$\mathbf{E} + \mathbf{u} \times \mathbf{B} = -\nabla \Phi \qquad (1)$$

The addition of a potential Φ to the expression for the ideal Ohm's law does not modify the evolution of the magnetic field for $\partial \mathbf{B}/\partial t = -\nabla \times \mathbf{E}$. The validity of Equation (1) in the vicinity of any point at which $\mathbf{B}$ is non-zero is easily demonstrated, for an arbitrary electric field can be written in this form. To show this, choose Φ so $\mathbf{B} \cdot \nabla \Phi = -\mathbf{E} \cdot \mathbf{B}$. Once such a Φ is found, choose the components of $\mathbf{u}$ perpendicular to the magnetic field so $\mathbf{u}_\perp = (\mathbf{E} + \nabla\Phi) \times \mathbf{B}/B^2$. That is, Φ and the two components of $\mathbf{u}_\perp$ give three functions which can be used to balance the three components of an arbitrary electric field. Equation (1) holds for generic fields, which have nulls, but the proof will be postponed until the importance of the equation is explained.

Although generic magnetic fields can have a non-ideal evolution, Equation (1) says that near every point the evolution appears ideal. In the Lagrangian coordinates defined by the flow $\mathbf{u}(\mathbf{x},t)$, Boozer [1992] has shown that neither the magnetic field lines nor the quantity $\mathbf{A} \cdot \mathbf{B}$ depend on time.

How can the paradox between a locally ideal and a globally non-ideal evolution be resolved? The resolution is in the equation $\mathbf{B} \cdot \nabla \Phi = -\mathbf{E} \cdot \mathbf{B}$, which gives the potential Φ. This equation is equivalent to $d\Phi/d\ell = -\mathbf{E} \cdot \mathbf{B}/B$ with $d\ell$ the differential distance along a field line. Although the equation for the potential Φ can always be solved locally, Newcomb [1959] has shown that a problem can arise along closed magnetic field lines. On a closed field line of length L, the potential must satisfy the single-valuedness constraint that $\Phi(\ell+L) = \Phi(\ell)$. This constraint is satisfied if and only if the loop voltage along the closed line $\mathcal{V} \equiv \oint (\mathbf{E} \cdot \mathbf{B}/B) d\ell$ is zero. One obtains the important result that the topology of the magnetic field lines is fixed if the loop voltage along every closed magnetic field line is zero. In astrophysical problems, magnetic field lines are often open, which means they enter and leave the spatial region of interest. Topology changes of open field lines are also controlled by the loop voltage when appropriately defined [Boozer 1990]. The appropriate definition is the integral of $\mathbf{E} \cdot \mathbf{B}/B$ along a line from one crossing of the bounding surface to another.

The requirement of a parallel electric field for a non-ideal evolution of a generic magnetic field has a number of important implications. For example, the usual Ohm's law of plasma physics, $\mathbf{E} + \mathbf{v} \times \mathbf{B} = \eta_\| \mathbf{j}_\| + \eta_\perp \mathbf{j}_\perp$, implies the perpendicular resistivity $\eta_\perp$ has nothing to do with whether the evolution of the magnetic field is ideal. But, the perpendicular resistivity can give a velocity difference between the plasma and the field $\mathbf{v} - \mathbf{u}$. A type II superconductor with a melted flux lattice has the same form of Ohm's law as a plasma except $\eta_\|$ is zero, so topology changes cannot occur. Such superconductors, especially in toroidal form, should be a fascinating basis for experiments, but this has never been exploited.

The validity of the ideal Ohm's law, Equation (1), near a null of the magnetic field remains to be demonstrated. Equation (1) is valid if a $\mathbf{u}(\mathbf{x},t)$ exists such that $\nabla \times (\mathbf{E} + \mathbf{u} \times \mathbf{B}) = 0$. At the null $\nabla \times (\mathbf{E} + \mathbf{u} \times \mathbf{B}) = -\partial \mathbf{B}/\partial t - \mathbf{u} \cdot \nabla \mathbf{B}$. Near the null the magnetic field has the form $\mathbf{B} = \mathcal{B} \cdot \mathbf{x}$ with $\mathcal{B}$ a matrix and $\mathbf{x}$ the distance from the null. The required $\mathbf{u}$ is, therefore, the solution to the equation $\mathcal{B} \cdot \mathbf{u} = -\partial \mathbf{B}/\partial t$, which can be solved when the determinant of $\mathcal{B}$ is non-zero.

Why must $\mathcal{B}$, the matrix of derivatives of $\mathbf{B}$ near a field null, have a non-zero determinant for a generic magnetic field? The equation $\mathbf{B}=0$ is really three equations $B_x=0$, $B_y=0$, $B_z=0$ for the three Cartesian coordinates x,y,z. Three equations with three unknowns is a well-posed problem. If the determinant of $\mathcal{B}$ must vanish at the point $\mathbf{B}=0$, there are four equations with three unknowns, which generically does not have a solution.

Some familiar models of fields with nulls do not satisfy the constraint of being of a generic form and are presumably of little physical relevance. The most familiar example of a non-generic field with nulls is $\mathbf{B} = -\hat{\mathbf{z}} \times \nabla A(x,y)$ with ∇A having nulls. For this field, the addition of an arbitrarily small field in the z direction removes all nulls. In such non-generic fields, peculiar effects, such as topology changes with $\mathbf{E} \cdot \mathbf{B} = 0$, frequently arise but are presumably of little physical significance.

3. IMPLICATIONS FOR DYNAMOS

Magnetic helicity and related conservation laws place a number of constraints on the theory of dynamos [Boozer 1993]. By a dynamo, we mean the creation or maintenance of a magnetic field within a bounded, simply connected region of space (such a sphere) by a fluid moving with velocity $\mathbf{v}(\mathbf{x},t)$ that is coupled to the electric field by the Ohm's law $\mathbf{E} + \mathbf{v} \times \mathbf{B} = \eta \mathbf{j}$. Since the derivations of the results that will be given in this section have been published, the emphasis will be on the results.

Possibly the oldest, but not yet fully accepted, implication is that the well-known α effect of mean-field dynamo theory cannot be physically correct. The usual Ohm's law of mean-field dynamo theory [Krause and Rädler 1980] is $\mathbf{E} + \mathbf{v} \times \mathbf{B} = (\eta + \beta)\mathbf{j} - \alpha \mathbf{B}$ with α and β given by small scale properties of the fluid velocity

$v(x,t)$. Boozer [1986] has shown that this Ohm's law cannot be a physically correct description for a conducting fluid with small spatial scales, $1/k$, for it changes magnetic energy and helicity on the same time scale. Bhattacharjee and Yuan [1995] as well as Gruzinov and Diamond [1996] have shown that the terms retained in the derivation of the α-effect form for Ohm's law are subdominant. A physically correct mean field theory has a parallel Ohm's law of the form $\mathbf{E}\cdot\mathbf{B} = \eta\mathbf{j}\cdot\mathbf{B} + \nabla\cdot\mathbf{h}$ with $2\mathbf{h}$ the flux of helicity due to the small scale flow field [Boozer 1986]. If the fluid has negligible energy, as in the reversed field pinch, the helicity flux can be written as $\mathbf{h} = -\lambda\nabla(j_\parallel/B)$ with λ positive and known as the hyperresistivity. The helicity flux $\mathbf{h}$ can have additional terms that represent the transfer of energy from the small scale fluid motion to the large scale magnetic field [Boozer 1993].

The concepts that were explained in Section (2). imply that the region occupied by a single steady-state dynamo must have a non-zero helicity [Boozer 1993]. However, one can have a double dynamo, which consists of two dynamos of equal but opposite helicity. The helicity may be the best measure of the strength of a dynamo produced field. The magnetic energy can be pathological, for the bulk of the energy can be located at very small spatial scales if the magnetic Reynolds number R_m is large, $R_m \equiv (\mu_o/\eta)vL$ with L a characteristic spatial scale. The minimum magnetic energy associated with a given helicity occurs when $j_\parallel/B$ is constant, so the helicity is a useful measure of the large scale magnetic field. A related result brings into question the physical relevance of so-called fast dynamos. In a fast dynamo, the mean magnetic field $\langle B \rangle$ is produced on a fast, or L/v, time scale compared to the characteristic dissipative time scale $(\mu_o/\eta)L^2$. For the Finn-Ott [1988, 1990] fast-dynamo models, Boozer [1992] found that the power dissipated resistively while producing a mean field energy $\langle B \rangle^2/2\mu_o$ is approximately $(\langle B \rangle^2/2\mu_o) R_m^{1/2} (v/L)$. Since fast-dynamo theory is supposed to apply in the limit as $R_m \to \infty$, the Finn-Ott fast dynamos can produce mean fields of only infinitesimal strength.

4. IMPLICATIONS FOR THE SOLAR CORONA

A star that has an outer convective zone with an embedded magnetic field must develop a corona-like structure. This result follows from three points: (a) The strength of the current along each field line, $j_\parallel/B$, is a constant in the upper atmosphere of a star. The constant is determined by the lower atmosphere or the convective zone. (b) The parallel current, or $j_\parallel/B$, that flows out of the photosphere is expected to be large with little spatial correlation from field line to field line. That is, the spatial distribution of $j_\parallel/B$ is expected to be fractal. (c) If the parallel current, $j_\parallel$, flows in a plasma of ever decreasing density, the plasma eventually breaks down. By plasma breakdown we mean the current carriers change from being near-Maxwellian electrons to being high energy, or runaway, electrons. The current carried by near-Maxwellian electrons is limited to $j_\parallel \ll en_e (T_e/m_e)^{1/2}$. The transition from Maxwellian to non-Maxwellian current carriers should occur over a very narrow region. These three points are discussed in the three subsections that follow.

The properties of a corona produced by runaway electrons is largely determined by the strength of $j_\parallel$ just below the radius at which the corona is formed. The strength and the spatial distribution of this current are determined by equations associated with those of helicity conservation. Working backwards from the known properties of the solar corona, $j_\parallel$ must reach a peak value of roughly $100 A/m^2$. The peak current density arises only on small bundles of field lines due the expected fractal spatial distribution of $j_\parallel/B$.

A. CONSTANCY OF $j_\parallel/B$

The variation of the parallel current along a field line is determined by two equations: (1) force balance between the magnetic field and the plasma and (2) Ampere's law, $\nabla\times\mathbf{B} = \mu_o\mathbf{j}$ with its associated constraint $\nabla\cdot\mathbf{j} = 0$. Let $\mathbf{f}$ be the force exerted on the field by the plasma; a typical expression is $\mathbf{f} = \rho(\partial\mathbf{v}/\partial t + \mathbf{v}\cdot\nabla\mathbf{v}) + \nabla p$. The force exerted by the magnetic field balances $\mathbf{f}$, so $\mathbf{f} = \mathbf{j}\times\mathbf{B}$. Vector identities imply that the variation of $j_\parallel/B$ is

$$\mathbf{B}\cdot\nabla\left(\frac{j_\parallel}{B}\right) = 2\mathbf{j}\cdot\boldsymbol{\kappa} + \frac{\mathbf{B}}{B^2}\cdot\nabla\times\mathbf{f} \quad (2)$$

with the field line curvature $\boldsymbol{\kappa} \equiv \hat{\mathbf{b}}\cdot\nabla\hat{\mathbf{b}} = -\hat{\mathbf{b}}\times(\nabla\times\hat{\mathbf{b}})$ and $\hat{\mathbf{b}} \equiv \mathbf{B}/|B|$ a unit vector along $\mathbf{B}$.

Even in the solar photosphere, it is difficult to remove a large $j_\parallel$ using the curvature term of Equation (2). The curvature term, which can also be written as $2\mathbf{j}_\perp\cdot\boldsymbol{\kappa}$, implies a variation in $j_\parallel$ of approximately $\Delta j_\parallel \approx j_\perp L/R$ with L the distance along a field line over which the perpendicular current has a value $j_\perp$ and R the radius of curvature of the line. For field lines that penetrate the photosphere and reach the corona, one would expect L/R to be no larger than roughly unity, so only a parallel current comparable to $j_\perp$ can be removed.

The $\nabla \times \mathbf{f}$ term in Equation (2) is, at least for a small plasma velocity, determined by the shear-Alfvén wave. The plasma pressure makes no contribution. Shear Alfvén waves propagate along magnetic field lines at the Alfvén speed, $B/\sqrt{\mu_o \rho}$. The parallel current at a point along a field line is determined by the region that can be reached by shear Alfvén waves during the time scale set by the magnetic field evolution.

Calculations of the magnetic field above the photosphere require two boundary conditions: the magnetic field normal to the photosphere $\mathbf{B} \cdot \hat{\mathbf{n}} = B_n(\theta, \varphi)$ and the parallel current $j_\parallel/B = k(\theta,\varphi)/\mu_o$ at the photosphere. The propagation of shear Alfvén waves is sufficiently rapid to insure a close coupling of the parallel currents above the photosphere with those below.

B. FRACTAL NATURE OF $j_\parallel/B$

The parallel current, or $j_\parallel/B$, that flows out of the photosphere is expected to be large with little spatial correlation from field line to field line. Although $k \equiv \mu_o j_\parallel/B$ has very large values, the magnetic field variation need not be large. A relatively small variation in the field occurs when the correlation distance δ in $j_\parallel/B$ from field line to field line is short, $(k\delta)^2 \ll 1$.

The formation of a fractal $j_\parallel/B$ distribution, in the photosphere or convective zone, will be explained using a simple two dimensional model. Boozer [1992] has given a more complicated three dimensional model that makes the argument for a fractal current distribution stronger. The magnetic field in a small region will be modelled by $\mathbf{B} = B_o \hat{\mathbf{z}} - \hat{\mathbf{z}} \times \nabla A(x,y,t)$ with B_o locally constant. The vertical or radial direction is $\hat{\mathbf{z}}$, and $A(x,y,t)$ is the z component of the vector potential. The local plasma flow will be assumed to be divergence free and in the x-y plane, so $\mathbf{v} = -\hat{\mathbf{z}} \times \nabla H(x,y,t)$. The standard Ohm's law, $\mathbf{E} + \mathbf{v} \times \mathbf{B} = \eta \mathbf{j}$, and $\mathbf{E} = -\hat{\mathbf{z}} \partial A/\partial t - \nabla \Phi$, imply

$$\partial A/\partial t + \mathbf{v} \cdot \nabla A = (\eta/\mu_o) \nabla^2 A, \quad (3)$$

which is the advection-diffusion equation. The streamlines of the flow obey Hamilton's equations $dx/dt = v_x = \partial H/\partial y$ and $dy/dt = v_y = -\partial H/\partial x$. The flow field $\mathbf{v}$, or equivalently $H(x,y,t)$, is set by the physics of the outer convective zone.

In naturally occurring flows, Equation (3) will be shown to imply a current, $j_z(x,y,t) = -\nabla^2 A/\mu_o$, that increases exponentially in time with little correlation from one (x,y) point to another. The features of the solutions of Equation (3) are largely determined by the properties of the streamlines of naturally occurring flows. The streamlines of the flow are the general solution to Hamilton's equations. The general solution has the form $x = x(x_o, y_o, t)$ and $y = y(x_o, y_o, t)$ with x_o and y_o the initial conditions. An element of fluid can be identified with an (x_o, y_o) point, and one can use (x_o, y_o) as coordinates, so-called Lagrangian coordinates. The expression for the distance ds between two infinitesimally separated fluid elements is given by $(ds)^2 = (dx)^2 + (dy)^2$ with $dx = (\partial x/\partial x_o) dx_o + (\partial x/\partial y_o) dy_o$, so $(ds)^2$ can be written as $(ds)^2 = g_{x_o x_o}(dx_o)^2 + 2 g_{x_o y_o}(dx_o)(dy_o) + g_{y_o y_o}(dy_o)^2$. The matrix g_{ij} with $x^1 = x_o$ and $x^2 = y_o$ is called the metric tensor, since it gives the metric, or measure, of distances. The metric tensor can be diagonalized, and $\nabla \cdot \mathbf{v} = 0$ implies that the two eigenvalues are reciprocals of one another. These eigenvalues can be written as $\exp[\pm 2\lambda t]$ with $\lambda(x_o, y_o, t)$ known as the finite-time Liapunov exponent [Boozer 1992]. If $\lambda(x_o, y_o, t \to \infty)$ is non-zero, the fluid point (x_o, y_o) is said to be in a region of chaotic flow. Otherwise, it is said to be in a region of regular flow. In a generic, or natural, flow most fluid points lie in chaotic regions and $\lambda(x_o, y_o, t)$ takes on a fractal spatial structure for large t [Tang and Boozer 1996, 1997].

Exponential growth in the metric tensor causes an exponential growth in j_z. Coordinate transformation theory implies that the z component of the current density, $\mu_o j_z = -\nabla^2 A$, can be written in Lagrangian coordinates as $\mu_o j_z = -\sum \partial_i (g^{ij} \partial_j A)$ and that Equation (3) can be written as $(\partial A/\partial t)_{x^1 x^2} = -\eta j_z$ with g^{ij} the matrix inverse of g_{ij} and $\partial_i \equiv \partial/\partial x^i$. If η were zero, the vector potential, A, would be independent of time in Lagrangian coordinates. However, the current density j_z would increase exponentially in chaotic regions, roughly as $\exp(2\lambda t)$. In a generic, and presumably any natural, flow the current density increases exponentially with time until either (1) the term ηj_z becomes sufficiently large to break the constancy of A in Lagrangian coordinates or (2) the magnetic forces $\mathbf{j} \times \mathbf{B} = j_z \nabla A \propto \exp(3\lambda t)$ become sufficiently large to modify the flow. The short spatial correlation distance for $j_\parallel/B$ from field line to field line, δ, comes from the fractal structure of $\lambda(x_o,y_o,t)$, the finite time Liapunov exponent [Tang and Boozer 1996, 1997].

C. FORMATION OF RUNAWAYS

Just above the photosphere, the solar density drops exponentially with a scale height of approximately 100 km while $j_\parallel/B$ remains constant. At the base of the corona, a sudden transition occurs from a plasma in which the current is carried by Maxwellian electrons with a thermal energy of about 0.5 eV to a plasma with a typical electron energy of order 100 eV. This sudden transition from cold to hot electrons may be explained by the runaway electron effect.

The breakdown of a plasma into runaway electrons is a familiar phenomena in toroidal plasmas [Connor and Hastie 1975, Wesson 1997]. The cause is that the collision frequency ν_e scales as $1/v^3$ for electrons with energy above the thermal average. For sufficiently high energy electrons, the electric field is a larger force than collisions, and these electrons are accelerated to arbitrarily high energies. A near Maxwellian approximation yields a rate of runaway electron production that depends exponentially on the electric field along the magnetic field lines. Let $\varepsilon \equiv E/E_D$ with the so-called Dreicer field $E_D = 4.6 \times 10^{-16} n_e/T_e$. The electric field is in Volts per meter, the electron density n_e is in particles per meter cubed, and the electron temperature, T_e, is in electron volts. The runaway production rate is then

$$dn_r/dt = \nu_e \alpha(\varepsilon) n_e \quad (4)$$

with ν_e the thermal electron collision frequency and $\alpha(\varepsilon) \approx \varepsilon^{1/2} \exp[-(1/4\varepsilon) - (2/\varepsilon)^{1/2}]$.

At plasma breakdown, the current flowing along the magnetic field, $j_\parallel$, switches from being carried by near-Maxwellian electrons to being carried by high energy, runaway electrons. The electric field before breakdown is $E = \eta j_\parallel$ with the resistivity η depending only on the electron temperature, $\eta \propto 1/T_e^{3/2}$. For simplicity, assume the temperature T_e is relatively constant below the corona but the density n_e drops with a scale height L_n, which is about 100 km on the sun. The current $j_\parallel$ is also almost constant, so the electric field E is. However, the Dreicer field, $E_D \propto n_e$, is dropping exponentially with altitude, so ε is increasing exponentially. Breakdown occurs when ε becomes sufficiently large for the runaway electrons to carry the current, $j_\parallel = e n_r v_r$. Equation (4) implies $(v_r/L_b)n_r = \alpha \nu_e n_e$, with v_r the typical velocity of runaways and L_b the radial distance over which breakdown occurs. Consequently, $\alpha(\varepsilon) n_e L_b = \varepsilon E_D/(\eta \nu_e) \approx 2.1 \times 10^{16} \varepsilon T_e^2$. The vertical scale over which breakdown occurs is given by $1/L_b \approx d\ln(\alpha)/dr$, so $L_b \approx \varepsilon L_n$. Typically ε is about 1/10 at breakdown and $L_n \approx 100$ km, so $L_b \approx 10$ km. The typical energy of runaways in electron volts is $u_r \equiv m_e v_r^2/2 \approx e \varepsilon E_d L_b$, so u_r and ε are related by

$$u_r \approx 16 \, T_e \, \varepsilon^2/\alpha(\varepsilon) \quad (5)$$

The typical energy of runaways u_r, which is the energy of electrons in the corona, is determined by the parallel current density $j_\parallel$ and the length of coronal field lines L_c. The coronal density $n_c = n_r$ can only rise to the level at which runaway electrons just reach the other end of field lines in the corona without being slowed by collisions. This implies $n_c L_c \approx 4.1 \times 10^{15} u_r^2$. This equation plus $j_\parallel = e n_c v_r \approx 9.5 \times 10^{-14} n_c u_r^{1/2}$ give the desired relation

$$j_\parallel L_c \approx 3.9 \times 10^2 \, u_r^{5/2}. \quad (6)$$

The current is in Amperes per square meter, the length of coronal field lines L_c is in meters, and the typical runaway energy is in electron-volts. Given $j_\parallel$ and L_c, one knows u_r, Equation (6), and the typical energy of electrons in the corona, $m_e u_r^2/2$. Equation (5) then determines the value of ε at breakdown. However due to the exponential dependence of $\alpha(\varepsilon)$ on ε this value is always about 1/10.

The power that can be transferred to the corona by a large but fractal distribution of $j_\parallel/B$ depends on the fraction of the surface the photosphere that is occupied by field lines with $j_\parallel/B$ large. This issue could be addressed through models but this has not yet been done.

Plasma breakdown into runaway electrons merits further study as the cause of the solar corona. One can say that if a star has no corona, it must either have very weak convection in the lower atmosphere or have no embedded magnetic field.

Acknowledgements: The author wishes to acknowledge support for this work received from grant DE-FG02-97ER54441 under the NSF/DoE Partnership in Plasma Science and Engineering.

REFERENCES

Berger, M. A. and Field, G. B.: 1984, *The topological properties of magnetic helicity*, J. Fluid Mech. *147*, 133.

Bevir, M. and Gray, J.: 1981, *Relaxation, flux consumption, and quasi steady state pinches*, in *Proceedings of the Reverse Field Pinch Theory Workshop*, edited by H. R. Lewis and R. A. Gerwin (Los Alamos National Laboratory, Los Alamos, NM), Vol. III, p. A-3.

Bhattacharjee, A. and Yuan, Y.: 1995, *Self-consistency constraints on the dynamo mechanism*, Astrophys. J. *449*, 739.

Boozer, A. H.: 1986, *Ohm's law for mean magnetic fields*, J. Plasma Phys. *35*, 133.

Boozer, A. H.: 1990, *The evolution of magnetic fields and plasmas in open field line configurations*, Phys. Fluids B *2*, 2300.

Boozer, A. H.: 1992, *Dissipation of magnetic energy in the solar corona*, Astrophys. J. *394*, 357.

Boozer, A. H.: 1993, *Magneitc helicity and dynamos*, Phys. Fluids B *5*, 2271.

Connor, J. W. and Hastie, R. J.: 1975, *Relativistic limitations on runaway electrons*, Nucl. Fusion *15*, 415.

Elsässer, W. M.: 1956, *Hydrodynamic dynamo theroy*, Rev. Mod. Phys. *28*, 135.

Finn, J. M. and Ott, E.: 1988, *Chaotic flows and magnetic dynamos*, Phys. Fluids *31*, 2992.

Finn, J. M. and Ott, E.: 1990, *The fast kinematic dynamo and the dissipationless limit*, Phys. Fluids B*2*, 916.

A. V. Gruzinov, A. V., and Diamond, P. H.: 1996, *Nonlinear mean field electrodynamics of turbulent dynamos*, Phys. Plasmas *3*, 1853.

Krause, F. and Rädler, K.-H.: 1980, *Mean-Field Magnetohydrodynamics and Dynamo Theory* (Pergamon, Oxford).

Newcomb, W. A.: 1959, *A plasma equilibrium constraint*, Phys. Fluids *2*, 362.

Tang, X. Z. and Boozer, A. H.: 1996, *Finite time Liapunov exponent and advection-diffusion equation*, Physica D *95*, 283.

Tang, X. Z. and Boozer, A. H.: 1997, *Hamiltonian structure of Hamiltonian chaos*, Phys. Lett. A *236*, 476.

Taylor, J. B.: 1974, *Relaxation of toroidal discharges*, Phys. Rev. Lett. *33*, 1139.

Wesson, J.: 1997, *Tokamaks* (Clarendon Press, Oxford) p. 72.

Woltjer, L.: 1958, *A theorem on force-free magnetic fields*, Proc. Natl. Acad. Sci. USA *44*, 489.

Influence of Geometry and Topology on Helicity

Jason Cantarella, Dennis DeTurck, Herman Gluck, Mikhail Teytel

Department of Mathematics, University of Pennsylvania, Philadelphia, PA 19104

The *helicity* of a smooth vector field defined on a domain in 3-space is the standard measure of the extent to which the field lines wrap and coil around one another; it plays important roles in fluid mechanics, magnetohydrodynamics and plasma physics. In this report we show how the relation between energy and helicity of a vector field is influenced by the geometry and topology of the domain on which it is defined. In particular, we will see that the standard model for the magnetic field in the Crab Nebula (equivalently, the *spheromak field* of plasma physics) is the unique energy-minimizing divergence-free vector field of given nonzero helicity, defined on and tangent to the boundary of a round ball, and that the essential features of this energy-minimizing field persist even as the domain changes topological type. We will also see that when volume-preserving deformation of domain is permitted, the spheromak field is not the absolute energy-minimizing field with given helicity; instead, the round ball on which it is defined can be dimpled in at the poles and expanded out at the equator to further decrease the field energy while preserving helicity. Our numerical computations suggest that this volume-preserving, helicity-preserving, energy-decreasing deformation of domain and field converges to a singular domain, in which the north and south poles have been pressed together at the center, along with a corresponding singular field.

1. TWO FUNDAMENTAL PROBLEMS

We organize this report by focusing on two fundamental problems:

1. Minimize energy among all divergence-free vector fields of given nonzero helicity, defined on and tangent to the boundary of a given domain.

2. Find the above minimum over all domains of given volume.

Such energy-minimizing vector fields provide models for stable force-free magnetic fields in gaseous nebulae and laboratory plasmas, while the search for them seems to bring out some of the deepest and most useful mathematics connected with helicity.

2. HELICITY AND WRITHING NUMBER

The *helicity* $H(V)$ of a smooth (meaning C^∞) vector field V on the domain Ω in 3-space, defined by the formula

$$H(V) = \frac{1}{4\pi} \int_{\Omega \times \Omega} V(x) \times V(y) \cdot \frac{x-y}{|x-y|^3} \, d\text{vol}_x \, d\text{vol}_y,$$

is the standard measure of the extent to which the field lines wrap and coil around one another. See Figure 1.

It was introduced in [*Woltjer*, 1958] and it was named in [*Moffatt*, 1969].

The *writhing number* Wr(K) of a smooth, arc-length-parametrized curve K in 3-space, defined by the formula

$$\text{Wr}(K) = \frac{1}{4\pi} \int_{K \times K} \frac{dx}{ds} \times \frac{dy}{dt} \cdot \frac{x-y}{|x-y|^3} \, ds \, dt,$$

is the standard measure of the extent to which the curve wraps and coils around itself. See Figure 2. It was introduced in [*Călugăreanu*, 1959-61] and named in [*Fuller*, 1971], and has proved its importance for molecular biologists in the study of knotted duplex DNA, and of the enzymes which affect it.

Clearly, writhing number for knots is the analogue of helicity for vector fields. Both formulas above are variants of the integral formula of [*Gauss*, 1833] for the linking number of two disjoint closed space curves.

3. RELATION BETWEEN HELICITY AND WRITHING NUMBER

A useful formula connecting the helicity of vector fields to the writhing of knots appears in [*Berger and Field*, 1984].

Let V be a vector field defined in a tube about a knot K, orthogonal to the cross-sectional disks, with length depending only on distance from K. Such a vector field is always divergence-free. See Figure 3.

Then

$$\text{H}(V) = \text{Flux}(V)^2 \, \text{Wr}(K).$$

Here, Flux(V) denotes the flux of V through any of the cross-sectional disks. We also refer the reader to the two papers of [*Moffatt and Ricca*, 1992] for related results.

4. HOW THE GEOMETRY OF THE DOMAIN INFLUENCES HELICITY

All the numbered theorems in this report are due to the authors, and may be found, together with their proofs in the papers cited.

THEOREM 1. *Let V be a smooth vector field defined on the compact domain Ω with smooth boundary. Then the helicity* H(V) *of V is bounded by*

$$|\,\text{H}(V)| \leq R(\Omega) \, \text{E}(V),$$

where $R(\Omega)$ is the radius of a ball with the same volume as Ω and $\text{E}(V) = \int_\Omega V \cdot V \, d\text{vol}$ *is the energy of V.*

This upper bound is not sharp, but it is the right order of magnitude. For example, the model for the magnetic field in the Crab Nebula is a vector field V

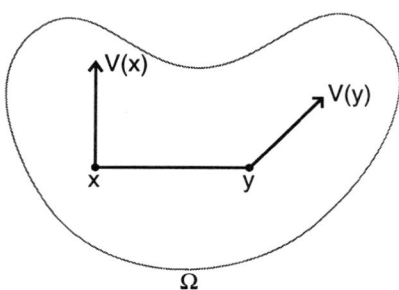

Figure 1. The helicity integrand.

on a round ball Ω with helicity greater than one-fifth of the asserted upper bound. Sharp upper bounds will be obtained by the spectral methods discussed below.

THEOREM 2. *The helicity of a unit vector field V defined on the compact domain Ω is bounded by*

$$|\,\text{H}(V)| \leq \frac{1}{2} \text{vol}(\Omega)^{4/3}.$$

This theorem, together with the formula of Berger and Field given above, yields an upper bound for the writhing of a DNA strand in terms of its length L and thickness $2R$:

THEOREM 3.

$$|\,\text{Wr}(K)| \leq \frac{1}{4}\left(\frac{L}{R}\right)^{4/3}.$$

Similar bounds have been obtained independently in [*Buck and Simon*, 1998], and also in [*Freedman and He*, 1991].

5. MAGNETIC FIELDS AND HELICITY

Start with a vector field V on the domain Ω, regard it as a current distribution, and use the Biot-Savart Law to compute its magnetic field:

$$\text{BS}(V)(y) = \frac{1}{4\pi} \int_\Omega V(x) \times \frac{y-x}{|y-x|^3} \, d\text{vol}_x.$$

See Figure 4.

The helicity of V can then be expressed as the integrated dot product of V with its magnetic field BS(V):

$$\begin{aligned} \text{H}(V) &= \frac{1}{4\pi} \int_{\Omega \times \Omega} V(x) \times V(y) \cdot \frac{x-y}{|x-y|^3} \, d\text{vol}_x \, d\text{vol}_y \\ &= \int_\Omega V(y) \cdot \frac{1}{4\pi} \int_\Omega V(x) \times \frac{y-x}{|y-x|^3} \, d\text{vol}_x \, d\text{vol}_y \\ &= \int_\Omega V \cdot \text{BS}(V) \, d\text{vol}. \end{aligned}$$

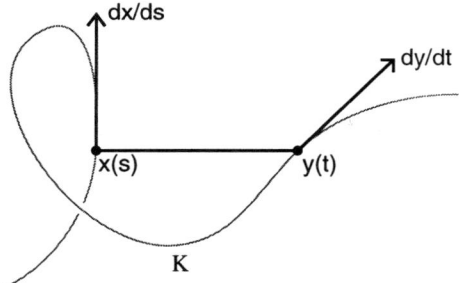

Figure 2. The writhing number integrand.

6. A GENERAL POINT OF VIEW

Let Ω be a compact domain in 3-space with smooth boundary. Let $\text{VF}(\Omega)$ denote the set of all smooth vector fields V on Ω. Then $\text{VF}(\Omega)$ is itself an infinite-dimensional vector space.

Define an inner product on $\text{VF}(\Omega)$ by the formula

$$\langle V, W \rangle = \int_\Omega V \cdot W \, d\text{vol}.$$

Although the magnetic field $BS(V)$ is well-defined throughout all of 3-space, we will restrict it to Ω; thus the Biot-Savart Law provides an operator

$$BS : \text{VF}(\Omega) \to \text{VF}(\Omega).$$

Using the above inner product notation, our formula for the helicity of V can be written

$$H(V) = \langle V, BS(V) \rangle.$$

7. THE MODIFIED BIOT-SAVART OPERATOR

Let $K(\Omega)$ denote the subspace of $\text{VF}(\Omega)$ consisting of all smooth divergence-free vector fields defined on Ω and tangent to its boundary.

Start with a vector field V in $K(\Omega)$ and compute its magnetic field, $BS(V)$. Restrict $BS(V)$ to Ω and subtract a gradient vector field so as to keep it divergence-free while making it tangent to $\partial\Omega$. Call the resulting vector field $BS'(V)$. The Hodge Decomposition Theorem in the Appendix tells us that the gradient vector fields on Ω form the orthogonal complement of $K(\Omega)$; hence $BS'(V)$ can be viewed as the orthogonal projection of $BS(V)$ back into $K(\Omega)$.

The *modified Biot-Savart operator*

$$BS' : K(\Omega) \to K(\Omega),$$

will play a leading role in our story.

The helicity of a vector field V in $K(\Omega)$ is given by

$$H(V) = \langle V, BS'(V) \rangle,$$

since $BS(V)$ and $BS'(V)$ differ by a gradient vector field, which as we just noted is orthogonal in the inner product structure of $\text{VF}(\Omega)$ to any vector field V in $K(\Omega)$.

8. SPECTRAL METHODS

From now on, we focus on vector fields which are divergence-free and tangent to the boundary of their domain, that is, on the subspace $K(\Omega)$ of $\text{VF}(\Omega)$, and on the modified Biot-Savart operator $BS' : K(\Omega) \to K(\Omega)$. A standard functional analysis argument yields

THEOREM 4. *The modified Biot-Savart operator BS' is a bounded operator, and hence extends to a bounded operator on the L^2 completion of its domain; there it is both compact and self-adjoint.*

The *Spectral Theorem* then promises that BS' behaves like a real self-adjoint matrix: the L^2 completion of its domain admits an orthonormal basis of eigenfields, in terms of which the operator is "diagonalizable". The eigenfields corresponding to the eigenvalues $\lambda(\Omega)$ of maximum absolute value have maximum helicity for given energy, and we obtain the sharp upper bound

$$|H(V)| \le |\lambda(\Omega)| \, E(V),$$

for all V in $K(\Omega)$.

This approach to the study of helicity was initiated in [*Arnold*, 1974] for the setting of closed orientable 3-manifolds. For a corresponding approach via the curl operator on domains in Euclidean space, see [*Yoshida and Giga*, 1990] and [*Laurence and Avellaneda*, 1991].

Figure 3. The vector field V.

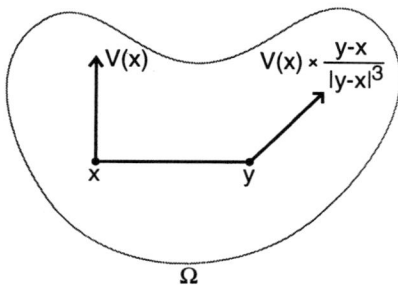

Figure 4. The Biot–Savart integrand.

9. CONNECTION WITH THE CURL OPERATOR

If the vector field V is divergence-free and tangent to the boundary of its domain Ω, that is, if V is in $K(\Omega)$, then
$$\nabla \times BS(V) = V.$$
Since $BS(V)$ and $BS'(V)$ differ by a gradient vector field, we also have
$$\nabla \times BS'(V) = V.$$
If V is an eigenfield of BS',
$$BS'(V) = \lambda V,$$
then
$$\nabla \times V = \frac{1}{\lambda} V.$$

Thus the eigenvalue problem for BS' can be converted to an eigenvalue problem for curl on the image of BS', which means to a system of partial differential equations. Even though we extended BS' to the L^2 completion of $K(\Omega)$ in order to apply the spectral theorem, the eigenfields are smooth vector fields in $K(\Omega)$; this follows, thanks to elliptic regularity, because on divergence-free vector fields, the square of the curl is the negative of the Laplacian. Hence these vector fields can be (and are) discovered by solving the above system of PDEs.

10. EXPLICIT COMPUTATION OF ENERGY-MINIMIZING VECTOR FIELDS

We solve $\nabla \times V = (1/\lambda)V$ on the flat solid torus $D^2(a) \times S^1$, where $D^2(a)$ is a disk of radius a and S^1 is a circle of any length; see [*Cantarella et al*, 1997a]. Although this is not a subdomain of 3-space, the solution here is so clear-cut and instructive as to be irresistible.

The eigenvalues of BS' of largest absolute value are
$$\lambda(D^2(a) \times S^1) = \pm \frac{a}{2.405...},$$
where the denominator is the first positive zero of the Bessel function J_0, and the corresponding eigenfields, discovered in [*Lundquist*, 1951], are
$$V_\lambda = J_1(r/\lambda)\,\hat{\varphi} + J_0(r/\lambda)\,\hat{z},$$
expressed in terms of cylindrical coordinates (r, φ, z) and the Bessel functions J_0 and J_1. See Figure 5.

It follows that for any V in $K(D^2(a) \times S^1)$,
$$|\,H(V)| \leq \frac{a}{2.405..}\, E(V),$$
with equality for the eigenfield V_λ.

We solve $\nabla \times V = (1/\lambda)V$ on the round ball $B^3(a)$ of radius a in terms of spherical Bessel functions in [*Cantarella et al*, 1998b].

The eigenvalues of BS' of largest absolute value are
$$\lambda(B^3(a)) = \pm \frac{a}{4.4934...},$$
where the denominator is the first positive zero of the function $(\sin x)/x - \cos x$. The corresponding eigenfields V_λ are Woltjer's models for the magnetic field in the Crab Nebula. In spherical coordinates (r, θ, φ) on a ball of radius $a = 1$,
$$V_\lambda(r, \theta, \varphi) = u(r,\theta)\,\hat{r} + v(r,\theta)\,\hat{\theta} + w(r,\theta)\,\hat{\varphi},$$
where
$$u(r,\theta) = \frac{2\lambda}{r^2}\left(\frac{\sin(r/\lambda)}{r/\lambda} - \cos(r/\lambda)\right)\cos\theta,$$
$$v(r,\theta) = -\frac{1}{r}\left(\frac{\cos(r/\lambda)}{r/\lambda} - \frac{\sin(r/\lambda)}{(r/\lambda)^2} + \sin(r/\lambda)\right)\sin\theta,$$
$$w(r,\theta) = \frac{1}{r}\left(\frac{\sin(r/\lambda)}{r/\lambda} - \cos(r/\lambda)\right)\sin\theta.$$

The values $\lambda = \pm 1/4.4934...$ make both $u(r,\theta)$ and $w(r,\theta)$ vanish when r=1, that is, at the boundary of the ball. As a consequence, the vector field V_λ is tangent to the boundary of the ball, and directed there along the meridians of longitude. See Figure 6.

It follows that for any V in $K(B^3(a))$,
$$|\,H(V)| \leq \frac{a}{4.4934...}\, E(V),$$
with equality for the eigenfield V_λ.

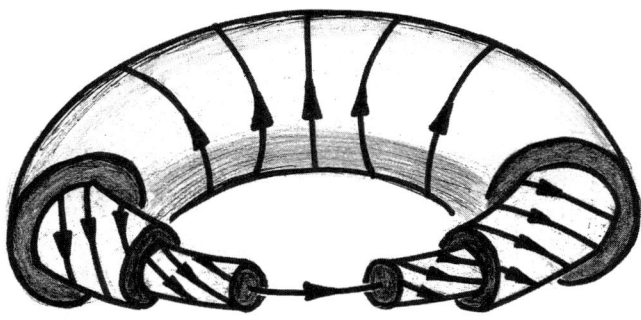

Figure 5. The Lundquist field.

Compare this with the rough upper bound from Theorem 1:
$$|H(V)| \leq a\, E(V).$$

11. THE ISOPERIMETRIC PROBLEM

We focus now on our second fundamental problem. A special case of this problem was considered in [*Chui and Moffatt*, 1995]:

Minimize energy among all divergence-free vector fields of given nonzero helicity, defined on and tangent to the boundary of all domains of given volume in 3-space.

When the domain Ω is fixed, the largest eigenvalue $\lambda(\Omega)$ of the modified Biot-Savart operator BS' is the largest possible value of the Rayleigh quotient:
$$\lambda(\Omega) = \max_V \frac{H(V)}{E(V)} = \max_V \frac{\langle V, BS'(V)\rangle}{\langle V, V\rangle}.$$

To maximize $\lambda(\Omega)$ among all domains of given volume, we want to take the "first derivative" of this quotient as the domain varies and set it equal to zero. This leads us to seek first variation formulas for helicity and energy.

12. FIRST VARIATION FORMULAS

Suppose the domain Ω is subject to a smooth volume-preserving deformation $h_t : \Omega \to \Omega_t$, with h_0 the identity, and initial velocity the divergence-free vector field W defined by $W(x) = \frac{d}{dt}|_{t=0} h_t(x)$.

Choose a vector field V in $K(\Omega)$, and let $V_t = (h_t)_* V$ be its push-forward to a vector field on the domain Ω_t. In other words, let V_t be frozen into the domain Ω_t as it deforms.

THEOREM 5. *The helicity $H(V_t)$ is independent of t.*

THEOREM 6. *The first variation of the energy of V_t, calculated at $t = 0$, is given by*
$$\delta E(V) = 2\langle V \times (\nabla \times V), W\rangle - \int_{\partial\Omega} |V|^2\, (W\cdot n)\, d\,\text{area}.$$

THEOREM 7. *The first variation of the largest eigenvalue $\lambda(\Omega_t)$ of BS' on Ω_t, calculated at $t = 0$, satisfies the inequality*
$$\delta\lambda(\Omega) \geq \lambda(\Omega)\frac{\int_{\partial\Omega} |V_\lambda|^2\, (W\cdot n)\, d\,\text{area}}{\int_\Omega |V_\lambda|^2\, d\,\text{vol}}.$$

where V_λ is a corresponding eigenfield.

The inequality appears only in the case that the largest eigenvalue has multiplicity > 1. This can certainly happen: when Ω is a round ball the largest eigenvalue has multiplicity 3. When this eigenvalue is simple, the inequality can be replaced by an equality.

13. CONSTRAINTS ON ANY OPTIMAL DOMAIN

The first variation formula in Theorem 7 leads to

THEOREM 8. *Suppose the vector field V defined on the compact, smoothly bounded domain Ω minimizes energy among all divergence-free vector fields of given nonzero helicity, defined on and tangent to the boundary of all such domains of given volume in 3-space. Then*

1. *$|V|$ is a nonzero constant on $\partial\Omega$.*

2. *All the components of $\partial\Omega$ are tori.*

3. *The orbits of V are geodesics on $\partial\Omega$.*

Thus, *no* smooth simply connected domain is optimal in the above sense. In principle, one could have

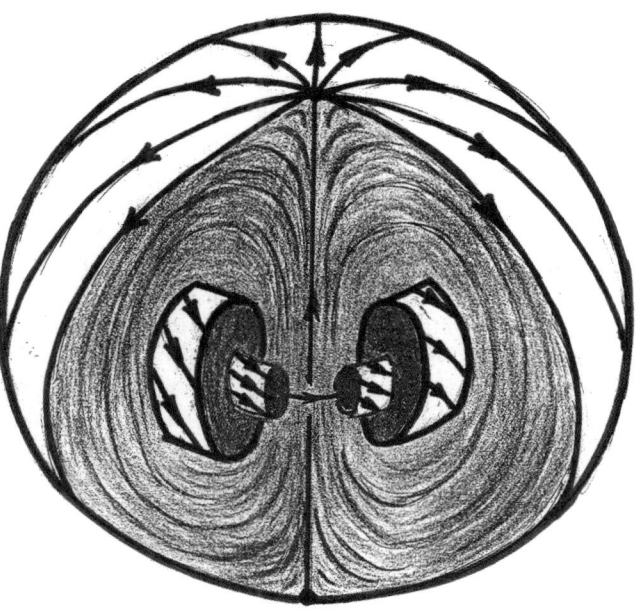

Figure 6. The Woltjer field.

a smooth optimal domain in the shape, say, of a solid torus. But we believe that *there are no smooth optimal domains at all,* regardless of topological type, and that the true optimizer looks like the singular domain shown in the next section.

14. THE SEARCH FOR OPTIMAL DOMAINS

Suppose we begin with the spheromak field V_λ which minimizes energy for given nonzero helicity on a round ball Ω, as discussed and pictured in section 10.

We seek a volume-preserving deformation of Ω which increases $\lambda(\Omega)$, guided by our inequality

$$\delta\lambda(\Omega) \geq \lambda(\Omega) \frac{\int_{\partial\Omega} |V_\lambda|^2 (W \cdot n) \, d\,\text{area}}{\int_\Omega |V_\lambda|^2 \, d\,\text{vol}}.$$

We maximize the right hand side by choosing

$$W \cdot n = |V_\lambda|^2 - \text{average value of } |V_\lambda|^2 \text{ on } \partial\Omega.$$

Then we imagine a volume-preserving deformation of Ω whose initial velocity field W has this preassigned normal component along the boundary. The deformation begins by dimpling Ω inwards near the poles and bulging it outwards near the equator, making the ball look somewhat like an apple.

At each stage Ω_t of the deformation, consider a vector field V_t which minimizes energy for given helicity on Ω_t, with $V_0 = V_\lambda$, and which determines the normal component $W_t \cdot n$ of the deformation velocity field W_t along the boundary in the same way as at the beginning:

$$W_t \cdot n = |V_t|^2 - \text{average value of } |V_t|^2 \text{ on } \partial\Omega_t.$$

Such a deformation tries to follow a path of steepest ascent for the largest eigenvalue $\lambda(\Omega_t)$ of the modified Biot-Savart operator.

We believe that this procedure will continue to dimple the apple inwards at the poles and bulge it outwards at the equator, until it reaches roughly the shape pictured in Figure 7, which then maximizes the largest eigenvalue $\lambda(\Omega)$ of the modified Biot-Savart operator among all domains of given volume. We can think of this singular domain either as an extreme apple, in which the north and south poles have been pressed together, or as an extreme solid torus, in which the hole has been shrunk to a point. We also show the expected appearance of the energy-minimizing vector field. The domain curiously resembles the NSTX (National Spherical Torus Experiment) containment device currently under construction at the Princeton Plasma Physics Laboratory.

Comparison of this picture with those of the energy-minimizers on the flat solid torus and on the round ball, given earlier, shows that we expect the common underlying pattern to persist even as the domain becomes singular, with the field in each case tangent to a family of nested tori with a single core curve.

A computational search for this singular optimal domain and the energy-minimizing vector field on it is at present under way, guided by a discrete version of the evolution described above.

APPENDIX. THE HODGE DECOMPOSITION THEOREM

1.1. *How Domain Topology Influences Vector Calculus*

In a multivariable calculus course, we are taught that the topology of the underlying domain affects the calculus of vector fields defined on it. For example, we learn that to test whether a vector field is the gradient of a function, we must take its curl and see if it is zero. If the curl is not zero, then the vector field is certainly not a gradient. If the curl is zero and the domain is simply connected, we learn that the vector field is a gradient. But if the curl is zero and the domain is not simply connected, then we learn that the vector field may or may not be a gradient, and that further tests are required.

The Hodge Decomposition Theorem for vector fields on domains in 3-space provides a more sophisticated level of control over this same subject.

The following two questions help to set the mood.

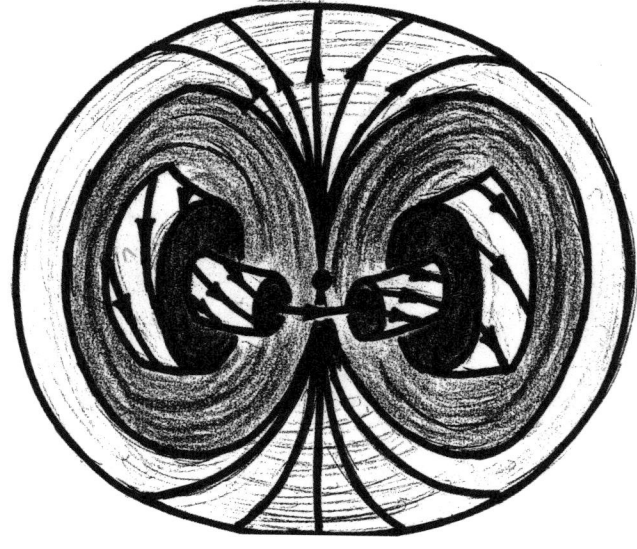

Figure 7. The expected optimal domain and field.

Question 1. *Is there a nonzero vector field V on the domain which is divergence-free, curl-free and tangent to the boundary?*

Question 2. *Is there a nonzero gradient vector field V on the domain which is divergence-free and orthogonal to the boundary?*

Domain	Answers to Question	
	1	2
Ball	No	No
Solid torus	Yes	No
Spherical shell	No	Yes
Toroidal shell	Yes	Yes

1.2. The Hodge Decomposition Theorem

Let Ω be a compact domain with smooth boundary in 3-space.

The following is arguably the single most useful expression of the interplay between the topology of the domain Ω, the traditional calculus of vector fields defined on this domain, and the inner product structure on VF(Ω) defined in section 6 by the formula $\langle V, W \rangle = \int_\Omega V \cdot W \, d\text{vol}$.

[Blank-Friedrichs-Grad, 1957] and [Schwarz, 1995] are good references; a detailed treatment and proof of this theorem in the form given below appears in our paper [Cantarella et al, 1997b].

HODGE DECOMPOSITION THEOREM. *We have a direct sum decomposition of VF(Ω) into five mutually orthogonal subspaces,*

$$\text{VF}(\Omega) = \text{FK} \oplus \text{HK} \oplus \text{CG} \oplus \text{HG} \oplus \text{GG},$$

with

$$\begin{aligned}
\ker \text{curl} &= \quad\quad\quad\;\; \text{HK} \oplus \text{CG} \oplus \text{HG} \oplus \text{GG}\\
\text{image grad} &= \quad\quad\quad\quad\quad\quad\;\; \text{CG} \oplus \text{HG} \oplus \text{GG}\\
\text{image curl} &= \text{FK} \oplus \text{HK} \oplus \text{CG}\\
\ker \text{div} &= \text{FK} \oplus \text{HK} \oplus \text{CG} \oplus \text{HG}
\end{aligned}$$

where

$\text{FK} = \{\nabla \cdot V = 0, V \cdot n = 0, \text{all interior fluxes } = 0\}$,
$\text{HK} = \{\nabla \cdot V = 0, \nabla \times V = 0, V \cdot n = 0\}$,
$\text{CG} = \{V = \nabla \varphi, \nabla \cdot V = 0, \text{all boundary fluxes } = 0\}$,
$\text{HG} = \{V = \nabla \varphi, \nabla \cdot V = 0, \varphi \text{ loc. constant on } \partial\Omega\}$,
$\text{GG} = \{V = \nabla \varphi, \varphi|_{\partial\Omega} = 0\}$,

and furthermore,

$$\begin{aligned}
\text{HK} &\cong H_1(\Omega; \mathbf{R}) \cong H_2(\Omega, \partial\Omega; \mathbf{R})\\
&\cong \mathbf{R}^{\text{genus of } \partial\Omega}.\\
\text{HG} &\cong H_2(\Omega; \mathbf{R}) \cong H_1(\Omega, \partial\Omega; \mathbf{R})\\
&\cong \mathbf{R}^{(\# \text{ components of } \partial\Omega) - (\# \text{ components of } \Omega)}.
\end{aligned}$$

We need to explain the meanings of the conditions which appear in the statement of this theorem.

The outward pointing unit vector field orthogonal to $\partial\Omega$ is denoted by n, so the condition $V \cdot n = 0$ indicates that V is tangent to the boundary of Ω.

Let Σ stand generically for any smooth surface in Ω with $\partial\Sigma \subset \partial\Omega$. Orient Σ by picking one of its two unit normal vector fields n. Then, for any vector field V on Ω, the *flux* of V through Σ is the value of the integral $\Phi = \int_\Sigma V \cdot n \, d\text{area}$.

If V is divergence-free and tangent to $\partial\Omega$, then the value of this flux depends only on the homology class of Σ in the relative homology group $H_2(\Omega, \partial\Omega; \mathbf{R})$. For example, if Ω is an n-holed solid torus, then there are disjoint oriented cross-sectional disks $\Sigma_1, \ldots, \Sigma_n$, positioned so that cutting Ω along these disks will produce a simply-connected region. The fluxes $\Phi_1, \ldots, \Phi_n$ of V through these disks determine the flux of V through any other cross-sectional surface.

If the flux of V through every smooth surface Σ in Ω with $\partial\Sigma \subset \partial\Omega$ vanishes, we say *all interior fluxes* $= 0$. Thus the subspace of vector fields V in VF(Ω) which have

$$\nabla \cdot V = 0, V \cdot n = 0, \text{ and all interior fluxes} = 0,$$

is called the subspace FK of *fluxless knots*.

The subspace HK of vector fields V in VF(Ω) with

$$\nabla \cdot V = 0, \nabla \times V = 0, V \cdot n = 0,$$

called *harmonic knots*, is isomorphic to the absolute homology group $H_1(\Omega; \mathbf{R})$ and also by Poincaré duality to the relative homology group $H_2(\Omega, \partial\Omega; \mathbf{R})$. It is thus a finite-dimensional vector space, with dimension equal to the (total) genus of $\partial\Omega$.

The orthogonal direct sum of these two subspaces,

$$K(\Omega) = \text{FK} \oplus \text{HK},$$

is the subspace of VF(Ω) mentioned earlier, consisting of all divergence-free vector fields defined on Ω and tangent to its boundary.

If V is a vector field defined on Ω, we will say that *all boundary fluxes of V are zero* if the flux of V through each component of $\partial\Omega$ is zero. The subspace of V in VF(Ω) with

$$V = \nabla \varphi, \nabla \cdot V = 0, \text{all boundary fluxes} = 0$$

is called the subspace CG of *curly gradients*, because these are the only gradients which lie in the image of curl.

The subspace HG of *harmonic gradients* consists of all V in VF(Ω) such that

$$V = \nabla\varphi, \nabla \cdot V = 0, \varphi \text{ locally constant on } \partial\Omega,$$

meaning that φ is constant on each component of $\partial\Omega$. This subspace is isomorphic to the absolute homology group $H_2(\Omega; \mathbf{R})$ and also, via Poincaré duality, to the relative homology group $H_1(\Omega, \partial\Omega; \mathbf{R})$, and is hence a finite-dimensional vector space, with dimension equal to the number of components of $\partial\Omega$ minus the number of components of Ω.

The definition of the subspace GG of *grounded gradients*, which consists of all V in VF(Ω) such that

$$V = \nabla\varphi, \varphi|_{\partial\Omega} = 0,$$

is self-explanatory.

We refer the reader to [*Cantarella et al*, 1997b] for a thorough treatment of the Hodge Decomposition Theorem and a variety of applications to boundary value problems for vector fields.

REFERENCES

Arnold, V.I., The asymptotic Hopf invariant and its applications, English translation in *Selecta Math. Sov.*, 5(4), 327-342, 1986; original in Russian, Erevan, 1974.

Berger, M.A. and Field, G.B., The topological properties of magnetic helicity, *J. Fluid Mech.* 147, 133-148, 1984.

Blank, A.A., Friedrichs, K.O., and Grad, H., Theory of Maxwell's Equations without Displacement Current. Theory on Magnetohydrodynamics V., *AEC Research and Development Report*, MHS, NYO-6486, 1957.

Buck, G. and Simon, J., Thickness and crossing number of knots, to appear in *Topology and its Applications*, 1998.

Călugăreanu, G., L'intégral de Gauss et l'analyse des noeuds tridimensionnels, *Rev. Math. Pures Appl.* 4, 5-20, 1959.

Călugăreanu, G., Sur les classes d'isotopie des noeuds tridimensionnels et leurs invariants, *Czechoslovak Math. J.* 11(86), 588-625, 1961.

Călugăreanu, G., Sur les enlacements tridimensionnels des courbes fermees, *Comm. Acad. R.P. Romine* 11, 829-832, 1961.

Cantarella, J., DeTurck, D., and Gluck, H., Upper bounds for the writhing of knots and the helicity of vector fields, preprint, 1997a.

Cantarella, J, DeTurck, D., and Gluck, H., Hodge decomposition of vector fields on bounded domains in 3-space, preprint, 1997b.

Cantarella, J., DeTurck, D., and Gluck, H., The Biot-Savart operator for application to knot theory, fluid dynamics and plasma physics, preprint, 1997c.

Cantarella, J., DeTurck, D., Gluck, H., and Teytel, M., Isoperimetric problems for the helicity of vector fields and the Biot-Savart and curl operators, preprint, 1998a.

Cantarella, J., DeTurck, D., Gluck, H., and Teytel, M., Eigenvalues and eigenfields of the Biot-Savart operator on spherically symmetric domains, preprint, 1998b.

Chui, A.Y.K., and Moffat, H.K., The energy and helicity of knotted magnetic flux tubes, *Proc. R. Soc. Lond. A*, 451, 609-629, 1995.

Freedman, M., and He, Z.-X., Divergence-free fields: Energy and asymptotic crossing number, *Annals of Math.* 134, 189-229, 1991.

Fuller, F.B., The writhing number of a space curve, *Proc. Nat. Acad. Sci. USA* 68(4), 815-819, 1971.

Gauss, C.F., Integral formula for linking number, in *Zur mathematischen theorie der electrodynamische wirkungen*, Collected Works, Vol. 5, Königlichen Gesellschaft des Wissenschaften, Göttingen, 2nd edition, 605, 1833.

Laurence, P., and Avellaneda, M., On Woltjer's variational principle for force-free fields, *J. Math Phys.* 32(5), 1240-1253, 1991.

Lundquist, S. Magneto-hydrostatic fields, *Arkiv Fysik*, 2 (35), 361-365, 1951.

Moffatt, H.K., The degree of knottedness of tangled vortex lines, *J. Fluid Mech.* 35, 117-129 and 159, 359-378, 1969.

Moffatt, H.K., and Ricca, R., Helicity and the Călugăreanu invariant, *Proc. R. Soc. Lond. A*, 439, 411-429, 1992.

Ricca, R. and Moffatt, H.K., The helicity of a knotted vortex filament, in *Topological Aspects of the Dynamics of Fluids and Plasmas* (H. K. Moffatt, ed.), 225-236, 1992.

Schwarz, G., *Hodge Decomposition— A Method For Solving Boundary Value Problems*, Lecture Notes in Mathematics, No. 1607, Springer Verlag, 1995.

Woltjer, L., The Crab Nebula, *Bull. Astr. Netherlands* 14, 39-80, 1958a.

Woltjer, L., A theorem on force-free magnetic fields, *Proc. Nat. Acad. Sci. USA* 44, 489-491, 1958b.

Yoshida, Z. and Giga, Y., Remarks on spectra of operator rot, *Math Z.* 204, 235-245, 1990.

J. Cantarella, D. DeTurck, H. Gluck and M. Teytel, Department of Mathematics, University of Pennsylvania, 209 S. 33rd Street, Philadelphia, PA 19104-6395. cantarel@math.upenn.edu; deturck@math.upenn.edu; gluck@math.upenn.edu; teytel@math.upenn.edu

Magnetic Energy And Helicity In Open Systems

B. C. Low[1]

High Altitude Observatory, National Center for Atmospheric Research, Boulder, Colorado

This paper summarizes the basic ideas of magnetic relaxation in the low-β, high-conductivity limit, taking the Taylor theory from contained magnetic fields to magnetic fields that thread across the boundary of the plasma domain such as encountered in astrophysical atmospheres. Several issues relating to magnetic helicity, gravitational confinement, magnetic flux and helicity ejection in open domains are analyzed within a coherent hydromagnetic framework.

1. INTRODUCTION

Many astrophysical circumstances involve long-lived magnetic fields which penetrate into the dense base of an open atmosphere. In the solar atmosphere between eruptions, magnetic structures anchored in this manner may persist for periods of time long compared to the characteristic time scales of hydromagnetic instabilities. Such structures present some basic hydromagnetic issues which are the subject of this paper. In particular, we will address the turbulent relaxation of anchored magnetic fields in the limit of low-β and high electrical conductivity; and, the self-confinement and gravitational confinement of anchored magnetic fields in open domains. We emphasize elementary physical properties and lead the discussion to an intriguing aspect of mass expulsion out of the solar corona.

2. TAYLOR RELAXATION

We begin with the Taylor theory [1974, 1986] for the turbulent relaxation of a low-β hydromagnetic plasma,

---

[1]The National Center for Atmospheric Research is sponsored by the National Science Foundation

Magnetic Helicity in Space and Laboratory Plasmas
Geophysical Monograph 111
Copyright 1999 by the American Geophysical Union

to see intuitively how it has worked well and where some physical limitations may lie. In this paper, we limit our attention to the single-fluid picture with isotropic pressure and electrical conductivity, and, motivated by astrophysical concerns, we treat only simply connected domains.

A magnetic field $\mathbf{B} = \nabla \times \mathbf{A}$, expressed in terms of its vector potential $\mathbf{A}$, has the magnetic helicity density $h = \mathbf{A} \cdot \mathbf{B}$ which, summed over a finite volume V, gives

$$H(\mathbf{B}; V) = \int_V dV \mathbf{A} \cdot \mathbf{B}. \quad (1)$$

Both h and H are unphysical because of their dependence on the free gauge of $\mathbf{A}$, except where V completely contains $\mathbf{B}$ as expressed by the condition

$$B_n|_{\partial V} = 0 \quad (2)$$

on the normal component B_n on the boundary ∂V. In this case, H is a physical measure of the twist in the magnetic field wholly contained in V [*Moffatt* 1978].

For a plasma with perfect electrical conductivity, the magnetic flux threading through any parcel of gas is conserved. This implies that if equation (2) is true for a volume of plasma V_m at any instant, it will remain true for all time, where V_m is defined to contain the same plasma particles. Only the plasma particles have physical identity, not the field. The conservation of magnetic flux may then be stated to require the magnetic helicity in every magnetic volume V_m to be invariant in time. If the plasma is not a perfect electrical conductor, the first thing to go is the identity of the magnetic volume V_m because of the resistive diffusion of magnetic field across the plasma. The many topological invariants described by *Kruskal and Kulsrud* [1958] that define an equilibrium state in the ideal hydromagnetic limit are lost and, among them, the magnetic-helicity invariants.

The difference between a highly resistive plasma and a highly conducting plasma is that, in the latter, resistivity is important only at sites of extreme magnetic gradients where magnetic reconnection may occur. Outside of these sites, resistive diffusion is negligible. Reconnection in a highly conducting plasma may thus be viewed intuitively to produce a cutting and joining of magnetic volumes with the consequence that individual magnetic volumes V_m defined at any one time have no permanent identity. If the whole magnetic field is contained in the volume V_0 of a rigid, perfectly conducting container, V_0 is the only magnetic volume with a permanent identity. Moreover, the integrand in equation (1), applied to $V = V_0$, will have suffered resistive changes only for brief moments and over spatially isolated regions where resistivity matters. If we accept the plausibility that this resistive change is small compared to the total integral, it follows that $H(\mathbf{B}; V_0)$ is approximately conserved. On the other hand, reconnection can give rise to a significant change in magnetic energy; see *Berger* [1984] for a formal treatment. The change in field topology allows the execution of plasma displacements that drain magnetic energy, displacements which are forbidden if resistivity is completely absent.

Based on these premises, the Taylor theory identifies the end state of relaxation in the low-β limit to be a magnetic field of minimum energy subject to the conservation of the total magnetic helicity $H(\mathbf{B}; V_0)$. This is the variational problem of *Woltjer* [1958] leading to the eigenvalue problem:

$$\nabla \times \mathbf{B} = \alpha_0 \mathbf{B}, \qquad (3)$$

subject to the homogeneous boundary condition

$$B_n|_{V_0} = 0, \qquad (4)$$

for a force-free magnetic field. The eigenvalue α_0 is to be selected to give the lowest magnetic energy that has the conserved total magnetic helicity.

In general, a force-free magnetic field satisfies:

$$\nabla \times \mathbf{B} = \alpha \mathbf{B}, \ \mathbf{B} \cdot \nabla \alpha = 0, \qquad (5)$$

where α may vary in space but by virtue of $\nabla \cdot \mathbf{B} = 0$ is a constant along a line of force. The function α is also a measure of the twist in each layer of magnetic flux, and is related in a non-trivial way, within each magnetic volume V_m, to the magnetic helicity contained in it. The Taylor theory describes a relaxation process which spreads the total helicity throughout the contained plasma. The spread is even in the sense of a homogeneous distribution of α.

The Taylor theory is quantitatively successful in explaining certain laboratory experiments, notably, the Reversed-Field Pinch device. Application to other devices has been less successful, leading to suggestions that relaxation to energy minimum under some circumstances may involve constraints additional to the conservation of the total magnetic helicity, e.g., [*Bhattacharjee and Déwar* 1982]. In the discussion to follow, the same suggestion will arise, that the relaxation to a constant-α force-free field may be thwarted even if the total helicity is conserved. We will also see that for the process in open domains, the conservation of total helicity may be vitiated by the possibility of an ideal (astrophysical) transport of helicity to infinity.

3. ERGODIC LINES OF FORCE

There is a topological property of turbulent magnetic fields which has a fundamental role in magnetic relaxation. In the absence of symmetry, most magnetic fields in three-dimensional space, not necessarily in equilibrium, have ergodic lines of force [*Grad* 1967; *Parker* 1979; *White* 1983]. Most fields as well behaved mathematical functions of space, upon direct integration, give lines of force which are not closed but are such that a single line may wind endlessly to fill up a sub-volume of space [*Dombre et al.* 1986].

It is instructive to consider the following elementary problem on magnetic lines of force. A line in any locality of space is an intersection between two surfaces. This geometric fact lends readily to the representation of the divergence-free field $\mathbf{B}$ in terms of suitably constructed Euler (or Clebsch) potentials U and V:

$$\mathbf{B} = \nabla U \times \nabla V. \qquad (6)$$

Then, for constant values of U_0 and V_0, $U = U_0$ and $V = V_0$ generate two families of flux surfaces intersecting along the lines of force of $\mathbf{B}$ [*Stern* 1966]. The pair (U, V) are just the two integration constants of the pair of ordinary differential equations $\frac{dx}{B_x} = \frac{dy}{B_y} = \frac{dz}{B_z}$ for a line of force, written in Cartesian coordinates, with $\mathbf{B}$

being a known function of space. It is a mathematical fact that this representation of solutions cannot be made global for most prescribed $\mathbf{B}$, in the sense of integrability in the language of analytical mechanics. The pairs of integrals (U, V) developed at different localities cannot, in general, be replaced by a single pair which has global validity.

Only those fields which are untwisted may be represented by equation (6) with globally defined (U, V) [*Moffatt* 1978]. Equation (6) implies $\mathbf{B} = \nabla \times (U \nabla V)$ leading to the vector potential $\mathbf{A} = U \nabla V$ that renders $\mathbf{A} \cdot \mathbf{B} = 0$ everywhere. In other words, a gauge exists for the magnetic helicity density to vanish everywhere if a global pair (U, V) can be found. For fields which are twisted, any attempt to construct a global pair (U, V) will result in mathematical anomalies such as multivalued functions and related singularities, e.g. [*Rosner et al.* 1989].

It is important to distinguish between a line of force and a flux bundle, the former carries no flux and the latter carries a definite amount of flux threading across an area. Central to the idea of a flux bundle is that a magnetic flux surface exists to define the boundary of the local rope-like bundle. A flux bundle can always be so defined locally but, for twisted fields, runs into topological problems when extended globally along the lengths of the lines of force contained in the flux bundle. For example, starting with the representation in equation (6) in some locality, the flux surfaces U and V could be suitably continued out of the locality by following the line of force along its path. If the line is ergodic filling a subvolume, we would have the anomaly that every point of this subvolume lies on the two "surfaces" of constant U and V, pointing to the absence of a globally valid (U, V)-pair.

To keep some familiar objects in mind, think of an axisymmetric poloidal field to be an example of an untwisted field, and an axisymmetric field with a ϕ component an example of a twisted field. The latter has only one family of global flux surfaces, namely, the nested tori containing winding lines of force. Except for a subset of measure zero, these lines of force are ergodic in the torus flux surfaces in the sense that a single line winds progressively and endlessly to fill up an entire torus surface. If the symmetry about the axis is broken topologically (as opposed to deforming an axisymmetric field under the frozen-in condition), the lines of force generally become ergodic in a subvolume [*White* 1983].

Recall equations (5) describing the general force-free magnetic field. If such an equilibrium state is approached via states with volume-filling ergodic lines of force, the end state is likely to have a spatially uniform α, in particular, if the end state is a single ergodic line of force which propagates the same value of α in equations (5) to every point in the volume. In some situations, the device containing the magnetic field imposes some symmetry, say, axisymmetry. The magnetic field would then start as an axisymmetric unstable initial state, undergo turbulent disruption which breaks its symmetry, and then settle to an end state which may regain axisymmetry. At least one family of flux surfaces exist in both the initial and final states, although the loss of symmetry during the disruption would result in ergodic fields. Reconnection would need to destroy the volumetric ergodicity and establish the flux surfaces of the axisymmetric end state. If these flux surfaces form together and rapidly, it is conceivable that they would take on values of α close to each other. If they form in succession with time lapses, then it is not compelling to assert that the values of α trapped in different end-state flux surfaces would have to be the same. Then, some additional topological constraints must apply to forbid the Taylor end-state, in favor of a force-free field with a spatially variable α of some determined distribution. It is unclear whether such constraints exist. It is possible that outside of the Taylor theory one needs to solve the full hydromagnetic problem because the end state is history dependent and is not constrained by topological factors alone.

4. ANCHORED MAGNETIC FIELDS

Consider a magnetic field which threads across the boundary ∂V_0 of its domain V_0 with some given flux distribution F:

$$B_n|_{\partial V_0} = F(\partial V_0), \quad (7)$$

To keep matters simple, ∂V_0 is a rigid wall with no flow across it. Take V_0 to be finite for the present. The total helicity $H(\mathbf{B}; V_0)$ is not gauge invariant.

Berger and Field [1984] proposed the construction of a relative helicity H_R to replace $H(\mathbf{B}; V_0)$ as a gauge-invariant measure of magnetic topology in V_0. They first noted that boundary condition (7) defines a unique potential field $\mathbf{B}_{pot}$ satisfying

$$\nabla \times \mathbf{B}_{pot} = 0. \quad (8)$$

Denote the exterior of V_0 by V_0' defined so that the combined volume $V_T = V_0 + V_0'$ contains the entire magnetic field $\mathbf{B}$ extended in some unspecified manner from the volume V_0 to its exterior V_0'. Then, the total helicity $H(\mathbf{B}; V_T)$ is gauge invariant. Take the same magnetic field in V_0' to be the extension of $\mathbf{B}_{pot}$ and calculate the total helicity $H(\mathbf{B}_{pot}; V_T)$. The difference between the two total helicities is the relative helicity $H_R = H(\mathbf{B}; V_T) - H(\mathbf{B}_{pot}; V_T)$ given by

$$H_R(\mathbf{B}, \mathbf{B}_{pot}; V_0) = \int_{V_0} (\mathbf{A} - \mathbf{A}_{pot}) \cdot (\mathbf{B} + \mathbf{B}_{pot}) dV + \int_{\partial V_0} \chi (\mathbf{B} + \mathbf{B}_{pot}) \cdot \hat{\mathbf{n}} dS \qquad (9)$$

involving only integration in V_0 and on boundary ∂V_0. The quantity χ is expressible in terms of $\mathbf{B}$ and $\mathbf{B}_{pot}$ defined in V_0 [*Berger and Field* 1984]. This relative helicity is completely independent of the form of $\mathbf{B}$ taken in the exterior region V_0', and H_R is invariant to changes of gauge in either $\mathbf{B}$ or $\mathbf{B}_{pot}$. A criticism by this author that H_R is only gauge invariant in a restricted manner is in error and should be ignored [*Low* 1996].

The Berger-Field construction reduces naturally to the regular magnetic helicity when the boundary flux at ∂V_0 vanishes, in which case $\mathbf{B}_{pot} = 0, \chi = 0$ for simply connected domains. But, it is not a trivial question whether the relative total helicity in V_0 is conserved in hydromagnetic processes.

The conceptual involvement of $\mathbf{B}_{pot}$ may be avoided by considering a specific situation in which the external magnetic field in V_0' is explicitly known, to show how relative helicity makes complete physical sense. Suppose V_0' is a rigid perfect conductor containing an explicit extension of the magnetic field in V_0. As the magnetic field in V_0 undergoes relaxation by magnetic reconnection, the field in V_0' does not change. The total magnetic helicity in the combined volume V_T is conserved over the time scale of energy decay for the same reasons having to do with high conductivity given for the wholly contained plasma. Hence the end state must contain the same total helicity in V_T it started with. Having assured that this total helicity is conserved, it follows that the relative helicities of the initial and end states, obtained by adjusting the total helicity by the fixed amount $H(\mathbf{B}_{pot}; V_T)$, must be conserved.

If the conservation of relative helicity is the only constraint, Woltjer's variational problem for the minimum energy state again leads to equation (3) for a constant-α_0 force-free field but subject to the inhomogeneous boundary condition (7) [*Berger* 1984]. Despite the involvement of $\mathbf{B}_{pot}$, the Euler-Lagrangian equation, namely, equation (3), does not contain $\mathbf{B}_{pot}$.

Two questions about this extension of the Taylor theory merit attention. The original Taylor theory had rested on its remarkable success in explaining specific laboratory experiments, but there is no comparable experimental verification of the theory for anchored fields.

The magnetic lines of force anchored to the boundary ∂V_0 have finite footpoint-to-footpoint lengths in V_0. In the course of relaxation, some parts of the magnetic field may, by reconnection, completely disconnect from the boundary to lie entirely within V_0. Only such disconnected parts of the field may have ergodic lines of force filling up a subvolume and rendering the field-aligned currents compatible with a spatially uniform α in the end state. On the other hand, the anchored lines of force may trap a spatially variable α in the end state. The end state is then not a solution of equation (3), suggesting that topological constraints additional to the conservation of relative total magnetic helicity may apply. Future work needs to resolve this fundamental issue, perhaps by direct numerical simulation.

5. OPEN DOMAINS

The boundary value problem posed by equations (3) and (7) has a different character when V_0 is unbounded, with a part of ∂V_0 located at infinity. Equation (7) now describes a given flux on the finite part of the boundary to which we add the requirement that $\mathbf{B}$ vanishes at infinity. Typically we are thinking of the space outside a unit sphere in some astrophysical context. This boundary value problem has mathematically regular solutions for $\alpha_0 \neq 0$, but the solutions give magnetic fields vanishing at large distances so slowly that its total energy is generally unbounded. No finite-energy solution exists, except for the $\alpha_0 = 0$ potential field [*Seehafer* 1978; *Berger* 1985; *Aly* 1992; *Laurence and Avellaneda* 1993; *Low* 1996].

This result has a simple physical interpretation. The Woltjer variational problem seeks a minimum-energy state by spreading the conserved total magnetic helicity throughout the domain with no concentration anywhere. In an unbounded domain, spreading any finite quantity with no concentration must result in a density everywhere vanishingly small. A force-free field of vanishingly small helicity density is indistinguishable from a potential field. Thus the solution to the Woltjer variational problem for the infinite domain is the potential field irrespective of the magnitude of the prescribed total magnetic helicity. A force-free field with a non-zero α_0 is a field with a finite magnetic helicity density and thus can only exist in the infinite domain if the magnetic field has infinite total energy and helicity.

Extending the idea of relaxation to an unbounded domain is fundamentally problematical. A basic feature of high-conductivity relaxation processes is that outside of spatially isolated resistive regions, the plasma sends information as an ideal, low-β hydromagnetic fluid limited by the propagation speeds of Alfven waves. These speeds are very high for a typical laboratory device of meter size but are unimpressive over astronomical distances. Distant parts of an infinite system would take unboundedly long times to interact, and, for such a system, it is not generally meaningful to take the Taylor relaxation process to its logical conclusion. A different physical consideration is needed.

6. MAGNETIC SELF-CONFINEMENT

In a finite domain, magnetic helicity is transported from sub-volume to sub-volume by magnetic reconnection. In the open domain, there is the additional freedom to lower magnetic energy by having a part of the magnetic flux take some amount of magnetic helicity with it to be ejected to infinity. Consider the virial theorem for a force-free magnetic field:

$$\int_{V_0} \frac{B^2}{8\pi} dV = \frac{1}{4\pi} \int_{\partial V_0} \left\{ \frac{B^2}{2} (\mathbf{r} \cdot d\mathbf{S}) - (\mathbf{B} \cdot \mathbf{r})(\mathbf{B} \cdot d\mathbf{S}) \right\}, \tag{10}$$

where $\mathbf{r}$ is the position vector from the origin of the coordinate system used, and $d\mathbf{S}$ denotes the directed area element on the boundary ∂V_0. Thus, the total energy of a force-free field is determined entirely in terms of its vector boundary values. This theorem shows that force-free magnetic fields cannot be self-confining without being anchored to some non-force-free region [*Chandrasekhar* 1961]. The magnetic tension force counts on such an anchoring in order to confine the positive pressure of the magnetic field.

For our purpose, we look at two simple but instructive applications. First take V_0 to be the the spherical volume $r < r_0$, a constant radius. Equation (10) gives

$$\int_{r<r_0} \frac{B^2}{8\pi} dV = \frac{r_0^3}{8\pi} \int_{r=r_0} \left\{ B_\theta^2 + B_\phi^2 - B_r^2 \right\} \sin\theta d\theta d\phi, \tag{11}$$

in spherical coordinates. Any force-free field in $r < r_0$ will of course ensure that the surface integral on the right side is positive. The magnetic field can have any amount of energy depending on how highly twisted the field is. The case of a wholly contained field with $B_r = 0$ at $r = r_0$ is simplest: solutions to equations (3) and (4) allow the force-free field in $r < r_0$ to have any amplitude.

Now consider the case of V_0 being the infinite space $r > r_0$. For any physical electrical currents, the field at infinity disappears as fast as a dipole potential field, so that the surface integral at infinity in equation (10) has no contribution and we obtain

$$\int_{r>r_0} \frac{B^2}{8\pi} dV = \frac{r_0^3}{8\pi} \int_{r=r_0} \left\{ B_r^2 - B_\theta^2 - B_\phi^2 \right\} \sin\theta d\theta d\phi. \tag{12}$$

Equilibrium requires field anchoring at $r = r_0$ where $B_r \neq 0$. No force-free field can exist in $r > r_0$ that does not thread across $r = r_0$. Any field with $B_r = 0$ at $r = r_0$ would simply expand to zero field density in the unbounded space. With anchoring, there is an obvious upper bound E_{ub} on the energy of the force-free field, given by equation (12) retaining only the positive term B_r^2 in the integrand on the right. The least upper bound is significantly smaller than E_{ub} since no force-free field is likely to have vanishing mean square tangential components at $r = r_0$. By implication, when an anchored field with some fixed B_r distribution at $r = r_0$ is pumped to an excessive level of magnetic helicity associated with an extremely large amount of energy; a dynamical transition must disconnect a part of the magnetic flux from the inner boundary and eject the excessive helicity to infinity, if the bound on magnetic energy is not to be violated.

7. SOLAR CORONAL MASS EJECTIONS

Recent interest in magnetic flux ejection arose in the study of Coronal Mass Ejections, or CMEs [*Hundhausen* 1998; *Crooker, Joselyn and Feynman* 1997]. These are large-scale, rapid reconfigurations of the solar corona, each event ejecting a significant amount of mass to disturb the steady solar wind flowing out of the corona into interplanetary space. Observation supports the interpretation that a CME is the forceful outward stretching of an anchored bipolar magnetic field to release the mass trapped in the field. Once the trapped mass has been ejected, the opened field recloses by magnetic reconnection to produce a characteristic flare heating of the low corona.

A majority of CMEs originate in quiescent magnetic structures that have persisted for some length of time prior to eruption. If we take the view that the quiescent state still contains a large amount of magnetic energy associated with magnetic helicity accumulated over time, the CME can be interpreted as a two-step hydromagnetic process: The CME sets in as an ideal hydromagnetic outflow when the confinement of both field and helicity fails, and the magnetic disconnection of the ejected field returns the anchored part of the field to a state of lowered energy and helicity [*Low* 1994].

Out of this solar research came several basic hydromagnetic results relevant to our discussion, which we briefly describe to conclude this paper. First is the Aly conjecture which states that a force-free field in $r > r_0$ containing only lines of force anchored to $r = r_0$ cannot have more energy than an open state [*Aly* 1991]. The open state is meant to be one produced by forcefully opening every line of force of the anchored field to infinity, leaving B_r unchanged at $r = r_0$. Aly's conjecture removes a naive but otherwise attractive idea that a progressively sheared force-free field could, by its stored energy, spontaneously transit to a fully open state [*Barnes and Sturrock* 1972].

The characteristic amounts of energy involved are illustrated by the axisymmetric case of a potential dipole field in $r > r_0$ with a $\cos\theta$ profile for B_r on $r = r_0$ [*Low and Smith* 1993]. Denote its total energy in $r > r_0$ by E_{pot}. If this poloidal field is given a force-free current associated with a shear without changing the boundary flux at $r = r_0$, there is no way to bring its energy in $r > r_0$ above about $1.7 E_{pot}$. This threshold is the minimum energy lodged in the stretched-out field when the tops of all the lines of force are pulled to infinity with the footpoints of the lines rigidly anchored to $r = r_0$. Aly's conjecture says nothing against the sheared force-free field having an energy in excess of this threshold provided that, in addition to the anchored field, a bundle of magnetic flux closed and lying entirely in $r > r_0$ is present. Such a force-free field has enough energy to spontaneously open up but the excess energy available to drive the medium is small. Equation (12) sets an obvious upper limit of $E_{ub} = 2E_{pot}$ for the energies of all possible force-free fields with the same boundary dipolar flux at $r = r_0$. This upper limit is a clear over estimate. The least upper bound of the energies lies much closer to $1.7 E_{pot}$ than $2 E_{pot}$.

Of course the field is not compelled to open up all of its lines of force. Energy is always adequate for the modest opening up of only a fraction of the magnetic flux anchored to $r = r_0$ [*Low* 1986; *Wolfson and Low* 1992]. It is in the opening up of all the lines of force, the case of relevance to CMEs, that incurs the constraint of Aly's conjecture. In this case the CMEs compels us to abandon the force-free field model. Observation shows that a CME typically carries away a significant energy in comparable amounts of bulk kinetic energy and gravitational potential energy. The CME energy is of the same order of and sometimes even larger than the flare energy liberated by the reclosing of the opened magnetic field. This implies that as much or even more energy is spent in driving the coronal material as is left in the open field whose energy is subsequently liberated as a flare. Since the lower atmosphere moves far too slowly to directly drive a CME at its coronal hydromagnetic speeds, the combined energies of the CME and its associated flare must be stored largely in the pre-eruption coronal magnetic field. From the above analysis, this large amount of energy is unlikely to be stored in a force-free field.

The low-β approximation is not uniformly valid in the solar corona. The field declines with heliocentric distance more rapidly than the decline of the two-million degree hydrostatic, nearly isothermal, plasma pressure. Above about one solar radius from the coronal base, the plasma is able to take the frozen-in magnetic field out with the quasi-steady solar wind. Moreover, observation has shown that gravitational potential energy is significant for CMEs. This alone suggests that departure from the force-free assumption is essential to an understanding of CMEs.

Abandoning the force-free approximation, static equilibrium is described by the balance of forces:

$$\frac{1}{4\pi}(\nabla \times \mathbf{B}) \times \mathbf{B} - \nabla p - \rho \frac{GM}{r^2}\hat{\mathbf{r}} = 0, \quad (13)$$

where p and ρ are the plasma pressure and density, respectively, G is Newton's gravitational constant, and M the solar mass. The virial theorem applied to equation (13) then gives:

$$\int_{r>r_0} \left\{ \frac{B^2}{8\pi} + 3p - \rho\frac{GM}{r} \right\} dV =$$
$$\frac{r_0^3}{8\pi} \int_{r=r_0} \left\{ B_r^2 - B_\theta^2 - B_\phi^2 - 8\pi p \right\} \sin\theta d\theta d\phi . \quad (14)$$

The application of equation (14) to the corona treated as a static atmosphere is of course at variance with the existence of the global solar wind outflow everywhere above about a solar radius from the coronal base. For our discussion, this simplification is tolerated in order to make the following instructive point.

The bound E_{ub} on the magnetic energy encountered in equation (12) with force-free fields no longer applies in the gravitationally stratified atmosphere. The point is especially clear when there is cold material whose weight is significant and largely supported by magnetic tension force as opposed to pressure gradients [*Low* 1999]. Neglecting the pressure p in equation (14) in such a case, the term for the total gravitational potential energy allows the magnetic energy to exceed the bound E_{ub} on the energies of force-free fields. The unloading of the anchoring cold material would be a simple way of releasing the huge store of magnetic energy to drive a CME.

This issue of energy budget for the CME brings to focus the nature of force-free fields in the open domain. A force-free field expands under its positive magnetic pressure when progressively stressed. A rigid wall counters this expansion and allows an unbounded magnetic energy to build up in the highly compressed field; see equation (11) for a contained field. In an open domain with no external walls, a force-free field expands outward with progressive stressing to self-confine by its tension force. Such an expansion eventually leads to confinement failure to eject excessive twists so that the permissible states of equilibrium have energies with a stringent absolute bound. The weight of an atmosphere approximates a rigid wall to confine the magnetic field, and provides the only means of building up magnetic energies to, in principle, unlimited values. Direct modeling has shown that gravitational confinement combined with a detached flux rope embedded in an anchored coronal field can explain the energetics of the CME and its associated flare [*Low and Smith* 1993; *Low* 1994, 1996].

The interaction among the magnetic field, the plasma pressure and gravity, in addition to allowing for an open domain for astrophysical situations such as the solar corona, takes the physical problem to a complexity not likely to be fruitfully handled without dealing with the hydromagnetic equations directly. The elegance of the original Taylor theory is that it avoids the complexity of hydromagnetic turbulence to topologically find the end state in a remarkably forthright procedure. ¿From a solar physics perspective, observation still has much to tell us about how natural systems really work. Coupled with the high power of numerical simulation currently available, observation must be relied upon to get some phenomenology clearly understood before theoretical ideas could germinate and provide insight into the physics as the Taylor theory in its own way has.

8. CONCLUSION

This has been an essay to discuss the hydromagnetic issues of the turbulent relaxation of anchored magnetic fields such as encountered in astrophysical atmospheres. We have provided a coherent hydromagnetic framework within which these issues hopefully stand out clearly in physical terms. Not all the issues can be resolved at the present but theory may be guided by future work in observation and in high-conductivity numerical simulation.

Three conclusions in the discussion should be repeated here as a summary.

The first is that the low-β relaxation to a constant α force-free magnetic field in a finite domain bounded by rigid walls, may not be taken for granted even if the total magnetic helicity is conserved in the limit of high electrical conductivity. The need to impose additional constraints merits further investigation, especially for fields anchored to rigid walls.

The second is that there are stringent requirements for the self-confinement of force-free magnetic fields in open domains, complicated by magnetic disconnection and ejection to remove excessive helicity and energy, in quite interesting ways as seen in solar CMEs.

The third is the richness of physics, as encountered with CMEs, obtaining in an atmosphere with finite-β effects. In such a system, there is the possibility of building up magnetic energy, through gravitational confinement, well in excess of bounds which apply stringently only to force-free fields.

Acknowledgment. Bob Kerr and the referees provided helpful comments on the article. The author thanks C. Z. Cheng and Princeton Plasma Physics Laboratory for hospitality during a sabbatical visit.

REFERENCES

Aly, J. J., How much energy can be stored in a three-dimensional force-free magnetic field?, *Astrophys. J. Lett.*, *375*, L61, 1991.

Aly, J. J., Some properties of finite energy constant-α force-free magnetic fields in a half space, *Solar Phys.*, *138*, 133, 1992.

Barnes, C. W. and P. A. Sturrock, Force-free magnetic-field structures and their role in solar activity, *Astrophys. J.*, *174*, 659, 1972.

Berger, M. A., Rigorous new limits on magnetic helicity dissipation in the corona, *Geophys. Astrophys. Fluid Dyn.*, *30*, 79, 1984.

Berger, M. A., Structure and stability of constant α force-free fields, *Astrophys. J. Supp.*, *59*, 433, 1985.

Berger, M. A. and G. B. Field, The topological properties of magnetic helicity, *J. Fluid Mech.*, *147*, 133, 1984.

Bhattacharjee, A. and R. L. Dewar, Energy principle with global invariants, *Phys. Fluids*, *25*, 887, 1982.

Chandrasekhar, S., *Hydrodynamic and hydromagnetic stability*, (Oxford University Press), 1961.

Crooker, N., J.-A. Joselyn and J. Feynman (Eds.), *Coronal*

mass ejections, (American Geophys. Union: Washington, DC), 1997.

Dombre, T., et al., Chaotic streamlines of the ABC flows, *J. Fluid Mech.*, *167*, 353, 1986.

Grad, H., Toroidal containment of a plasma, *Phys. Fluids*, *10*, 137, 1967.

Hundhausen, A. J., Coronal mass ejections: A summary of SMM observations from 1980 and 1984 - 1989, in *The many faces of the sun*, edited by K. Strong, J. Saba and B. Haisch, in press, 1998.

Kruskal, M. D. and R. M. Kulsrud, Equilibrium of a confined plasma in a toroid, *Phys. Fluids*, *1*, 265, 1958.

Laurence, P. and M. Avellaneda, Woltjer's variational principle, II: The case of unbounded domain, *Geophys. Astrophys. Fluid Dyn.*, *69*, 201, 1993.

Low, B. C., Blowup of force-free magnetic fields in the infinite region of space, *Astrophys. J.*, *307*, 205, 1986.

Low, B. C., Magnetohydrodynamic processes in the solar corona: Flares, coronal mass ejections, and magnetic helicity, *Phys. Plasma.*, *1*, 1684, 1994.

Low, B. C., Solar activity and the corona, *Solar Phys.*, *167*, 217, 1996.

Low, B. C., The hydromagnetic structure of solar quiescent prominences, preprint, 1999.

Low, B. C. and D. F. Smith, The free energies of partially open coronal magnetic fields, *Astrophys. J.*, *410*, 413, 1993.

Moffatt, H. K., *Magnetic field generation in electrically conducting fluids*, (Cambridge U. Press), 1978.

Parker, E. N., *Interplanetary dynamical processes*, (Interscience: New York), 1963.

Parker, E. N., *Cosmical magnetic fields*, (Oxford University Press), 1979.

Rosner, R., B. C. Low, K. Tsinganos and M. A. Berger, On the relationship between the topology of magnetic field lines and flux surfaces, *Geophys. Astrophys. Fluid Dyn.*, *48*, 251, 1989.

Seehafer, N., Determination of constant α force-free magnetic fields from magnetograph data, *Solar Phys.*, *58*, 215, 1978.

Stern, D. P., The motion of magnetic field lines, *Space Sci. Rev.*, *6*, 147, 1966.

Taylor, J. B., Relaxation of toroidal plasma and generation of reverse magnetic fields, *Phys. Rev. Lett.*, *33*, 1139, 1974.

Taylor, J. B., Relaxation and magnetic reconnection in plasmas, *Rev. Mod. Phys.*, *58*, 741, 1986.

White, R. B., Resistive instabilities and field line reconnection, in *Handbook of Plasma Physics*, edited by M. N. Rosenbluth and R. Z. Sagdeev, Vol. 1, 611, (North Holland Pub. Co.), 1983.

Wolfson, R. and B. C. Low, Energy buildup in sheared force-free magnetic fields, *Astrophys. J.*, *391*, 353, 1992.

Woltjer, L., A theorem on force-free magnetic fields, *Proc. Natl. Acad. Sci. USA*, *44*, 489, 1958.

B. C. Low, High Altitude Observatory, National Center for Atmospheric Research, P. O. Box 3000, Boulder, CO 80301. (e-mail: low@ucar.edu)

Helicity and Its Role in the Varieties of Magnetohydrodynamic Turbulence

David C. Montgomery and Jason W. Bates[1]

Department of Physics and Astronomy, Dartmouth College, Hanover, New Hampshire

Magnetic helicity has appeared as an important but slippery quantity in the theory of magnetohydrodynamic (MHD) turbulence in two contexts: (1) as a slowly-decaying ideal invariant that can control to some extent the formation of a "relaxed" MHD state — one far from thermal equilibrium — in laboratory confinement devices such as the toroidal pinch; and (2) as a potentially inversely-cascadable global quantity in driven, homogeneous MHD turbulence. In the former case, the origin of helicity is straightforwardly clear: electric current is forced to flow along a dc magnetic field, generating poloidal magnetic flux and causing the magnetic field lines to kink up, helically. In the latter, helicity's origins and physical interpretation are more obscure, sometimes having to do with mechanically driven helical motions which supposedly generate magnetic helicity that, however, no longer has any obvious "linked flux" interpretation. In both cases, its usefulness and even its definition sometimes depend sensitively on boundary conditions in a way that, say, those for energy do not. We will examine what the utility of the concept of magnetic helicity has so far been shown to be in discussing turbulent MHD, and comment on some of the ways it differs from other global ideal invariants that have been discussed, such as kinetic energy in 2D Navier-Stokes flows, and mean-square magnetic vector potential in 2D MHD. Attention will be devoted to the evidence for variational principles such as "maximal helicity," or "minimum energy," conjectured to predict various relaxation processes and late-time laminar states in evolving MHD situations. What is believed to be an important distinction between applications of the principles to decaying and driven situations will be stressed. Our discussion will be confined to the cases of small but non-zero transport coefficients, and will not deal with any possible role of helicity in ideal MHD.

---

[1] Applied Theoretical and Computational Physics Division, Los Alamos National Laboratory, Los Alamos, New Mexico

Magnetic Helicity in Space and Laboratory Plasmas
Geophysical Monograph 111
Copyright 1999 by the American Geophysical Union

1. INTRODUCTION

This article will focus on two rather different contexts in which the ideal global invariant, magnetic helicity, has appeared in the theory of magnetohydrodynamic (MHD) turbulence. Ideal global invariants in continuum mechanics appear in two rather different varieties. First, there are very general conservation laws, affecting quantities like total energy, total momentum, or total angular momentum, whose constancy may be implied for an isolated mechanical system by a fundamental symmetry of the dynamics, such as invariance of the Lagrangian under translations, rotations, etc. (Noether's theorem [*Goldstein*, 1981]). Then there are invariants that are peculiar to a particular system and may only be invariant under a particular approximation or in a particular dimensionality, not connected with any obvious fundamental symmetry: e.g., enstrophy, mean square vector potential, or (in incompressible flow) pointwise mass density and its moments. One of the major directions of investigations into turbulent continuum behavior that have characterized the last forty or fifty years has been determining how, why, and under what circumstances these ideally-conserved invariants influence the development of turbulent fields for fluids, magnetofluids, and plasmas. The activity itself was largely initiated by *Kraichnan* [1958, 1967], following up on a statistical mechanical theorem of *Lee* [1952]. One such invariant is magnetic helicity.

Magnetic helicity lies in the second class of ideal invariants; it is conserved in ideal three dimensional magnetohydrodynamics (3D MHD) under a variety of boundary conditions, but is not even well defined in two dimensional magnetohydrodynamics (2D MHD), ideal or not. Its simplest definition is the volume integral

$$H_m = \int_V \mathbf{A} \cdot \mathbf{B} \, d^3x, \qquad (1)$$

where $\mathbf{B}$ is the magnetic field, and $\mathbf{A}$ is a vector potential which gives rise to it ($\mathbf{B} = \nabla \times \mathbf{A}$). Under sufficiently simple geometric circumstances, H_m has the well-known interpretation [*Moffatt*, 1978; *Krause and Raedler*, 1980] of being the total "linked magnetic flux," but under other circumstances, where the complexity of the magnetic field structure prohibits any useful interpretation as "linked flux," it is still a conserved quantity that amounts to an ideal global invariant; it can be computed any time the fields are available. It is a quantity which is created whenever an electric current is forced to flow along the local magnetic field, forcing the magnetic field lines to "kink up" into helical shapes because of the new magnetic field contributions perpendicular to the original direction that the current creates. Its existence is not limited to MHD, but as far as we know, most of its dynamical significance is.

Magnetic helicity owes its significance, as far as we can see, to its constancy in the absence of Ohmic dissipation and to its generic preference for being transferred to long wavelengths in nonlinear time-dependent processes of the kind that characterize turbulence. The origin of the former property is easy to see, but the latter we still understand mainly on a formal level, with our beliefs mostly reinforced by computer solutions of the MHD equations, though it is at this point a result that it would be hard to deny. However, we shall argue below that the theoretical and computational framework in which these conclusions about the role of magnetic helicity in MHD turbulence have been reached requires re-examination and perhaps modification.

The two different classes of processes in which helicity appears so far to play a significant role are highly dissimilar, but not totally unrelated. There is first its appearance in the statistical theory of homogeneous 3D MHD turbulence, formulated in the framework pioneered by *Kolmogorov* [1941a,b], *Obukhov* [1941], *Batchelor* [1953], and others [*Tennekes and Lumley*, 1972; *Orszag*, 1977; *Frisch*, 1995] for Navier-Stokes fluids. Though one might argue about one or two peripheral details, it seems established beyond doubt that there is a robust tendency for 3D MHD turbulence to transfer H_m to long wavelengths in a spectral decomposition, in a turbulent MHD flow. It is thus a prime candidate for a process that can generate large scale magnetic fields from small scale turbulence, mechanical or magnetic. This has been very much a French subject, led by Pouquet and Frisch, and many of the references are Nicoise; we will list here only a fraction of them [Frisch et al., 1975; *Pouquet and Paterson*, 1978; *Pouquet*, 1978; Meneguzzi et al., 1981; Pouquet et al., 1988; *Pouquet*, 1993, 1996]. While it is not reasonable to expect nature ever to present us with a situation that is as cleanly symmetric as the theory assumes, and while the intense interest in such theoretical niceties as the numerical values of exponents, for example, may remain largely aesthetic in character, it is also clear that a qualitatively new physical process has been identified in this way, one that may be broadly responsible for many of the magnetic fields that are observed in nature, even when the symmetric conditions under which it has been identified are inapplicable.

The other context in which helicity has appeared in the last twenty-five years (there were early precursors in the astrophysics literature [*Woltjer*, 1958, 1959a,b,c, 1960; *Wells and Norwood*, 1969]) is in connection with the "pinch effect" from the general area of thermonuclearly-intended laboratory confinement experiments. The toroidal pinch, in particular, is a magnetofluid confinement device whose dynamical behavior under many circumstances experiences an initially turbulent phase and then "relaxes" for awhile to a much more quiescent one that can be approximately characterized as a state of maximum H_m, for a given value of magnetic energy, defined by the volume integral

$$W = \int_V \frac{B^2}{8\pi} d^3x \,. \qquad (2)$$

More precisely, what is observed is a near conservation of H_m with an attendant decay of energy, to the point where the ratio W/H_m is nearly minimal, a state which is essentially laminar. This observation, due to Taylor [*Taylor*, 1974, 1975, 1976, 1986], spawned a great amount of activity in the 1970s and 1980s, starting with attempts to explain the operation of the British "ZETA" toroidal pinch device and then moving on to the operation of spheromaks and even tokamaks as well as solar prominences and arcades. The apparent success of the hypothesis was accompanied by numerical tests of it and related Navier-Stokes decay principles, using solutions of the Navier-Stokes and MHD equations [e.g., *Bretherton and Haidvogel*, 1976; *Matthaeus and Montgomery*, 1980; Riyopoulos et al., 1982; Ting et al., 1986; Dahlburg et al., 1986] of increasing degrees of realism. When certain features of the computations failed to fit in entirely with the details of the state obtained by minimizing (1) subject to (2) — in particular, there were residual velocity fields which refused to go away in the "relaxed" state — considerable interest arose in formulating alternative variational principles which would simplify the description of late-time evolution of turbulent MHD and Navier-Stokes systems, the testing of those hypotheses against numerical data and, where available, laboratory experiments. "Selective decay" hypotheses [*Bretherton and Haidvogel*, 1976; *Matthaeus and Montgomery*, 1980; Riyopoulos et al., 1982; Ting et al., 1986; Dahlburg et al., 1986, 1987, 1988], in which some global ideal invariant would be transferred to small spatial scales and dissipated there while a second one was transferred to long wavelengths and preserved, led to a wide variety of suggestions for minimizing one turbulent quantity while approximately conserving another; such processes are clearly related to "inverse cascade" behavior, but are not the same thing. The results of the various conjectures were almost invariably interesting, even when they were not physically correct. At various times, hypotheses of "minimum energy," "minimum energy dissipation rate," "minimum entropy production rate," and "maximum entropy" have all been candidates for a variational principle that would predict late-time relaminarization of a turbulent MHD flow field. Additional constraints that were suggested to be included in the variational formalism were, for instance, constant total current, constant magnetic flux (toroidal or poloidal), flux tube topology, and so on.

In Sec. 2 of this article, we will offer some discussion of the first of these two classes of processes, formulated in the homogeneous turbulence framework. These have been widely discussed, and our emphasis here will be on the internal consistency of some symmetries and approximations which have been made within the framework of rectangular periodic boundary conditions. These assumptions are crucial to many of the conclusions that have been drawn, and seem to require reexamination. In Sec. 3, we will attempt some similar reconsiderations for the case of laboratory MHD confinement devices, with what is in practice usually a toroid being idealized as a straight, periodic cylinder. In both Sec. 2 and Sec. 3, space limitations will mandate against the extensive reproduction of computer data and graphics that have led to some of the conclusions, and the reader will be referred to the cited references for the required documentation. In Sec. 4, some general observations will be provided with an emphasis on moving to true toroidal geometry to reconsider the same class of questions; this, we believe, is the single area of investigation most ripe at present for theoretical investigation.

2. CONSERVATION OF H_m; SPATIAL PERIODICITY

When H_m obeys a conservation law is first of all an electromagnetic question. We start with Faraday's law, expressed in cgs units:

$$\frac{\partial \mathbf{B}}{\partial t} = -c \nabla \times \mathbf{E} \,, \qquad (3)$$

where $\mathbf{E}$ is the electric field and c is the speed of light. We will also assume an Ohm's law of the form

$$\mathbf{E} + \mathbf{v} \times \mathbf{B}/c = \mathbf{j}/\sigma \,, \qquad (4)$$

where $\mathbf{v}$ is the magnetofluid velocity, $\mathbf{j}$ is the electric current density, and σ is the electrical conductivity. We may further neglect the displacement current, in perhaps the most fundamental MHD approximation (what Grad called the "pre-Maxwell" approximation),

$$\nabla \times \mathbf{B} = 4\pi \mathbf{j}/c \,, \qquad (5)$$

as the connection between $\mathbf{B}$ and $\mathbf{j}$.

Pulling a curl off Eq. (3) provides an equation of evolution for $\mathbf{A}$ at the price of including a scalar potential Φ:

$$\frac{\partial \mathbf{A}}{\partial t} = \mathbf{v} \times \mathbf{B} - c\nabla\Phi - c\mathbf{j}/\sigma \,. \quad (6)$$

The scalar potential is not fixed until a gauge is chosen. Frequently, this has been the Coulomb gauge; then taking the divergence of Eq. (6) provides a Poisson equation for Φ, with a source involving $\mathbf{v}$ and $\mathbf{B}$ and the derivatives of $\mathbf{B}$. It is already clear that questions of electromagnetic boundary conditions, required to determine a solution for Φ, will be sensitive.

If we dot Eq. (3) with $\mathbf{A}$, Eq. (6) with $\mathbf{B}$, add them together, integrate over a simply-connected volume V bounded by a closed surface S, and carry out a few vector manipulations, we find upon using the relations written down that, without further assumptions or approximations,

$$\begin{aligned}\frac{dH_m}{dt} &= c\int_S d\mathbf{s} \cdot [\mathbf{A} \times \mathbf{j}/\sigma] - 2c\int_V \mathbf{j}\cdot\mathbf{B}/\sigma\, d^3x \\ &+ \int_S d\mathbf{s} \cdot [(\mathbf{v}\times\mathbf{B})\times\mathbf{A} - c\mathbf{B}\Phi] \,.\end{aligned} \quad (7)$$

We defer until the next section questions associated with finite bounding conductors and non-periodic geometries. If spatial periodicity is assumed on all field variables, so that the surface integrals vanish, it is clear that Eq. (7) will collapse to

$$\frac{dH_m}{dt} = -2c\int_V \frac{\mathbf{j}\cdot\mathbf{B}}{\sigma} d^3x \,. \quad (8)$$

If the conductivity is taken as infinite for finite $\mathbf{B}$ and $\mathbf{j}$, it is clear that Eq. (8) amounts to a conservation law for H_m. Note that the MHD equation of motion, taken here to be for the incompressible case,

$$\rho\left(\frac{\partial \mathbf{v}}{\partial t} + \mathbf{v}\cdot\nabla\mathbf{v}\right) = -\nabla p + \mathbf{j}\times\mathbf{B}/c + \rho\nu\nabla^2\mathbf{v}, \quad (9)$$

has not been used in demonstrating Eq. (8). Here, p is the scalar pressure, ρ is the mass density, and ν is the kinematic viscosity.

Equation (8) establishes the conservation law for H_m, and two other conservation laws of a similar character are readily demonstrable from the equations of motion if the transport coefficients $1/\sigma$ and ν are set equal to zero [Frisch et al., 1975; Pouquet, 1993]; the total energy and the "cross helicity," defined respectively by the expressions

$$W_{tot} = \int_V \left(\frac{\rho\mathbf{v}^2}{2} + \frac{\mathbf{B}^2}{8\pi}\right) d^3x \,, \quad (10)$$

and

$$H_c = \int_V \rho\, \mathbf{v}\cdot\mathbf{B}\, d^3x \,. \quad (11)$$

The statistical mechanical story, which uses H_m, W_{tot}, and H_c to construct absolute equilibrium Gibbs ensembles in a Fourier-series phase space, and then predict a condensation of H_m in the longest wavelengths allowed by the boundary conditions in the limit of an infinite number of degrees of freedom, has been well told [Frisch et al., 1975; Pouquet, 1996], and need not be repeated here. Likewise, the cascade predictions that follow for homogeneous, isotropic turbulence when $1/\sigma$ and ν are reinstated, and an external injection mechanism is added to the right hand sides of Eqs. (3) and/or (9) are also familiar, as are the various conjectured power law results for the cascades, inverse cascades, and dynamo processes that result. This may be described now as the standard set of theoretical predictions of 3D MHD that have been developed in scores of papers, most importantly by Pouquet and collaborators [Frisch et al., 1975; Pouquet and Paterson, 1978; Pouquet, 1978; Meneguzzi et al., 1981; Pouquet et al., 1988; Pouquet, 1993, 1996; Stribling and Matthaeus, 1990; Stribling and Matthaeus, 1991; Shebalin, 1989].

What will be remarked upon now are some troublesome points that underlie the conceptual framework in which these theories and computations have developed, in particular those associated with the simultaneous neglect of displacement current [leading to Eq. (5)], and the assumption of spatial periodicity in all three coordinates that makes many of the manipulations possible. To introduce the question, let us re-instate the displacement current, not making the "pre-Maxwell" approximation, and have, instead of Eq. (5),

$$\nabla\times\mathbf{B} = \frac{4\pi}{c}\mathbf{j} + \frac{1}{c}\frac{\partial \mathbf{E}}{\partial t} \,. \quad (12)$$

This does not alter the possibility of a magnetohydrodynamics, it just reinstates a term the net effect of which is to advance $\mathbf{E}$, $\mathbf{v}$, and $\mathbf{B}$ all on a parallel footing, with the Ohm's law, Eq. (4), now regarded as determining $\mathbf{j}$.

Notice now that if we Fourier-analyze Eq. (12) term by term, expanding all fields in the solenoidal orthogonal periodic functions that vary spatially as $\exp(i\mathbf{k}\cdot\mathbf{x})$, and assuming spatially periodic behavior for **B**, **j**, and **E**, the left hand side vanishes, and we are left with the statement that

$$4\pi <\mathbf{j}> +\frac{\partial}{\partial t}<\mathbf{E}> = 0, \quad (13)$$

where $<\mathbf{E}>$ and $<\mathbf{j}>$ represent the spatial averages of **E** and **j**. This is an exact result and so far does not assume any approximations that are peculiar to MHD. To be able to "neglect" a term in a partial differential equation, there must be at least three of them, so that one of them can be considered "small" relative to the non-trivial pair which survives to provide the "approximate" differential equation. It is mathematically unacceptable to attempt to satisfy a differential equation by "approximating" every term by zero. That amounts simply to discarding whatever physical content the differential equation might have provided. Clearly, for the $\mathbf{k} = 0$ part of Eq. (12) as realized in Eq. (13), no "neglects" are permissible.

Note that if the Ohm's law, Eq. (4), is similarly Fourier-decomposed, the result for the $\mathbf{k} = 0$ component of it is

$$<\mathbf{E}> + <\mathbf{v}\times\mathbf{B}/c> = <\mathbf{j}>/\sigma, \quad (14)$$

where the angle bracket $<>$ always indicates a spatial average over the basic cube of volume V. In general, the term $<\mathbf{v}\times\mathbf{B}/c>$ is composed of Fourier components from non-zero **k**. Substituting Eq. (14) into Eq. (13) gives

$$\frac{\partial}{\partial t}<\mathbf{E}> +4\pi\sigma<\mathbf{E}> = \frac{-4\pi\sigma}{c}<\mathbf{v}\times\mathbf{B}>. \quad (15)$$

In Eq. (15), the displacement current accounts for the first term, and has not been neglected. But neither neglecting it or not neglecting it will lead to the conclusion that $<\mathbf{E}>$ vanishes. We conclude that $<\mathbf{E}>$ must be a non-vanishing function of time, since very large numbers of configurations exist for which spatially periodic **v** and **B** fields will lead to a non-vanishing value of $<\mathbf{v}\times\mathbf{B}>$. And since $<\mathbf{E}>$ is non-zero and time-dependent, Eq. (13) guarantees a non-zero value of $<\mathbf{j}>$.

The conclusion emerges that we must consider both $<\mathbf{E}>$ and $<\mathbf{j}>$ as non-zero and time dependent, whether or not the displacement current is neglected; moreover, if one does neglect the displacement current, then the Ampère's law which follows from Eq. (5) shows that **B** itself cannot be spatially periodic (for a spatially periodic **B**, $<\nabla\times\mathbf{B}> = 0$). The only way to preserve spatial periodicity with internal consistency is to permit non-zero $<\mathbf{E}>$ and $<\mathbf{j}>$. It may to be harmless to neglect contributions from $\partial\mathbf{E}/\partial t$ for the $\mathbf{k}\ne 0$ components and hereafter we shall assume it is all right to do so. However, it appears that $\partial<\mathbf{E}>/\partial t$ must be retained in the (non-trivial) $\mathbf{k} = 0$ terms.

Infinite periodic systems in space are only a meaningful construct if they can be imagined as an idealization or limit of finite systems, because finite systems are what Maxwell's equations apply to and are derived for; they have no meaning otherwise. One decision that has to be made for any finite system is whether a net electric current is to be allowed to pass through it or not: the two choices are "open circuit" or "closed circuit" boundary conditions. Clearly, the only option available in MHD under periodic boundary conditions is to have imagined ourselves with "closed circuit" boundary conditions that permit a net current to flow through the basic box. And if we do that, we are driven to the conclusion immediately that if we try to neglect the displacement current including the $\mathbf{k} = 0$ components, spatial periodicity of the field variables will no longer be possible! It may be hoped that somehow the effect we have noted is "small" or "negligible," but no one has as yet shown in what sense that might be true. Numerical investigations permitting participation of $<\mathbf{E}>$ and $<\mathbf{j}>$ would be required. Note that we are at this point not saying anything yet about sophisticated derived concepts such as H_m or gauge invariance; rather, we are only asking if it makes sense to treat 3D MHD in the same "homogeneous turbulence" framework that is conventionally used in the Navier-Stokes case. And we have found that the preservation of spatial periodicity over time depends on our specifically *not* neglecting the displacement current entirely, if Maxwell's equations are given their full authority. This amounts to a revision or enlargement of MHD itself.

Similar troubling questions may be raised in terms of the equation of motion, Eq. (9), where the flow is incompressible (**v** is divergenceless) and the mass density ρ is spatially uniform. The problem we now mention is associated with the possible presence of a non-zero, but spatially periodic mean magnetic field $<\mathbf{B}>$ — a case often considered and one of the greatest physical importance. The fluid velocity evolves as a consequence of the combination of pressure forces, $\mathbf{j}\times\mathbf{B}/c$ Lorentz forces, and viscous forces (for simplicity, we will assume that both $1/\sigma$ and ν are constant and uniform). The pressure p is to be determined from the Poisson equation that results upon dividing Eq. (9) by ρ and taking its divergence. The question is now whether it is internally consistent to assume **v** is spatially periodic with zero spatial average, as is conventionally done, at the same time that p is assumed to be spatially periodic. Let us spatially average Eq. (9). Both $<$ **j** $>$ and $<$ **B** $>$, if there is a non-zero $<$ **B** $>$, will be part of the total **j** and **B** which contribute to the average $<$ **j**$\times$**B** $>$ Lorentz force term in the result. Since $<$ **j** $>$ is definitely non-zero for periodic geometry ("closed-circuit" boundary conditions), then a non-zero $<$ **B** $>$ will imply a non-zero spatial average to $<$ **j**$\times$**B** $>$, except in the exceptional circumstance in which $<$ **j** $>$ and $<$ **B** $>$ are parallel ("force free states"). Specific values of the spatially-varying parts of the fields might cancel the total overall $<$ **j**$\times$**B** $>$, but in general they will not. The terms containing velocity, according to the periodicity hypothesis we are examining, will all have zero spatial averages. Thus the gradient of p will in general have a non-zero spatial average, and p cannot be spatially periodic. In short, we can only find spatially periodic solutions to Poisson equations, such as that for the pressure, when the "source" term for those Poisson equations have certain integral properties: their spatial averages must vanish. In both the examples we have considered, we find that this is difficult to justify unless $<$ **B** $>= 0$.

Even with $<$ **B** $>= 0$, a non-zero $<$ **j** $>$ will cause one additional term to survive on the right hand side of the equation of motion, an extra term $<$ **j** $>\times$**B**, which typically has not appeared in discussions of 3D MHD turbulence. It remains to be investigated how important its effect will be. It does not violate the invariance, under periodic boundary conditions, of two of the familiar "rugged invariants," energy and cross helicity, but it may affect plane sinusoidal waves in an unfamiliar and inconvenient way because of its time-dependence. Moreover, it adds an extra term into the conserved total energy, one associated with the energy density of the electric field. The new, conserved energy is, for periodic boundary conditions,

$$W_{tot} = \int_V \left[\frac{1}{2}\rho \mathbf{v}^2 + \frac{\mathbf{B}^2}{8\pi}\right] d^3x + \int_V \frac{\langle \mathbf{v}\times\mathbf{B}\rangle^2}{8\pi c^2} d^3x. \quad (16)$$

For fluctuation amplitudes **v** and **B** that are not too large, the quartic nature of the last term in Eq. (16) means that it may sometimes be considered negligible compared to the standard energy integral, which amounts to only the first two terms in Eq. (16).

We recapitulate, before turning our attention to H_m. It appears to be possible to affirm the internal consistency and temporal preservation of spatially periodic boundary conditions for a finite $<$ **j** $>$, but not for a finite $<$ **B** $>$. The prominent role played by the displacement current makes for a somewhat different looking version of MHD than the familiar one without the $<$ **j** $>$. It does seem quite possible to reformulate MHD without neglecting the displacement current, but the implications of that remain to be explored. We remark that in 2D, a finite $<$ **B** $>$ becomes consistent, as long as it points in the direction perpendicular to the two directions of spatial variation. We also remark that in some sense not yet suggested, the apparent contradictions with the assumed symmetries that we have noted may appear to be "small." This seems an inquiry worth pursuing.

We turn now to the question of the definition of H_m. In the modified $\mathbf{k} = 0$ Eq. (15), we may define H_m straightforwardly from Eq. (1) and it still obeys the conservation law. There will in general be a (benign) non-zero $<$ **A** $>$ as long as $<$ **B** $>= 0$, but it will affect nothing in the dynamics and will not contribute to H_m. It can be identified as a time dependent quantity that contributes to the spatially averaged electric field, but not to the magnetic field. As soon as we want to talk about a mean non-zero $<$ **B** $>$, however, the previous problems all arise, plus one more: a non-zero $<$ **B** $>$ has a non-zero $<$ **A** $>$ which cannot be spatially periodic. Redefinitions of H_m in terms of a magnetic field from which $<$ **B** $>$, and/or its associated vector potential contribution, has been subtracted off seem less than satisfactory, for the reasons already stated: the periodicity assumption itself seems to cease to be internally consistent, however H_m is redefined.

There seems to be a conceptual difficulty, not yet entirely clear in its origins, with assuming spatially periodic MHD boundary conditions at the same time that a mean dc magnetic field is present. It may inhere in the fact that no finite-sized spatial system can be imagined, including those currents that are ultimately necessary and responsible for the mean dc magnetic field, that would lead in any clear way to such a triply-periodic configuration with a non-zero mean $<$ **B** $>$ in the limit of an infinite system. It is considered by us an open question as to whether homogeneous turbulence theory can be consistently formulated with a non-zero $<$ **B** $>$, to any approximation. This is alarming, since in almost all physical cases of interest, a mean $<$ **B** $>$ can be identified and is important.

Stribling et al., [1994] have suggested a possible aperiodic statistics which can nonetheless be expected to behave locally periodically for a restricted time, even with a non-zero $<$ **B** $>$, but containing a negligible

$< \mathbf{j} >$. Details remain to be proved, however, concerning this intriguing possibility, which leads to a rather different dynamics for helicity.

We should also note the related discussion of *Berger* [1997], which is concerned with the question of how to define magnetic helicity while retaining its interpretation as linked magnetic flux, in the presence of spatially periodic boundary conditions. Our concern here is rather with what we perceive as an inconsistency in rectangular periodic boundary conditions themselves at the same time the displacement current is neglected in MHD. This is a prior concern to any definition of magnetic helicity. What seems plausible to us would seem to be a need for an exploration of an enlarged version of MHD itself, with three independent fields ($\mathbf{v}$, $\mathbf{B}$, and $\mathbf{E}$), obeying three independent partial differential equations, and with j given by the Ohm's law.

3. FINITE BOUNDED GEOMETRIES; MHD "RELAXATION" PROCESSES

The typical geometry in which magnetic helicity has been of interest is that of toroidal confinement devices: pinches, spheromaks, sometimes tokamaks. However, for technical reasons, toroidal MHD is mathematically very difficult to deal with at a global level. In true toroidal coordinates, the Helmholtz equation is not separable, which results in extraordinary technical difficulties for various vector operations that are straightforward in rectangular, cylindrical, or spherical geometry. These technical difficulties have been surmounted by Bates and Lewis [*Bates and Lewis*, 1996; *Bates*, 1997], but so far only in time-independent cases. In cylindrical coordinates, toroidal boundaries are difficult to represent and implement boundary conditions over, and in effect we are limited to rectangular toroidal cross sections, if we want to carry out numerical manipulations comparable in complexity with the cylindrical computations that have been done. For these reasons, attention has been directed toward the straight cylindrical boundary with periodically identified ends, thought to have some features of the torus at a much lower computational price. There are certainly some features of MHD that change qualitatively from cylinder to toroid [*Montgomery and Shan*, 1994; Montgomery et al., 1997a,b; Kamp et al., 1998; *Bates and Montgomery*, 1998], but that is a story for another occasion. There is a further complication that the actual boundaries of confinement devices are themselves complex, highly asymmetric, and usually involve regions (such as the "scrape off layer" in a tokamak) where there is no chance that MHD itself is even applicable. This has led to a variety of approximations and compromises with reality in order to obtain even any tractable problems. Some of these are as follows.

First, we assume no vacuum region separating the magnetofluid from the mechanical wall. This is not realistic, but seems to be necessary at this stage to avoid intractable boundary-layer analyses involving plasmas with large gradients in density and temperature near any surface at which the boundary conditions are to be enforced (the "zero pressure plasma"). In the present treatment, that will mean a uniform mass density right up to the cylindrical walls. The cylinders are assumed to be able to withstand normal stresses that in a confinement device might spell disaster.

The cylinder walls are usually taken as unbroken, even though the real life version of the toroidal containers must contain slits and slots cut in them to permit the applied fields — electric and magnetic — to penetrate into the magnetofluid. An unbroken perfectly conducting toroidal shell would exclude all electromagnetic activity from its interior. The cuts destroy the axisymmetry that one would like to use (translational axial symmetry for the straight cylinder), and stand as permanent embarrassments in all the theories so far. We have no choice here but to assume unbroken cylinder walls as well.

What are the walls to be made of? Perfectly conducting metal has been a popular assumption, though in fact many of the plasmas of interest have conductivities by no means small compared to that of copper. One complication of a perfectly conducting wall for an even slightly resistive magnetofluid is that it demands the vanishing of the tangential component of $\mathbf{E}$ there. From Ohm's law, and assuming $\mathbf{v} \cdot \hat{\mathbf{n}} = 0$, this demands the vanishing of the tangential component of $\mathbf{j}$ at the wall as well, in any magnetofluid with a finite conductivity. This greatly complicates the participation of many of the more-or-less tractable current profiles, such as that of the celebrated "Taylor state" itself (to be discussed presently). However, a perfectly conducting, mechanically impenetrable boundary will demand the vanishing of the normal components of $\mathbf{v}$ and $\mathbf{B}$, and the tangential component of $\mathbf{j}$, at the walls of a periodic cylinder, and it as at once apparent from Eq. (7) that that combination of boundary conditions again results in Eq. (8) for the cylinder. To the extent that the conductivity becomes infinite, H_m again defines a conserved quantity for that case.

A second idealization of the wall, more convincing in some respects, is that of a perfect conductor coated with a very thin layer of insulating dielectric. This removes the restriction that the tangential electric field must vanish at the wall, but leaves in place the vanishing of the normal component of **B**. It permits a current profile whose tangential **j** component need not vanish at the wall, but whose normal component must. From the point of view of the Ohm's law, with no normal velocity at the wall (see below), this amounts to a Neumann boundary condition on the scalar potential Φ.

For mechanical boundary conditions on **v**, it is natural to assume that $\mathbf{v} \cdot \hat{\mathbf{n}} = 0$. The remaining viscous boundary condition can be either the vanishing of the tangential component of **v** (the rough wall) or the vanishing of the tangential component of viscous stress (the perfectly slippery wall).

The Taylor assumption [*Taylor*, 1974, 1975, 1976, 1986] that energy decays to the minimum value it can have subject to the constraint of approximately conserved H_m implies the disappearance of any kinetic energy of fluid motion that may have been there (the minimum-energy conjecture is likely not to be true in any sense for compressible flow, where energy must include the thermal energy of particle motions), and the simultaneous decay of magnetic energy to its minimum value. The result of such a conjecture leads through the calculus of variations to the Euler-Lagrange equation

$$\nabla \times \mathbf{B} = 4\pi \mathbf{j}/c = \lambda \mathbf{B}, \qquad \lambda = const. \qquad (17)$$

Equation (17) has been arrived at by considering the class of all infinitesimal variations of **A** such that their tangential components vanish at the wall, and periodic boundary conditions in the axial coordinate z have been assumed. The quantity λ is a Lagrange multiplier. The solutions are the functions of *Chandrasekhar and Kendal* [1957], suitably modified for cylindrical geometry. Typically the values of the eigenvalues λ must be determined from boundary conditions, or by imposing (in the case of the axisymmetric solutions) some other constraint such as fixed values of the axial magnetic flux and/or current.

Equation (17) defines an infinite sequence of ideal equilibria which are "force free" (i.e., $\mathbf{j} \times \mathbf{B} = 0$) magnetohydrodynamic states, one for each λ. Only one of these minimizes the energy absolutely for a given H_m and boundary. The axisymmetric one, earlier called [*Voslamber and Callebaut*, 1962] the "Bessel function model," has become known as the "Taylor state." Its properties have been much calculated and discoursed upon, and there seems to be no doubt that it can reproduce, in some rudiments, the essential features of the phenomenon of "field reversal" in the toroidal pinch [*Taylor*, 1974, 1975, 1976, 1986]. (It is noteworthy, however, that "reversed current" pinches are rarely discussed.)

It is tempting to give the decay of energy relative to magnetic helicity a universal significance as one of a class of similar "selective decay" processes comparable to the decay of enstrophy relative to energy in two-dimensional Navier-Stokes flow, and decay of energy relative to mean-square vector potential in 2D MHD. At this stage, whether it happens in detail as a consequence of the MHD equations is primarily a question to be answered numerically. Laboratory data is suggestive but far too incomplete for a convincing case; moreover, the limitations of MHD itself in describing magnetized plasma processes in regimes of recent interest are severe.

Numerical evidence accumulated in decaying turbulence situations has supported this "selective decay" hypothesis, within limits [*Matthaeus and Montgomery*, 1980; Riyopoulos et al., 1982; Ting et al., 1986; Dahlburg et al., 1986, 1987, 1988]. There are basically three shortcomings of the hypothesis that have disappointed the initially high hopes that were held for it. (1) The effect can only occur as a consequence of finite resistivity and viscosity; otherwise, nothing would decay. But if a finite resistivity is postulated, the late-time Taylor state does not obey the Ohm's law internally with any curl-free electric field present, in the absence of flow velocity. (2) Boundary conditions are difficult to impose without disrupting the essential nature of the Taylor state. One can postulate a strangely non-uniform electrical conductivity going to zero at the wall which will make $\mathbf{E} = \mathbf{j}/\sigma$ everywhere curl-free while **j** is definitely not curl-free, but to do so in the absence of any deeper reason for such a spatial dependence seems desperate. (3) Perhaps most importantly, numerical attempts to follow the decay of energy relative to magnetic helicity, while observing the relative decay of energy and clearly seeing the anticipated field reversal near the outer boundary of the magnetofluid [*Riyopoulos et al.*, 1982; *Dahlburg et al.*, 1986, 1987, 1988], always stopped short of reaching an energy minimum. In particular, a residual kinetic energy of flow, with energies of the order of a small percentage of the magnetic energy, always remained in the form of helical vortices. The current channel, as well, remained helically distorted. The "Taylor state" has not been accurately achieved in any numerical computation, and its utility seems therefore limited.

Another difficulty arose when one began to try to sort out the differences between driven and decaying MHD steady states. In some cases, such as tokamak operation, toroidal voltages are maintained and the toroidal discharge is held approximately steady as long as possible in a quiescent state, in times that are now sometimes measured in minutes. In toroidal pinches, on the other hand, there is sometimes an initial voltage pulse that initiates the discharge, but the voltage is then "crow-barred" to allow the discharge to proceed on its own, largely independent of any driving mechanism, with the current maintained by the system's own inductance. It is unlikely that the same variational principle would suffice for both cases. The toroidal pinch case might be idealized as a "decaying" situation, but the tokamak never could; it is inherently "driven."

¿From pondering the numerical results, it appeared to many of us that some other variational principle might be operative that was similar to, but distinct in some important respect from, the simple energy decay principle described above. The decay phenomena showed many regularities, but they just were not all ones implied by the "minimum energy" principle. (A related principle of "minimum energy," had long been used as a means of testing the ideal stability of ideal MHD equilibria [Bernstein et al., 1958; *Bateman*, 1978; *Wesson*, 1997], but seemed difficult to connect to turbulent processes where resistive and viscous decays were of the essence.) The easiest way to modify the implications of the Taylor conjecture is to discover additional constraints that for one reason or another might be thought to represent a conserved quantity as robust as helicity seemed to be. For a period of close to ten years, it would be difficult to suggest anything plausible that was *not* explored by someone as an additional constraint to be included, often with strikingly interesting results [*Lewis*, 1982; *Turner and Christensen*, 1981; Aydemir et al., 1985]. But the level of experimental precision and stability of support was far below the level that would have allowed any decisive tests of any of these conjectures, which were, by necessity not considered "main stream."

The statistical-mechanical principle of "minimum rate of entropy production" has received great formal attention in the decades since its enunciation by *Onsager* [1931a,b]. Influential monographs [*Glansdorff and Prigogine*, 1971; *de Groot and Mazur*, 1962; *Keizer*, 1987] have been devoted to it, but it can seem less general than some of the claims that have been suggested for it. For example, it cannot apparently deal with the simple problem of Fourier heat conduction between two planes held at different temperatures in the steady state, except for materials whose thermal conductivity falls off as the inverse square of the temperature, and no known substance has this property [*Jaynes*, 1980]. Progressive and more elaborate re-definitions of entropy production followed the appearance of system after system in which the naive consequences of the principle were shown to be at variance with what was known or desired.

One antecedent for the principle of minimum entropy production rate, though not mentioned by Onsager, was an observation made by *Kirchoff* [1848]. Kirchoff asked what electric current profile would result, in a rigid non-uniform conductor, if its boundaries were held at a given spatially-varying electrostatic potential. Current is allowed to enter and leave locally through the surface of the conductor however it likes. He was able to solve the problem, both rigorously and simply, then to show that the current would distribute itself in such a way as to minimize the total rate of Ohmic dissipation of energy, subject to the boundary condition. It is this principle that is operative, in fact, when a current divides itself among parallel branches in a dc network involving resistors. Later in the century, this principle of minimum energy dissipation appeared in quite a different (fluid) context: various wall-bounded hydrodynamic shear flows (pipe flow, plane Poiseuille flow, plane Couette flow, for example) were seen to be states that minimized the rate of viscous energy dissipation, subject to boundary conditions [*Lamb*, 1930; *Montgomery and Phillips*, 1988], but with no proof attached that had the generality given by Kirchoff for the electromagnetic case.

How are these energy dissipation principles related to "minimum rate of entropy production?" The answer is simple for the uniform temperature case. For if one adopts the most pedestrian definition of entropy, $dS = \delta Q/T$, where δQ is the amount of heat transferred at temperature T, it is clear that in any dissipation process which dissipates energy into heat, entropy will be produced at a proportional rate and the factor of proportionality is just $1/T$. The Fourier heat conduction problem can be seen as the simplest case in which the two alleged principles differ at all. *Jaynes* [1980] has given an ingenious example to show that energy dissipation rate is to be favored over entropy production rate as a basic principle. Consider two different resistors in parallel, through which a total current is forced to flow. Put one of the resistors in a heat bath which maintains it at a different temperature from the other. It is now straightforward to minimize either the energy dissipation rate or the entropy production rate for the whole system. Try it and see which one leads to the correct answer as defined by standard undergraduate circuit theory!

It is unclear that the principle of minimum entropy production rate has produced any knowledge that is simultaneously correct and not already contained in the more simple-minded energy dissipation rate principle of *Kirchoff* [1848]. It occurred to us at one point [*Montgomery and Phillips,* 1988] that it might be useful to seek MHD states as states of minimum energy dissipation rate. We thought we had done so and found a number of very plausible-looking predictions for the periodic cylinder, but then realized that we had not found the true minimum-dissipation states that we had sought. Any minimization of viscous plus Ohmic dissipation must be carried out with the variational field functions constrained to obey the Ohm's law. Techniques from the calculus of variations simply do not at present seem adequate to be able to include such pointwise vector constraints, involving solenoidal fields, on the variational functions in a manageable way. While we still believe the principle of minimum energy dissipation rate is likely applicable to the driven case of incompressible flow, it has not yet been successfully implemented, in the writer's opinion. We do believe it is consistent with what we have more recently learned [*Montgomery and Shan,* 1995] about steady states into which turbulent magnetofluids evolve.

A brief description of those investigations follows; reference must be made to the literature for a more complete presentation [Shan et al., 1991; *Shan and Montgomery,* 1993a,b, 1994a,b; *Montgomery and Shan,* 1995]. Most of the computations and the cleverest analytical innovations are due to X. Shan. We became interested for the reasons noted above in doing 3D MHD computations for a uniform-density magnetofluid inside a periodic straight cylinder. The computations were done spectrally, and since the fields were all solenoidal, they could be expanded in eigenfunctions of the curl operator. That is, all the eigenfunctions of the type given in Eq. (17) (properly called Chandrasekhar-Kendall functions [*Chandrasekhar and Kendall,* 1957; Montgomery et al., 1978] are believed to be a complete set [*Yoshida,* 1992] in which to expand the solenoidal functions $\mathbf{v}$, $\mathbf{B}$, $\mathbf{j}$, and the vorticity $\boldsymbol{\omega}$. In addition to the axial periodicity, all four fields were chosen to have vanishing normal components at the cylinder radius. This is less than entirely satisfactory for the velocity field; since most hydrodynamicists would probably say that while $\boldsymbol{\omega} \cdot \hat{\mathbf{n}} = 0$ is a consequence of no-slip boundary conditions at the wall ($\mathbf{v} = 0$ there), it is less than a complete specification of them. This brings up unsolved problems in the simpler context of the theory of Navier-Stokes fluids which have been troublesome for quite some time (see, e.g., *Gresho,* [1991]), and we can say no more about them at this point; their resolution will probably not occur first in MHD.

But assuming that Chandrasekhar-Kendall functions are an acceptable basis for spectral computations, we find that they can be made orthonormal and used in a spectral decomposition of the 3D MHD equations. Using Galerkin methods, they can be reduced to a set of nonlinear ordinary differential equations, first order in the time, that advance the expansion coefficients of the various fields. The usual questions of resolution present in any numerical treatment of turbulence have to be dealt with and provide a limitation on the Reynolds-like numbers that can be accommodated. Steady state current profiles can be found (they depend upon the assumption about the spatial variation of the resistivity) and perturbed arbitrarily at $t = 0$ in order to explore their stability and nonlinear behavior. Since the Chandrasekhar-Kendall functions bear a close geometrical correspondence to the linearly unstable normal modes when they appear, convergence of the Galerkin expansion of the fields is rather rapid as long as one does not get very far above the stability thresholds. The stability thresholds are curves in a plane whose axes are the pinch ratio and the Hartmann number. The pinch ratio is the mean poloidal or azimuthal magnetic field at the cylinder wall divided by the mean axial magnetic field, and the Hartmann number is a geometric mean of two Reynolds-like dimensionless numbers: each one has an Alfvén speed and a length scale upstairs, and a kinematic viscosity or a magnetic diffusivity downstairs. This method is limited to cases that are not too turbulent, since the number of degrees of freedom that can be followed with a few tens of hours of Cray C-90 time is limited to about 2000. An interested reader is referred to the original papers [Shan et al., 1991; *Shan and Montgomery,* 1993a,b, 1994a,b; *Montgomery and Shan,* 1995] for details that cannot be dwelt on here. Both stable and unstable cases, and laminar and turbulent ones, have been computed in detail.

There is very little for a variational principle to do in such a context, where every MHD variable is being specified at $t = 0$ and followed in time regardless of whether the evolution is turbulent or not. We would like to be able to argue that the voltage-driven magnetofluid, after its initial phase of instability growth, had sought out the state of minimum energy dissipation rate. But since we have been unable to say what that state might be, this has not been possible. What we can note is the following. As the axisymmetric state first becomes unstable, the magnetofluid seeks out a helically distorted steady state with flow. In effect, the solution bifurcates into two time-independent steady states, as the current is raised, before it becomes turbulent. One state remains axisymmetric and quiescent, the other becomes helical and involves vortices. Inevitably, the state chosen, the stable one, has been one of lower energy dissipation rate (despite the viscous dissipation occasioned by the velocity field). This much of the variational formulation survives, but only this much.

Our belief (and at this point, that is all it can be) is that the Taylor hypothesis, of rapid energy decay relative to magnetic helicity is broadly correct for the decaying initially-turbulent pinch. But while it correctly predicts such features as field reversal and gives qualitative agreement with the measured "F-theta" diagrams [*Taylor*, 1975, 1976; *Lewis*, 1982], it does not predict accurately the late-time nature of the turbulent evolution; in particular, it misses the velocity fields associated with helical vortices, and the helical distortion of the current channel. These vortices are actually responsible for lowering the energy dissipation rate below its zero-flow value for the same applied voltage [*Montgomery et al.*, 1989; *Phillips*, 1996]. It is not possible to formulate the hypothesis for the driven steady state in a meaningful way. On the other hand, minimum energy dissipation rate is a compelling principle which is consistent with everything we know for the driven case, describes the steady states below and just above the stability thresholds consistently, but has not been made to yield any recognizable predictions for the steady-state turbulent situation.

Suppose minimum energy dissipation rate subject to boundary conditions were a universal characteristic property of driven steady states in continua — what would it mean? About this we can only speculate, though perhaps there is no harm in doing so on such an occasion. Consider that what nature would like to do with any non-equilibrium system would be to bring it to uniform, smooth, thermal equilibrium. But for driven problems of the kind under consideration, boundary conditions permanently prop the system away from thermal equilibrium. In a sense, the system might be thought to prefer to get as close to thermal equilibrium as possible, compatible with the boundary conditions. If one were to write down an integral measure of the *smoothness* of a mandated non-zero viscous or resistive profile, the integrals defining the Ohmic and viscous dissipation,

$$\int_V \frac{\mathbf{j}^2}{\sigma} d^3x = \int_V \frac{c^2 (\nabla \times \mathbf{B})^2}{4\pi^2 \sigma} d^3x \quad (18a)$$

$$\rho \int_V \nu (\nabla \times \mathbf{v})^2 d^3x = \rho \int_V \nu \omega^2 d^3x, \quad (18b)$$

would be the first things, for the solenoidal fields $\mathbf{B}$ and $\mathbf{v}$, that one might write down. A similar expression for a material of thermal conductivity κ and temperature profile T,

$$\int_V \kappa (\nabla T)^2 d^3x, \quad (19)$$

would be the corresponding quantity for a temperature profile, and in fact minimizing the expression (19) deals quite accurately with the Fourier heat conduction problem mentioned previously, though it is difficult to relate (19) to anything to do with rate of entropy production. What actually seems to be involved is a principle of "maximum smoothness" of the relevant fields, as measured by expressions like (18) and (19), in the presence of mandated boundary conditions. This smells perhaps like a "maximum entropy" principle, for a non-equilibrium entropy that has not been fully defined yet.

4. DISCUSSION AND SUMMARY

Magnetic helicity H_m is an intriguing MHD quantity, potentially of great significance, but its conceptual foundations are in need of clarification. We have considered two situations here in both of which promising beginnings have been made, but where the physical consequences and significance of the quantity H_m need to be further articulated.

We considered first the much-studied problem of incompressible homogeneous MHD turbulence. Questions about the internal consistency of conveniently-assumed symmetries and approximations with simultaneous assumptions of rectangular periodic boundary conditions have been raised but not answered. In detail, it seems impermissible to neglect uniformly the displacement current relative to the conduction current at the same time that rectangular periodic boundary conditions are being assumed. Inclusion of the spatially-uniform part of the displacement current is accompanied by the need for a spatially uniform contribution to the conduction current. These will modify the content of MHD turbulence theory and computations to an uncertain extent, even in the absence of a mean dc magnetic field $<\mathbf{B}>$. Inconsistencies appear in 3D MHD if a nonzero mean dc magnetic field $<\mathbf{B}>$ is included. These arise even before questions of the definition and evolution of H_m come up, and raise troubling uncertainties about the whole propriety of treating MHD turbulence in the same symmetric framework that Navier-Stokes turbulence has become accustomed to.

Secondly, we have remarked on the problems of relaxing MHD turbulence in straight cylinders with periodically identified ends intended to mimic toroidal situations in a more tractable framework. The results discussed appear mostly in lengthy computational papers the graphical results of which have not been possible to reprint here. We have observed certain shortcomings in the predictions of the "Taylor state," sometimes thought to represent a rigorous and well-established consequence of reversed-field pinch dynamics. In addition to noting that it does not obey the Ohm's law, it is also noted that in computations simulating it, while axial field reversal and a decreasing energy to magnetic helicity ratio has been observed, so also has a residual vortical kinetic energy and an inevitable helical distortion of the current channel; axial magnetic field reversal, moreover, is not typically accompanied by axial current reversal. For a variety of reasons, and particularly in the driven steady state, a principle of "minimum rate of energy dissipation" has seemed to be a more accurate predictor of what transpires. However, technical calculus of variations problems associated with imposing the Ohm's law as a pointwise constraint in the variational procedures has limited our ability to predict relaxed steady states variationally when they include velocity fields.

It may be premature to advertise helicity as a mechanical quantity that will have the organizing predictive power of some other ideal invariants, such as energy or angular momentum, for turbulent systems, astrophysical or laboratory. Certainly it seems worthwhile to attempt to sharpen up the conclusions and deductions associated with magnetic helicity despite the difficulties that arise in doing so.

REFERENCES

Aydemir, A. Y., D. C. Barnes, E. J. Caramana, A. A. Mirin, R. A. Nebel, D. D. Schnack, and A. G. Sgro, Compressibility as a feature of field reversal maintenance in the reversed-field pinch, *Phys. Fluids 28*, 898-902, 1985.

Batchelor, G. K. *Theory of Homogeneous Turbulence,* 197 pp., Cambridge University Press, Cambridge, 1953.

Bateman, G., *MHD Instabilities*, 263 pp., MIT Press, Cambridge, Mass., 1978.

Bates, J. W., and H. R. Lewis, A toroidal boundary-value problem in resistive magnetohydrodynamics, *Phys. Plasmas 3*, 2395-2400, 1996.

Bates, J. W., On toroidal Green's functions, *J. Math. Phys. 38*, 3679-3691, 1997.

Bates, J. W., and D. C. Montgomery, Toroidal visco-resistive magnetohydrodynamic steady states contain vortices, *Phys. Plasmas 5*, 2649-2653, 1998.

Berger, M. A., Magnetic helicity in a periodic domain, *J. Geophys. Res. 102*, 2637-2644, 1997.

Bernstein, I. B., E. A. Frieman, M. D. Kruskal, and R. M. Kulsrud, An energy principle for hydromagnetic stability problems, *Proc. Roy. Soc. (London) A223*, 17-40, 1958.

Bretherton, F.P., and D. B. Haidvogel, Two-dimensional turbulence above topography, *J. Fluid Mech. 78*, 129-154, 1976.

Chandrasekhar, S., and P. C. Kendall, On force-free magnetic fields, *Astrophys. J. 126*, 457-460, 1957.

Dahlburg, J. P., D. Montgomery, G. D. Doolen, and L. Turner, Turbulent relaxation to a force-free field-reversed state, *Phys. Rev. Lett. 57*, 428-431, 1986.

Dahlburg, J. P., D. Montgomery, G. D. Doolen, and L. Turner, Turbulent relaxation of a confined magnetofluid to a force-free state, *J. Plasma Phys. 37*, 299-321, 1987.

Dahlburg, J. P., D. Montgomery, G. D. Doolen, and L. Turner, Driven, steady state RFP computations, *J. Plasma Phys. 40*, 39-68, 1988.

de Groot, S. R., and P. Mazur *Non-equilibrium Thermodynamics*, 510 pp., North-Holland, Amsterdam, 1962.

Frisch, U. *Turbulence, the Legacy of A.N. Kolmogorov,* 296 pp., Cambridge University Press, Cambridge, 1995.

Frisch, U., A. Pouquet, J. Leorat, and A. Mazure, Possibility of an inverse cascade of magnetic helicity in magnetohydrodynamic turbulence, *J. Fluid Mech. 68*, 769-778, 1975.

Glansdorff, P., and I. Prigogine *Thermodynamic Theory of Structure, Stability, and Fluctuations*, 306 pp., Wiley-Interscience, New York, 1971.

Goldstein, H. *Classical Mechanics*, 2nd edition, 672 pp., Addison-Wesley, Reading, Mass., 1981.

Gresho, P. M., Incompressible fluid dynamics: some fundamental formulation issues, *Annu. Rev. Fluid Mech. 23*, 413-453, 1991.

Jaynes, E. T., The minimum entropy production principle, *Annu. Rev. Phys. Chem. 31*, 579-601, 1980.

Kamp, L. P. J., D. Montgomery, and J. W. Bates, Toroidal flows in resistive magnetohydrodynamic steady states, *Phys. Fluids 10*, 1757-1766, 1998.

Keizer, J. *Statistical Thermodynamics of Nonequilibrium Processes*, 506 pp., Springer-Verlag, New York, 1987.

Kirchoff, G. D., Uber die Anwendigbarkeit der Formeln für die Intensitäten der galvanische Ströme in Einem Systeme linearer Leiter auf Systeme, die zum Theil aus nicht linearen Leitern bestehen, *Ann. Phys. und Chemie (Leipzig) 75*, 189-205, 1848.

Kolmogorov, A. N., The local structure of turbulence in incompressible viscous fluid for very large Reynold's number, *Dokl. Akad. Nauk SSSR 30*, 9-13, 1941a.

Kolmogorov, A. N., On degeneration (decay) of isotropic turbulence in an incompressible viscous liquid, *Dokl. Akad. Nauk SSSR 31*, 538-540, 1941b.

Kraichnan, R. H., Irreversible statistical mechanics of incompressible hydromagnetic turbulence, *Phys. Rev. 109*, 1407-1422, 1958.

Kraichnan, R. H., Inertial ranges in two-dimensional turbulence, *Phys. Fluids 10*, 1417-1423, 1967.

Krause, F., and K.-H. Raedler *Mean-Field Magnetohydrodynamics and Dynamo Theory*, 271 pp., Pergamon Press, Oxford, 1980.

Lamb, H. *Hydrodynamics*, 6th ed., 738 pp., Dover, New York, 1932.

Lee, T. D., On some statistical properties of hydrodynamic and magnetohydrodynamic fields, *Q. Appl. Math. 10*, 69-74, 1952.

Lewis, H. R. (Ed.), Proceedings of the reversed-field pinch theory workshop, April 29-May 2, 1980, *Report LA-8944-C*, 298 pp., Los Alamos National Laboratory, Los Alamos, New Mexico, 1982.

Matthaeus, W. H., and D. Montgomery, Selective decay hypothesis at high mechanical and magnetic Reynolds numbers, *Ann. N.Y. Acad Sci. 357*, 203-222, 1980.

Meneguzzi, M., U. Frisch, and A. Pouquet, Helical and nonhelical turbulent dynamos, *Phys. Rev. Lett. 47*, 1060-1064, 1981.

Moffatt, H. K. *Magnetic Field Generation in Electrically Conducting Fluids*, 343 pp., Cambridge University Press, Cambridge, 1978.

Montgomery, D., L. Turner, and G. Vahala, Three dimensional magnetohydrodynamic turbulence in cylindrical geometry, *Phys. Fluids 21*, 757-764, 1978.

Montgomery, D., and L. Phillips, Minimum dissipation rates in magnetohydrodynamics, *Phys. Rev. A38*, 2953-2964, 1988.

Montgomery, D., L. Phillips, and M. L. Theobald, Helical dissipative, magnetohydrodynamic states with flow, *Phys. Rev. A40*, 1515-1523, 1989.

Montgomery, D., and X. Shan, Toroidal resistive MHD equilibria, *Comments on Plasma Phys. & Contr. Fusion 15*, 315-320, 1994.

Montgomery, D., and X. Shan, Magnetohydrodynamic turbulence with net currents, in *Small-Scale Structures in Three Dimensional Hydrodynamic and Magnetohydrodynamic Turbulence*, edited by M. Meneguzzi, A. Pouquet and P.-L. Sulem, pp. 241-254, Springer-Verlag, Berlin, 1995.

Montgomery, D., J. W. Bates, and H. R. Lewis, Resistive magnetohydrodynamic equilibria in a torus, *Phys. Plasmas 4*, 1080-1086, 1997a.

Montgomery, D., J. W. Bates, and S. Li, Toroidal vortices in resistive magnetohydrodynamic equilibria, *Phys. Fluids 9*, 1188-1193, 1997b.

Obukhov, A. M., Spectral energy distribution in a turbulent flow, *Izv. Akad. Nauk. SSSR Ser. Georg. Geofiz. 5*, 453-466, 1941.

Onsager, L., Reciprocal relations in irreversible processes. I, *Phys. Rev. 37*, 405-426, 1931a.

Onsager, L., Reciprocal relations in irreversible processes. II, *Phys. Rev. 38*, 2265-2279, 1931b.

Orszag, S. A., Statistical theory of turbulence, in *Fluid Dynamics: Les Houches 1973*, edited by R. Balian and J.-L. Peube, pp. 237-374, Gordon and Breach, New York, 1977.

Phillips, L., States of minimum dissipation in magnetohydrodynamics: a review, *J. Plasma Phys. 56*, 531-551, 1996.

Pouquet, A., and G. S. Patterson, Numerical simulation of helical magnetohydrodynamic turbulence, *J. Fluid Mech. 85*, 305-323, 1978.

Pouquet, A., On two-dimensional magnetohydrodynamic turbulence, *J. Fluid Mech. 88*, 1-16, 1978.

Pouquet, A., P. L. Sulem, and M. Meneguzzi, Influence of velocity-magnetic field correlations on decaying magnetohydrodynamic turbulence with neutral X points, *Phys. Fluids 31*, 2635-2643, 1988.

Pouquet, A., Magnetohydrodynamic turbulence, in *Proc. Les Houches Summer School on Astrophysical Fluid Dynamics, July, 1987*, edited by J. P. Zahn and J. Zinn-Justin, pp. 139-227, Elsevier, Amsterdam, 1993.

Pouquet, A., Turbulence, statistics, and structures: an introduction, in *Plasma Astrophysics: Proc. VIIth European School in Astrophysics, San Miniato, Italy, 1994*, edited by C. Chieuderi and G. Einaudi, pp. 163-212, Springer-Verlag, Berlin, 1996.

Riyopoulos, S., A. Bondeson, and D. Montgomery, Relaxation toward states of minimum energy in a compact torus, *Phys. Fluids 25*, 107-115, 1982.

Shan, X., D. Montgomery, and H. Chen, Nonlinear magnetohydrodynamics by Galerkin-method computation, *Phys. Rev. A44*, 6800-6818, 1991.

Shan, X., and D. Montgomery, On the role of Hartmann number in magnetohydrodynamic activity, *Plasma Phys. & Contr. Fusion 35*, 619-631, 1993a.

Shan, X., and D. Montgomery, Global searches of Hartmann-number dependent stability boundaries, *Plasma Phys. & Contr. Fusion 35*, 1019-1032, 1993b.

Shan, X., and D. Montgomery, Magnetohydrodynamic stabilization through rotation, *Phys. Rev. Lett. 73*, 1624-1627, 1994a.

Shan, X., and D. Montgomery, Rotating magnetohydrodynamics, *J. Plasma Phys. 52*, 113-128, 1994b.

Shebalin, J. V., Broken ergodicity and coherent structures in homogeneous turbulence, *Physica D 37*, 173-191, 1989.

Stribling, T., and W. H. Matthaeus, Statistical properties of ideal three-dimensional magnetohydrodynamics, *Phys. Fluids B 2*, 1979-1988, 1990.

Stribling, T., and W. H. Matthaeus, Relaxation processes in a low-order three-dimensional magnetohydrodynamic model, *J. Phys. Fluids B 3*, 1848-1864, 1991.

Stribling, T., W. H. Matthaeus, and S. Ghosh, Nonlinear decay of magnetic helicity in magnetohydrodynamic turbulence with a mean magnetic field, *J. Geophys. Res. 99*, 2567-2576, 1994.

Taylor, J. B., Relaxation of toroidal plasma and generation of reverse magnetic fields, *Phys. Rev. Lett. 33*, 1139-1141, 1974.

Taylor, J. B., Relaxation of toroidal discharges to stable states and generation of reverse magnetic fields, in *Plasma Physics and Controlled Fusion, Tokyo, 1974*, pp. 161-167, International Atomic Energy Agency, Vienna, 1975.

Taylor, J. B., Relaxation of toroidal discharges, in *Pulsed, High-Beta Plasmas*, edited by D.E. Evans, pp. 59-67, Pergamon Press, Oxford, 1976.

Taylor, J. B., Relaxation and magnetic reconnection in plasmas, *Revs. Mod. Phys. 58*, 741-763, 1986.

Tennekes, H., and J. L. Lumley *A First Course in Turbulence*, 300 pp., MIT Press, Cambridge, Mass., 1972.

Ting, A. C., W. H. Matthaeus, and D. Montgomery, Turbulent relaxation processes in magnetohydrodynamics, *Phys. Fluids 29*, 3261-3274, 1986.

Turner, L., and J. P. Christensen, Incomplete relaxation of pinch discharges, *Phys. Fluids 24*, 893-898, 1981.

Voslamber, D. K., and D. K. Callebaut, Stability of force-free magnetic fields, *Phys. Rev. 128*, 2016-2021, 1962.

Wells, D. R., and J. Norwood, Jr., A variational approach to the dynamic stability of high-density plasmas in magnetic containment devices, *J. Plasma Phys. 3*, 21-46, 1969.

Wesson, J. *Tokamaks*, 680 pp., Clarendon Press, Oxford, 1997.

Woltjer, L., A theorem on force-free magnetic fields, *Proc. Natl. Acad. Sci. USA 44*, 489-491, 1958.

Woltjer, L., Hydromagnetic equilibrium. II Stability in the variational formulation, *Proc. Natl. Acad. Sci. USA 45*, 769-771, 1959a.

Woltjer, L., Hydromagnetic equilibrium. III Axisymmetric incompressible media, *Astrophys. J. 130*, 400-404, 1959b.

Woltjer, L., Hydromagnetic equilibrium. IV Axisymmetric compressible media, *Astrophys. J. 130*, 405-413, 1959c.

Woltjer, L., On the theory of hydromagnetic equilibrium, *Revs. Mod. Phys. 32*, 914-915, 1960.

Yoshida, Z., Eigenfunction expansions associated with the curl derivatives in cylindrical geometries: Completeness of Chandrasekhar-Kendall eigenfunctions, *J. Math. Phys. 33*, 1252-1256, 1992.

D. C. Montgomery, Department of Physics & Astronomy, Dartmouth College, Hanover, NH 03755 (e-mail: david.c.montgomery@dartmouth.edu)

J. W. Bates, Los Alamos National Laboratory, X-HM, D 413, Los Alamos, NM 87545. (e-mail: batesj@lanl.gov)

Planetary Dynamos and Helicities

K.-H. Rädler

Astrophysikalisches Institut Potsdam, Germany

After a short survey on observational facts on planetary magnetic fields the fundamentals of planetary dynamo theory are reviewed. Particular attention is paid to the mean–field approach to dynamo models, which assumes irregular small–scale fluid motions, in particular turbulence, and gives some basic insight into possibilities of dynamo mechanisms. In addition, dynamo models with regular large–scale flow patterns are briefly mentioned. A few details are given for a model of this kind which may provide some explanation for the high degree of axisymmetry of the field observed at Saturn. The crucial point in the mean–field approach is the α–effect, that is, the occurrence of a mean electromotive force with a component in the direction of the mean magnetic field, as a rule a consequence of irregular small–scale motions under the influence of Coriolis forces. The α–effect is closely connected with the kinetic helicity of the motions. Results are reviewed which show connections between the α–effect and the current helicities of the mean and the small–scale parts of the magnetic field. They lead to the conclusion that in dynamos of α^2–type the signs of the kinetic helicity and the total current helicity essentially coincide. For the model mentioned above developed with a view to Saturn some numerical results are presented showing that the distributions of kinetic and current helicities follow the same rule.

1. INTRODUCTION

The best known planetary magnetic field is that of the Earth. The dominant part of the field outside the rigid Earth corresponds to a dipole, creating south magnetic polarity around the geographic north pole and vice versa, but being not completely aligned with the rotation axis. The magnetic moment is given by $M = 8.2 \cdot 10^{15}\,\text{Tm}^3$, the corresponding magnetic flux density at the poles by $B_p = 0.6 \cdot 10^{-4}$ T and the inclination of the dipole against the rotation axis by $\psi = 11.4°$. The contributions of quadrupole, octupole and higher multipoles lack in general symmetry with respect to both dipole and rotation axis. Their share of the magnetic energy density averaged over the Earth surface is about 5%. The magnetic field of the Earth shows a broad spectrum of time variations, out of which two secular ones should be mentioned. Firstly, there are drifts of magnetic field patterns non–symmetric with respect to the rotation axis, preferably westward, with drift rates up to the order of tenths of a degree per year. Secondly, as we know from paleomagnetic studies, the Earth had a magnetic field with about today's intensity over at least $3.5 \cdot 10^9$ years. However, it reversed its polarity after intervals of irregularly distributed durations between 10^5 and 10^7 years; the characteristic time of the reversals is only 10^4 years.

The past three decades of exploration of the planetary system by spacecrafts brought us information about the

Table 1. Planetary Magnetic Fields

Planet	$M[\text{Tm}^3]$	$\psi[°]$	P	$B_p[\text{nT}]$
Mercury	$5.0 \cdot 10^{12}$	14.0°	–	660
Venus	$< 4.0 \cdot 10^{11}$	...	...	< 4
Earth	$8.2 \cdot 10^{15}$	11.4°	–	6200
Mars	$< 2.0 \cdot 10^{12}$	...	...	< 120
Jupiter	$1.6 \cdot 10^{20}$	9.6°	+	856000
Saturn	$4.7 \cdot 10^{18}$	0.0°	+	42400
Uranus	$3.8 \cdot 10^{17}$	58.6°	+	46000
Neptune	$2.0 \cdot 10^{17}$	46.8°	+	28000
Moon	$< 1.0 \cdot 10^{9}$	...	...	< 0.4

M denotes the magnetic dipole moment, ψ the inclination between dipole and rotation axis, P the magnetic polarity (+ or − depending on whether the angle between dipole moment and angular momentum is < or > 90°, respectively), B_p the magnetic flux density resulting from the dipole only at the intersection of its axis and the planetary surface. Data taken from *Ness* [1994].

magnetic fields of all planets except Pluto; see, e.g., *Ness* [1994]. As far as the terrestrial planets are concerned, only at Mercury has the existence of an intrinsic magnetic field been proved; it is, however, much smaller than that of the Earth. At Venus no such field could be found. The situation at Mars is still unclear; if any intrinsic field exists it is quite small. For Jupiter, Saturn, Uranus and Neptune not only the existence of intrinsic magnetic fields but also details concerning their intensities and structures, in particular multipole representations including dipole, quadrupole and octupole are known. The magnetic flux density at the surface of Jupiter is more than a factor 10 larger than that at the Earth, but the corresponding values for Saturn, Uranus and Neptune are comparable to that for the Earth. Whereas the geometrical structure of the Jovian magnetic field including the inclination of the dipole axis against the rotation axis is very similar to that of the Earth's field, the situation with the other three planets is clearly different. In the best-fitted model of the Saturnian field the latter appears to be completely symmetric about the rotation axis of the planet, and with Uranus and Neptune there are drastic deviations from this symmetry. Some more details are given in Table 1; see also *Rädler and Ness* [1990].

It is generally believed that the magnetic fields of the Earth and some of the planets like Jupiter, Saturn, Uranus and Neptune are generated and maintained by dynamo processes in the electrically conducting fluid regions of their interiors. Some elementary understanding of these processes has been gained from kinematic dynamo models, in particular from those based on the mean–field approach. Much progress has been made during the last decades towards dynamically consistent models for the Earth and also for planets, reflecting essential features of their magnetic fields. In particular we note the impressive results of extensive numerical simulations of the geodynamo, e.g., by *Glatzmaier and Roberts* [1996], and of numerical studies of planetary dynamos, e.g., by *Busse, Grote and Tilgner* [1998].

We want to review here a few aspects of the dynamo theory relevant for the Earth and the planets with particular attention to aspects connected with the helicities in the fluid motions and the magnetic fields. The role of the kinetic helicity in dynamo processes is well–known and has often been discussed. It seems however that in geodynamo and planetary dynamo studies the magnetic helicity has never been considered, and the current helicity only in a few cases. So we have to restrict our explanations and discussions to the kinetic and the current helicities. Since their relevance for dynamo processes can be most easily demonstrated in the kinematic mean–field dynamo theory we will mainly deal with this approach and add only a few further considerations.

2. PLANETARY DYNAMO MODELS

2.1. Fundamentals

Let us briefly describe the fundamentals of planetary dynamo theory. We consider the region responsible for dynamo action, that is, the Earth's core or a corresponding part of a planetary interior, as a spherical body of electrically conducting fluid in electrically insulating surroundings. We further assume that the magnetic flux density $\boldsymbol{B}$ is governed by the induction equation,

$$\eta \Delta \boldsymbol{B} + \nabla \times (\boldsymbol{u} \times \boldsymbol{B}) - \frac{\partial \boldsymbol{B}}{\partial t} = \boldsymbol{0}, \quad (1)$$
$$\text{div } \boldsymbol{B} = 0,$$

inside the fluid and continues as a solenoidal potential field in outer space. As usual, η denotes the magnetic diffusivity, for simplicity assumed to be constant, and $\boldsymbol{u}$ the velocity of the fluid motion. We speak of a dynamo if there are solutions $\boldsymbol{B}$ of (1) which do not decay to zero as $t \to \infty$.

The kinematic dynamo problem, which we will deal with first, consists in finding fluid motions, that is, velocity fields $\boldsymbol{u}$, which allow for such non–decaying solutions $\boldsymbol{B}$. On this level the fluid–dynamic equations governing the motions and describing in particular the back–reaction of the magnetic field on the motions are ignored.

We note that there are theorems saying that a dynamo cannot work with simple geometrical structures of the magnetic or the velocity field; see, e.g., *Moffatt* [1978] or *Krause and Rädler* [1980]. For instance, according to Cowling's theorem an axisymmetric magnetic field can never be generated or maintained by a dynamo. Another theorem tracing back to Bullard and Gellman excludes in the case of constant magnetic diffusivity any dynamo action of a purely toroidal motion.

2.2. The Mean–Field Approach

In order to give a first idea on how a dynamo may work let us imagine fluid motions of convective or turbulent nature with components varying on small space or time scales. Then the magnetic field too has such components. When considering situations of this kind it is useful to follow the lines of mean–field magnetohydrodynamics; see, e.g., *Krause and Rädler* [1980].

In that sense we split $\boldsymbol{u}$ and $\boldsymbol{B}$ into mean fields $\overline{\boldsymbol{u}}$ and $\overline{\boldsymbol{B}}$, defined by a proper averaging procedure that maintains large–scale components only, and "fluctuating" fields $\boldsymbol{u}'$ and $\boldsymbol{B}'$, which consist of the small–scale components. Starting then from the induction equation (1) and accepting the usual Reynolds averaging rules we arrive immediately at the mean–field induction equation

$$\eta \Delta \overline{\boldsymbol{B}} + \nabla \times (\overline{\boldsymbol{u}} \times \overline{\boldsymbol{B}} + \boldsymbol{\mathcal{E}}) - \frac{\partial \overline{\boldsymbol{B}}}{\partial t} = \boldsymbol{0}, \quad (2)$$

$$\operatorname{div} \overline{\boldsymbol{B}} = 0,$$

where $\boldsymbol{\mathcal{E}}$ is an electromotive force due to the fluctuations $\boldsymbol{u}'$ and $\boldsymbol{B}'$,

$$\boldsymbol{\mathcal{E}} = \overline{\boldsymbol{u}' \times \boldsymbol{B}'}. \quad (3)$$

It turns out that $\boldsymbol{\mathcal{E}}$ is a functional of $\overline{\boldsymbol{u}}$, $\boldsymbol{u}'$ and $\overline{\boldsymbol{B}}$ which is linear in $\overline{\boldsymbol{B}}$. If the variations of $\overline{\boldsymbol{B}}$ within the length and time scales of the fluctuations are sufficiently weak we may conclude that

$$\mathcal{E}_i = a_{ij} \overline{B}_j + b_{ijk} \partial \overline{B}_j / \partial x_k, \quad (4)$$

with tensors a_{ij} and b_{ijk} depending on $\overline{\boldsymbol{u}}$ and $\boldsymbol{u}'$ only.

It is instructive to consider first the very special case in which $\overline{\boldsymbol{u}} = \boldsymbol{0}$ and $\boldsymbol{u}'$ corresponds to a locally isotropic turbulence. We call turbulence isotropic with respect to a given point if all averaged quantities depending on $\boldsymbol{u}'$ are invariant under arbitrary rotations of the $\boldsymbol{u}'$–field about arbitrary axes through this point. For all points in which $\boldsymbol{u}'$ shows this property, relation (4) takes the form

$$\boldsymbol{\mathcal{E}} = \alpha \overline{\boldsymbol{B}} - \beta \nabla \times \overline{\boldsymbol{B}}, \quad (5)$$

with coefficients α and β depending on $\boldsymbol{u}'$. The first term describes the α–effect, more precisely the isotropic α–effect, that is, the occurrence of an electromotive force parallel or antiparallel to the mean magnetic field, and the second one gives rise to introduce a mean–field diffusivity or a mean–field conductivity different from the the corresponding values in the absence of fluctuating motions.

In this context, the behavior of quantities depending on $\boldsymbol{u}'$ under reflections of the $\boldsymbol{u}'$–field deserves special interest. We call a turbulence reflectionally symmetric with respect to a point if all averaged quantities depending on $\boldsymbol{u}'$ are invariant under reflections of the $\boldsymbol{u}'$–field about arbitrary planes containing this point. Since a reflection converts a right–handed helical motion into a left–handed one and vice versa, reflectional symmetry implies an equipartition of right and left–handed motions. It turns out that the coefficient α changes its sign under reflection of the $\boldsymbol{u}'$–field at planes containing this point. Thus it can be non–zero only for a turbulence lacking reflectional symmetry. The coefficient β is invariant under such reflections. In a wide range of reasonable assumptions it proved to be positive so that the mean–field diffusivity is higher and the mean–field conductivity smaller than the values in the absence of turbulence.

In a spherical rotating body such as the Earth's or a planetary core, even in a proper frame of reference in which $\overline{\boldsymbol{u}} = \boldsymbol{0}$ applies, $\boldsymbol{u}'$ can no longer be considered as isotropic turbulence. One preferred direction is already given by the radial gravitational force, another one by the rotation axis, which is relevant for the Coriolis forces too. These forces in particular produce helical structures in the turbulent motions and thus disturb any reflectional symmetry. Then instead of (5) another relation with additional terms on the right–hand side applies, a part of which can be described by α and β–tensors. As long as $\boldsymbol{\mathcal{E}}$ shows a component parallel or antiparallel to $\overline{\boldsymbol{B}}$ we speak again of α–effect or, if other components exist, more precisely of anisotropic α–effect. Furthermore, we have now in general anisotropic mean–field diffusivities or conductivities.

The magnitude of the α–effect and the structure of the α–tensor depends on a parameter $\Omega \tau_c$, where Ω is the angular velocity of the rotation, which is also responsible for the Coriolis forces, and τ_c a characteristic time scale of the turbulent motions.

2.3. Kinematic Mean–Field Dynamo Models

With this background simple kinematic dynamo models have been developed; see, e.g., *Krause and Rädler* [1980] or *Rädler* [1980,1986]. In these models rotating fluid bodies are assumed showing small–scale convection or turbulence subject to Coriolis–forces. As a consequence, an α–effect occurs. The α–effect, alone or in combination with a differential rotation, that is, a dependence of the angular velocity on radius or latitude, is able to generate or maintain magnetic fields. Dynamo action always consists in an interplay of the poloidal and the toroidal part of the magnetic field. The regeneration of the poloidal from the toroidal part is in all cases mainly due to α–effect. Depending on whether the regeneration of the toroidal from the poloidal part is again mainly due to α–effect or due to differential rotation, we speak of α^2– or $\alpha\omega$–mechanisms.

It seems natural to assume some symmetries of the distribution of the fluid motion. Let us formulate them by saying that all averages depending on $\boldsymbol{u}$ are invariant under rotation of the $\boldsymbol{u}$–field about the rotation axis and under its reflection about the equatorial plane of the fluid body. This implies, of course, axisymmetry and equatorial symmetry of $\overline{\boldsymbol{u}}$, but also some symmetries of quantities like α and β. For instance, the components of the α–tensor with respect to spherical or cylindrical polar co–ordinates are again symmetric about the rotation axis but some of them, in particular the diagonal ones, are antisymmetric about the equatorial plane. With this symmetry assumptions it can be shown that the solutions of (2) are modes of the form

$$\overline{\boldsymbol{B}} = \Re(\hat{\boldsymbol{B}} \exp(im\varphi + \lambda t)), \qquad (6)$$

or superpositions of such modes. $\hat{\boldsymbol{B}}$ is a complex steady vector field being symmetric about the rotation axis and either antisymmetric or symmetric about the equatorial plane, m is an integer, φ the azimuthal co–ordinate, and λ a complex constant, the real part of which describes the growth rate of the solution. We denote the modes by Am or Sm corresponding to their antisymmetry or symmetry about the equatorial plane, with m indicating the azimuthal variation.

A number of numerical investigations of dynamo models of that kind has been carried out; see again, e.g., *Krause and Rädler* [1980], *Rädler* [1980,1986], *Rüdiger and Elstner* [1994], *Moss and Brandenburg* [1995]. It turned out that for the α^2–mechanism in the case $\Omega \tau_c \ll 1$ the excitation condition for the axisymmetric modes, $A0$ and $S0$, and the first non–axisymmetric ones, $A1$ and $S1$, are rather close together; all other modes are less easier to excite. In the case $\Omega \tau_c \gg 1$, however, there is a preference of the first non–axisymmetric modes, $A1$ or $S1$, over all others. For the $\alpha\omega$–mechanism always one of the axisymmetric modes, $A0$ or $S0$, is preferred over all others; the selection between them depends on the details of the differential rotation.

These results provide us at least with a rough idea on the planetary dynamo processes. They also show that the magnetic fields need not be essentially symmetric about the rotation axis of the planet but may well deviate drastically from this symmetry.

2.4. Models beyond the Mean–field Approach

By several reasons kinematic mean–field models are only a first step toward more or less realistic models of planetary dynamos.

Firstly, depending on the nature of the fluid motions in a planetary interior the mean–field approach, which works well only with some separation of large and small scales of these motions, may provide us only with a rough approximation. In this context also other kinds of kinematic dynamo models with large–scale flow patterns are of interest. We refer here to models as elaborated by *Pekeris, Accad and Shkoller* [1973], *Gubbins* [1973], *Kumar and Roberts* [1975], *Dudley and James* [1989], *Sarson* [1994], *Holme* [1996] or *Love and Gubbins* [1996].

Secondly, kinematic dynamo models ignore any back–reaction of the magnetic field on the fluid motion. Therefore they do not say anything about intensities of the magnetic fields generated or the behavior of a dynamo in the nonlinear regime. By this and other reasons we have to proceed to dynamically consistent models, based on the induction equation (1) in combination with the fluid–dynamic equations including the Lorentz forces. Examples for models of that kind are those by *Glatzmaier and Roberts* [1996] or by *Busse, Grote and Tilgner* [1998].

2.5. A Saturnian Dynamo Model

We do not discuss the dynamo models mentioned in the last section but give only a few details of a model of some interest for Saturn, which we want refer to later in the helicity context too. At the first glance, because of Cowling's theorem the high degree of axisymmetry of the Saturnian magnetic field seems to be in conflict with the assumption of a dynamo. The model explained in the following, proposed by *Rädler* [1990] and elaborated by *Rheinhardt* [1997], demonstrates that a dynamo may well work with a magnetic field whose first multipoles are completely axisymmetric.

To describe the kinematic version of this model let us consider $\boldsymbol{u}$ and $\boldsymbol{B}$ as Fourier series with respect to the azimuthal coordinate φ. More precisely, each of their components should have an expansion after the pattern of $F = \sum_m F^m$ with $F^m = \Re\{\hat{F}^m e^{im\varphi}\}$. Suppose now that $\boldsymbol{u}$ is periodic in φ with a period $2\pi/k$, $k > 1$, so that

$$\boldsymbol{u} = \boldsymbol{u}^0 + \boldsymbol{u}^k + \boldsymbol{u}^{2k} + \cdots. \tag{7}$$

Then it can be easily concluded that the induction equation (1) has independent solutions $\boldsymbol{B}^{(0)}, \boldsymbol{B}^{(1)}, \cdots \boldsymbol{B}^{(l)}$, $l \leq k/2$, with the structures

$$\begin{aligned}
\boldsymbol{B}^{(0)} &= \boldsymbol{B}^0 + \boldsymbol{B}^k + \boldsymbol{B}^{2k} + \cdots, \\
\boldsymbol{B}^{(1)} &= \boldsymbol{B}^1 + \boldsymbol{B}^{k-1} + \boldsymbol{B}^{k+1} + \boldsymbol{B}^{2k-1} + \boldsymbol{B}^{2k+1} + \cdots, \\
&\vdots \\
\boldsymbol{B}^{(l)} &= \boldsymbol{B}^l + \boldsymbol{B}^{k-l} + \boldsymbol{B}^{k+l} + \boldsymbol{B}^{2k-l} + \boldsymbol{B}^{2k+l} + \cdots.
\end{aligned} \tag{8}$$

For each of these fields the usual multipole representation can be given for the outer space. For the solution $\boldsymbol{B}^{(0)}$ the contributions of the first $k-1$ multipoles are axisymmetric. This results from the facts that the multipole field of order n is described by spherical harmonics Y_n^m, and that $Y_n^m = 0$ if $|m| > n$. If, e.g., a flow pattern according to (7) with $k \geq 4$ is given, a magnetic field of $\boldsymbol{B}^{(0)}$-type has axisymmetric dipole, quadrupole and octupole parts, and the necessary deviations from axisymmetry occur only in higher multipoles.

Numerical investigations of dynamo models of that kind have been carried out on the basis of the induction equation (1) with motions defined by

$$\begin{aligned}
\boldsymbol{u} = -U\{ &\nabla \times [\boldsymbol{r} \times \nabla(f_1(r) Y_{2k}^{kc}(\vartheta, \varphi))] \\
&+ \boldsymbol{r} \times \nabla(f_2(r) Y_{2k+1}^{kc}(\vartheta, \varphi))\},
\end{aligned} \tag{9}$$

where r, θ, φ are the usual spherical polar co-ordinates, r measured in units of the radius of the conducting body, U is a constant, f_1 and f_2 are functions of r only, and $Y_l^{mc}(\vartheta, \varphi)$ stands for $P_l^m(\cos\vartheta)\cos m\varphi$, with P_l^m being associated Legendre polynomials. As for specifications of f_1, f_2 and k we restrict our attention here to two cases. The first one, denoted by (a), is defined by $f_1 = 0.067 f_2 = r^2(1 - r^2)$ and $k = 4$. In the second case, denoted by (b), f_1 and f_2 are more complex functions which change their signs three times when r runs from 0 to 1, and it holds $k = 16$, that is, compared to (a) the length scales of the motions are smaller both in r-direction and perpendicular to it. Flow patterns for these two cases are shown in Figures 1 and 2.

Let us first deal only with case (a), which was studied in more detail; a result for case (b) is given in the helicity context. Indeed, solutions with the structures described by (8) have been found. It turned out that in this case solutions of type $\boldsymbol{B}^{(0)}$ are less easily excitable than those of type $\boldsymbol{B}^{(1)}$. If, however, $\boldsymbol{u}$ is changed by adding a term corresponding to a differential rotation $\boldsymbol{B}^{(0)}$ is preferred.

A dynamically consistent version of this model with prescribed buoyancy forces and with Coriolis forces has been considered too. With a proper distribution of buoyancy forces a flow pattern similar to that given by (9) in the case (a) is generated. In addition automatically a differential rotation occurs, and so a magnetic field of $\boldsymbol{B}^{(0)}$-type, that is, with completely axisymmetric dipole, quadrupole and octupole.

3. DYNAMO ACTION AND HELICITIES

3.1. α-effect and Kinematic Helicity

The α-effect is closely connected with the kinetic helicity. We demonstrate this here for the simple case in which $\overline{\boldsymbol{u}} = \boldsymbol{0}$. In addition we restrict ourselves on the second-order correlation approximation, or first-order smoothing approximation, which is often used in mean-field magnetohydrodynamics but can be easily justified only under the assumption $\min(u_c \lambda_c/\eta, u_c \tau_c/\lambda_c) \ll 1$, where u_c, λ_c and τ_c are characteristic values of the magnitude and the length and time scales of $\boldsymbol{u}'$. Within this framework a straightforward calculation provides us with

$$a_{ij} = \int\!\!\int_0^\infty\!\!\int G(\xi, \tau) K_{ij}(\boldsymbol{x}, t; \boldsymbol{\xi}, \tau) d^3\xi\, d\tau, \tag{10}$$

where G is a Green's function,

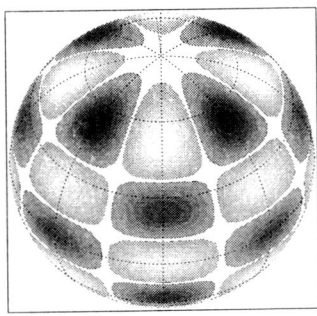 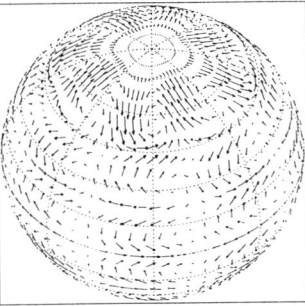

Figure 1. Dynamo model defined by the fluid velocity (9), case (a). Normal (left) and tangential (right) components of the velocity on the surface $r = 0.25$. Light areas show upward, dark areas downward flow.

$$G(\xi, \tau) = (4\pi\eta\tau)^{-3/2} \exp(-\xi^2/4\eta\tau), \quad (11)$$

K_{ij} a two–point correlation function,

$$K_{ij}(\boldsymbol{x}, t; \boldsymbol{\xi}, \tau) = -\epsilon_{ilm}\epsilon_{mnp}\epsilon_{pqj}\,\overline{u'_l(\boldsymbol{x}, t)\frac{\partial u'_q(\boldsymbol{x} - \boldsymbol{\xi}, t - \tau)}{\partial \xi_n}}, \quad (12)$$

and the ξ–integration is over all space; see *Krause and Rädler* [1980].

We further specify $\boldsymbol{u'}$ so that it corresponds to an isotropic turbulence. Then (5) applies, and we find

$$\alpha = -\frac{1}{3}\int\!\!\int_0^\infty G(\xi, \tau)\, h(\boldsymbol{x}, t; -\boldsymbol{\xi}, -\tau)\, d^3\xi\, d\tau \quad (13)$$

with

$$h(\boldsymbol{x}, t; \boldsymbol{\xi}, \tau) = \overline{\boldsymbol{u'}(\boldsymbol{x}, t) \cdot (\nabla \times \boldsymbol{u'}(\boldsymbol{x} + \boldsymbol{\xi}, t + \tau))}. \quad (14)$$

Clearly, the coefficient α occurring with the isotropic α–effect as given by (5) is determined by the kinetic–helicity two–point correlation function $h(\boldsymbol{x}, t; \boldsymbol{\xi}, \tau)$.

In the low–conductivity limit, $\tau_c \gg \lambda_c^2/\eta$, we may reduce (13) and (14) to

$$\alpha = -\frac{1}{3\eta}\,\overline{\boldsymbol{a} \cdot (\nabla \times \boldsymbol{a})}, \quad (15)$$

where $\boldsymbol{a}$ is vector potential of $\boldsymbol{u'}$ so that $\nabla \times \boldsymbol{a} = \boldsymbol{u'}$ and div $\boldsymbol{a} = 0$.

For the high–conductivity limit, $\tau_c \ll \lambda_c^2/\eta$, we find

$$\alpha = -\frac{1}{3}\int_0^\infty \overline{\boldsymbol{u'}(\boldsymbol{x}, t) \cdot (\nabla \times \boldsymbol{u'}(\boldsymbol{x}, t - \tau))}\, d\tau, \quad (16)$$

or

$$\alpha = -\frac{1}{3}\,\overline{\boldsymbol{u'} \cdot (\nabla \times \boldsymbol{u'})}\,\tau_c^*, \quad (17)$$

with a correlation time τ_c^* just defined by equating the right–hand sides of (16) and (17).

We note that in the high–conductivity limit α approaches a non–zero value. By the way, the same applies to β. These facts imply the possibility of fast dynamos working on the basis of the α–effect.

The results presented here show that, in the framework of our approximations, the coefficient α depends on the fluid velocity $\boldsymbol{u'}$ via the kinetic–helicity two–point correlation function $h(\boldsymbol{x}, t; \boldsymbol{\xi}, \tau)$. Statements saying that α essentially coincides with the kinetic helicity in the usual sense, that is with $\overline{\boldsymbol{u'} \cdot (\nabla \times \boldsymbol{u'})}$, must be considered with caution. Their validity is, apart from the second–order correlation approximation and the assumption of isotropic turbulence, restricted to the high–conductivity limit, and even then α is primarily given by (16), which can be written in the form (17) only for non–zero $\overline{\boldsymbol{u'} \cdot (\nabla \times \boldsymbol{u'})}$. Incidentally, if the assumption of isotropic turbulence is relaxed and thus an α–tensor occurs, relations (13) to (17) apply with α replaced by $\frac{1}{3}$ trace(α), but it is not simply this quantity which is responsible for dynamo action in models as envisaged here.

3.2. Current Helicity

We remain within the mean–field concept and consider now the average of the current helicity,

$$\overline{\boldsymbol{B} \cdot (\nabla \times \boldsymbol{B})} = \overline{\boldsymbol{B}} \cdot (\nabla \times \overline{\boldsymbol{B}}) + \overline{\boldsymbol{B'} \cdot (\nabla \times \boldsymbol{B'})}. \quad (18)$$

In the following we will discuss the two contributions separately.

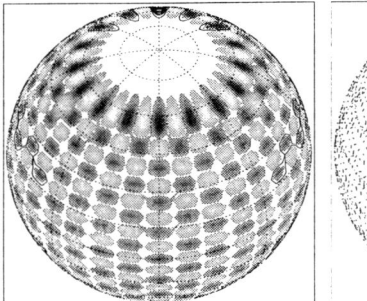 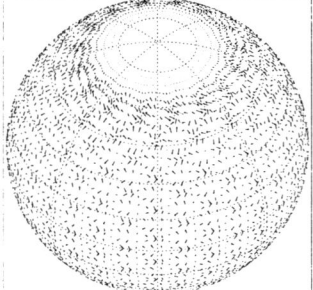

Figure 2. Dynamo model defined by the fluid velocity (9), case (b). Explanations given with Figure 1 apply, but with $r = 0.75$.

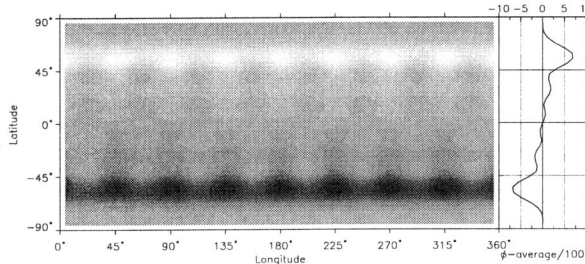

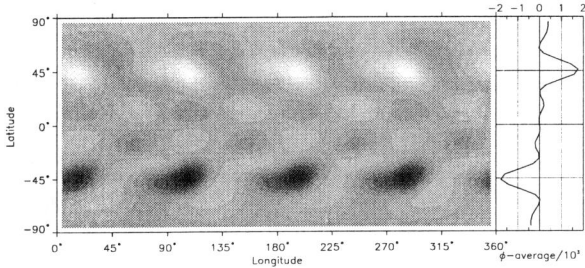

Figure 3. Dynamo model defined by the fluid velocity (9), case (a). Kinetic (above) and current (below) helicity in arbitrary units on the surface $r = 0.25$. Light areas correspond to positive, dark to negative values.

Let us start with the second contribution, that is, $\overline{B' \cdot (\nabla \times B')}$. Under the assumptions made above, that is, $\overline{u} = 0$ and second–order correlation approximation, it can be shown that

$$\overline{B' \cdot (\nabla \times B')} = -\frac{1}{\eta} a_{ij} \overline{B_i} \cdot \overline{B_j} \ ; \qquad (19)$$

see *Rädler and Seehafer* [1990]. In the case of isotropic turbulence this turns into the form

$$\overline{B' \cdot (\nabla \times B')} = -\frac{\alpha}{\eta} \overline{B}^2 \ , \qquad (20)$$

which was already given by *Keinigs* [1983], see also *Keinigs and Gerwin* [1986], and by *Matthaeus, Goldstein and Lantz* [1986]. A derivation of (20) which avoids the assumption $\overline{u} = 0$ and the second–order correlation approximation was proposed by *Seehafer* [1994], see also *Seehafer* [1996].

We see that the signs of $\overline{B' \cdot (\nabla \times B')}$ and α are opposite to each other. As long as the sign of α is opposite to that of the kinetic helicity, the sign of the kinetic helicity coincides with that of the current helicity of the fluctuations of the magnetic field.

In order to find a condition for the other contribution, $\overline{B} \cdot \nabla \times \overline{B}$, we first remark that for a mean–field dynamo model where $\overline{B}$ is governed by equations (2) inside a fluid body and continues as a solenoidal potential field in outer space, the balance of the energy stored in the mean magnetic field is given by

$$\frac{d}{dt} \int_\infty \frac{\overline{B}^2}{2\mu} dv = -\int \eta (\nabla \times \overline{B})^2 dv$$
$$+ \frac{1}{\mu} \int \overline{u} \cdot ((\nabla \times \overline{B}) \times \overline{B}) \, dv \quad (21)$$
$$+ \frac{1}{\mu} \int \mathcal{E} \cdot (\nabla \times \overline{B}) \, dv \ .$$

Note that the integral at the left hand side is over all space, the other integrals over the fluid body only; μ is the permeability which is supposed to be the same in the fluid and in free space.

Consider now a dynamo of α^2–type for which, in a proper frame of reference, $\overline{u} = 0$. A necessary condition for dynamo action is then that the last integral on the right–hand side of (21) is positive. If $\mathcal{E}$ is given by (5) with a non–negative β, this leads to

$$\int \alpha \left(\overline{B} \cdot (\nabla \times \overline{B}) \right) dv > 0 \ . \qquad (22)$$

That is, the signs of α and $\overline{B} \cdot (\nabla \times \overline{B})$ must essentially coincide in the sense that the regions with this coincidence determine the sign of the integral.

With (20) and (22) in mind we write now

$$\overline{B' \cdot (\nabla \times B')} = -f \, \overline{B} \cdot (\nabla \times \overline{B}) \qquad (23)$$

and conclude that f is essentially positive in a similar sense. Moreover it can be shown that f is essentially larger than unity; see again *Rädler and Seehafer* [1990]. So the sign of $\overline{B} \cdot (\nabla \times \overline{B})$ is essentially given by $\overline{B' \cdot (\nabla \times B')}$, which was already noted by *Keinigs and Gerwin* [1986].

Consequently the sign of the average of the total current helicity, $\overline{B} \cdot (\nabla \times \overline{B})$ should coincide with that of the kinetic helicity.

3.3. Kinematic and Current Helicity in a Saturnian Dynamo Model

Let us now leave the mean–field approach and return to the Saturnian dynamo model described in Section 2.5. Guided by the results on the mean–field level the distribution of the kinetic and current helicities has been studied on the basis of a numerical solution of the induction equation (1) for B with u given by (9). Figures 3 and 4 show results for the solutions of $B^{(0)}$–type for the cases (a) and (b) defined above. As to

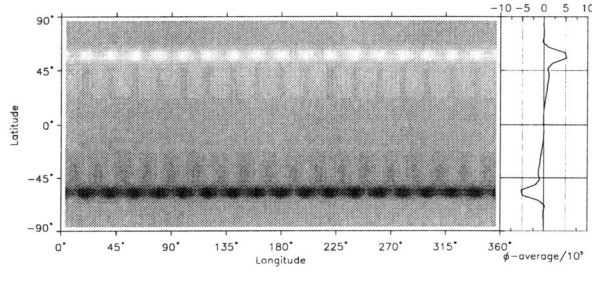

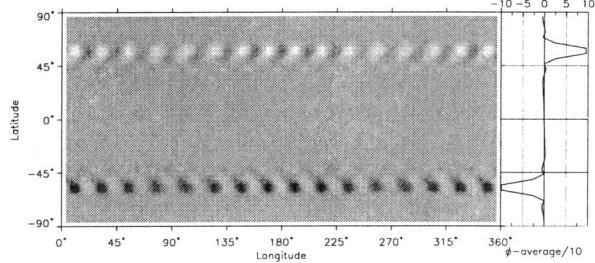

Figure 4. Dynamo model defined by the fluid velocity (9), case (b). Explanations given with Figure 3 apply, but with $r = 0.75$.

be expected from the above mean–field considerations, striking correlations between the kinetic and current helicities can be seen.

Acknowledgments. The author thanks Dr. M. Rheinhardt for providing him with the unpublished numerical results on the helicities presented in the last paragraph.

REFERENCES

F. H. Busse, E. Grote, and A. Tilgner. On convection driven dynamos in rotating spherical shells. *Studia geoph. et geod.*, 42:211–223, 1998.

M. L. Dudley and R. W. James. Time–dependent kinematic dynamos with stationary flows. *Proc. R. Soc. London A*, 425:407–429, 1989.

G. A. Glatzmaier and P. H. Roberts. An anelastic evolutionary geodynamo simulation driven by compositional and thermal convection. *Physica D*, 97:81–94, 1996.

D. Gubbins. Numerical solutions of the kinematic dynamo problem. *Phil. Trans. Roy. Soc. London A*, 274:493–521, 1973.

R. Holme. Three–dimensional kimematic dynamos with equatorial symmetry: Application to the magnetic fields of Uranus and Neptune. *Phys. Earth Planet. Inter.*, 102:105–122, 1997.

R. K. Keinigs. A new interpretation of the alpha effect. *Phys. Fluids*, 26:2558–2560, 1983.

R. K. Keinigs and A. Gerwin. The alpha effect: The connection between cyclonic events and current helicity. *IEEE Transact. Plasma Sci.*, PS-14:858–861, 1986.

F. Krause and K.-H. Rädler. *Mean–Field Magnetohydrodynamics and Dynamo Theory.* Akademie-Verlag Berlin and Pergamon Press Oxford, 1980.

S. Kumar and P.H. Roberts. A three–dimensional kinematic dynamo. *Proc. R. Soc. Lond. A.*, 344:235–258, 1975.

J. J. Love and D. Gubbins. Optimized kinematic dynamos. *Geophys. J. Int.*, 124:787–800, 1996.

W. H. Matthaeus, M. L. Goldstein, and S. R. Lantz. The alpha dynamo parameter and the measurability of helicities in magnetohydrodynamic turbulence. *Phys. Fluids*, 29:1504–1508, 1986.

H. K. Moffatt. *Magnetic Field Generation in Electrically Conducting Fluids.* Cambridge University Press, 1978.

D. Moss and A. Brandenburg. The generation of nonaxisymmetric magnetic field in the giant planets. *Geophys. Astrophys. Fluid Dynamics*, 80:229–240, 1995.

N. F. Ness. Intrinsic magnetic fields of the planets: Mercury to Neptune. *Phil. Trans. R. Soc. Lond.*, A 349:249–260, 1994.

C. L. Pekeris, Y. Accad, and B. Shkoller. Kinematic dynamos and the Earth's magnetic field. *Phil. Trans. Roy. Soc. London A*, 275:425–442, 1973.

K.-H. Rädler. Mean–field approach to spherical dynamo models. *Astron. Nachr.*, 301:101–129, 1980.

K.-H. Rädler. Investigations of spherical kinematic mean-field dynamo models. *Astron. Nachr.*, 307:89–113, 1986.

K.-H. Rädler. Can the highly axisymmetric magnetic field of saturn be maintained by a dynamo? *Adv. Space Research*, 12:281–284, 1992.

K.-H. Rädler and N. F. Ness. The symmetry properties of planetary magnetic fields. *J. Geophys. Research*, 95:2311–2318, 1990.

K.-H. Rädler and N. Seehafer. Relations between helicities in mean-field dynamo models. In H. K. Moffatt and A. Tsinober, editors, *Topological Fluid Mechanics*, pages 157–163. Cambridge University Press, 1990.

M. Rheinhardt. *Untersuchungen kinematischer und dynamisch konsistenter Dynamomodelle in sphärischer Geometrie.* PhD thesis, Universität Potsdam, 1997.

G. Rüdiger and D. Elstner. Non-axisymmetry vs. axisymmetry in dynamo-excited stellar magnetic fields. *Astron. Astrophys.*, 281:46–50, 1994.

G.R. Sarson. *Kinematic Dynamo Calculations for Geomagnetism.* PhD thesis, The University of Leeds, 1994.

N. Seehafer. Current helicity and the turbulent electromotive force. *Europhys. Lett.*, 27:353–357, 1994.

N. Seehafer. Nature of the alpha-effect in magnetohydrodynamics. *Phys. Rev. E*, 53:1283–1286, 1996.

K.-H. Rädler, Astrophysikalisches Institut Potsdam, An der Sternwarte 16, D-14482 Potsdam, Germany. (e-mail:khraedler@aip.de)

Helicity, Relaxation, and Dynamo in a Laboratory Plasma

S.C. Prager

Department of Physics, University of Wisconsin, Madison, Wisconsin

The toroidal laboratory plasma known as the reversed field pinch (RFP) is understood through magnetohydrodynamics to undergo an approximately helicity-conserving relaxation to a state which is near a minimum energy "Taylor state." Experimental observation is consistent with this view. Measurements of the MHD dynamo (the alpha effect), obtained through various experimental measurements or inferences of the fluctuating plasma flow velocity, indicate that the MHD dynamo is active. Additional dynamo effects beyond the standard MHD model, such as pressure-driven or kinetic effects, are also under study experimentally, with some experimental support. However, the standard MHD model has motivated control experiments in which external adjustment of the mean fields leads to reduction in the fluctuation-induced relaxation and dynamo activity.

1. INTRODUCTION

Magnetic helicity is a quantity which is invariant in a perfectly conducting plasma. However, plasmas in nature are not perfectly conducting. Thus, the significance of helicity arises in large part from the conjecture that it is approximately conserved in real plasmas, and that this conservation is an important constraint on the plasma dynamics. It has been suggested that helicity conservation regulates magnetic field turbulence, magnetic field relaxation, and dynamo behavior in various plasma situations in nature. To study these issues in a laboratory experiment requires a highly conducting plasma undergoing magnetic relaxation. The laboratory plasma known as the reversed field pinch [*see, for example, Bodin and Newton,* 1980] satisfies this criterion. The RFP is a toroidal plasma with toroidal and poloidal magnetic fields which superpose to form a helical magnetic field. Whereas part of the magnetic field is imposed by the experimenter, the RFP spontaneously produces a magnetic field structure in which the toroidal component of the field reverses direction near the wall. Taylor showed that if one assumes that helicity is conserved as the magnetic energy relaxes (is minimized) then the final state is indeed one in which the field reverses direction with radius, as is observed. The helicity-conserving Taylor state approximates the observed RFP state, although the two differ in detail. It appears that the plasma does not relax fully to the Taylor state, and can be considered to be partially relaxed, as described in Section 2.

Whereas the Taylor conjecture provides a very useful framework for understanding relaxation, recent research has focused on the dynamics which produce the relaxation. Nonlinear MHD theory and computation provide a detailed portrayal of the relaxation of the magnetic field through the action of plasma fluctuations. Similar to many natural plasmas, the RFP plasma spontaneously generates plasma current which drives the plasma toward a relaxed magnetic field. This dynamo effect arises from the correlation between velocity and magnetic fluctuations. MHD provides a self-consistent treatment in which the fluctuations are affected by the mean fields and the mean fields are altered by the fluctuations. The magnetic fluctuations responsible for the dynamo also cause magnetic reconnection, which is thereby an intrinsic part of the plasma relaxation. The

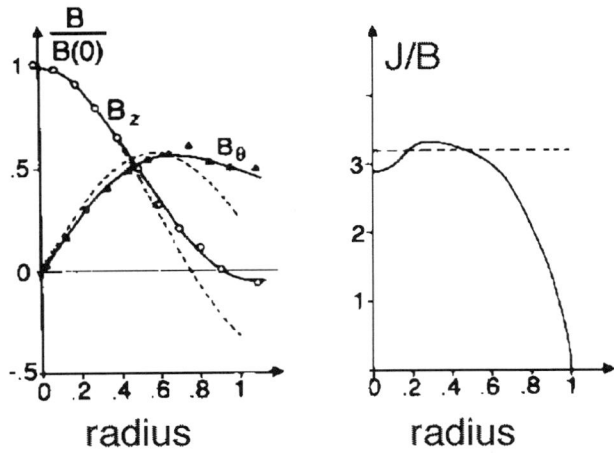

Figure 1. (a) Experimentally measured magnetic field vs radius (solid lines) and theoretical prediction of equation (1), (dashed lines) (b) experimentally measured profile of j/B. Figure taken from *Ortolani and Schnack*, 1993.

MHD understanding is summarized in Section 3. Despite the obvious accessibility of laboratory plasmas, experimental study of the dynamo effect challenges our ability to measure the fluctuating plasma flow velocity with high time resolution. The MHD dynamo effect has been measured in the edge of the plasma with electrostatic Langmuir probes to infer the fluctuating plasma velocity from **E** x **B** drifts, and is being measured in the hot plasma core with Doppler spectroscopy to detect the flow velocity of tracer ions. Each technique demonstrates that the MHD dynamo effect is large under certain plasma conditions, as described in Section 4.

Theoretical treatment of the dynamo has evolved from kinematic theories to studies of the backreaction - the effect of the growing mean magnetic field on the plasma fluctuations. Whereas some theories indicate that the backreaction is strongly quenches the dynamo, in the RFP both nonlinear MHD computation and experiment indicate a robust local dynamo effect in steady state. The computation is self-consistent and fully incorporates all backreaction effects (as, of course, does the experiment). In Section 5 we discuss the weakness of the backreaction for the RFP in the context of recent theoretical discussions. Although the MHD model for the dynamo is compelling, with some experimental support, there are numerous possible dynamo mechanisms which extend beyond the standard model, either through two -fluid effects or through kinetic (non-fluid) effects. In section 6 we discuss three such effects: a dynamo arising fluctuating electron pressure (correlated with fluctuating magnetic field), a dynamo-like effect arising from electron streaming along fluctuating magnetic field lines, and a two-fluid prediction of the individual Taylor-like relaxation which occurs in the separate electron and ion fluids.

Attempts are underway to actively control the relaxation and dynamo effects in the RFP, employing the understanding accumulated through MHD. Such an experiment serves to test and to extend the understanding. To date, through rather coarse external control of the spatial structure of the mean current density we have reduced the amplitude of magnetic fluctuations which induce relaxation by a factor of two. This effect is as expected from the understanding of the relationship between the mean field and the fluctuations. In addition, the fluctuation reduction in turn diminished the transport of energy across the mean field five-fold, revealing that the magnetic fluctuations which produce relaxation also yield energy transport (presumably from electrons following chaotic field lines). The active control experiments are described in Section 7.

2. HELICITY-CONSERVING RELAXATION IN THE LABORATORY

It is well-known that if magnetic energy ($\int(B^2/2\mu_0)dV$) in a plasma is minimized while the magnetic helicity ($\int \mathbf{A} \cdot \mathbf{B} dV$) is held constant, then the magnetic field will obey the equation $\nabla \times \mathbf{B} = \lambda \mathbf{B}$, where λ is a constant. Thus, the ratio j/B is a spatial constant. For a one-dimensional cylindrical plasma this relation yields the simple Bessel function solution [*Taylor*, 1986], known as the "Taylor state,"

$$B_z(r) = B_0 J_0(\lambda r) \quad (1)$$
$$B_\theta(r) = B_0 J_1(\lambda r)$$

where B_0 is a constant. The remarkable feature of this solution is that it captures key properties of the magnetic field in the RFP (comparing the field in the toroidal direction to B_z and the field in the poloidal direction to B_θ). It displays the monotonic decay of the toroidal magnetic field with radius and its reversal of direction at large radius. Figure 1a compares the theoretical solution, eqn (1), to the fields measured in experiment . The agreement is good, although the minimum energy solution is more deeply reversed than the experiment. The deviation between experiment and the simple minimum energy state is also evident in the radial profile of the ratio j/B, shown in Figure 1b. The ratio is roughly constant over about two-thirds of the radius; thus, the plasma can be characterized as partially relaxed.

The approximate Taylor state is also recovered through nonlinear MHD computation [see *Ortolani and Schnack*, 1980, and references therein]. Extensive study of the relaxation process has been achieved through computational solution of the three-dimensional, nonlinear, resistive MHD equations. Typically, these computations have been performed with zero plasma pressure since the relaxation is mediated by instabilities driven by the plasma current. A simple Ohm's law ($E + v \times B = \eta j$) is employed which ignores inertial and Hall effects. However, it retains the electromotive force provided by the plasma flow which is key to the relaxation, as discussed in the next section. The computations to date have employed a cylindrical plasma. It is expected that the toroidal effects present in experiments are small since the toroidal magnetic field is relatively weak. The figure of merit for the importance of toroidicity is the field line pitch rB_z/RB_θ, where R is the major radius of the torus. This parameter is small (<<1) for the RFP.

A typical final state which evolves from the solution is illustrated in Figure 2. With a conducting wall boundary, as is provided in most RFP experiments, the plasma is only partially relaxed. This is evident in the radial profile of j/B which, similar to experiments, deviates from the Taylor relaxed state in the outer third of the radius (Figure 2a). However, if one removes the conducting wall boundary condition, by imposing a vacuum region between the plasma and a distant conducing wall, then the plasma indeed approaches a fully relaxed state in which j/B is nearly constant over the entire plasma (Figure 2b). This arises since the fluctuations which produce relaxation (as described in the next section) are enhanced by removing the stabilizing conducting wall. The increase in relaxation by removal of the conducting wall is also observed in experiment [*Alper et al*, 1989]

The relaxation toward a minimum energy state often occurs in experiments as events which are discrete in time [*Watt and Nebel*, 1983; *Hokin et al*, 1991]. For example, the toroidal magnetic flux often increases suddenly, as shown in Figure 3. The toroidal flux is generated spontaneously in repetitive events which persist for about 100 μsec, sometimes referred to as discrete dynamo events. Between the discrete events, the flux decays. The cyclic nature of the dynamo is understood through the evolution of the radial profile of j/B. The applied electric field causes the j/B profile to become centrally peaked, as discussed in Section 3. As the j/B profile peaks, the deviation from the Taylor state increases, as do the plasmas fluctuations, also described in Section 3. At sufficiently large fluctuation amplitude, the fluctuation-induced relaxation resets the j/B profile to one closer to the Taylor state, after which the cycle repeats. The precise trigger for the relaxation is not fully understood. Nonetheless, these sudden events are convenient experimental signatures of the relaxation process.

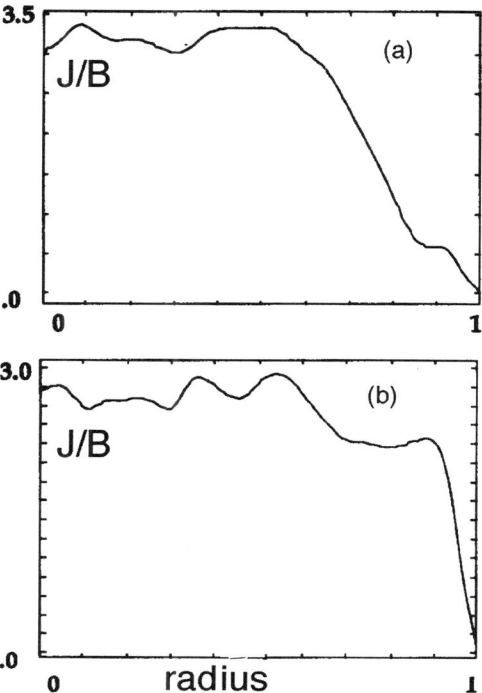

Figure 2. Radial profile of j/B, calculated from nonlinear MHD computation with (a) a conducting boundary and (b) a resistive boundary.

A further test of the Taylor conjecture is to measure the change in the volumetric magnetic helicity and magnetic energy during a discrete relaxation event. This is an extremely difficult experimental measurement. It requires measurement of the vector magnetic field over a full poloidal cross-section. However, an approximate measurement of helicity and energy has been accomplished by combining measurement of the magnetic field at the plasma surface with modeling of the instantaneous plasma state as a slowly varying MHD equilibrium [*Ji, Prager, and Sarff*, 1995]. It is reasonable to assume that through the discrete events the plasma mean fields obey the equilibrium force balance equation, $j \times B = \nabla p$. Although the plasma changes rapidly, the change is slow compared to the Alfven time scale on which the equilibrium equation is satisfied. Hence, the plasma can be viewed as evolving, via relaxation, through a series of quasi-static equilibrium states. Solution of the equilibrium equation, constrained by surface field measurements and experimental information on the plasma

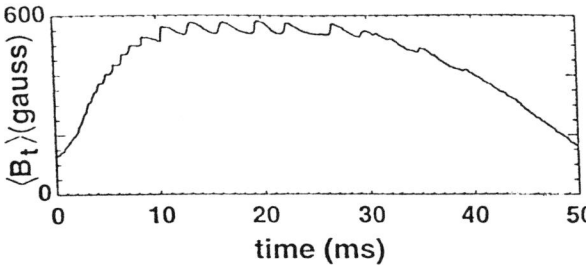

Figure 3. Toroidal magnetic flux within the plasma volume vs time (measured in MST experiment).

pressure, permits inference of the magnetic field within the plasma. In this way, it is concluded that during a discrete event the magnetic energy decreases by about 8%, while the magnetic helicity changes by about 3% (Figure 4). Hence, the discrete events indeed minimize energy relative to the helicity. Moreover, the change in the radial profile of j/B brings the plasma closer to a Taylor state.

The decrease of energy relative to helicity is also observed in MHD computation. Under certain conditions the computational plasma exhibits an oscillatory behavior in which relaxation occurs cyclically, somewhat similar to the experimental behavior. In such cases, the magnetic energy decreases more rapidly than the helicity, again indicating that the plasma relaxes approximately in accord with the Taylor conjecture.

3. DYNAMICS OF RELAXATION

The Taylor conjecture provides a useful framework to depict approximately the final state of the relaxation process. However, it provides no information on the dynamics which underlie relaxation. A complete, self-consistent treatment of the dynamics is provided by nonlinear MHD, summarized in this section. Insight into the MHD physics is obtained from analytical quasilinear studies and simplified treatment of nonlinear mode coupling. However, computational solution of the MHD equations is necessary to calculate both the fluctuating and mean fields. These solutions are self-consistent in that the fluctuations influence the mean fields which, in turn, determine the fluctuations. The relaxation toward an approximate Taylor state magnetic field structure requires the self-generation, or redistribution, of current in the plasma. A current generation mechanism, within MHD, is evident from the mean field Ohm's law

$$<E> + <\delta v \times \delta B> = \eta <j> \qquad (2)$$

where δv and δB are the fluctuating flow velocity and magnetic field, and $<>$ denotes an ensemble average over the fast fluctuation time or spatial scale (or in the cylindrical description of the RFP, an average over the axial and azimuthal directions). The current generation term $<\delta v \times \delta B>$ is considered to be a dynamo effect since it provides a local force to drive plasma current. This is the same alpha effect term considered for natural dynamos, although the cause of the fluctuations is situation dependent. In the RFP it is well understood that the fluctuations are generated by "tearing instabilities" - magnetic instabilities associated with magnetic reconnection [*Ortolani and Schnack*, 1993]. These instabilities arise from the gradients in the current density.

The strength of the dynamo effect in the RFP can be judged through MHD computation by comparing the dynamo term to the other terms in Ohm's law. In Figure 5 we plot the radial dependence of the parallel component of each term in Ohm's law. A spatially constant mean axial electric field is imposed in the calculation, similar to the toroidal electric field which is induced in experiments. The parallel component of the applied axial electric field decreases with radius as a result of the strong radial variation of the direction of the magnetic field. It is seen (Figure 5a) that there is a nonzero current at the radius where the electric field is zero, and across the entire plasma the current cannot be attributed to the electric field. The edge requires an extra current drive mechanism, whereas the center requires spontaneous current generation counter to the electric-field-driven current. The computed $<\delta v \times \delta B>$ dynamo, shown in Figure 5b, provides the additional current drive, and balances Ohm's law. The local dynamo effect is large, driving all the current at the radius where the parallel electric field vanishes. However, from a more global view, the dynamo effect in the RFP with a conducting boundary merely

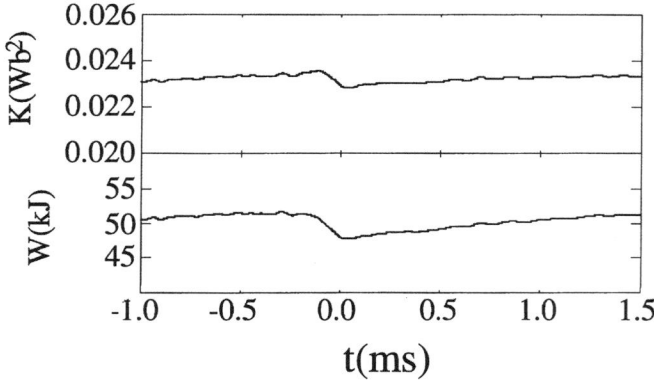

Figure 4. Magnetic helicity and magnetic energy vs time, throughout a magnetic relaxation event, computed from equilibrium modeling of an MST plasma.

redistributes the plasma current from small radius to large radius. With a resistive boundary the dynamo drives net current in the plasma.

The MHD fluctuations do indeed conserve magnetic helicity. The dissipation of magnetic helicity is given by $dK/dt = -2\int \mathbf{E}\cdot\mathbf{B}\,dV$. The contribution from the mean electric field, using Ohm's law, is

$$\frac{dK}{dt} = -2\int \eta \langle\mathbf{j}\rangle\cdot\mathbf{B}\,dV - 2\int \langle\delta\mathbf{v}\times\delta\mathbf{B}\rangle\cdot\mathbf{B}\,dV \quad (3)$$

The first term on the RHS represents resistive dissipation, while the second term is the magnetic helicity change arising from fluctuations. It is observed from Figure 5b that the second term is approximately zero. Fluctuations dissipate helicity at small radii, but generate helicity at large radii. An alternative view is that fluctuations transport helicity from the center to the edge. Similarly, the dynamo effect can be viewed either as local current generation, or a redistribution of current from the center to the outer region of the plasma.

4. EXPERIMENTAL MEASUREMENTS OF THE MHD DYNAMO

A clear test of the MHD dynamo theory is to measure each term in Ohm's law. Efforts to devise methods to measure the dynamo term, $\langle\delta\mathbf{v}\times\delta\mathbf{B}\rangle$, have been underway for several years. The task is aided by the cyclic nature of the dynamo, which yields an identifiable time variation to the quantities to be measured. It is notable that plasma diagnostics have not yet advanced to the point that the fluctuating flow velocity of the plasma can be measured directly. However, the MHD theory has been examined experimentally by two techniques. First, the velocity fluctuation specifically driven by the fluctuating E x B drift (from the fluctuating electric field which accompanies the tearing fluctuations) has been measured in the edge of the plasma in the MST experiment [Ji et al, 1994]. This component of the velocity $\delta\mathbf{v} = \delta\mathbf{E}\times\langle\mathbf{B}\rangle/\langle\mathbf{B}\rangle^2$, is not necessarily the total plasma velocity; however, it is the dominant contribution to the velocity as calculated by the pressureless, MHD computations described above. The electric field was measured in the plasma edge by Langmuir probes, and the magnetic field was measured with magnetic field pickup loops. The mean electric field and the mean current density were also measured, so that each of the three terms in Ohm's law were estimated from experimental data.

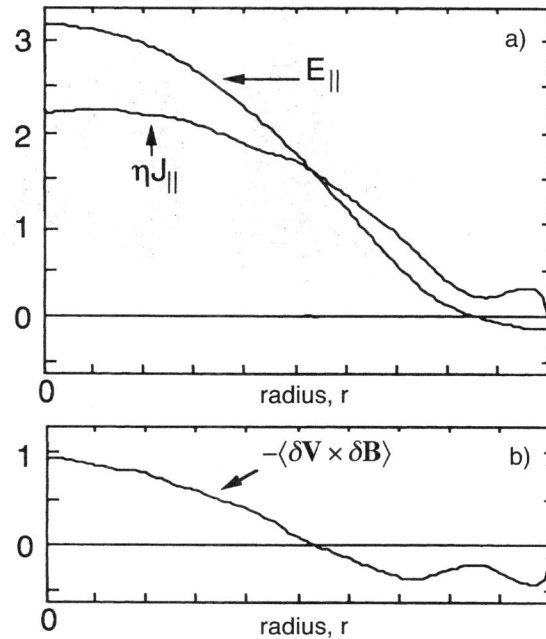

Figure 5. Radial profile of each term in Ohm's law, equation (1), showing (a) the mean electric field and current density terms and (b) the fluctuation-induced dynamo term. Each term is evaluated from nonlinear MHD computation.

The result is that the Ohm's law is satisfied, both during the discrete dynamo event, and between events.

The second technique to infer the MHD dynamo term is spectroscopic detection of the Doppler emission from impurity ions which are embedded within the hydrogen plasma [D. J. Den Hartog et al, 1998, J.T. Chapman et al, 1997]. For example, carbon ions originating from the vacuum vessel wall penetrate the plasma and function as tracer elements. The velocity of the carbon ions is determined from the Doppler shift of the emission. In this way, the fluctuating flow velocity was measured to high time resolution within the hot core of the plasma. As seen in Figure 6 the dynamo effect surges during a discrete dynamo event. The effect is quite large, rising to about 10 Volts/meter. This measurement yields the full dynamo only if the impurity ion velocity is equal to the majority ion velocity. This is a plausible assumption for two reasons: (1) we expect that the dominant flow arises from the E x B drift, which is identical for all charged particles, and (2) the collisional equilibration time between the impurity and majority ions is relatively short. Nonetheless, techniques are being developed to detect the majority ion flow velocity, as well as to achieve greater spatial resolution than the present chord-averaging spectroscopy. The other terms in Ohm's law have not yet been measured in the plasma core.

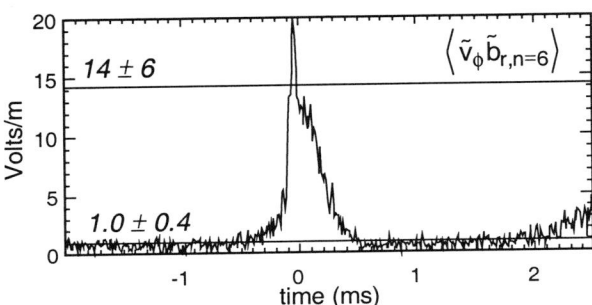

Figure 6. One component of the MHD dynamo $\langle\delta v_\phi \delta B_r\rangle$, contributed by the toroidal mode number n = 6, vs time throughout a magnetic relaxation event. Velocity fluctuations were measured by Doppler spectroscopy in the MST core, and magnetic fluctuations with an edge coil array.

5. THE BACKREACTION

Early dynamo theories were kinematic - the magnetic field was calculated for an assumed velocity field. More recently, the nonlinear "backreaction" of the growing magnetic field on the velocity has received attention [*Cattaneo and Vainshtein*, 1991; *Kulsrud and Anderson*, 1992; *Gruzinov and Diamond*, 1994]. Some results suggest that the backreaction will strongly quench dynamo growth of the magnetic field. In the RFP, the dynamo is visible as a large, saturated, steady-state effect. Clearly, the experimental plasmas incorporate the backreaction, as does the nonlinear MHD computation which predicts a strong dynamo. Thus, the question arises as to whether the observed dynamo is consistent with calculations of the backreaction.

Although the assumptions of most backreaction theories (such as isotropic turbulence) do not apply to the RFP, a sense of the resolution of the apparent discrepancy can be obtained from those calculations. From the parallel component of Ohm's law, $\mathbf{E} + \mathbf{v} \times \mathbf{B} = \eta\mathbf{j}$, the alpha effect can be expressed as

$$\alpha = \frac{-\eta\langle\delta\mathbf{j}\cdot\delta\mathbf{B}\rangle + \langle\delta\mathbf{E}\cdot\delta\mathbf{B}\rangle}{\langle B\rangle^2} \quad (4)$$

where $\alpha\langle B\rangle^2 = \langle\delta\mathbf{v}\times\delta\mathbf{B}\rangle_\parallel$. From the induction equation for the evolution of magnetic field, alpha may also be written as (*Pouquet et al.*, 1976)

$$\alpha = \alpha_0 - \frac{\tau}{3\rho}\langle\delta\mathbf{j}\cdot\delta\mathbf{B}\rangle \quad (5)$$

where $\alpha_0 = \frac{\tau}{3\rho}\langle\delta\mathbf{v}\cdot\delta\mathbf{w}\rangle$ is the alpha term in the absence of the backreaction, $\delta\mathbf{w}$ is the vorticity, and τ is the correlation time of the turbulence (assumed to to be the same for the velocity and magnetic field fluctuations). The second term represents the effect of the Lorentz force on the velocity field. Combining the above two equations yields [*Bhattacharjee and Yuan*, 1995]

$$\alpha = \frac{\alpha_0 + \frac{\tau}{3\rho\eta}\langle\delta\mathbf{E}\cdot\delta\mathbf{B}\rangle}{1 + \frac{\tau}{3\rho\eta}\langle B\rangle^2} \quad (6)$$

We see that there are two terms which arise from the backreaction and can cause α to deviate from α_0. The term in the denominator is positive definite, and therefore is always in the direction to reduce the dynamo. The term in the numerator, which depends on electric field fluctuation, can be either positive or negative. It may be an alpha-enhancing backreaction. In the limit of high electrical resistivity, or weak electric field, we can ignore the backreaction term in the numerator, to yield [*Gruzinov and Diamond*, 1994]

$$\alpha = \frac{\alpha_0}{1 + \frac{\tau}{3\rho\eta}\langle B\rangle^2} \quad (7)$$

illustrating the quenching effect. However, in the limit of low resistivity we find

$$\alpha = \frac{\langle\delta\mathbf{E}\cdot\delta\mathbf{B}\rangle}{\langle B\rangle^2} \quad (8)$$

In this limit, the second term in equation (2) dominates the resistive term. The strength of the dynamo requires solution for the fluctuations. For the RFP processes described in the previous section, this term is dominant and provides the dynamo effect.

6. DYNAMO EFFECTS BEYOND THE STANDARD MODEL

The standard MHD model described above is compelling theoretically and enjoys some measure of experimental support. However, there are dynamo effects beyond that model, some of which also have some experimental support, and which may also prove to be important. In this section we discuss a dynamo effect generated by electron plasma pressure (the "diamagnetic dynamo"), a dynamo effect generated by electron streaming along chaotic magnetic field lines (the "kinetic dynamo"), and a two-fluid version of helicity-conserving magnetic relaxation.

The Diamagnetic Dynamo: Langmuir probe measurements of the dynamo were also conducted in the edge of an RFP in Japan (TPE1RM-20), with a surprising result [*Ji et al*, 1995]. The MHD dynamo generated by the fluctuating E x B drift was measured to be small. However, a pressure-driven dynamo term in Ohm's law was measured to account roughly for the internal current drive. This can be understood by including the electron pressure term in the generalized, parallel mean-field Ohm's law,

$$\langle E \rangle_\parallel - \eta \langle j \rangle_\parallel = -\langle \delta E \cdot \delta B \rangle - \frac{\langle \nabla(\delta p) \cdot \delta B \rangle}{ne} \quad (9)$$

The first term on the right hand side is the standard dynamo effect obtained from the $\langle \delta v \times \delta B \rangle$ effect including only the contribution to δv from the $\delta E \times B$ drift. The second term represents the contribution to the $\langle \delta v \times \delta B \rangle$ term from the fluctuating electron diamagnetic drift which arises from a fluctuating electron pressure. If the pressure fluctuation correlates with the magnetic field fluctuation, a "diamagnetic" dynamo effect results. The edge conditions in TPE-1RM20 are somewhat different than in MST (for example, TPE-1RM20 is more collisional), but the cause for the different results is not yet known.

The Kinetic Dynamo: A second dynamo effect outside the standard MHD model is the adjustment of the current density profile by the transport of electron parallel momentum (or current) in the radial direction by magnetic fluctuations. This transport process is known as the "kinetic dynamo" [*Jacobson and Moses*, 1984]. It is observed in many RFP experiments that fast electrons exist at the plasma edge with parallel energies characteristic of the central temperature. This implies that the electrons may originate from the plasma center and are rapidly transported to the plasma edge. However, definitive measurement of the kinetic dynamo requires measurement of new fluctuating quantities associated with the electrons. In addition, the inclusion of the self-consistency constraint of Ampere's law may inhibit the kinetic dynamo mechanism [*Terry and Diamond*, 1990].

Although the theoretical basis for the MHD dynamo is compelling, and some experimental tests are supportive, it still remains to be determined which of the several possible dynamo mechanisms dominate the variety of plasma conditions found in the various RFP experiments. We conjecture that, although the relative influence of various dynamo mechanisms may be situation dependent, the mechanisms will always sum to yield the natural current density profile of the partially relaxed Taylor-like state. The MHD velocity and magnetic field fluctuations can adjust, in the presence of other dynamo contributors, to maintain the relaxed profile. This conjecture is consistent with the similarity of j/B profiles observed in a wide variety of RFP experiments, and with the dynamic changes of the fluctuations to plasma conditions described in Section 4.

Two Fluid Relaxation: Whereas the Taylor conjecture, and MHD theory, describe the plasma as a single fluid, experimental measurements can distinguish the behavior of the electrons and the ions. Hence, a full understanding of relaxation requires a two-fluid treatment. The Taylor conjecture predicts that j/B is a spatial constant, where j is the total electrical current composed of electron current and ion current. Recently, an analogous approach has been applied to each individual species [*Avinash and Taylor*, 1991; *Steinhauer and Ishida*, 1997; *Hegna*, 1998]. Some calculations suggest that each species will *individually* approach a Taylor-like state in which both j_e/B and j_i/B are spatial constants, where j_e and j_i are the electron and ion current density (or momentum density), respectively. This result follows from various arguments. A generalization to the Taylor argument follows by defining a generalized species-dependent helicity which includes both magnetic and flow effects. It also follows from two-fluid linear theory of tearing fluctuations. In MST it has indeed been observed that the ion flow velocity changes suddenly during a discrete dynamo event [*Den Hartog et al.*, 1998]; modeling is underway to determine whether the sudden changes are consistent with the two-fluid relaxation model.

7. ACTIVE CONTROL OF RELAXATION

In the plasma relaxation described above, the helicity-conserving fluctuations reduce the gradient in the normalized current density, j/B. The gradient in j/B drives the fluctuations, which then generate local current to bring the plasma closer to a stable, Taylor-like state. An interesting test of the model is to employ this understanding to reduce the plasma fluctuations. For example, if the gradient in the current density gradient was reduced by some external means, we would expect the fluctuations to also diminish. To test this idea, a poloidal electric field was induced in the outer region of the MST plasma (by varying the toroidal magnetic flux in the plasma), in a direction to reduce the current density gradient [*Sarff, Lanier, Prager, and Stoneking,* 1997].

It was indeed observed that the fluctuations in the magnetic field were reduced, by a factor of two from $|\delta B|/<B> \approx 1.3\%$ to 0.7%. The externally driven current produces a plasma closer to the Taylor state, thereby diminishing the need for fluctuation-induced relaxation. A rather remarkable consequence of the reduced fluctuations is an increase in the electron temperature by about 50% (from 400 eV to 600 eV) and a decrease in the energy flux in the radial direction by a factor of five. The decrease in energy loss was an expected consequence of the experiment. The magnetic fluctuations which enable the dynamo also cause energy transport. The magnetic fluctuations are of an amplitude and wave number spectrum to cause the magnetic field lines to follow a chaotic trajectory. In the absence of fluctuations the field lines are confined on approximate circles. The fluctuations introduce radial wander of the field lines. The particles travel along the field lines, and thereby experience an increased transport in the radial direction. Hence, a reduction of fluctuations diminishes energy loss from the plasma, leading to an elevation of the temperature.

8. CONCLUSIONS

The RFP laboratory plasma displays relaxation of the magnetic field, dynamo generation of plasma current, and magnetic fluctuations which underlie these effects. Nonlinear MHD theory and computation provides a self-contained description, including the mutual interaction between plasma fluctuations and mean fields. Magnetic and velocity fluctuations, driven by gradients in the mean current density, produce a dynamo effect which redistributes the plasma current and magnetic field, driving the plasma to a state which approximates the Taylor state. The plasma does not develop a perfect Taylor state, presumably because the fluctuations are insufficient to establish the requisite current density profile in the outer region of the plasma.

In experiment, relaxation also drives the plasma partially towards a Taylor state, and there is experimental evidence that the MHD alpha-effect dynamo is active. In addition, the fluctuation-induced relaxation is able to be controlled, in part, by the experimental alteration of the mean fields, providing additional consistency with the nonlinear MHD description.

The development of the MHD-based understanding has led to additional intriguing physics questions. Equations beyond MHD suggest a variety of dynamo and relaxation mechanisms outside the standard model. For example, a pressure-driven dynamo effect has been observed under certain plasma conditions, and kinetic and Hall term dynamos have been explored. An understanding of the conditions under which various dynamo effects dominate is not in hand. The mean field behavior of the RFP suggest that perhaps the MHD dynamo will adjust to whatever other dynamo effects are active, so that the mean fields resemble a Taylor state. This conjecture requires both experimental and theoretical test.

The experiments to control or reduce relaxation and dynamo will permit us to examine the relationship between fluctuations and relaxation, as each is controllably varied. Moreover, it raises the question of whether fluctuations, dynamo, magnetic chaos, and energy transport can be suppressed to nearly zero. The control of the mean current profile, and thereby of the fluctuations and relaxation, has so far been relatively coarse. Finer techniques under development are required to address these questions.

Acknowledgments The author would like to acknowledge useful discussions with many members of the MST research group. In particular, the synopsis presented in this paper reflects many contributions from D. Den Hartog, C. Hegna, H. Ji, J. Sarff, and P. Terry.

REFERENCES

Alper B et al., in *Plasma Physics and Controlled Nuclear Fusion Research* (Proc. 12th In. Conf.) Vol. 2, International Atomic Energy Agency, p. 431, 1989.

Avinash K., and Taylor, J.B., *Comments Plasma Physics 14*, 1267, 1991.

Bhattacharjee A., and Yuan, Y, *ApJ 449*, 739 1995.

Bodin H.A.B., and Newton A.A., *Nucl. Fusion 20*, 1255, 1980.

Cattaneo, F., and Vainshtain, S.I., *ApJ 376*, L21 1991.

Chapman J.T. et al., *Bull. Am. Phys. Soc. 42*, 2046, 1997.

Den Hartog D.J. et al., *Plasma Physics Reports 24*, 2169, 1998.

Gruzinov A.V., and Diamond P.H., *Phys. Rev. Lett. 72*, 1651 (1994).

Hegna C.C., *Phys. Plasmas 5*, 2257, 1998.

Hokin S.A. et al., *Phys. Fluids B3*, 2241, 1991.

Jacobson, A.R., and Moses, R.W., *Phys. Rev. A 29*, 3335, 1984.

Ji H. Almagri A., Prager S.C., and Sarff, J.S., *Phys. Rev. Lett. 73*, 668, 1994.

Ji H., Prager S.C., and Sarff, J.S., *Phys. Rev. Lett. 74*, 2945, 1995.

Ji H. et al., *Phys. Rev. Lett. 75*, 1086 (1995).

Kulsrud R.M., and Anderson S.W., *ApJ 396*, 606 (1992).

Ortolani, S. and Schnack, D.D., *Magnetohydrodynamics of Plasma Relaxation*, World Scientific Publishing Co. Singapore, 1993.

Pouquet A, Frisch U., Leorat J., *J. Fluid Mechanics* 77, 321 1976.

Sarff J.S., Lanier N.E., Prager S.C., and Stoneking, M.R., *Phys. Rev. Lett.* **62**, 1997.

Steinhauer L.C., and Ishida A., *Phys. Rev. Lett.* 79, 3423, 1997.

Taylor J.B., *Rev. Mod. Phys.*, 58, 741, 1986.

Terry P.W., and Diamond, P.H., *Phys Fluids B 2*, 1128 (1990).

Watt R., and Nebel. R., *Phys. Fluids 26*, 1168, 1983..

Helicity in Large-Scale Dynamo Simulations

Axel Brandenburg

Department of Mathematics, University of Newcastle upon Tyne, NE1 7RU, UK

Various hydromagnetic turbulence simulations exhibiting large scale dynamo action are analysed: rotating convection with shear, rotating shear flow, and isotropically forced helical turbulence. The signs and magnitudes of the various helicities are compared and related to the effective dynamo alpha parameter. In isotropically forced helical flows the alpha parameter is found to be a negative multiple of the residual helicity, which is the difference between kinetic and current helicity. The convection simulations are consistent with this, but the rotating shear flow simulations are not. In the latter case shear is responsible for reversing the sign of the stress, and it is the sign of the magnetic stress that determines the sign of the magnetically driven dynamo alpha. Finally, the inverse magnetic cascade is related to the alpha effect and attempts are shown to evaluate the magnitudes of alpha and turbulent diffusivity in a simulation exhibiting an inverse cascade.

1. INTRODUCTION

The concept of helicity is central to all theories of large scale dynamos. In the early work by Parker (1955) the concept of cyclonic convection was introduced as a means of producing poloidal magnetic field from a toroidal field by twisting rising flux tubes via the Coriolis force. This was later quantified with the development of the α-effect (Steenbeck, Krause & Rädler 1966), which measures the magnitude of the mean electromotive force in the direction of the mean magnetic field. The books by Moffatt (1978) and Krause & Rädler (1980) give a comprehensive account of the kinematic mean-field dynamo theory. The main result is that when α is large enough a dynamo instability sets in and a large scale magnetic field is generated. For isotropic turbulence the α parameter is a negative multiple of the kinetic helicity.

An important discovery was made by Frisch et al. (1975) and Pouquet et al. (1976), who found that the presence of magnetic helicity can give rise to an inverse cascade, by which magnetic helicity and energy are being transferred from small to large scales. The growth of the large scale field depends here on the *residual* helicity, i.e. the difference between kinetic and (electric) current helicities. The involvement of magnetic fields, or rather magnetically driven fluid motions, could be crucial, especially in those circumstances where the magnetic field is strong. This is the case in practically *all* astrophysically interesting applications (stars, accretion discs, and galaxies).

There are now many different simulations displaying large scale dynamo action in astrophysically relevant systems. The purpose of this paper is to compare the helicities in some of those simulations. We begin with recent simulations of overshooting convection with imposed shear. We then discuss simulations without convection and just shear, relevant to accretion discs, and finally compare with simulations of isotropically forced flows.

2. DYNAMOS FROM OVERSHOOTING CONVECTION WITH SHEAR

The overshoot layer beneath the solar convection zone proper is often thought to be the place where the dynamo operates. This is the main reason why it is useful to include overshoot in convective dynamo simulations. Results of Nordlund et al. (1992) and Brandenburg et al. (1996) suggest that dynamo action occurs actually throughout the entire convection zone proper, but that the field is then transported downwards into the overshoot layer by turbulent pumping of magnetic fields via rapidly spinning downdrafts. Recently, those simulations have been extended to include the effects of shear (Brandenburg et al. 1999). Shear takes the role of the omega-effect, although here the concept of alpha-omega dynamos is not explicitly invoked. The main result is the generation of large scale fields on the scale of the box. Those fields are of significant strength and can exceed the equipartition field strength by an order of magnitude.

For orientation we give the basic parameters of the simulation. The simulation is carried out at 30° northern latitude and the resulting inverse Rossby number, $2\Omega L/u_{\rm rms}$, is around 5. Here, $u_{\rm rms}$ is the turbulent rms velocity, L is the depth of the unstable layer, and Ω is the angular velocity. Uniform latitudinal shear is imposed by a body force throughout the convection zone proper, but it vanishes towards the radiative interior, resulting in vertical shear around the lower overshoot layer. 'Sliding-periodic' boundary conditions (Hawley et al. 1995) are used in the cross-stream direction and ordinary periodic boundary conditions in the streamwise direction. The ratio between shear gradient and angular velocity is 0.5 and the velocity difference across the box is $\Delta U \approx \pm 0.4 u_{\rm rms}$. The resolution is $63 \times 63 \times 64$ meshpoints, the ordinary and magnetic Prandtl numbers are $\text{Pr} = \nu/\chi = 0.2$ and $\text{Pr}_M = \nu/\eta = 0.5$, i.e. the kinematic viscosity ν is smaller than the magnetic and thermal diffusivities (η and χ). In the sun the two Prandtl numbers are much smaller than unity, but this is impossible to simulate in a simulation of only modest resolution. The Reynolds number is $\text{Re} = u_{\rm rms} L/\nu = 240$, the Rayleigh and Taylor numbers are $\text{Ra} = gL^4 s'_0/(c_p \chi \nu) = 5 \times 10^5$ and $\text{Ta} = (2\Omega L^2/\nu)^2 = 10^6$. Here, g is gravity and s'_0 is the entropy gradient of the associated unstable hydrostatic solution.

The orientation of the cartesian box is as follows: x points north, y points east in the toroidal direction, and z points downwards. The top and bottom boundaries are stress free and the horizontal field vanishes, so there

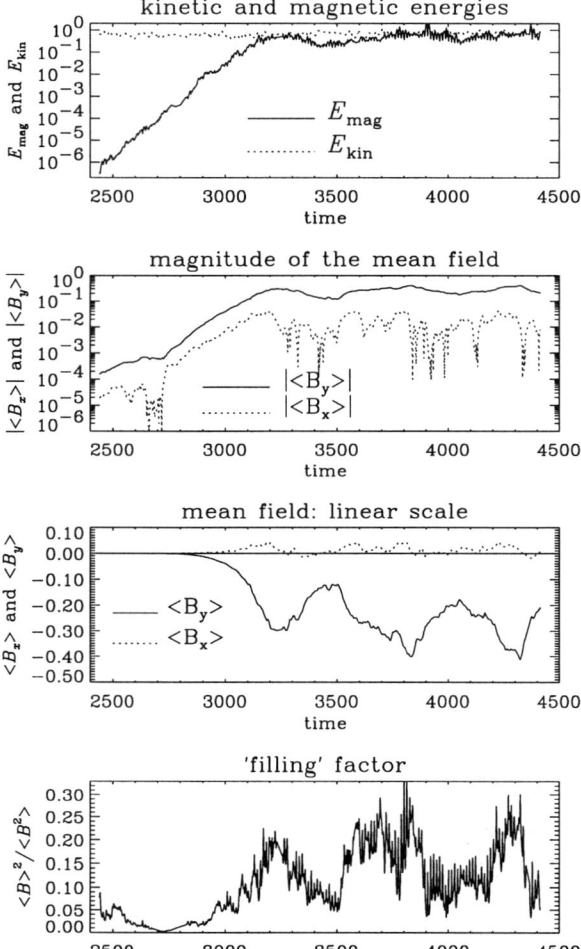

Figure 1. Evolution of magnetic and kinetic energies, mean magnetic field, and $\langle B \rangle^2 / \langle B^2 \rangle$ (which may be interpreted as a 'filling' factor) in a convection simulation with imposed shear.

is no vertical Poynting flux through the boundaries. Initially there is no net flux through the box.

In figure 1 we show the evolution of the total magnetic energy and the mean magnetic field in such a simulation. The magnetic energy increases by 6 orders of magnitude and then saturates. There is also an exponential growth of the *mean* field (averaged over the entire box), which increases by 3 orders of magnitude until saturation is reached. (This is at around $t = 3200$, approximately the same time when the magnetic energy saturates; the time unit is $\sqrt{L/g}$.) Note that the energy in the mean magnetic field can be as large as 20% of the total magnetic energy.

The main effect of the shear is the generation of strong ordered toroidal fields, $\langle B_y \rangle$. There is also a much weaker poloidal field component. The component in the latitudinal direction, $\langle B_x \rangle$, is about 10 times weaker and oriented mostly in the opposite direction, i.e. $\langle B_x \rangle \langle B_y \rangle < 0$ for most of the time. This is simply a consequence of the shear, $\partial U_y/\partial x < 0$, which turns a positive $\langle B_x \rangle$ into a negative $\langle B_y \rangle$.

In figure 2 we show the evolution of the various helicities for this run: kinetic helicity $\langle \boldsymbol{\omega} \cdot \boldsymbol{u} \rangle$, current helicity $\langle \boldsymbol{J} \cdot \boldsymbol{B} \rangle$, cross helicity $\langle \boldsymbol{u} \cdot \boldsymbol{B} \rangle$, and the magnetic helicity, $\langle \boldsymbol{A} \cdot \boldsymbol{B} \rangle$. Here, $\boldsymbol{\omega} = \operatorname{curl} \boldsymbol{u}$ is the vorticity, $\boldsymbol{u}$ the velocity, $\boldsymbol{J} = \operatorname{curl} \boldsymbol{B}/\mu_0$ the current density, $\boldsymbol{B} = \operatorname{curl} \boldsymbol{A}$ the magnetic field, $\boldsymbol{A}$ the magnetic vector potential, and μ_0 the vacuum permeability.

We find that the kinetic helicity is negative, and that its magnitude *increases* as the dynamo becomes saturated, i.e. when the magnetic energy levels off. This suggests that not only the current helicity, but also the kinetic helicity is driven (at least partly) by the magnetic field. In other words, the part of the velocity that contributes mostly to the helicity integral is caused mainly by the Lorentz force. Note also that current and kinetic helicities have the *same* sign. This is in contrast to some simulations of magnetoconvection with imposed magnetic field and at smaller magnetic Reynolds number (Brandenburg et al. 1990), where the two helicities have opposite sign. This may hint at an important difference between more-or-less passive magnetic field evolution on the one hand and dynamo-generated magnetic fields on the other.

Looking at the third panel of figure 2 we note that there is also some cross helicity being generated. The cross helicity, which is perhaps more sensibly written as $\langle \boldsymbol{B} \cdot \operatorname{curl}^{-1} \boldsymbol{\omega} \rangle$, measures the linkage between $\boldsymbol{B}$-tubes and $\boldsymbol{\omega}$-tubes. Here, $\operatorname{curl}^{-1} \boldsymbol{\omega} = \boldsymbol{u}$. Significant magnetic helicity, $\langle \boldsymbol{B} \cdot \operatorname{curl}^{-1} \boldsymbol{B} \rangle$, which measures the linkage of $\boldsymbol{B}$-tubes with themselves, is also being generated at the time when the large scale field reaches saturation. That too is negative, so all three fields, $\boldsymbol{\omega}$, $\boldsymbol{J}$ and $\boldsymbol{B}$, have the same sign of the linkage number after the time the large scale field saturates. Only the linkage between $\boldsymbol{\omega}$ and $\boldsymbol{B}$ tubes has the opposite sign.

The helicities given in figure 2 are all calculated using the full velocity and magnetic fields. In mean-field dynamo theory one often needs the helicity calculated with respect to the fluctuations about the mean value, i.e. $\langle \boldsymbol{\omega}' \cdot \boldsymbol{u}' \rangle = \langle \boldsymbol{\omega} \cdot \boldsymbol{u} \rangle - \langle \boldsymbol{\omega} \rangle \cdot \langle \boldsymbol{u} \rangle$ and $\langle \boldsymbol{J}' \cdot \boldsymbol{B}' \rangle = \langle \boldsymbol{J} \cdot \boldsymbol{B} \rangle - \langle \boldsymbol{J} \rangle \cdot \langle \boldsymbol{B} \rangle$. The difference is negligible, however, because the large scale kinetic and current helicities are small; see the dotted lines in figure 2a and

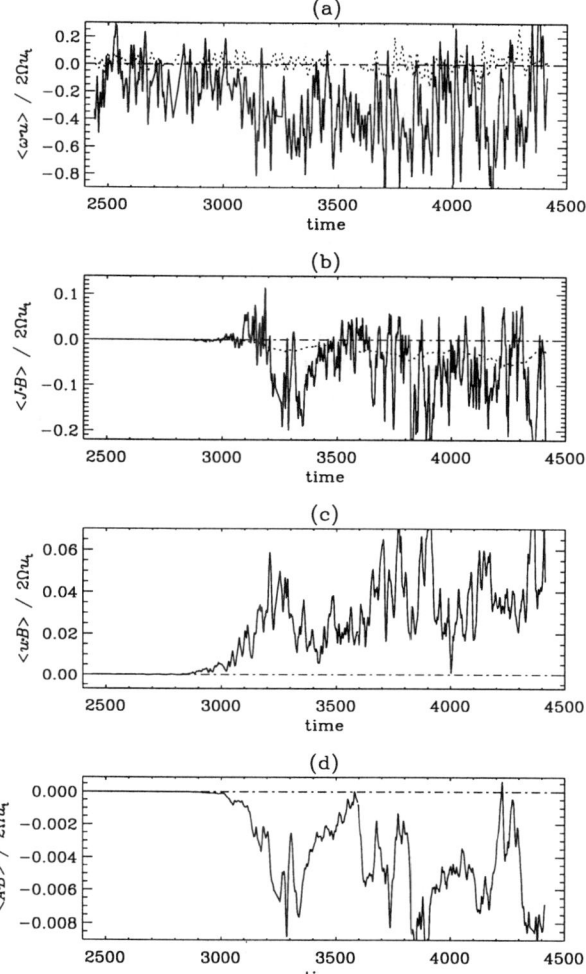

Figure 2. Evolution of kinetic helicity, current helicity, cross helicity, and magnetic helicity, in a convection simulation with imposed shear. The dotted lines in (a) and (b) give the large scale kinetic and current helicities, multiplied by a factor of 10.

b. The fluctuations in the toroidal component of the electromotive force, $\langle \boldsymbol{u}' \times \boldsymbol{B}' \rangle_y$, are large, and there is only a very weak positive correlation with the mean toroidal magnetic field, $\langle B_y \rangle$. Such a correlation would be suggestive of a positive (but very noisy) alpha-effect, if $\langle \boldsymbol{u}' \times \boldsymbol{B}' \rangle_y = \alpha_{yy} \langle B_y \rangle +$ other terms. It may be surprising or even implausible that a very noisy effect could explain a strong and well-defined mean magnetic field as seen in figure 1. The reason is perhaps that even a very noisy or an incoherent alpha-effect (Vishniac & Brandenburg 1997; see also Vishniac's chapter) could give rise to a large scale magnetic field.

3. HELICITY IN SHEAR-DRIVEN ACCRETION DISC DYNAMOS

A somewhat different situation is encountered in accretion discs, where there is no direct source of turbulence, because discs are hydrodynamically stable. Only in the presence of a magnetic field there is a linear instability (Velikhov 1959, Chandrasekhar 1960, 1961). This instability is now often called the magnetorotational or Balbus-Hawley (1991) instability. However, the flows generated by this instability would tend to destroy the magnetic field via turbulent diffusion. Nevertheless, at the same time the turbulence can also amplify the magnetic field via dynamo action. Simulations unanimously point towards the possibility of a cycle where the field generates turbulence and the turbulence generates more magnetic fields (Brandenburg et al. 1995, Hawley et al. 1996, Stone et al. 1996). In particular, in simulations of Brandenburg et al. (1995) there is a large scale magnetic field, which is oscillatory and varies on a time scale of about 30 orbits, $T_{\rm rot} = 2\pi/\Omega_0$, where Ω_0 is the angular velocity. Many quantities vary cyclically with the mean field, of which the toroidal component $\langle B_y \rangle$ is the strongest.

In figure 3 we plot the kinetic and current helicities in the upper disc plane as functions of the mean toroidal field, $\langle B_y \rangle$. Note that $\langle \boldsymbol{J} \cdot \boldsymbol{B} \rangle$ is approximately proportional to $\langle B_y \rangle^2$, as indicated by the solid line. On the other hand, $\langle \boldsymbol{\omega} \cdot \boldsymbol{u} \rangle$ shows strong scatter and is independent of $\langle B_y \rangle$. In contrast to the case of convection with shear the kinetic and current helicities have now opposite signs. The negative sign of $\langle \boldsymbol{\omega} \cdot \boldsymbol{u} \rangle$ is in agreement with the interpretation in terms of cyclonic motions, but the sign of $\langle \boldsymbol{J} \cdot \boldsymbol{B} \rangle$ is not. At large scales the signs of the two helicities are actually different (see the lower two panels of figure 3), but their magnitudes are small, so this does not explain the result. So, the origin of the sign of $\langle \boldsymbol{J} \cdot \boldsymbol{B} \rangle$ remains unclear. It is probably connected with the strong effects of shear, which can give rise to unusual signs of the α-effect. This will be discussed next. The connection between α and $\langle \boldsymbol{J} \cdot \boldsymbol{B} \rangle$ will be discussed in the following section.

In the case of the accretion disc simulations it is possible to estimate the magnitude and sign of the effective dynamo α parameter by correlating at different time steps the mean electromotive force with the resulting mean magnetic field and to establish a fit of the form $\langle \boldsymbol{u}' \times \boldsymbol{B}' \rangle_y = \alpha \langle B_y \rangle$ (Brandenburg et al. 1995, Brandenburg & Donner 1997). Here primes denote fluctuations. The α measured in that way is found to be *negative* in the upper disc plane. Therefore, the sign of α is in

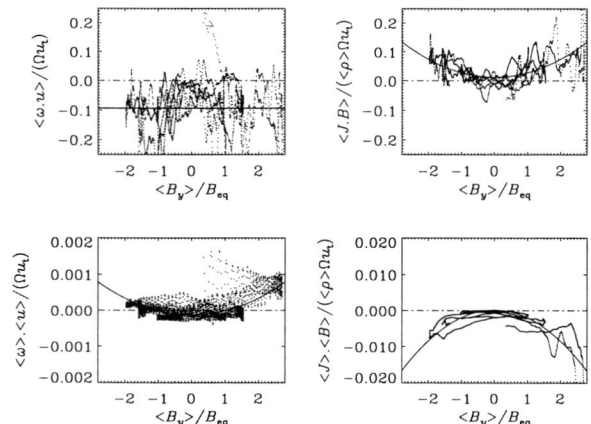

Figure 3. Kinetic helicity and current helicity in the upper disc plane of in an accretion disc dynamo simulation. The lines give a fit through the data. The last two panels show that the two helicities of the mean field are small and of opposite sign.

disagreement with that expected form kinetic and current helicities (see the next section). The perhaps most convincing explanation for this negative sign is that the shear twists buoyant magnetic structures in the opposite sense as the Coriolis force (Brandenburg & Campbell 1997, Brandenburg 1997, 1998, 1999). The mean toroidal electromagnetic force, $\langle \boldsymbol{u}' \times \boldsymbol{B}' \rangle_y$, is then governed by the vertical velocity fluctuations u'_z, and radial magnetic field fluctuations, B'_x, so $\langle \boldsymbol{u}' \times \boldsymbol{B}' \rangle_y = \langle u'_z B'_x \rangle$. If u'_z originates mainly from magnetic buoyancy then $u'_z \sim -(\rho'/\langle\rho\rangle)g\tau$, where τ is some relevant timescale and $\rho'/\langle\rho\rangle - \approx \langle B_y \rangle B'_y/\langle \mu_0 \rho c_s^2 \rangle$, so $\langle \boldsymbol{u}' \times \boldsymbol{B}' \rangle_y = \alpha \langle B_y \rangle$ with $\alpha \sim \langle B'_x B'_y \rangle g\tau/\langle \mu_0 \rho c_s^2 \rangle$. This would explain the negative sign of α, because $\langle B_x B_y \rangle < 0$ (in agreement with the sign of the mean shear, $\partial U_y/\partial x = -\frac{3}{2}\Omega_0$. This is an example where a negative α results from a flow that is driven exclusively by the Lorentz force and not, like in the case of convection, by thermal buoyancy of other nonmagnetic forces.

4. THE RELATION BETWEEN ALPHA-EFFECT AND HELICITY

There have been attempts to estimate α from forced MHD turbulence. Simulations of Tao et al. (1993), for example, verify that α is a negative multiple of the kinetic helicity. However, there has so far been no verification that α is related to the *residual* helicity (Pouquet et al. 1976),

$$H_{\rm res} = \langle \boldsymbol{\omega} \cdot \boldsymbol{u} \rangle - \langle \boldsymbol{J} \cdot \boldsymbol{B} \rangle/\rho. \tag{1}$$

The $\langle \boldsymbol{J} \cdot \boldsymbol{B} \rangle$ term in this expression can lead to significant modifications once the magnetic energy is strong compared with the kinetic energy. This is likely to be case in accretion discs. In order to access this parameter regime we now discuss another model (Brandenburg & Bigazzi 1999) where the flow is magnetically driven. In that case we adopt some random forcing, $\boldsymbol{E}_f$, directly in the induction equation, which then takes the form

$$\frac{\partial \boldsymbol{A}}{\partial t} = \boldsymbol{u} \times \boldsymbol{B} + \eta \nabla^2 \boldsymbol{A} - \boldsymbol{E}_f, \quad (2)$$

where $\boldsymbol{E}_f$ consists of plane Beltrami waves of maximum (positive) helicity. The spatial pattern is renewed in regular time intervals, Δt_f.

An explicit forcing in the induction equation is adopted mainly for mathematical convenience, rather than physical reality. In the case of accretion discs there is actually an extra term, $\boldsymbol{E}_f = SA_y \hat{\boldsymbol{x}}$, where S is the shear parameter, but this corresponds to a multiplicative forcing, not to an additive forcing as in the present model, because $\boldsymbol{E}_f$ is proportional to the y-component of $\boldsymbol{A}$. We comment further on this forcing term in the next section, where we adopt a forcing at high wavenumbers.

In the present model a large scale vertical magnetic field is imposed, so the field in Eq. (2) consists of two parts: $\boldsymbol{B} = \hat{\boldsymbol{z}} B_0 + \boldsymbol{\nabla} \times \boldsymbol{A}$. The flow is driven exclusively by the Lorentz force in the momentum equation. The resulting current helicity is large and positive, and the kinetic helicity has now the *same* sign, i.e. $\langle \boldsymbol{\omega} \cdot \boldsymbol{u} \rangle > 0$ and $\langle \boldsymbol{J} \cdot \boldsymbol{B} \rangle > 0$, but with $|\langle \boldsymbol{\omega} \cdot \boldsymbol{u} \rangle| \ll |\langle \boldsymbol{J} \cdot \boldsymbol{B} \rangle|$.

We measure the dynamo α by dividing the z-componen of the resulting electromotive force, $\mathcal{E} = \langle \boldsymbol{u} \times \boldsymbol{B} \rangle$, by B_0. We find that to a good approximation α is a negative multiple of H_{res}; see figure 4. In the present case with magnetic forcing the residual helicity is always negative, because the current helicity dominates over the kinetic helicity and both are positive. A somewhat different situation arises when the forcing is applied in the momentum equation instead. In that case the kinetic helicity is larger than the current helicity and so the sign of the residual helicity is positive. The corresponding sign of α is then also reversed. This is how we obtained the points on the right hand side of figure 4 for positive values of H_{res}. The data points for both magnetic and hydrodynamic forcing match the linear fit equally well.

Although the approach in this section is enlightening as far as the connection between α and the various helicities is concerned, is remains unsatisfactory for a number of reasons. Firstly, for small imposed fields and sufficiently large magnetic Reynolds numbers there is a dynamo effect that causes the mean electromotive

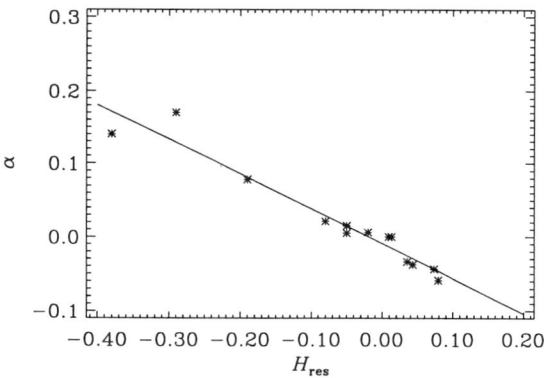

Figure 4. Alpha effect as a function of residual helicity.

force to grow to large values until saturation occurs. At the same time the mean magnetic field is conserved. Thus there can be no casual relation between the mean electromotive force and the mean magnetic field. It is therefore no longer possible to estimate α by just dividing the electromotive force by B_0. The result would have been arbitrary and therefore meaningless. This difficulty does not arise for small magnetic Reynolds numbers, although that case is of course less interesting. Secondly, when measuring α using averages defined by projections onto the $k = 0$ wavenumber the results may be spurious, because the field in the $k = 0$ wavenumber (i.e. the flux through the box) is conserved for periodic boundary conditions and therefore not affected by the dynamics. A more satisfactory approach is therefore to measure α by projecting onto the wavenumber $k = 1$ and forcing at sufficiently small scale to have some sort of scale separation. This is done in the next section. One could still use an initial field at wavenumber $k = 1$, but now this field can evolve. We find that it grows to appreciable field strengths due to dynamo action, and it is such a state that will be used for estimating α. First, however, we look at the growth of the field starting from random initial conditions.

5. THE INVERSE CASCADE EFFECT IN ISOTROPICALLY FORCED SYSTEMS

Following the early work on inverse cascades (Pouquet et al. 1976) we now adopt a *high* wavenumber forcing in the induction equation. Apart from the higher forcing wavenumber and the absence of an imposed field everything else is like in the previous section. Because the forcing is at high wavenumbers only ($k = 10$) the magnetic field evolution at the large scales ($k = 1$) is not immediately affected, except of course for the inverse

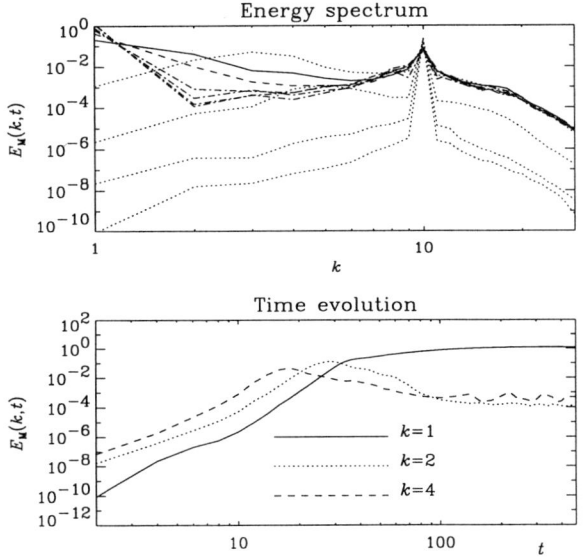

Figure 5. The inverse cascade seen in power spectra of the magnetic field taken at different times (upper panel). The four dotted curves are for $t = 2, 4, 10, 20$, the solid and dashed curves are for $t = 40$ and 60, respectively, and the dash-dotted curves are $t = 80, 100, 200$, and 400. The lower panel shows the evolution of the spectral power in the $k = 1, 2$, and 4 modes in a double-logarithmic plot.

cascade effect which governs the evolution on wavenumbers smaller than the forcing wavenumber. We also point out that the general behavior is similar, regardless of whether the forcing is applied in the induction equation or in the momentum equation.

Looking at power spectra of the magnetic field at subsequent times (figure 5) we see that the energy at the largest possible scale in the system ($k = 1$) grows until some saturation level is reached at around $t = 40$; see the lower panel of figure 5. Curiously enough, at the time when the $k = 1$ mode reaches saturation the power in the $k = 2$ mode begins to be suppressed (see the dotted line). Looking more carefully at this plot reveals that at the time when the $k = 2$ mode began to saturate (at around $t = 20 - 30$) the power in the next higher modes, $k = 3$ and $k = 4$, was suppressed. This has also been observed in similar calculations of low Reynolds number flows (Gilbert & Sulem 1990, Galanti et al. 1991, Galanti & Sulem 1991). In our case the Reynolds numbers (ordinary and magnetic), based on the box size and the rms velocity, are around 140. However, the Reynolds number based on the wavenumber 10 is only 14. In that sense our simulation too is rather diffusive.

The orientation of the magnetic field is not determined a priori and depends on chance and on initial conditions. Sometimes we found a field that varied mostly in the x-direction, while for other simulations the field varied mostly in the y or z-directions. If the mean field varies only in the x-direction, for example, then $\partial_y \langle \boldsymbol{B} \rangle = \partial_z \langle \boldsymbol{B} \rangle = 0$ and only $\partial_x \langle \boldsymbol{B} \rangle$ is nontrivial. Then, however, because

$$0 = \boldsymbol{\nabla} \cdot \langle \boldsymbol{B} \rangle = \partial_x \langle B_x \rangle, \qquad (3)$$

we have $\langle B_x \rangle = \mathrm{const} = 0$, so $\langle \boldsymbol{B} \rangle = (0, \langle B_y \rangle, \langle B_z \rangle)$. In other words, the field vector lies in a plane whose normal is parallel to the direction in which it varies, but it has no component in that direction. Once the large scale field has selected a preferred direction, it will stick to it for all times. We note, however, that we never encountered a case where the field is oblique to any of the coordinate planes. An oblique mean field would diffuse faster, because the turbulent diffusion operator, $\eta_t(k_x^2 + k_y^2) = 2\eta_t k_{\min}^2$, is always larger than just $\eta_t k_x^2 = \eta_t k_{\min}^2$. This is probably the reason why diagonal fields are not being generated.

In figure 6 we show the resulting mean magnetic field from a simulation in which the preferred direction of the mean field is the z-direction. Note that there is a $90°$ phase difference between the x and y-components of the mean magnetic field.

The approach just described allows us to study the evolution and saturation of the large scale magnetic field. An obvious question is then whether the field produced by the inverse cascade resembles qualitatively and perhaps even quantitatively the field generated by an α^2 dynamo, and if so, what are then the corresponding values of α and turbulent diffusivity, η_t.

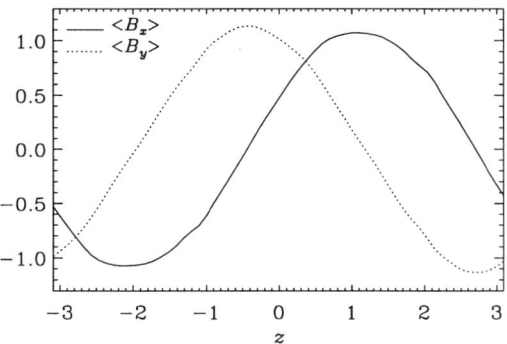

Figure 6. The mean magnetic field components, $\langle B_x \rangle$ and $\langle B_y \rangle$, as functions of z in a simulation where the mean field varies only in the z-direction.

6. CONNECTION WITH AN ALPHA-SQUARED DYNAMO

The mean magnetic field found in the previous section resembles in many ways an α^2 dynamo. In such a dynamo the large scale field is governed by the equations

$$\frac{\partial \langle B_x \rangle}{\partial t} = -\alpha \frac{\partial \langle B_y \rangle}{\partial z} + (\eta + \eta_t) \frac{\partial^2 \langle B_x \rangle}{\partial z^2}, \quad (4)$$

$$\frac{\partial \langle B_y \rangle}{\partial t} = +\alpha \frac{\partial \langle B_x \rangle}{\partial z} + (\eta + \eta_t) \frac{\partial^2 \langle B_y \rangle}{\partial z^2}, \quad (5)$$

where we have assumed that the mean field varies only in the z-direction, which is the situation in the particular solution displayed in figure 6 (section 5). The averages are taken over the x and y directions. In the saturated case the field is dominated by the smallest waver number $k = 1$; see the inset of figure 5. Therefore we now take the solution to be of the form

$$\langle B_x \rangle = \hat{B}_x(t) \sin(z - z_0), \quad \langle B_y \rangle = \hat{B}_y(t) \cos(z - z_0), \quad (6)$$

where z_0 is a constant (phase factor). With this, Eqs. (4) and (5) take the form

$$\frac{\partial \hat{B}_x}{\partial t} = \alpha \hat{B}_y - (\eta + \eta_t) \hat{B}_x, \quad (7)$$

$$\frac{\partial \hat{B}_y}{\partial t} = \alpha \hat{B}_x - (\eta + \eta_t) \hat{B}_y. \quad (8)$$

In the steady state we have $\alpha = \eta + \eta_t$. In order to estimate the value of α we modify the actual field in the simulation by setting momentarily the mean field in either the x or the y-direction to zero, i.e. we replace at some instance in time $B_x \to B_x - \langle B_x \rangle$ or $B_y \to B_y - \langle B_y \rangle$. Looking at eq. (7) we see that setting $\langle B_x \rangle = \hat{B}_x = 0$ means that immediately after this manipulation the $\hat{B}_x$ field should recover at a rate $\alpha \hat{B}_y$. This rate is approximately 0.02 (see figure 7), and since $\hat{B}_y \approx 1$ we have $\alpha \approx 0.02$. This value is already affected by the nonlinear feedback in the system (alpha-quenching, for example). Assuming that the value of α is the same before and after removing one of the two mean field components we have therefore $\eta + \eta_t \approx \alpha \approx 0.02$. Since in this simulation $\eta = 0.01$ we have $\eta_t \approx \eta$. Those values of α and η_t are rather small, suggesting again that the effective magnetic Reynolds number is small.

This method can in principle be applied to systems with different field strengths, different magnetic Reynolds numbers, and different amounts of helicity.

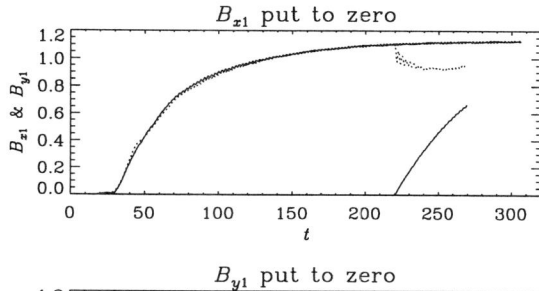

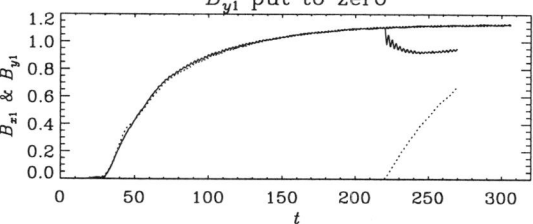

Figure 7. Response of the large scale field after removing the mean field from the B_x and B_y fields, respectively. After $t = 220$ the field component that was set to zero ($\hat{B}_x$ in the upper panel, $\hat{B}_y$ in the lower) began to grow at a rate ≈ 0.02.

7. CONCLUSIONS

Helicity is closely connected with large scale dynamos. In isotropically forced turbulence helicity leads to a growth of the large scale field in a way that is very similar to the case of α-effect mean-field dynamos. Measuring the value of α in such a case gave evidence that this α is a negative multiple of the residual helicity, as was expected some time ago by Frisch et al. (1975) and Pouquet et al. (1976). However, in the non-isotropic case (sections 2 and 3) the situation is not so obvious and there may no longer be a clear relation between α and helicity, or it may be more complicated and affected by other factors such as shear, for example. Another important point is that in the inverse cascade mechanism the large scale field appears to saturate by quenching the power in the next higher Fourier modes. This suggests that modelling this phenomenon in terms of the α-effect requires some modification such as a k-dependence of α in Fourier space. However, a multiplication of the form $\alpha(k)\hat{B}(k)$ corresponds, in real space, to a convolution with some α-kernel. Such possibilities were recently explored by Brandenburg and Sokoloff (1999), and may even be necessary to explain stellar cycle data (Brandenburg et al. 1998).

Looking at figure 5 we note that once the field has reached saturation at some wavenumber, the power in the next larger wavenumber begins to be suppressed.

This may be interpreted in terms of a wavenumber dependent quenching. However, only future work can show whether the apparent similarity with α-effect dynamos is coincidental or not. Whatever the result is, it is clear that in both approaches (inverse cascade and α-effect) helicity does play an important role.

There is another important issue that needs to be mentioned here. In all situations of practical relevance the relative kinetic and current helicities (normalized by the rms values of velocity and vorticity or magnetic field and current, respectively) are never close to 100%, as was the case in the cascade model. In the simulations of rotating convective and shear flow turbulence, for example, the relative helicities were at most around 3%-5%. However, it is important to realize that even for zero net helicity, and just helicity fluctuations, a large scale field can grow, provided there is shear; see Vishniac & Brandenburg (1997). The large scale field generated in such a case varies somewhat irregularly in time and may show reversals on a diffusive time scale, so this effect alone would be insufficient to explain the solar cycle, which is more regular. Nevertheless, it is quite plausible that even a small amount of net helicity suffices to produce mean fields with the spatial and temporal order seen on the sun. Therefore, when measuring the helicity in the sun observationally it is important to measure not only the mean helicity, but also its variance (see the chapter by Pevtsov, for example).

REFERENCES

Balbus, S. A. & Hawley, J. F. 1991 A powerful local shear instability in weakly magnetized disks. I. Linear analysis. *Astrophys. J.* 376, 214-222.

Brandenburg, A. 1997 Large scale turbulent dynamos. *Acta Astron. Geophys. Univ. Comenianae* XIX, 235-261.

Brandenburg, A. 1998 Theoretical Basis of Stellar Activity Cycles. In *Tenth Cambridge Workshop on Cool Stars, Stellar Systems, and the Sun* (ed. R. Donahue & J. Bookbinder), pp. 173-191. Astron. Soc. Pac. Conf. Ser., Col. 154.

Brandenburg, A. 1999 Disc Turbulence and Viscosity. In *Theory of Black Hole Accretion Discs* (ed. M. A. Abramowicz, G. Björnsson & J. E. Pringle), Cambridge University Press.

Brandenburg, A. & Bigazzi, A. 1999 "Nonlinear alpha-effect and inverse cascade in hydromagnetic turbulence," *Astrophys. J.* (to be submitted).

Brandenburg, A. & Campbell, C. G. 1997 Modelling magnetised accretion discs. In *Accretion disks – New aspects* (ed. H. Spruit & E. Meyer-Hofmeister), pp. 109-124. Springer-Verlag.

Brandenburg, A. & Donner, K. J. 1997 The dependence of the dynamo alpha on vorticity. *Monthly Notices Roy. Astron. Soc.* 288, L29-L33.

Brandenburg, A. & Sokoloff, D. 1999 "Local and nonlocal magnetic diffusion and alpha-effect tensors in shear flow turbulence," *Geophys. Astrophys. Fluid Dyn.* (submitted).

Brandenburg, A., Nordlund, Å., Pulkkinen, P., Stein, R.F., & Tuominen, I. 1990 3-D Simulation of turbulent cyclonic magneto-convection. *Astron. Astrophys.* 232, 277-291.

Brandenburg, A., Nordlund, Å., Stein, R. F., & Torkelsson, U. 1995 Dynamo generated turbulence and large scale magnetic fields in a Keplerian shear flow. *Astrophys. J.* 446, 741-754.

Brandenburg, A., Jennings, R. L., Nordlund, Å., Rieutord, M., Stein, R. F., & Tuominen, I. 1996 Magnetic structures in a dynamo simulation. *J. Fluid Mech.* 306, 325-352.

Brandenburg, A., Saar, S. H., & Turpin, C. R. 1998 Time evolution of the magnetic activity cycle period. *Astrophys. J. Letters* 498, L51-L54.

Brandenburg, A., Nordlund, Å., & Stein, R. F. 1999 "Simulation of a convective dynamo with imposed shear," *Astron. Astrophys.* (to be submitted).

Chandrasekhar, S. 1960 The stability of non-dissipative Couette flow in hydromagnetics. *Proc. Natl. Acad. Sci.* 46, 253-257.

Chandrasekhar, S. 1961 *Hydrodynamic and Hydromagnetic Stability*. Dover Publications, New York., pp. 384

Frisch, U., Pouquet, A., Léorat, J., & Mazure, A. 1975 Possibility of an inverse cascade of magnetic helicity in hydrodynamic turbulence. *J. Fluid Mech.* 68, 769-778.

Galanti, B. & Sulem, P.-L. 1991 Inverse cascades in three-dimensional anisotropic flows lacking parity invariance. *Phys. Fluids* A 3, 1778-1784.

Galanti, B., Sulem, P.-L. & Gilbert, A. D. 1991 Inverse cascades and time-dependent dynamos in MHD flows. *Physica* D 47, 416-426.

Gilbert, A. D. & Sulem, P.-L. 1990 On inverse cascades in alpha effect dynamos. *Geophys. Astrophys. Fluid Dyn.* 51, 243-261.

Hawley, J. F., Gammie, C. F., & Balbus, S. A. 1995 Local three-dimensional magnetohydrodynamic simulations of accretion discs. *Astrophys. J.* 440, 742-763.

Hawley, J. F., Gammie, C. F., & Balbus, S. A. 1996 Local three dimensional simulations of an accretion disk hydromagnetic dynamo. *Astrophys. J.* 464, 690-703.

Krause, F., & Rädler, K.-H. 1980 *Mean-Field Magnetohydrodynamics and Dynamo Theory*. Akademie-Verlag, Berlin; also Pergamon Press, Oxford.

Moffatt, H. K. 1978 *Magnetic Field Generation in Electrically Conducting Fluids*. Cambridge University Press, Cambridge.

Nordlund, Å., Brandenburg, A., Jennings, R. L., Rieutord, M., Ruokolainen, J., Stein, R. F., & Tuominen, I. 1992 Dynamo action in stratified convection with overshoot. *Astrophys. J.* 392, 647-652.

Parker, E. N. 1955 Hydromagnetic dynamo models. *Astrophys. J.* 122, 293-314.

Pouquet, A., Frisch, U., & Léorat, J. 1976 Strong MHD helical turbulence and the nonlinear dynamo effect. *J. Fluid Mech.* 77, 321-354.

Steenbeck, M., Krause, & F., Rädler, K.-H. 1966 Berechnung der mittleren Lorentz-Feldstärke $\overline{\mathbf{v} \times \mathbf{B}}$ für ein elektrisch leitendendes Medium in turbulenter, durch Coriolis-Kräfte beeinflußter Bewegung. *Z. Naturforsch.* 21a, 369-376. See also the translation in Roberts & Stix, The turbulent dynamo..., Tech. Note 60, NCAR, Boulder, Colorado (1971).

Stone, J. M., Hawley, J. F., Gammie, C. F., & Balbus, S. A. 1996 Three dimensional magnetohydrodynamical simulations of vertically stratified accretion disks. *Astrophys. J.* 463, 656-669.

Velikhov, E. P. 1959 Stability of an ideally conducting liquid flowing between cylinders rotating in a magnetic field. *Sov. Phys. JETP* 36, 1398-1404. (Vol. 9, p. 995 in English translation)

Vishniac, E. T. & Brandenburg, A. 1997 An incoherent $\alpha - \Omega$ dynamo in accretion disks. *Astrophys. J.* 475, 263-274.

Axel Brandenburg, Department of Mathematics, University of Newcastle upon Tyne, NE1 7RU, UK (email: Axel.Brandenburg@Newcastle.ac.uk)

Creation of Twist at the Core-Convection Zone Interface

Peter A. Gilman and Paul Charbonneau

High Altitude Observatory, National Center for Atmospheric Research,[1] Boulder, Colorado

Current helicity is a convenient measure of twist of magnetic flux at the base of the solar convection zone. Accurate estimates of twist at these depths are difficult: only surface observations are available, dynamos predict mean rather than total fields, and flux concentration is not well understood. We generate 'butterfly diagrams' for a variety of mean field dynamos, showing they produce a variety of current helicities. Not all have the predominance of left-handed twist in the northern hemisphere seen in modern observations. Current helicity therefore may be a useful discriminator among various solar dynamos. Convection influenced by rotation and shear should be the primary source of twist within the convection zone, while 'helical waves' driven by magnetic buoyancy, and the nearly 2D mhd instability of toroidal field and differential rotation, could be sources below the convection zone.

1. SOME THOUGHTS ON TWIST AND HELICITY

The subject of this Chapman Conference is magnetic helicity, but our specific assignment is to consider the 'twist' of magnetic field patterns. These two properties of magnetic fields are obviously related but they are clearly not the same. Twist implies field aligned currents, but not necessarily force free fields, which are generally twisted. Magnetic helicity, or $\mathbf{A} \cdot \mathbf{B}$, is conserved when integrated over a closed system in ideal mhd, but the vector potential $\mathbf{A}$ is hard to measure even with a vector magnetograph. By contrast, the 'current helicity,' or $\mathbf{j} \cdot \mathbf{B}$, can be measured, but is generally not conserved, even in closed systems in the ideal mhd limit. Dynamos generally predict $\mathbf{j} \cdot \mathbf{B}$, and magnetic fields at the core-convection zone interface are probably maintained by dynamo action. Therefore since current helicity can be both observed and predicted, we choose to focus on it in considering twist.

An observation of twist will be of the total field, but dynamo theory will predict twist in a 'mean' or average, large scale field. These need not be the same, but below the core-convection zone boundary, where presumably turbulence is weak, the total field and the mean field may be nearly the same. Clearly they will not be within the convection zone, where convective turbulence is strong.

We would like to know what the twist of the field is at the core-convection zone boundary, but we can observe twist only at the solar surface or above. We have to recognize that many processes occurring within the convection zone – meridional circulation, convective turbulence and turbulent diffusion, magnetic buoyancy, and others – may modify twist created at the base of the convection zone [*Longcope et al.*, this volume]. Toroidal and poloidal fields generated there, that together comprise the total twisted field, may be modified in different ways as each threads its way up through the convection

---

[1] The National Center for Atmospheric Research is sponsored by the National Science Foundation.

zone to the surface. Thus inferences from surface observations about twist at the base are uncertain. In what follows we will therefore stick to prediction of twist at the base.

Mean field dynamos will predict the twist of the *global* field at the base of the convection zone, but the existence of sunspots is interpreted as evidence of *concentrated* toroidal flux rings there. These concentrations should also have twist, but at present we do not know how the sun changes the global field there into concentrations of flux.

Changing an untwisted axisymmetric toroidal flux ring into a twisted one is not simple. No axisymmetric (longitude independent, or $m = 0$) motion can produce any twist: differential rotation does not shear the ring because it is everywhere parallel, and meridional circulation will transport toroidal field lines around in meridian planes, but impart no twist. A nonaxisymmetric motion (for example, periodic in longitude, or $m > 0$) can produce local twist, but the global average still vanishes. So for $m = 1$ motions, for example, the twist would be opposite on opposite sides of the sun. Nonaxisymmetric motions that displace flux perpendicular to the axis of the original ring can, when coupled with reconnection, produce twist in the mean field associated with that ring, which then becomes a combination of toroidal and poloidal fields. This is what happens in mean field dynamos, so we will examine results from them extensively.

In mean field dynamos, the field-aligned current essential to sustaining dynamo action and creating twist is created by the kinetic helicity $\mathbf{V} \cdot \nabla \times \mathbf{V}$ in which $\mathbf{V}$ is the small-scale, fluctuating velocity field, coupled with small but finite magnetic diffusion. This process is called the 'α effect' (which needs to be kept distinct from the α of force free fields discussed in other articles in these proceedings). Therefore, kinetic helicity tends to produce current helicity or twist, because the flow drags field lines with it. The same process also produces magnetic helicity.

We examine kinetic helicity in the context of the production of twisted magnetic flux at two levels. First, we assume the existence of kinetic helicity or α effect, and study what twist or current helicity is produced in mean field dynamos. We concentrate on the core-convection zone interface, since this is where the flux ropes that will become sunspots are likely formed. Second, we then take a step back and discuss the characteristics of various classes of motions as sources of kinetic helicity.

2. PRODUCTION OF TWISTED MAGNETIC FLUX IN MEAN FIELD DYNAMOS

2.1. Mean Field Dynamo Equations

We solve the standard kinematic mean field dynamo equations in the form

$$\frac{\partial \mathbf{B}}{\partial t} = \nabla \times (\mathbf{U} \times \mathbf{B} + \alpha \mathbf{B} - \eta \nabla \times \mathbf{B}) \qquad (1)$$

in which $\mathbf{B}$ is the vector magnetic field, $\mathbf{U}$ the assumed large-scale velocity field, including differential rotation and meridional circulation (if any), α is the 'α effect,' proportional to $-\mathbf{V} \cdot \nabla \times \mathbf{V}$, η is the total (molecular plus turbulent) diffusivity of the system.

In a spherical shell, we solve for the axisymmetric magnetic field $\mathbf{B} = \nabla \times (A \hat{e}_\phi) + B_\phi \hat{e}_\phi$, in which $A(r, \theta, t) \hat{e}_\phi$ is the vector potential of the poloidal field, and $B_\phi(r, \theta, t)$ is the toroidal field, r is radius and θ is colatitude. In component form, (1) then reduces to the following scalar equations

$$\frac{\partial A}{\partial t} = \left(\nabla^2 - \frac{1}{r^2 \sin^2 \theta} \right) A + C_\alpha \alpha B \qquad (2)$$

$$\begin{aligned}\frac{\partial B}{\partial t} &= \left(\nabla^2 - \frac{1}{r^2 \sin^2 \theta} \right) B \\ &+ r \sin \theta \, C_\Omega [\nabla \times (A \hat{e}_\phi)] \cdot \nabla \Omega\end{aligned} \qquad (3)$$

in which $\Omega \, (r, \theta)$ is the differential rotation and $C_\alpha = \alpha_0 R/\eta$, $C_\Omega = \Omega_0 R^2/\eta$, are dynamo numbers, in which α_0 and Ω_0 are suitable amplitudes for the α effect and differential rotation, and R is a radius scale.

For solar values, typically $C_\alpha/C_\Omega \sim 10^{-3}$ so that we omitted the α-effect term in (3). This defines the so-called $\alpha\Omega$ dynamo; the toroidal field in (3) is generated exclusively by the differential rotation $\nabla \Omega$. We then solve equations (2) and (3) as a linear eigenvalue problem, assuming a time-dependence of the form $e^{\lambda t}$. All solutions discussed below are slightly supercritical, in that $\text{Re}(\lambda) > 0$ but $\text{Re}(\lambda) \ll \text{Im}(\lambda)$. For given α and Ω spatial distributions, the eigenvalue λ is entirely determined by the product $D = C_\alpha \times C_\Omega$. The absolute magnetic field strength is left undetermined in such linear solutions, but $|B_r|/|B_\phi| \sim (C_\alpha/C_\Omega)^{1/2}$ at constant D. In all cases reported below, dipole symmetry is favored. These dynamo solutions are very similar to those computed by *Stix* [1976] and others in the 1970's.

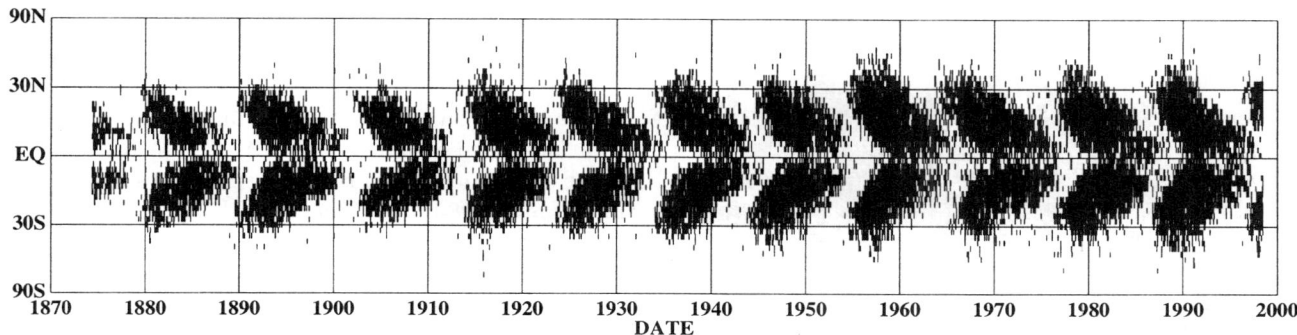

Figure 1. "Butterfly diagram" Plot of sunspot area as a function of date and latitude for the period 1875 to the present. Obtained from David Hathaway at NASA/MSFC.

2.2. Observations Compared with Dynamo Output

The primary observation that all solar dynamo model predictions are compared with is the so-called 'butterfly diagram,' shown in Figure 1, which is a plot of the latitude of sunspot occurrence as a function of time in successive solar cycles.

In addition, historically such models are also compared with evidence for the phase relation in time between the sun's toroidal and poloidal fields [*Stix*, 1976; *Sheeley*, 1991], as inferred from sunspot observations and magnetograms. *Stix* [1976] originally suggested a phase relationship of 180° between the radial and toroidal components. However, the magnetogram observations he used for the radial field are likely dominated by the decay of active regions. The polar faculae calibration of *Sheeley* [1991] on the other hand, indicate that the radial field at high latitudes lags the toroidal field by 90°.

In the past, mean field dynamos have not been compared to observations of the twist or 'handedness' of the large scale solar magnetic field, presumably because observations of this quantity were not developed until recently. However, estimates of twist or swirl in magnetic structures around sunspots had been the subject of a small study by *Richardson* [1941], who reported that Hα images showed that perhaps 35% of spots had twist, and 2/3 to 3/4 of these showed left-handed twist in the northern hemisphere, right-handed in the southern. Recent estimates of the sense of twist discussed elsewhere in this volume (see papers by Pevtsov and by Canfield) indicate a preponderance of left-handed twist in photospheric fields in the northern hemisphere and right-handed in the southern, between 60 and 80%, depending on the measure used.

2.3. Predicted Butterfly Diagrams and Twist for Various Dynamos

Figures 2–6 show multiple butterfly diagrams for five mean-field $\alpha\Omega$ dynamo solutions having different configurations of differential rotation and α effect. In all cases, $\alpha \sim \cos(\theta)$ consistent with most previous simple mean field solar dynamo models. All quantities plotted are calculated at the base of the convection zone, about $r = 0.7R$. In each plot, the uppermost panel is the toroidal field B_ϕ, the middle panel the radial field B_r, and the bottom panel the current helicity $j_\phi B_\phi$. In $\alpha\Omega$ dynamos for which $C_\alpha/C_\Omega \ll 1$, $\mathbf{j} \cdot \mathbf{B} \sim j_\phi B_\phi$; that is, the mean field is predominately toroidal, with the weak poloidal field adding twist about a local azimuthal axis.

Time-latitude diagrams for the toroidal field can be compared to the sunspot butterfly diagram (Figure 1) under the two additional assumptions that the sunspot-producing toroidal flux ropes (a) form in the regions of strongest toroidal field, and (b) rise radially to the photosphere.

Figures 2 and 3 show solutions with cylindrical isorotation contours approximately matched to the observed surface rotation, with negative (Figure 2) and positive (Figure 3) α-effect. The first solution reproduces reasonably well the sunspot butterfly diagram of Figure 1, and gives rises to dominantly left-handed twist (negative $j_\phi B_\phi$) in the northern hemisphere, right-handed in the southern, in agreement with current observations. However, the toroidal and radial fields are nearly in phase, contrary to the interpretation of surface observations of sunspot fields and synoptic magnetograms favored in the 1970's. On this basis *Stix* [1976] and others suggested that a solution having *negative* radial shear and *positive* α-effect should be preferred. Such a

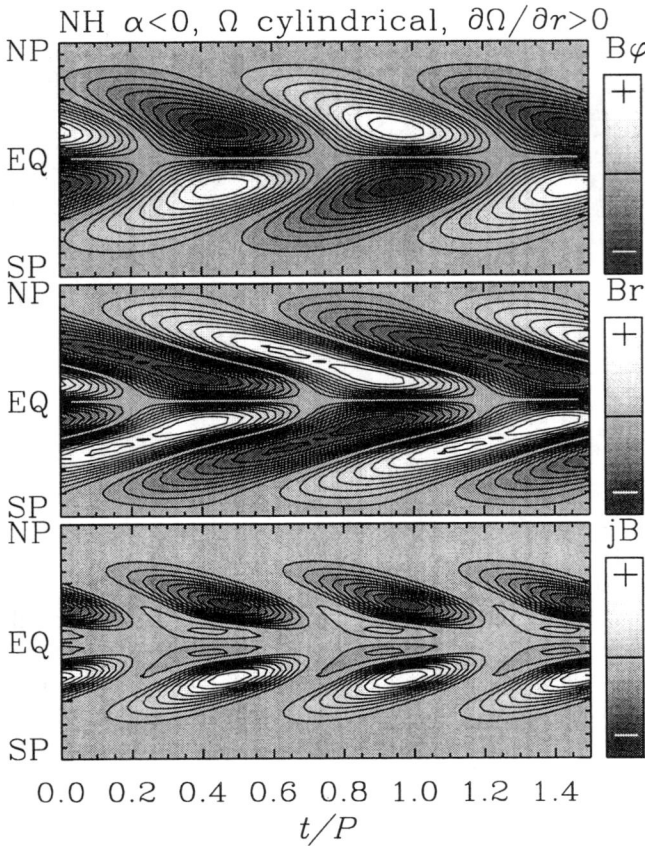

Figure 2. Butterfly diagrams for cylindrical differential rotation matched to solar surface values, and $\alpha < 0$ in the NH.

equatorward propagation materializes in models where the product of the radial shear and α effect is negative in the northern hemisphere. Note that the butterfly diagrams shown on Figure 3 —and underlying differential rotation— resemblance qualitatively the full mhd dynamo simulations for a rotating spherical shell achieved by *Gilman* [1983] and *Glatzmaier* [1985].

The solutions discussed up to now are primarily of pedagogical interest, since helioseismology has shown that the solar convective envelope is *not* in a state of cylindrical rotation. Instead, differential rotation in the envelope is primarily latitudinal, matching smoothly across a thin shear layer immediately beneath the core-envelope interface (the 'tachocline') to a core rotating (very nearly) rigidly at the rate of the surface mid-latitude. This implies that the radial shear is confined to the interface region, and is positive (negative) at low (high) latitudes. Figure 4 displays butterfly diagrams for a solution using such a differential rotation profile.

solution (not shown) has a toroidal butterfly diagram similar to Figure 2 (top panel), with the radial field (middle panel) reversed in sign but otherwise similar in shape, thus yielding what was then considered to be the correct phase relationship. However, the sign of the current helicity distribution (bottom panel) is also reversed, a direct consequence of flipping the sign of the radial field, so that right-handed twist now predominates in the northern hemisphere. If the handedness of the solar field had been well known at the time these mean field dynamos were developed, conclusions about how well they simulated the sun's dynamo would probably have been different! It is possible to maintain a 180° phase lag *and* have predominantly left-handed twist in the northern hemisphere by enforcing a positive radial shear and negative α-effect. Such a solution is shown on Figure 3. However, the toroidal field then migrates to high latitudes. This is expected from the classical propagation 'rule' for dynamo waves, which indicate that

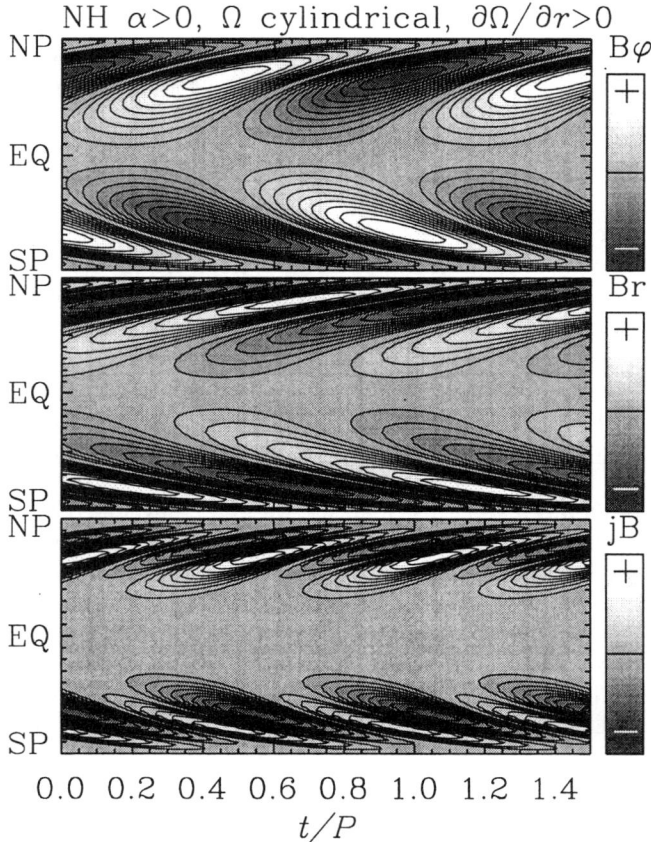

Figure 3. Butterfly diagrams for cylindrical differential rotation matched to solar surface values, and $\alpha > 0$ in the NH.

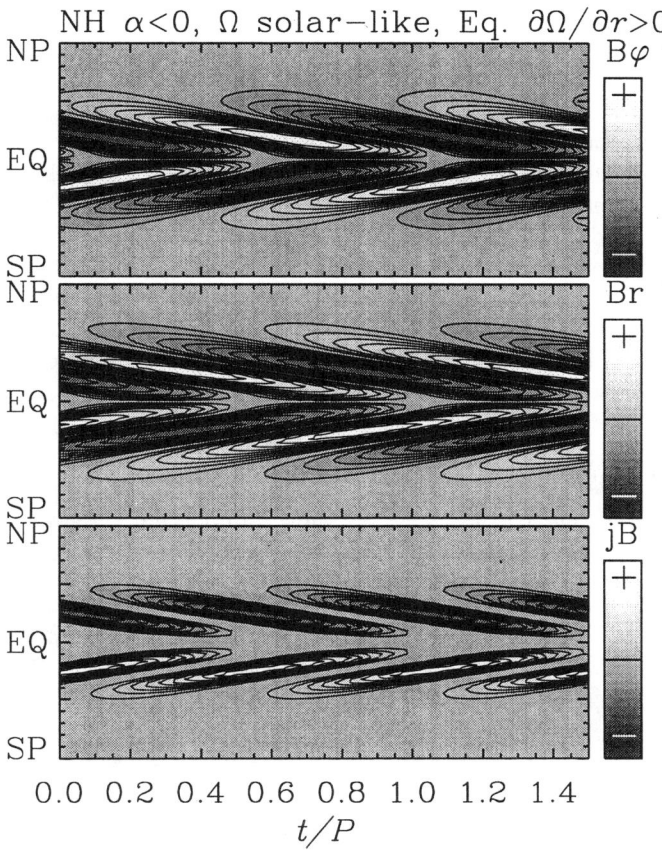

Figure 4. Butterfly diagrams for solar like differential rotation, equatorial radial gradient of rotation > 0, and $\alpha < 0$ in the NH.

Once again we have a reasonable toroidal butterfly diagram, this time with left-handed twist in the northern hemisphere. Careful examination of Figure 4 reveals that the radial field *leads* the toroidal component by about 90° at low latitudes. As in previous cases, since the plots are constructed at the core-envelope interface, there remain the possibility that processes acting within the envelope, such as meridional circulation, can bring the phase of the surface radial field is closer agreement. One serious difficulty is that it is hard to justify theoretically an $\alpha < 0$ in the northern hemisphere through the bulk of the convection zone [*Gilman*, 1983; *Glatzmaier*, 1985]. It is much more likely to have this sign near the base of the convection zone.

Focus in solar dynamo theory has shifted toward the base of the convection zone and below for several reasons. Most important is that the radiative zone and to a lesser degree the overshoot layer have the capacity to store magnetic flux of high amplitude long enough to be important on solar cycle time scales. Fields of peak strength 5×10^4 to 10^5 gauss seem to be necessary to ensure that when the flux enters the convection zone from below, it will rise to the photosphere at sunspot latitudes. Many studies, starting with *Choudhuri and Gilman* [1987], and including *Fan et al.* [1993], *Caligari et al.* [1995] and others, have shown that weaker flux concentrations will traverse the convection zone on a path parallel to the rotation axis rather than radially, which would imply emergence at latitudes poleward of 45°, in conflict with the observed emergence of sunspots and active regions. For dynamos seated below the convection zone base with such strong toroidal fields, it was thought initially that dynamo action, particularly the α effect, would be quenched and the dynamo would be unable to attain or sustain such large fields. But this difficulty has been overcome with the development of the so-called interface dynamos, first introduced by *Parker* [1993], and elaborated for spherical systems by *Charbonneau and MacGregor* [1997].

In these models, various parameters of the problem are assumed to change rapidly at the interface between the convection zone and the interior. Below the interface, the α effect and turbulent diffusivity for magnetic flux are assumed small, while the radial gradient in rotation is taken to be large. Above the interface, these amplitudes are reversed: α effect and diffusivity large, and differential rotation small. Some diffusion must occur across the interface to keep the magnetic field in the layers above and below linked.

Results from these models are promising, in that dynamo action is achieved in a system in which the induced toroidal field below the interface can exceed equipartition values (compared to the mixing length convection amplitude for the base of the convection zone), and plausible butterfly diagrams result. Therefore we show in Figure 5 the same set of butterfly diagrams for an interface dynamo solution as in Figures 2-4 for the bulk convection zone models. We use the same 'solar like' differential rotation as in Figure 4, with the radial gradient concentrated below the interface. *Charbonneau and MacGregor* [1997] find three distinct dynamo modes for a 'solar-like' differential rotation. Two of these three modes (their 'polar' and 'hybrid' modes) lead to toroidal field concentrated at too high latitudes as compared to the sunspot butterfly diagram. The 'equatorial' mode, on the other hand, is in qualitative agreement with the sunspot butterfly diagram, as can be seen on Figure 5. (see *Charbonneau and MacGregor* [1997] for further details).

As can be seen on the bottom pattern of Figure 5, the equatorial interface mode generates magnetic

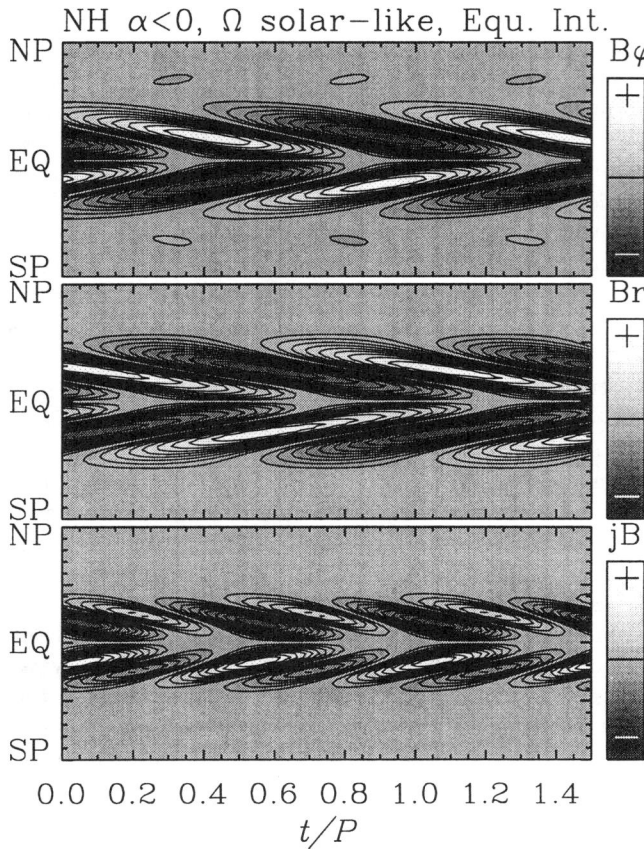

Figure 5. Butterfly diagrams for interface dynamo with solar like differential rotation, for equatorial dynamo mode, $\alpha > 0$ in the NH.

solar observations; and a nonlinear, nonlocal α effect arising from twist acquired by rising buoyant flux tubes acted upon by Coriolis forces. Figure 6 shows butterfly diagrams from one solution of this model. We see that the toroidal field migrates all the way from polar regions to the equator, but with the strongest toroidal fields confined to low latitudes where sunspots are found. The poloidal field is strong only near the poles at all phases of the cycle. The resulting butterfly diagram for the twist or current helicity is predominantly but not completely left-handed in the northern hemisphere. As discussed in Dikpati and Charbonneau, this model agrees well with other observational constraints.

2.4. Conclusions about Twist from Mean Field Dynamos

It is clear from the cases shown above that different assumptions about α and Ω can lead to very different patterns that exhibit mixed handedness in each hemisphere, though there is a definite bias toward a left-handed northern hemisphere. The hybrid mode behaves in a similar way, with positive handedness in the northern hemisphere during the first half of the cycle, followed by negative handedness in the second half. The polar mode is the only one showing negative handedness throughout the cycle. Even from this brief discussion it is clear that it is possible to get a wide range of dynamo behavior and associated field twist even among different modes of an interface dynamo.

The final dynamo example we show here is for a so-called 'flux transport' dynamo, as described in *Dikpati and Charbonneau* [1999]. The governing equations for this model are similar to those for other bulk convection zone and interface dynamos, but with meridional circulation added, in the form of a single cell that has poleward flow near the outer boundary, in agreement with

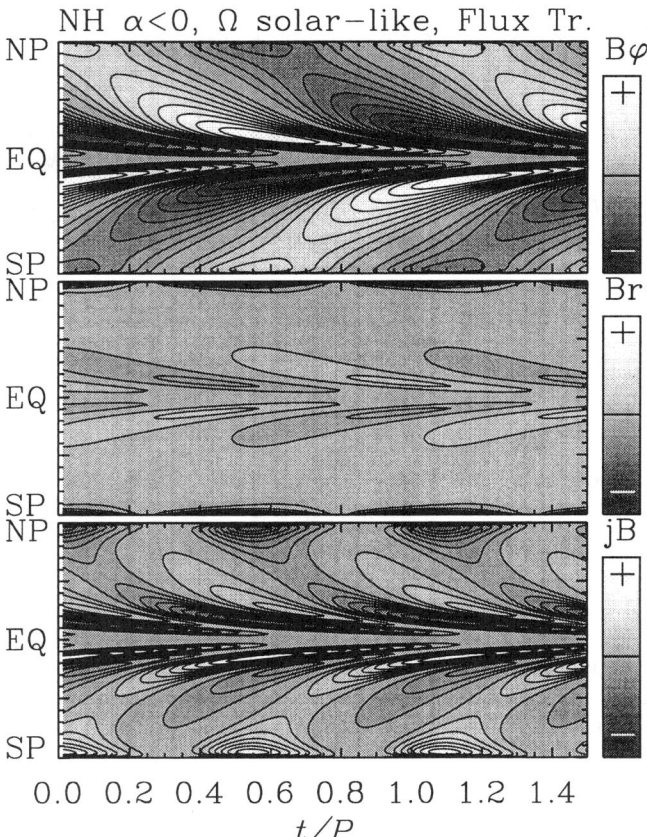

Figure 6. Butterfly diagrams for flux transport dynamo model of Dikpati and Charbonneau with solar like differential rotation, single celled meridional circulation (poleward flow near the outer boundary) and $\alpha < 0$ in the NH.

current helicity or 'handedness' patterns. This means that **j · B** may be a useful discriminator for evaluating the relative merit of various dynamo models applied to the sun. We have seen in particular that dynamos containing a 'solar' differential rotation as inferred from helioseismology, together with $\alpha \sim -\cos(\theta)$ predict nearly all left-handed twist in the northern hemisphere and right-handed in the southern, as does the flux transport model of *Dikpati and Charbonneau* [1999]. The interface dynamo solutions of the type displayed in *Charbonneau and MacGregor* [1997] mix left-handedness and right-handedness in each hemisphere much more, but with one sense of twist early in a half cycle, and the opposite sense later in the same half. So far, no observational evidence has been reported of twist sign dependence or sunspot cycle phase.

3. CHARACTERISTICS OF MOTIONS THAT LEAD TO TWIST

There are general characteristics of mhd fluids and flows that are either essential for producing twisted flux, or greatly enhance production. The first and most obvious necessary property is that at high conductivity the field lines tend to be frozen to the fluid particles as they move. Then if the fluid particle trajectories are long and nonperiodic so they do not close back on themselves, field lines are stretched and convolved, and twist production is enhanced. In addition, flow responding to forces that are perpendicular to the local velocity vector are also conducive to twist production, because the particle trajectories will tend to become spirals. The most obvious example of this type of force is the coriolis force of rotating fluids, which is probably why so many rotating highly conducting celestial bodies–both planets and stars–have magnetic fields that seem to be maintained by dynamo action.

These characteristics suggest some types of flows should be distinctly better than others at producing twist. For example, unstable flows are more likely to produce twist than are waves. Particle trajectories lengthen as the forces of the instability push the system further from equilibrium. Lengthening particle trajectories require that work be done against the electromagnetic body force. This energy comes from whatever energy reservoir is being tapped to drive the unstable flow. Thermal convection is an obvious relevant example. By contrast, waves involve restoring forces that tend to make fluid particles retrace trajectories. This works against producing twist, particularly in simple cases, such as a pure sound wave. An exception may be wave motions arising from coriolis forces themselves, called inertial oscillations ('Rossby waves' are a limiting case). This is because in such waves the particle trajectories tend to be epicyclic or spiral in form, since the coriolis force is perpendicular to the motion rather than opposing. Such waves are obviously more effective at producing twist than either sound or gravity waves.

Shear flow is obviously effective for stretching fluid elements, and therefore lengthening magnetic field lines, but needs to be coupled with other processes to produce twist. We see this in mean-field $\alpha\Omega$ dynamos, in which twist comes from the combined action of differential rotation and kinetic helicity.

Given the above, combinations of instability, shear, and inertial waves are good candidates for twist production. The solar convection zone, which is convecting, rotating, and generating differential rotation, should be an active producer of twist of magnetic fields found there. This combination can also occur in stably stratified layers adjacent to rotating stellar convection zones. For example, just below the solar convection zone is where most of the radial and latitudinal differential rotation of the solar 'tachocline' is found. If this layer is unstable to global nearly two-dimensional (longitude-latitude) disturbances, kinetic helicity and therefore twist could be produced there too. We discuss this possibility for twist in more detail below.

Magnetic fields are not an intrinsic part of the instabilities discussed so far as sources of twist. But there are at least two instabilities that do require magnetic fields that could also be effective sources of twist. One is magnetostrophic or helical waves [*Ferriz-Mas et al.*, 1994], and the other is the joint instability of differential rotation and toroidal magnetic fields [*Gilman and Fox*, 1997]. In the joint instability, when the magnetic field is weak compared to the differential rotation, the perturbation magnetic fields generate a Maxwell stress that transports angular momentum toward the poles, thereby extracting kinetic energy from a solar type differential rotation in latitude. Without the magnetic field, this differential rotation profile is stable to horizontal perturbations. For strong magnetic fields, the toroidal field itself becomes the energy source for the instability, but requires differential rotation to exist.

Helical waves involve magnetically buoyant flux tubes rising and twisting in an environment that is rotating and slightly subadiabatic. For these waves to grow, the stratification must be very close to the adiabatic gradient (within 10^{-5} or 10^{-6}), and so they can only occur in the (perhaps extremely thin) overshoot layer just under the convection zone. By contrast, the 2D instability of differential rotation and toroidal field is favored in the radiative layer below that, where the strong buoyancy restoring force tends to suppress global scale radial motions. To get kinetic helicity and therefore twist from this instability nevertheless requires some displacement in the radial direction. This appears easiest to achieve by allowing the whole tachocline to push up against or suck down the base of the convection zone. These two instabilities could both be present, with the magnetically buoyant helical waves excited above the global 2D mhd instability. We note that both favor low longitudinal wavenumbers, though their latitudinal structure is quite different.

Ferriz-Mas et al. [1994] show that the kinetic helicity of helical waves driven by magnetic buoyancy produce kinetic helicity and therefore something resembling an α effect that could, together with differential rotation, drive an $\alpha\Omega$-like dynamo. A particularly interesting feature is that the α effect is not quenched by strong magnetic fields, because the instability depends on them to exist. There is instead a lower limit to the magnetic field required for the regeneration of the poloidal field to take place, so the dynamos produced by this class of α-like effect are fundamentally nonlinear. This feature also means that dynamo action can still occur when the toroidal field is much larger than an equipartition value. This is necessary if the dynamo is to produce strong enough toroidal fields that buoyant flux tubes rising through the convection zone from where they are stored, will emerge at sunspot latitudes.

Potentially the same features could be produced by the 2D global instability of toroidal field and differential rotation, because this instability also occurs for magnetic fields much larger in energy than the differential part of the rotation. It appears that with some radial displacement allowed, the α effect produced will have the correct sign in each hemisphere to produce equatorial migration of toroidal field, as well as left-handedness in the northern hemisphere, right-handedness in the southern. As discussed in *Gilman and Fox* [1997] and *Dikpati and Gilman* [1999], the unstable modes have other properties that have potential to explain certain features of solar activity.

REFERENCES

Caligari, P., F. Moreno-Insertis, and M. Schussler, Emerging flux tubes in the solar convection zone, I, Asymmetry, tilt, and emergence latitude, *Astrophys. J.*, *441*, 886-902, 1995.

Charbonneau, P., and K. B. MacGregor, Solar interface dynamos, II, Linear kinematic models in spherical geometry, *Astrophys. J.*, *486*, 502-520, 1997.

Choudhuri, A. R., and P. A. Gilman, The influence of the coriolis force on flux tubes rising through the solar convection zone, *Astrophys. J.*, *316*, 788-800, 1987.

Dikpati, M., and P. Charbonneau, A Babcock-Leighton flux transport dynamo with solar-like differential rotation *Astrophys. J.*, in press, 1999.

Dikpati, M., and P. A. Gilman, Joint instability of latitudinal differential rotation and concentrated toroidal fields below the solar convection zone, *Astrophys. J.*, *512*, 1999.

Fan, Y., G. H. Fisher, and E. E. DeLuca, The origin of morphological asymmetries in bipolar active regions, *Astrophys. J.*, *405*, 390-401, 1993.

Ferriz-Mas, A., D. Schmitt, and M. Schussler, A dynamo effect due to instability of magnetic flux tubes, *Astron. Astrophys.*, *289*, 949-956, 1994.

Gilman, P. A., Dynamically consistent nonlinear dynamos driven by convection in a rotating spherical shell, II, Dynamos with cycles and strong feedbacks, *Astrophys. J. Suppl.*, *53*, 243-268, 1983.

Gilman, P. A., and P. A. Fox, Joint instability of latitudinal differential rotation and toroidal magnetic fields below the solar convection zone, *Astrophys. J.*, *484*, 439-454, 1997.

Glatzmaier, G. A., Numerical simulations of stellar convective dynamos, II, Field propagation in the convection zone, *Astrophys. J.*, *291*, 300-307, 1985.

Longcope, D., M. Linton, A. Pevtsov, G. Fisher, and I. Klapper, Twisted flux tubes and how they get that way, in *Magnetic Helicity in Space and Laboratory Plasmas*, edited by Pevtsov, Canfield, and Brown, 1999.

Parker, E. N., A solar dynamo surface wave at the interface between convection and nonuniform rotation, *Astrophys. J.*, *408*, 707-719, 1993.

Richardson, R. S., The nature of solar hydrogen vorticies, *Astrophys. J.*, *93*, 23-28, 1941.

Stix, M., Differential rotation and the solar dynamo, *Astron. Astrophys.*, *47*, 243-254, 1976.

Sheeley, N. R. Jr., Polar faculae: 1906-1990, *Astrophys. J.*, *374*, 386-389, 1991.

P. Charbonneau and P. A. Gilman, High Altitude Observatory, National Center for Atmospheric Research, P.O. Box 3000, Boulder, CO 80303. (e-mail: paulchar@ucar.edu; gilman@ucar.edu)

Flows in the Solar Convection Zone

A. G. Kosovichev

Stanford University, Stanford, California

The Solar Oscillation Investigation – Michelson Doppler Imager (SOI/MDI) experiment on SOHO and ground-based helioseismic networks (GONG, MWO, BBSO, TON) have provided new interesting results on the differential rotation and flows in the solar convection zone. Accurately measured frequency splitting has allowed us to detect the zonal shear flows associated with the 'torsional oscillations' in the upper convection zone and also study the evolution of these flows during the current solar minimum. The new data have also revealed anomalously slow rotation and other interesting variations of the differential rotation at high latitudes, and provided important insight into the structure and dynamics of the transition layer ('tachocline') at the bottom of the convection zone. With the new method of time-distance solar tomography it has become possible to study meridional circulation and asymmetries in the differential rotation in the convection zone. This method has also been employed for probing the structure of supergranulation and internal flows associated with sunspots and emerging magnetic flux. These and other new developments in helioseismology open important prospects for understanding the mechanisms of generation of solar magnetic fields.

1. HELIOSEISMIC MEASUREMENTS OF SOLAR ROTATION AND FLOWS

Helioseismology provides a unique tool for probing the interior structure and dynamics of the Sun. The information about the thermodynamic and magnetic properties and differential rotation and flows is obtained by inverting frequencies and travel times of solar oscillations. The oscillations are usually observed by measuring the Doppler shift of a solar absorption line formed in the lower part of the solar atmosphere. The MDI instrument (Scherrer *et al.*, 1995) on board Solar and Heliospheric Observatory (SOHO) has provided long time series of stable Doppler images of the solar surface of various degree of resolution. Because of the stochastic nature of solar oscillations substantial spatial and temporal averaging of data is required to measure the frequencies and travel times accurately.

The frequencies of solar eigenmodes, ω, are obtained from oscillation power spectra, an example of which is shown in Figure 1. Different ridges in the power spectrum represent mode multiplets of different radial order n, which corresponds to the number of radial nodes of mode eigenfunctions. The lowest ridge is the fundamental (f) mode which is essentially a surface gravity mode at high angular degree l. The other ridges represent acoustic (p) modes. The frequencies of the mode multiplets (n, l) are split with respect to the azimuthal order m because of rotation and asphericity.

Local wave dispersion relation $\omega(k_x, k_y)$, where k_x and k_y are the horizontal wave-numbers, are measured from three-dimensional power spectra (Figure 2), which

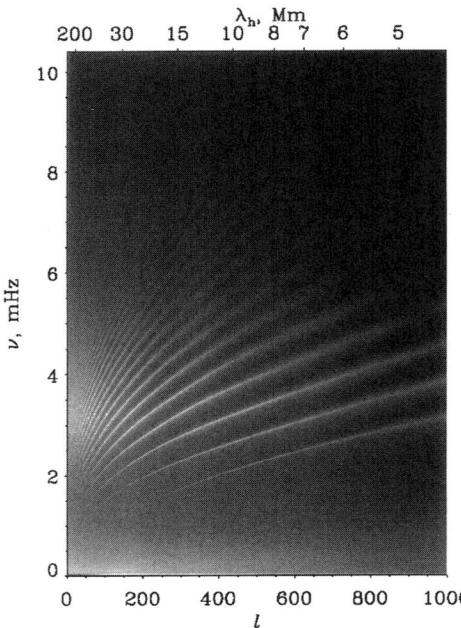

Figure 1. The power spectrum ($l - \nu$ diagram) obtained from the MDI data for solar modes averaged over azimuthal order m and plotted as a function of mode angular degree l (and the corresponding wave length, λ_h), and cyclic frequency $\nu = \omega/2\pi$.

are obtained by applying a 3D Fourier transform to series of Doppler images of small areas on the solar surface. For a given frequency the oscillation power is concentrated in rings in the (k_x, k_y)-plane. These rings correspond to groups of modes of different radial order n. Advection of the wave fronts by flows shifts the wave frequency: $\Delta\omega = \boldsymbol{k} \cdot \boldsymbol{U}$. This method which is called 'ring-diagram analysis' allows us to measure the horizontal components of the flow velocity, $\boldsymbol{U}$ (Hill, 1988).

The travel times are measured from a cross-covariance function (Figure 3) calculated for oscillation signal at different distances (Duvall et al., 1993). The lowest set of ridges in the time-distance diagram corresponds to waves propagated to a distance, Δ, without additional reflections from the solar surface. The middle ridge is produced by the waves arriving to the same distance after one reflection from the surface, and the upper ridge resulted from the waves arriving after two bounces from the surface.

These three approaches are complementary: the frequencies of solar modes are used to infer large-scale properties, such as the rotation rate as a function of radius and latitude, through the whole Sun, whereas the local dispersion relation and acoustic travel times have been used for determining local properties of convective and magnetic structures in the subsurface layers.

Most of the helioseismic analysis is based on a perturbation analysis which provides linear integral relations between the helioseismic measurements and the internal properties, flow velocities and sound-speed variations (e.g. Kosovichev, 1999). For instance, the relation between frequency splitting $\Delta\omega_{nlm}$ and rotation rate $\Omega(r, \theta)$, as a function of radius r and co-latitude θ is given by:

$$\Delta\omega_{nlm} = \int_0^R \int_0^\pi \Omega(r,\theta) K_{nlm}(r,\theta) d\theta dr, \quad (1)$$

where n, l, and m are the radial order, angular degree, and azimuthal order of a solar eigenmode, K_{nlm} is the integral kernel computed using the eigenfunctions of a spherically symmetrical static solar model. This method provides information only about the flow component which is axisymmetrical and symmetrical with respect to the equator.

Similarly, estimates of the horizontal components, U_x and U_y, of the flow velocity averaged in small areas as a function of depth are obtained from the wave dispersion relation (ring diagrams) by solving the following equation:

$$\Delta\omega(k_x, k_y) = \int_0^R K(r) \left[k_x U_x(r) + k_y U_y(r) \right] dr, \quad (2)$$

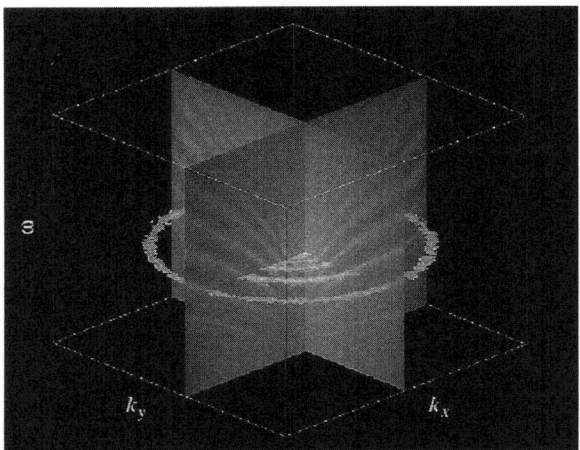

Figure 2. Three-dimensional power spectrum from MDI of a circular region on the Sun of about 15° in diameter. The range of horizontal wave numbers k_x and k_y is from 0 to 2.15 Mm^{-1}, and the range of frequency $\omega/2\pi$ is from 0 to 8.33 mHz. Power is shown projected along two the horizontal axes, and on a sigle cut in the vertical axis. (Bogart et al., 1997).

where $\Delta\omega(k_x, k_y)$ is the frequency shift due to advection, and $K(r)$ is the sensitivity kernel for the horizontally averaged flow.

In the time-distance method, the flow velocity, U, is measured from the difference between the reciprocal travel times, which in the ray approximation is given by:

$$\delta\tau \equiv \tau_+ - \tau_- = -2 \int_\Gamma \frac{(n \cdot U)}{c^2} ds, \quad (3)$$

where τ_+ and τ_- are acoustic travel times in the opposite directions along the same ray path, Γ, n is the unit vector targent to Γ, and s is the distance along the ray path (Kosovichev and Duvall, 1997). This method can provide 3D velocity maps of flows in the convection zone, however, it is less accurate than the frequency method because the travel times are measured less accurately than the frequency splittings, and, to some extent, because of the lack of theoretical interpretation based of wave theory.

Functions $\Omega(r, \theta)$ and $U(r)$ are inferred from equations (1)-(3) by linear inversion techniques such as the regularized least-squares method (Tikhonov and Arsenin, 1977) and the optimally localized averaging technique (Backus and Gilbert, 1968). The helioseismic inversions deal with large datasets of $\Delta\omega_{nlm}$ and $\delta\tau$, which typically consist of $10^5 - 10^6$ measurements. For correct interpretation of inversion results it is important to take into account that inversions provide estimates of some localized averages, and not point values of the solar properties.

2. FLOWS IN THE CONVECTION ZONE

2.1. Tachocline

Knowledge of the internal dynamics of the solar convection zone has important consequences for understanding where and how the solar dynamo operates. Dynamo theory predicts that the direction of the latitudinal propagation of dynamo waves is determined by the sign of the product of the radial gradient of angular velocity and flow helicity. This is important for explaining the 'butterfly diagram' of migration of the sunspot zone towards the equator in the course of the solar cycle. The magnitude of the gradient determines the propagation speed of this zone. In most models, the dynamo process is assumed to occur at the base of the convection zone, in a transition zone between the convection zone and the radiative interior, which is sometimes called 'tachocline'. Helioseismology provides estimates of the parameters of the tachocline.

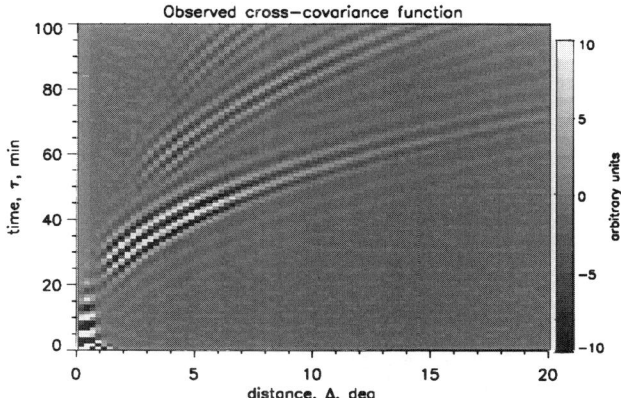

Figure 3. The observed cross-covariance function as a function of the distance on the solar surface, Δ, and the travel time, τ.

Figure 4 shows the rotation rate (in nHz) in the solar interior inferred by inverting frequency splittings obtained from a 144-day series of MDI Dopplergrams (Schou et al., 1998). This inversion result confirmed the previous findings that the solar differential rotation is confined mostly in the convection zone (the lower boundary of which is shown by the dashed curve), and that the radiative core rotates almost rigidly.

The results reveal two zones of strong radial gradient of the rotation rate at the lower and upper boundaries of the convection zone. For the dynamo theory, it is important to establish properties of these zones, and, in particular, the exact location of the lower transition region, 'tachocline', with respect to the boundary of the convection zone, because if most of the tachocline were located in the convection zone then the large scale magnetic field generated by the dynamo would be quickly destroyed by convection.

The spatial resolution of the 2D inversions is usually insufficient for determining the central location r_0 and width w of the tachocline. Therefore, Kosovichev (1996) has suggested determining these parameters by fitting an analytical model to a parameter of rotational splitting which characterizes the main component of the latitudinal differential rotation.

Figure 5 shows χ^2 contours of his fit to the BBSO data (Woodard and Libbrecht, 1993). The result of this fit shown in Figure 6 indicates that most of the tachocline is located below the adiabatic boundary of the convection zone, which is located at $0.713\,R$ according to Christensen-Dalsgaard et al. (1991) and Kosovichev and Fedorova (1991). The estimate of the width, w, (which was defined approximately as twice the FWHM) is less certain (see Fig. 5). It was estimated to

86 FLOWS IN THE SOLAR CONVECTION ZONE

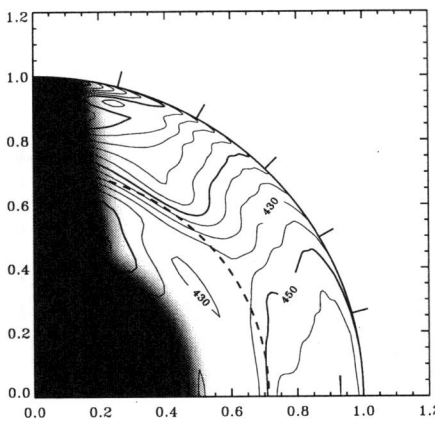

Figure 4. Inversion for rotation rate $\Omega/2\pi$ with radius and latitude. Some contours are labeled in nHz, and for clarity, selected contours are shown as bold. The dashed circle indicates the base of convection zone, and the tick marks at the edge of the outer circle are at latitudes $15°$, $30°$, $45°$, $60°$, $75°$. In such a quadrant display, the equator is the horizontal axis and the pole the vertical one, with the proportional radius labeled. The shaded area indicates the region in the Sun where the inversion results are not reliable (after Schou et al., 1998).

be $0.09\pm0.04\ R$. Charbonneau et al. (1998) have argued that the tachocline may be only $0.05\ R$ wide. Longer time series from the MDI and GONG will provide more accurate estimates of the tachocline.

2.2. Zonal Shear Flows

Observations of the surface flows by Howard and LaBonte (1980) revealed a pattern of zonal flows of fast and slow rotation, associated with the solar cycle, so-called, 'torsional oscillations'. However, the nature of these flows and their role in the dynamo are unknown. Recent helioseismic data have shown that these flows are probably quite deep, extending, at least, for 5% of the solar radius.

Figure 7 shows the rotation rate and its deviation from the three-term law:

$$\Omega(\theta) = A + B\cos^2\theta + C\cos^4\theta, \quad (4)$$

averaged over the depth 2-9 Mm. These results obtained from the f-mode splitting reveal the bands of slower and faster rotation. The f-mode splitting is measured more accurately than the splitting of p modes. However, the f mode provides the rotation rate only in a thin subsurface layer because this mode does not penetrate into the deep interior in the observed range of wavelengths.

The depth dependence of the zonal flows has been obtained by Schou et al. (1998) from the MDI p-mode data. These results (Figure 8) have provided evidence that the zonal flows are, at least, 35 Mm ($0.05\ R$) deep. An important problem is to establish whether these flows extend to the base of the convection zone or not. This would help in understanding the origin of the flows and their role in the solar dynamo.

2.3. Polar Vortex

A substantial deviation from the standard rotation law represented by equation (4) has been found in the

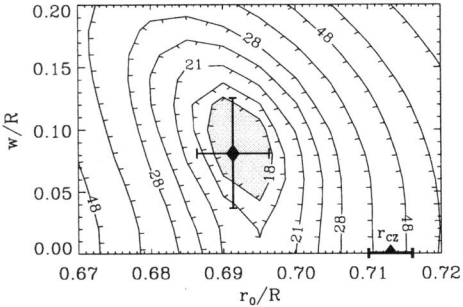

Figure 5. Contours of $\chi^2(w, r_0)$ of an analytical fit to the rotational splitting data at $\chi^2 = 18, 19, 21, 24, 28, 36, 48, 64$ and 96; w is the thickness of the tachocline, r_0 is its central radius. The shaded area corresponds to the increase of χ^2 by 1 from its minimum value, or 1σ uncertainty in the parameters. The error bars show the 1σ uncertainty estimated from statistical modeling by adding Gaussian noise to the data. (Kosovichev, 1996).

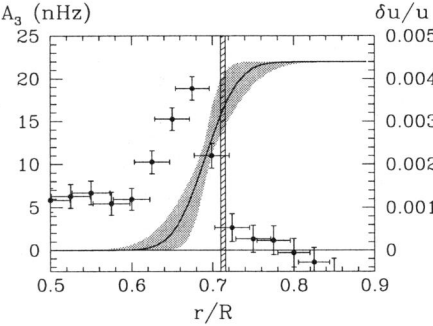

Figure 6. The solid curve with the shadow (indicating 1σ uncertainty) shows a parameter of the latitudinal differential rotation estimated from the BBSO data. The points with error bars represent the variations of the ratio of the pressure to the density, $u \equiv p/\rho$, relative to a standard solar model, inferred from the GONG data. The vertical hatched column shows the location of the base of the adiabatically stratified part of the convection zone. (Kosovichev, 1996).

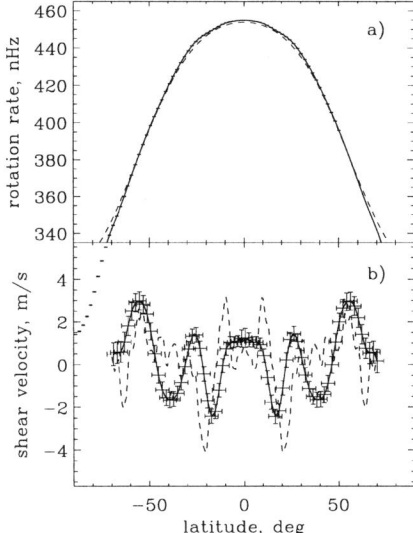

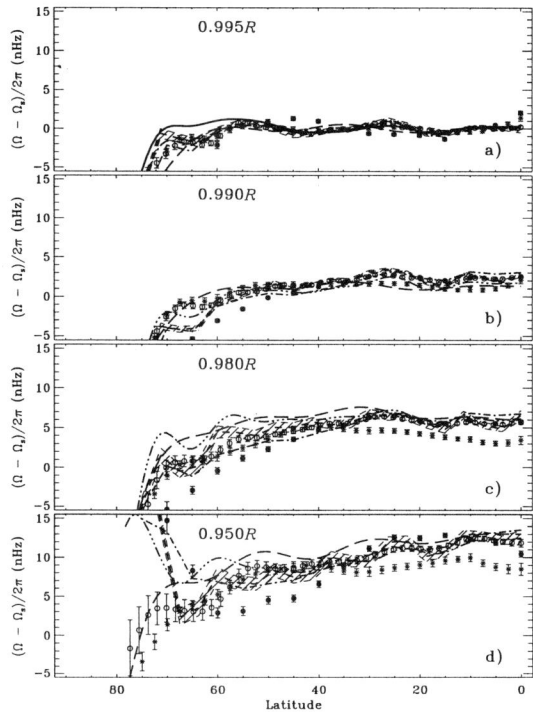

Figure 7. (a) The rotation rate, $\overline{U(\theta)}/2\pi R \sin\theta$, in the subsurface layer 2-9 Mm deep as a function of latitude as determined from the f-mode frequencies by Kosovichev and Schou (1997) (solid curve). The dashed curve shows the surface rotation rate obtained from Doppler measurements (Snodgrass, 1992). (b) The variations of the azimuthal velocity from the three terms rotation law. The error bars show 3σ random error estimates from the data. The horizontal bars show the latitudinal resolution. The dashed curve shows the symmetric component of the surface flows (Hathaway et al., 1996)

Figure 8. Residual $(\Omega-\Omega_s)/2\pi$ of the inferred rotation rate at different fixed radii (as labeled), plotted against latitude, after subtraction of fitted three-term rotation rate at $r = 0.995\,R$. The heavy solid curve shows the corresponding result obtained by Kosovichev & Schou (1997) from analysis of f-mode frequencies. Otherwise symbols and line styles represent the results of different inversion methods (Schou et al., 1998). The solution has been restricted to the region outside the shaded area in Fig. 4.

polar regions where the angular velocity turned out to be lower (Birch and Kosovichev, 1998; Schou et al., 1998). The mechanism of the slow rotation ('polar vortex') is not understood yet. However, the variation of the near-pole rotation is probably related to the mechanism of the solar cycle, because the helioseismic data from the BBSO (Woodard and Libbrecht, 1993) indicate that the rotation rate in the polar zone was higher at the cycle maximum (see Birch and Kosovichev, 1998).

Figure 9 shows the latitudinal variations of rotation, radially averaged in the range $0.96 < r/R < 0.99$, from two 72-day sets of MDI data of May 1 - July 12, 1996 (MDI A) and July 12 - September 22, 1996 (MDI B), and from a 108-day dataset from the GONG (June 6 - September 21, 1996). These results show approximately 5% decrease of the rotation rate at $\theta < 15°$.

Slow near-pole rotation was neither predicted by simulations nor expected from theory. It has been suggested that the sharp decrease in rotation is due to torque from the fast solar wind (e.g. Schou et al., 1998).

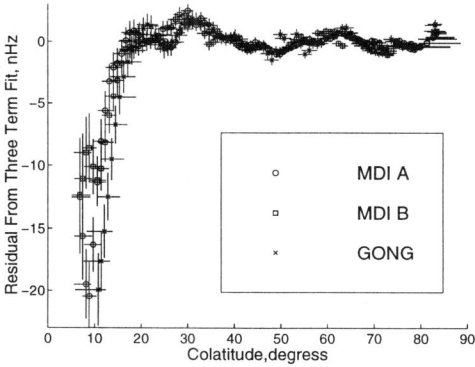

Figure 9. The residuals of the inversion results from three-term fits in even powers of $\cos\theta$ are shown as a function of colatitude. All three results show a sharp deviation from the three-term law at roughly 20° as well as small zonal flows. (Birch and Kosovichev, 1998).

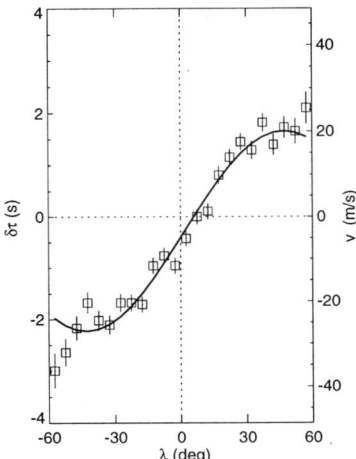

Figure 10. The average travel time difference (south minus north) as a function of latitude, λ, for surface separation of pairs of points in the range 12–73 Mm. The individual points are shown (squares) and the 1σ errors (vertical lines). The solid curve is the best fit 2-parameter model described in the text. The velocity scale on the right axis, in which 12.1 m/s flow corresponds to a 1 s time difference, is obtained from equation (3) (Giles et al., 1997).

However, this explanation is not compatible with previous discussions of convection zone dynamics. Gilman (1974) has argued that because of mixing the convection zone should be rigid to the solar wind torque.

2.4. Meridional Circulation

Meridional flows from the equator towards the north and south poles have been observed on the solar surface in direct Doppler-shift measurements (e.g. Duvall, 1979). The MDI observations by Giles et al. (1997) have provided the first evidence that such flows persist to great depths, and, thus, possibly play an important role in the 11-year solar cycle. The poleward flow can transport the magnetic remnants of sunspots generated at low latitudes to higher latitudes and, therefore, contribute to the cyclic polar field reversal.

The meridional flows in the solar interior were detected by the time-distance method. Figure 10 shows the differences between the travel times of acoustic waves propagating poleward and equatorward at different latitudes λ. These travel-time differences correspond to the mean meridional flow averaged over the penetration depth of the acoustic waves, which was 4-24 Mm in the measurements. By using equation (3) Giles et al. (1997) estimated that the maximum mean speed of the flow is $\simeq 20$ m s^{-1}. They have also found that the flow velocity is almost constant over the observed range of depth.

2.5. Giant Convective Cells

Giant convective cells which have been predicted to exist on the Sun (e.g. Simon and Weiss, 1965), could span the whole convection zone, thus, playing a substantial role in the global circulation in the Sun, and, possibly, in the processes of activity. Beck, Duvall and Scherrer (1998) have found new evidence for the giant cells by analyzing a 505-day series of MDI data.

They have detected long-lived velocity cells (Figure 11) extending over 40-50 degrees of longitude but less than 10 degrees of latitude. The large aspect is surprising, but may be a consequence of solar differential rotation.

2.6. Supergranulation

Supergranulation represents convective motions of an intermediate scale, $\simeq 20$ Mm, which falls between the scales of granules ($\simeq 1$ Mm) and giant cells ($\simeq 200$

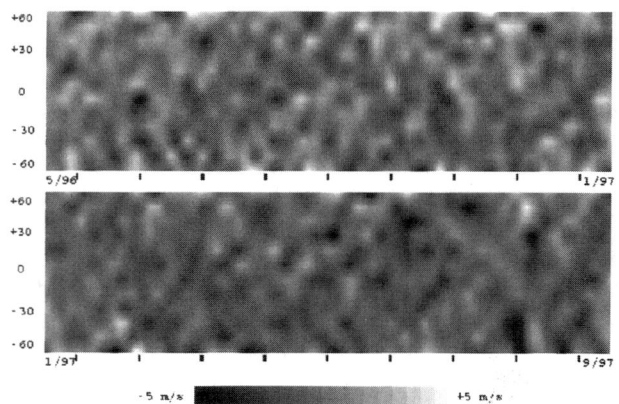

Figure 11. Maps of the east-west component of flow speed obtained by isolating the Doppler signal antisymmetrical across the central meridian and averaging over a disk passage. The signals are weighted by the sine of the longitude and are shifted in time before averaging to account for the solar rotation. A white signal corresponds to flow in the direction of solar rotation. Solar latitude is shown on the left scale. Time is on the horizontal axis with the first half of the 505 days covered in the top image and the second half in the image below. The range of dates covered is shown in the lower corners of the images. The small black boxes along the bottom of the images are separated by the rotation period defined by Carrington of 27.2753 days. About half the variance in these images is due to the long-lived structures and the other half of the variance is caused by shorter-lived smaller-scale supergranules (after Beck et al., 1998).

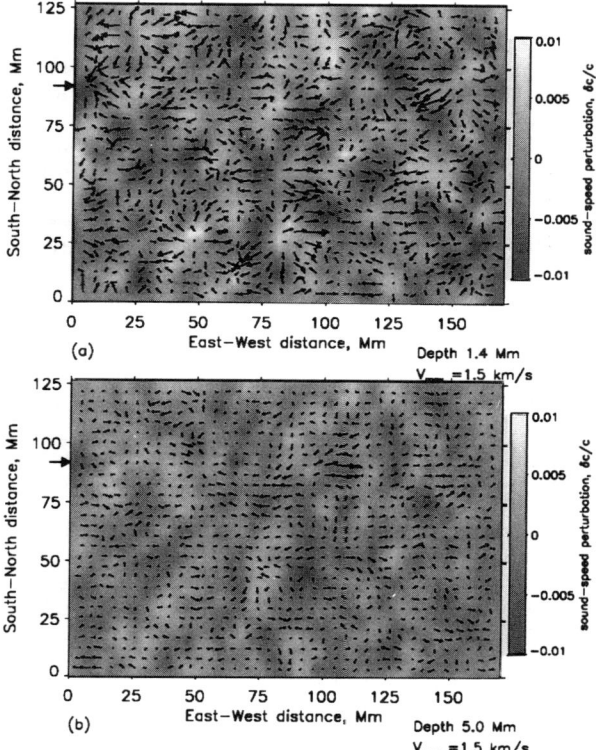

Figure 12. The horizontal flow velocity field (arrows) and the sound-speed perturbation (grey-scale background) at the depths of 1.4 Mm (a) and 5.0 Mm (b), as inferred from the SOHO/MDI high-resolution data of 27 January 1996. The arrows at the South-North axis indicate location of the vertical cut in East-West direction, which is shown in Fig. 13.

Mm). Supergranulation plays a fundamental role in active processes in the chromosphere and corona because most of the magnetic flux outside the active regions is concentrated at the boundaries of supergranular cells. The nature of supergranulation is not understood. It was originally believed that supergranulation is driven by convective instability in the HeII ionization zone located approximately 15 Mm below the solar surface. Using the time-distance technique Duvall et al.(1997) and Kosovichev and Duvall (1997) have attempted to determine the structure of the supergranular flows in the interior.

The results, an example of which is shown in Figures 12a and 12b and Figure 13, revealed that the supergranular outflow pattern previously observed on the surface is 2-3 Mm deep, and gradually disappears in deeper layers. The depth of supergranulation seems do not to exceed 8 Mm (Duvall, 1998).

Analyzing three-dimensional power spectra (ring diagrams) of small areas on the solar surface Hill (1990) and Patrón et al. (1995) found evidence for spiral flows on the scale of supergranulation (Figure 14), which could be important for generating magnetic helicity in the upper convection zone. However, their results are not fully consistent showing flow rotation in the opposite directions. Further investigations are required to determine whether the differences are due to temporal evolution of the flows. The spiral flows have not been detected by the time-distance analysis. A detailed study of the supergranulation flows is an important task of local helioseismology.

For understanding supergranulation and its role in solar activity it is also important to develop numerical models of this phenomenon and compare the models with the observation data. So far, supergranulation was not reproduced in numerical models of convection.

3. CONCLUSION

The helioseismic data from the MDI instrument on SOHO and from the ground-based observatories (BBSO, MWO, GONG, TON) have provided new information about the internal dynamics of the Sun. These data have shown a great variety of organized flows in the convection zone, such as the shear flow at the base of the convection zone (tachocline), migrating zonal flows in the sub-surface layers, slow rotation near the poles ('polar vortex'), meridional circulation, giant and supergranular cells. However, the relations among these flows and their role in the global circulation of the convection zone and generation of flow helicity are not understood. It will be of great importance to establish variations of the tachocline, the polar vortex and the zonal sub-surface flows with the phase of solar activity,

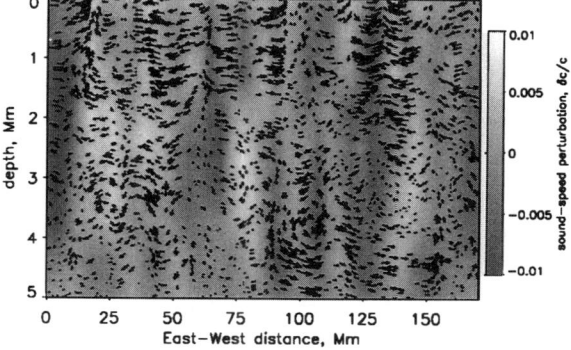

Figure 13. The vertical flow field (arrows) and the sound-speed perturbation (grey-scale background) at the North-South position indicated by arrows in Fig. 12.

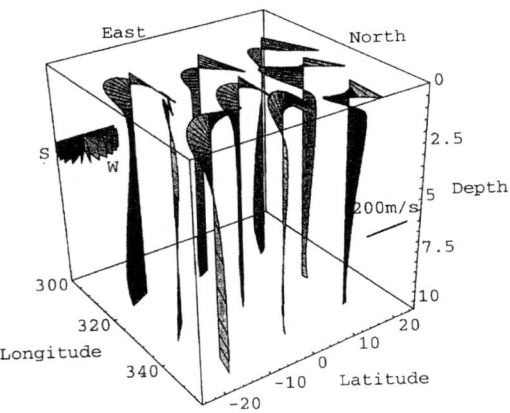

Figure 14. A three-dimensional pseudo-perspective plot of the velocity field inferred by Patrón et al. (1995) by analyzing three-dimensional oscillation power spectra from the Mount Wilson Observatory. Parallel lines to the latitude-longitude plane are proportional to the velocity vectors. The shading indicates the orientation of the flows. The numbers in the latitude-longitude plane are degrees (1 deg $\approx$ 12 Mm), and the units in depth are in Mm.

to determine the depth of meridional flows and giant cells, and to estimate the kinetic helicity of supergranulation. Future helioseismic studies will provide us with this information.

REFERENCES

Backus, G. E., and Gilbert, J. F., The resolving power of gross earth data, *Geophys. J. R. astr. Soc.*, 58, 631-654, 1979.

Beck, J. G., Duvall, T. L., Jr, and Scherrer, P. H., Long-lived giant cells detected at the surface of the Sun, *Nature*, 394, 653-655, 1998.

Birch, A. C., and Kosovichev, A. G., Latitudinal Variation of Solar Subsurface Rotation Inferred from p-Mode Frequency Splittings Measured with SOI-MDI and GONG, *Astrophys. J. Lett.*, 503, L187-L190, 1998

Bogart, R.S., Disher de Sá, L.A., González Hernández, I., Patrón Recio, J., Haber, D.A., Toomre, J., Hill, F., Rhodes, E.J., Xue, Y., Plane-Wave Analysis of the SOI Data, in: Sounding Solar and Stellar Interiors, Proc. IAU Symp. 181. Eds: J. Provost and F.-X. Schmider, Kluwer, 111-118, 1997.

Charbonneau, P., Christensen-Dalsgaard, J., Henning, R., Schou, J., Thompson, M.J., Tomczyk, S., Observational Constraints on the Dynamical Properties of the Shear Layer at the Base of the Solar Convection Zone, in: Sounding Solar and Stellar Interiors, Eds: J. Provost and F.-X. Schmider, Observatoire de la Côte d'Azur, Nice, France, 161-162, 1998.

Christensen-Dalsgaard, J., Gough, D.O., and Thompson, M.J., The depth of the solar convection zone, *Astrophys. J.*, 378, 413-437, 1991.

Duvall, T. L., Jr. Large-scale velocity fields, *Solar Phys.*, 63, 3-15, 1979.

Duvall, T. L., Jr, Jefferies, S. M., Harvey, J. W., and Pomerantz, Time-distance helioseismology, *Nature*, 362, 430-432, 1993.

Duvall, T. L., Jr., Recent Results and Theoretical Advances in Local Helioseismology, in: Proc. 6th SOHO Workshop, 1998, in press.

Giles, P. M., Duvall, T. L., Jr, and Scherrer, P. H., A subsurface flow of material from the sun's equator to its poles, *Nature*, 390, 52-54, 1997.

Gilman, P. A., Comments on 'Solar polar spindown', by Kenneth Schatten, *Solar Phys.*, 36, 61-64, 1974

Hathaway, D.H.; Gilman, P.A.; Harvey, J.W.; Hill, F.; Howard, R.F.; Jones, H.P.; Kasher, J.C.; Leibacher, J.W.; Pintar, J.A.; Simon, G.W. GONG observations of solar surface flows, *Science*, 272, 1306-1309, 1996.

Hill, F., Rings and Trumpets – Three Dimensional Power Spectra of Solar Oscillations, *Astrophys. J.*, 333, 996-1013, 1988.

Hill, F., A Map of the Horizontal Flows in the Solar Convection Zone, *Solar Phys.*, 128, 321-331, 1990.

Howard, R., and LaBonte, B. J., The Sun is observed to be a torsional oscillator with a period of 11 years, *Astrophys. J. Lett.*, 239, L33-L36, 1980.

Kosovichev, A. G., Helioseismic Constraints on the Gradient of Angular Velocity at the Base of the Solar Convection Zone, *Astrophys. J. Lett.*, 469, L61-66, 1996

Kosovichev, A. G., Inversion Methods in Helioseismology and Solar Tomography, *J. Comp. Appl. Math.*, 1999, in press.

Kosovichev, A.G., and Duvall, T.L., Jr. Acoustic tomography of solar convective flows and structures, in: SCORe'96 : Solar Convection and Oscillations and their Relationship, Eds.: F.P. Pijpers, J. Christensen-Dalsgaard, and C.S. Rosenthal, Kluwer Academic Publishers (*Astrophysics and Space Science Library*, Vol. 225), 241-260, 1997.

Kosovichev, A. G., and Fedorova, A. V., Construction of a seismic model of the Sun, *Sov. Astr.*, 35, 507-514, 1991.

Kosovichev, A. G., and Schou, J., Detection of Zonal Shear Flows beneath the Sun's Surface from f-Mode Frequency Splitting, *Astrophys. J. Lett.*, 482, L207-210, 1997.

Kosovichev, A. G., Schou, J., Scherrer, P.H. et al.., Structure and rotation of the solar interior: first results from the MDI medium-l program, *Solar Phys.*, 170, 43-61, 1997.

Patrón, J., Hill, F., Rhodes, E.J., Jr., Korzennik, S.G., and Cacciani, A., Velocity Fields Within the Solar Convection Zone: Evidence From Oscillation Ring Diagram Analysis of Mount Wilson Dopplergrams, *Astrophys. J.*, 455, 746-757, 1995.

Schou, J. Antia, H. M.; Basu, S.; Bogart, R. S.; Bush, R. I.; Chitre, S. M.; Christensen-Dalsgaard, J.; Di Mauro, M. P.; Dziembowski, W. A.; Eff-Darwich, A.; Gough, D. O.; Haber, D. A.; Hoeksema, J. T.; Howe, R.; Korzennik, S. G.; Kosovichev, A. G.; Larsen, R. M.; Pijpers, F. P.; Scherrer, P. H.; Sekii, T.; Tarbell, T. D.; Title, A. M.; Thompson, M. J.; Toomre, J., Helioseismic Studies of Differential Rotation in the Solar Envelope by the Solar Oscillations Investigation Using the Michelson Doppler Imager, *Astrophys. J.*, 505, 390-417, 1998.

Simon, G. W., and Weiss, N. O., Supergranules and the hydrogen convection zone, *Z. Astrophys.*, *69*, 435-450, 1968.

Snodgrass, H. B., Synoptic Observations of Large Scale Velocity Patterns on the Sun, in: The solar cycle; Proceedings of the National Solar Observatory/Sacramento Peak. 12th Summer Workshop, ASP Conference Series (ASP: San Francisco), *27*, p. 205-240, 1992.

Tikhonov, A. N., and Arsenin, V. Y., Solutions of ill-posed problems, Winston, Washington - New York, 1977.

Woodard, M. F., and Libbrecht, K. G., Observations of time variation in the sun's rotation, *Science*, *260*, 1778-1781, 1993.

A.G. Kosovichev, HEPL A204, Stanford University, Stanford, CA 94305-4085; ph. 650-723-7667; fax 650-725-2333; e-mail: sasha@khors.stanford.edu

Twisted Flux Tubes and How They Get That Way

Dana Longcope,[1] Mark Linton,[2] Alexei Pevtsov,[1] George Fisher,[2] and Isaac Klapper[3]

According to present theories, the Sun's magnetic field rises through the convection zone in the form of slender strands known as flux tubes, traditionally studied using "thin flux tube" models. While these models have been remarkably successful they have only recently begun to account for tubes with twisted magnetic flux, in spite of observational evidence for such twist. In this work we review the recent developments pertaining to twisted magnetic flux tubes and compare quantitative predictions to observations. Hydrodynamic theory predicts a role for twist in preventing fragmentation. Excessive twist can also lead to magnetohydrodynamic instability affecting the dynamics of the tube's axis. A thin tube model for a twisted tube suggests several possibilities for the origin of twist. The most successful of these is the Σ-effect whereby twist arises from deformation of the tube's axis by turbulence. Simulations show that the Σ-effect agrees with observations in magnitude as well as latitudinal dependence.

1. FLUX TUBES

Magnetic field appears at the solar surface in the form of isolated domains comprising active regions (ARs). These have been understood as the manifestation of slender, pressure-confined strands of magnetic field called *flux tubes* [Parker, 1955a]. It is believed that flux tubes rise buoyantly as arched Ω-loops, originating at the base of the convection zone (CZ) where they are generated by dynamo action. Tubes are thought to occur either because the dynamo generates magnetic field already in tube form [DeLuca et al., 1993], or in a smooth layer from which tubes break free [Cattaneo et al., 1990]. Equations describing the dynamical evolution of a buoyant flux tube were proposed by Spruit [1981] and have been employed in modified form by subsequent investigators [Moreno-Insertis, 1983; Choudhuri and Gilman, 1987; Chou and Fisher, 1989; D'Silva and Choudhuri, 1993; Fan et al., 1994; Caligari et al., 1995; Fan and Fisher, 1996]. Numerical solutions of such model equations have shown good quantitative agreement with sunspot data. This agreement includes the measured "tilt angle" ψ of a sunspot pair as it depends on solar latitude [D'Silva and Choudhuri, 1993], on magnetic flux [Fan et al., 1994; Fisher et al., 1995], and as its statistical dispersion depends on magnetic flux [Longcope and Fisher, 1996]. These models fit the data with very few free parameters, and offer our best estimates of field strengths at the base of the CZ.

The model flux tube is described by its axis, a space-curve $\mathbf{x}(\ell)$, parameterized by arclength ℓ. The tube's cross section is assumed to be a circle of radius $a(\ell)$ much smaller than all other scales — it is a *thin tube*. The plasma outside the tube is field-free and confines the tube by pressure ($\beta \gg 1$). Properties of the tube, such as the strength of its magnetic field $B_\parallel(\ell)$, are found from averages over the cross section: $B_\parallel(\ell) \equiv \Phi/\pi a^2(\ell)$, where Φ is the tube's total magnetic flux. The model equations describe the evolution of the axis

[1] Department of Physics, Montana State University
[2] Space Sciences Laboratory, UC Berkeley
[3] Department of Mathematics, Montana State University

due to magnetic tension, buoyancy and aerodynamic drag [Spruit, 1981; Choudhuri and Gilman, 1987]. These are most often derived by truncating expansions of the fields about the axis [Ferriz-Mas and Schüssler, 1990; Zhugzhda, 1996], however, the same equations can be found from integration of MHD forces over a section of differential length [Longcope and Klapper, 1997]. Conventional derivations have assumed that the magnetic field within the tube was everywhere parallel to its axis; the tube is *untwisted*. Currents occur at the tube's boundary (a surface current) and across the axis at a bend (giving rise to a curvature force); there is no axial current $J_\parallel$.

2. THE CASE FOR TWIST: OBSERVATIONS

Flux tubes are only truly observed where they cross the photospheric plane ($z = 0$) to form bipolar ARs. The total flux Φ is measured by integrating the vertical magnetic field, B_z, of one sign in a magnetogram (typically $\Phi = 10^{21}$ to 10^{22} Mx for an AR). The thin tube approximation is violated at the photosphere, along with several other simplifying assumptions. Nevertheless, it is frequently assumed that the tube's grossest characteristics, such as total flux and axis orientation, do not change over the top several Mm of the CZ and are therefore adequately measured at the photosphere.

Vector magnetograms often show vertical current density J_z coincident with vertical field B_z. Leka *et al.* (1996) used a sequence of vector magnetograms of emerging active region 7260 to show that the total vertical current I_z increased in proportion to the total flux, in several different bipoles (each $\sim 10^{20}$ Mx). This led them to the remarkable conclusion that each flux tube was carrying axial current prior to its emergence; the flux tubes were *twisted*. In the context of the Spruit model this would correspond to an azimuthal field component B_ϕ in addition to $B_\parallel$. Since the tube is still isolated both components vanish outside the tube and there is *per force* a cancelling axial return current flowing at the tube's surface, so that the tube carries no net current. The structure at the photosphere of such a tube, including its axial return current, is beyond any thin-tube model (there seems to be little evidence for these return currents in magnetograms [Leka et al., 1996]). Nevertheless, the observations of Leka *et al.* suggest that, like the total flux, the internal axial current is at most only slightly affected by this upper CZ boundary region.

Pevtsov, Canfield and Metcalf [1994, 1995] pioneered a method of quantifying the twist of an entire AR as a single value: α_{pcm}. Their technique is motivated by the constant α in a force-free field $\nabla \times \mathbf{B} = \alpha \mathbf{B}$, but does not actually assume the field to be force free (it is well known *not* to be force-free at the photosphere [Metcalf et al., 1995]). Measured horizontal magnetic field can be used to calculate the vertical current density $J_z(x,y) = \partial B_x/\partial y - \partial B_y/\partial x$. The ratio of current to field, J_z/B_z, can then be computed at each pixel of the magnetogram (though only where both quantities are accurately measured). Averaging this over the AR gives one global estimate of twist, α_{avg}. Alternatively, B_z can be used to compute a "force-free" version of B_x and B_y assuming a particular α. α is then varied until these vectors most closely approximate the measured values (in a least-squares sense). The minimizing value, which they call α_{best}, is a single global measurement of magnetic field twist, which we will refer to as α_{pcm}. Repeating the procedure on different magnetograms of the same AR provides one estimate of the intrinsic error in the measured value.

An extension of the original dataset to 203 ARs is shown in Figure 1 plotted against solar latitude [Longcope et al., 1998]. The typical magnitude is $\alpha_{\text{pcm}} \sim 2 \times 10^{-8}$ m^{-1}, comparable to calculations of α_{avg} [Pevtsov et al., 1995; Leka et al., 1996]. Substructures within an active region can have values of J_z/B_z an order of magnitude larger [Pevtsov et al., 1994; Leka et al., 1996], however, α_{pcm} reflects the AR as a whole, and thus is the most likely to reflect the twist of the active region flux tube at depth. There is a subtle, but statistically significant, trend for $\alpha_{\text{pcm}} < 0$ in the Northern hemisphere. The trend is statistically significant in the sense that the null hypothesis, that α_{pcm} is governed by identical distributions in the two hemispheres, can be ruled out definitively. A similar equally subtle trend has been found in a large set of magnetograms analyzed in a different manner [Bao and Zhang, 1998].

3. EFFECTS OF TWIST: THEORY

Consider a straight cylindrical flux tube with constant cross sectional radius a. If every field line in the tube has the same helical pitch $q = d\phi/d\ell$ then its field is given by

$$\mathbf{B}(r) = B_\parallel(r)[\hat{\mathbf{l}} + qr\hat{\phi}] , \qquad (1)$$

where $\hat{\mathbf{l}}$ and $\hat{\phi}$ are axial and azimuthal unit vectors respectively. Field lines wrap once around the axis over an axial distance $2\pi/q$. In principle the axial field profile $B_\parallel$ is arbitrary out to $r = a$, and vanishes beyond that.

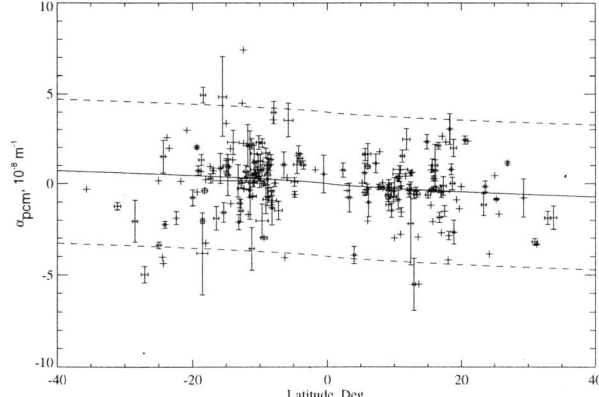

Figure 1. Values of $\alpha_{\rm pcm}$ measured in 203 ARs plotted against solar latitude. Error bars reflect variation in multiple measurements of the same AR. The solid line shows the mean value of $\alpha_{\rm pcm}$ generated by the theoretical Σ-effect in a flux tube of $\Phi = 10^{22}$ Mx. The intrinsic scatter in the Σ-effect is shown by the dashed lines.

A simple flat profile, $B_\| = \Phi/\pi a^2$, has the property that $J_\| = 2q\,B_\|$. Motivated by this we will henceforth make the association $q = \frac{1}{2}\alpha_{\rm pcm}$ and note that typical observed values correspond to $q = 0.01$ rad/Mm.

3.1. Integrity

By assuming $\hat{\mathbf{l}}$ symmetry in this straightened geometry the nonlinear MHD dynamics of the cross section can be studied in two-dimensions. In the absence of twist ($q = 0$) the magnetic field behaves as a gaseous phase with partial pressure $B_\|^2/8\pi$ and no mass. This is analogous to a *thermal* [Turner, 1973] and lacks any means of maintaining integrity [Parker, 1979; Tsinganos, 1980]. Numerical simulations have confirmed that a two-dimensional untwisted tube will spontaneously fragment under its own buoyancy-induced motion [Schüssler, 1979; Longcope et al., 1996].

Twist can prevent this fragmentation if the azimuthal magnetic tension is sufficient to overcome forces from the buoyant rise. A tube will rise at a terminal velocity $v_{\rm r}$ at which buoyancy is balanced by aerodynamic drag. Assuming a circular cross section, buoyancy of primarily magnetic origin and $qa \ll 1$ gives

$$v_{\rm r} \simeq \sqrt{\frac{ga|\delta\rho|}{\rho}} \simeq v_{\rm A\|}\sqrt{\frac{a}{H_{\rm p}}} \qquad (2)$$

where $H_{\rm p}$ is the local pressure scale height of the external atmosphere [Parker, 1975; Emonet and Moreno-Insertis, 1998]. In order for the azimuthal magnetic field to prevent its fragmentation its Alfvén speed $v_{\rm A\perp} = qa\,v_{\rm A\|}$ must be at least as large as $v_{\rm r}$ [Tsinganos, 1980]. Combining these two expression gives a criterion for flux tube integrity

$$q \gtrsim \frac{1}{\sqrt{aH_{\rm p}}}\;, \qquad (3)$$

[Linton et al., 1996; Emonet and Moreno-Insertis, 1998]. This lower limit is 0.1 rad/Mm for the typical values $a \sim 2$ Mm and $H_{\rm p} = 50$ Mm at the base of the CZ. Nonlinear two-dimensional simulations have shown the efficacy of twist at maintaining tube integrity [Fan et al., 1998b; Krall et al., 1998], and have confirmed that expression (3) is the amount of twist required [Emonet and Moreno-Insertis, 1998].

3.2. Instability

In the magnetic configuration (1) the axis is perfectly straight while the field lines are helical. Such equilibria can be susceptible to an instability, called the *helical kink*, whereby the axis spontaneously develops a helical pitch similar to that of the field lines. In low β contexts, such as fusion plasmas or the solar corona, the threshold for instability is given by a Kruskal-Shafranov criterion [Shafranov, 1957; Hood and Priest, 1981] $qL \gtrsim 2\pi$ where L is the axial length of the system. Infinitely long tubes are *always* unstable. In the high β pressure-confined case, however, it has been shown that instability requires

$$q \gtrsim \frac{1}{a}\;, \qquad (4)$$

even for an infinitely long tube [Linton et al., 1996]. For an active region flux tube to be stable at the base of the CZ therefore requires $q \lesssim 0.5$ rad/Mm.

A rising flux tube which undergoes a helical kink instability will develop a bend or concentrated kink in its axis, as various unstable helical modes interact. Figure 2 shows an example of such a deformation arising from an initial perturbation with a broad spectrum [Linton et al., 1998a]. Nonlinear three dimensional MHD simulations have confirmed this both for a straight tube [Linton et al., 1998b] and a rising Ω-loop [Fan et al., 1998a]. While the value of q will change little during the rise (see later discussion) the tube's radius a will increase dramatically. It is therefore possible for a tube to be initially stable, and become unstable during its rise [Linton et al., 1996].

Proper motion of emerging flux can reveal the shape of the tube's axis [Tanaka, 1991; Leka et al., 1996]. Motions of an island-δ spot analyzed by Tanaka [1991] provided evidence of a tightly knotted axis. Several of

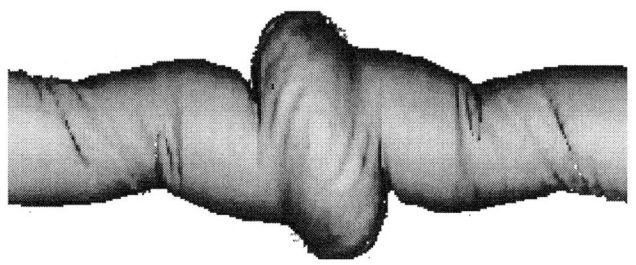

Figure 2. The saturated state of a helical kink instability. From nonlinear three-dimesnional simulations of Linton *et al.* [1998b]

the bipolar features within AR 7260, analyzed by Leka *et al.* [1996], also exhibited a twist-like deformation. If this resulted from the helical kink mode then the axis would be deformed *in the same sense* that the field lines are twisted (in contrast to the discussion below). This was shown to be the case for each of the small bipoles within the AR. Indeed, the values of α_{avg} for each bipole was sufficient to give $qa > 1$, thus exceeding the threshold for instability. While each of the features within the AR exceed the instability threshold, the AR as a whole does not ($qa = \ll 1$) nor does the entire AR exhibit proper motion characteristic of a kinked axis.

4. A MODEL FOR THIN TWISTED FLUX TUBES

Understanding the origin and evolution of twist in a flux tube requires a set of model equations applicable to general axis geometries. Several efforts have been made in this direction [Ferriz-Mas et al., 1989; Chui and Moffatt, 1995] of which a version by Longcope and Klapper [1997] most clearly extends the untwisted model of Spruit [1981]. As in Spruit's model the axis is a general space-curve $\mathbf{x}(\ell)$ whose local tangent vector is $\hat{\mathbf{l}}(\ell) \equiv \partial \mathbf{x}/\partial \ell$. There are two additional variables: $q(\ell)$ the field line twist, and $\omega(\ell)$ the angular velocity of internal material about $\hat{\mathbf{l}}$. An equation for the evolution of $\mathbf{x}(\ell)$ is found, exactly as in Spruit's model, by integrating the MHD forces acting on a differential tube segment [Longcope and Klapper, 1997]. This only changes the original Spruit equations where $q \gtrsim 1/a$. In particular, the evolution of the flux tube axis is unaffected in the *weakly twisted* limit, $qa \ll 1$. In light of the observed value $q \sim 0.01$ rad/Mm we will henceforth consider only this limit.

The moment of the MHD forces about the axis gives an equation for the evolution of spin

$$\frac{d\omega}{dt} = v_{A\|}^2 \frac{\partial q}{\partial \ell} - \frac{2}{a}\frac{da}{dt}\omega \ . \quad (5)$$

The second term on the right hand side (rhs) leads to "spin up" in a contracting tube (due to decreasing moment of inertia). Note also that to be in equilibrium a tube must be uniformly twisted: $q(\ell) = $ const. Finally, it is worth remarking that we have neglected possible torques exerted by the external fluid on the tube. If present such an effect might allow the fluid to "roll" the tube like a piece of string between two fingers [Rust, 1994]. Including such torques requires a model for the viscous coupling of the background plasma to the tube's outer boundary; no such model presently exists.

While spin and twist are easily defined about a straight axis, care must be taken in their definition on a general curve $\mathbf{x}(\ell, t)$ undergoing its own time evolution. Doing so results in a kinematic relation between the two quantities [Klapper and Tabor, 1994; Longcope and Klapper, 1997]

$$\frac{dq}{dt} = \frac{\partial \omega}{\partial \ell} - \left(\hat{\mathbf{l}} \cdot \frac{\partial \mathbf{u}}{\partial \ell}\right) q + \hat{\mathbf{l}} \cdot \left(\frac{\partial \hat{\mathbf{l}}}{\partial \ell} \times \frac{d\hat{\mathbf{l}}}{dt}\right) \ . \quad (6)$$

The second term on the rhs results in a decreased pitch q when the tube is stretched (the factor in parentheses is the rate of differential stretching). The first term shows the effect on twist if two portions of a tube are spinning differently; this effect is often invoked when "footpoints" are spun to impart twist. Retaining only the first rhs terms in Equations (5) and (6) leads to a wave equation for torsional Alfvén waves [Priest, 1982].

The final term on the rhs of Equation (6), sometimes called the Σ-term, is the most surprising: it represents a source of twist due to the evolution of the axis alone. As a result of this term it is possible to impart twist to a section of a tube by deforming its axis in a helical manner. The sense of twist imparted turns out to be *opposite* to the sense of the axial deformation. Figure 3 shows a tube section whose axis has been given a left-handed deformation (as if by a right-handed cyclonic event [Parker, 1970]), thus causing its field lines to develop right handed twist about the axis ($q > 0$).

The necessity of the Σ-term can be appreciated in the conservation of magnetic helicity. The helicity of a closed, thin tube can be written as a sum of two terms, $H = \Phi^2(Tw + Wr)$, called *twist* and *writhe* [Berger and Field, 1984; Moffatt and Ricca, 1992]. The individual terms are defined by the integrals

$$Tw = \frac{1}{2\pi}\oint q(\ell)\,d\ell \ , \quad (7)$$

$$Wr = \frac{1}{4\pi}\oint d\ell \oint d\ell' \frac{\hat{\mathbf{l}} \times \hat{\mathbf{l}}' \cdot (\mathbf{x} - \mathbf{x}')}{|\mathbf{x} - \mathbf{x}'|^3} \ , \quad (8)$$

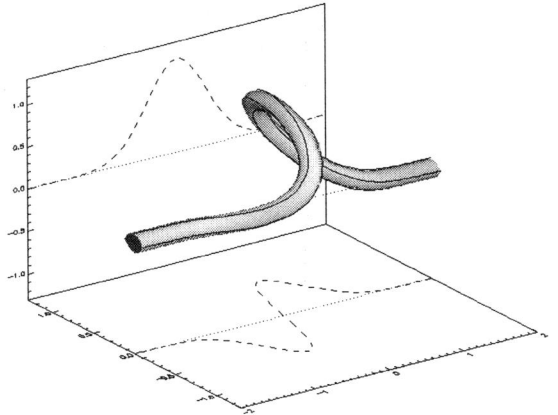

Figure 3. A flux tube (grey) on which several field lines are drawn. The axis of the tube has been deformed in a left-handed sense. The field lines twist about the axis to the right. The field lines at the ends of the tube have not moved during this deformation (from Longcope, Fisher & Pevtsov 1998).

where $\mathbf{x} \equiv \mathbf{x}(\ell)$, $\mathbf{x}' \equiv \mathbf{x}(\ell')$ and so on. Magnetic helicity H is strictly conserved under ideal motions and we are considering only ideal motion. Therefore, a change in writhe must be offset by an opposite change in twist. Writhe depends only on the axis while twist depends only on q, so that a deformation of the axis must be accompanied by a change in q. A left-handed helical deformation of the axis will add negatively to Wr, which will therefore cause a positive change in q, as seen in Figure 3.

5. ON THE POSSIBLE ORIGINS OF TWIST

To find possible sources of twist in a rising tube, we may seek instead sources of writhe. For the purpose of quantifying these we will begin by assuming flux tubes to be generated in an *untwisted* and *unwrithed* state ($Tw_0 = Wr_0 = 0$). Subsequent writhing of the axis will then result in an equal and opposite twist. Twist density q is introduced by the Σ-term in Equation (6) whence it is distributed along the axis by the propagation of torsional Alfvén waves.

5.1. Joy's law

One source of writhe is the tilting of the apex of a rising Ω-loop by the Coriolis force – Joy's Law. As it rises and expands a fluid parcel appears to rotate in a retrograde sense when viewed from the frame of the Sun. This causes the emerging sunspot pair to be tilted by an angle ψ, with the leading spot closer to the equator. Consider a flux tube which initially encircles the Sun in the Northern hemisphere (Figure 4). The rising and tilting gives the tube's axis a right handed pitch ($Wr > 0$) and thus a negative twist q in agreement with the tendency in Figure 1.

To calculate the writhe we begin with two limits in which Wr can be found readily. Without tilt ($\psi = 0$) the axis lies in a plane of constant latitude, even after rise, and therefore has $Wr = 0$. Alternatively, a tilt of $\psi = 180°$ deforms the axis into a figure-eight for which $Wr = +1$ [Berger and Field, 1984; Moffatt and Ricca, 1992]. We propose that intermediate values of tilt produce values of writhe $Wr \simeq \psi/180°$.

During the period of rise, torsional Alfvén waves will equalize $q(\ell)$ over some axial length L, giving an estimate of

$$q \simeq -\frac{2\pi Wr}{L} \simeq -\frac{2\pi \psi}{180° L} , \quad (9)$$

at the apex. Since the rise speed is always less than $v_{A\parallel}$, by Equation (2), we can deduce that $L > 200$ Mm, the depth of the CZ. For a typical active region tilt angle of $\psi = 6°$ [Howard, 1996] this gives an upper bound $|q| \lesssim 10^{-3}$ rad/Mm on the amount of twist possible from Joy's law tilt. This is an order of magnitude smaller than observed, and cannot be the primary source of twist. (It is, however, comparable to the magnitude of the mean trend.) It should be further noted that this mechanism would lead to a correlation between observed values of q (i.e. α_{pcm}) and tilt ψ. This correlation has not been observed [Pevtsov and Canfield, 1999].

5.2. Differential rotation

A second source of writhe is the well known differential rotation of the Sun. This process acts over extremely long times during which it is unlikley that a

Figure 4. A flux tube tilted by the Coriolis force acting on the rising apex.

single flux tube could survive. Even if the tube survived indefinitely, however, it can be shown that differential rotation would lead to $q \simeq 1/R_{CZ} \simeq 2 \times 10^{-3}$ rad/Mm, where $R_{CZ} \simeq 500$ Mm is the radius of the CZ base. This is well below observed values, and represents an upper bound on the twist possible from differential rotation.

5.3. The Σ-effect

The presence of two spatial derivatives (i.e. $\hat{\mathbf{l}} = \partial \mathbf{x}/\partial \ell$) in the definition of Wr (8) suggests that axis deformations on smaller scales may lead to more writhe than the large-scale mechanisms considered above. During its rise a tube will be buffeted by turbulent motions in the CZ. These motions are on the scale of the so-called mixing length, which decreases to several Mm towards the top of the CZ [Böhm-Vitense, 1958; Spruit, 1974]. Moreover, the turbulent velocity is thought to have a helical nature characterized by its kinetic helicity $\langle \mathbf{u} \cdot \nabla \times \mathbf{u} \rangle$. For the bulk of the CZ this quantity is believed to be negative in the Northern hemisphere [Steenbeck and Krause, 1966]. Fluid motions with negative kinetic helicity generate right-handed deformations in a flux tube [Parker, 1955b]. These right-handed deformations will give rise to a positive writhe, and thus negative twist in the North. This sign is consistent with the trend in the data (Figure 1) and the random character of the turbulence might explain its large statistical dispersion. This turbulent mechanism of twist generation is known as the Σ-effect [Longcope et al., 1998].

The Σ-effect generates twist of the *same* sign as the kinetic helicity. This differs from the sign relationship in the traditional alpha-effect of mean-field dynamo theory: $\alpha \sim -\langle \mathbf{u} \cdot \nabla \times \mathbf{u} \rangle$. This discrepency arises because the α-effect concerns motions *internal* to a magnetic field, while the Σ-effect concerns motions *external* to the field (i.e. outside the flux tube). Consequenlty, when using motions in the bulk of the CZ, the Σ-effect agrees with the observed sense of twist while the α-effect does not.

The magnitude of the Σ-effect has been estimated by numerical simulation [Longcope et al., 1998]. An initially untwisted, unwrithed flux tube was subjected to a random external velocity field representing the CZ turbulence. This velocity field had the amplitude, correlation length and kinetic helicity characteristic of mixing-length turbulence. The turbulence characteristics changed as the tube rose through the CZ. Equations for q and ω were solved until the end of the rise, yielding a single realization of q. To determine the statistical distribution of q, the procedure is then repeated 1000 times, each time with a different realization of the turbulent velocity. The distribution is then computed for tubes of different fluxes Φ at different solar latitudes. Figure 1 shows the mean (solid) and standard deviation (dashed) of the distribution for $\Phi = 10^{22}$ Mx. The typical values correspond to $q \sim 0.02$ rad/Mm, consistent with the data. Moreover, the latitudinal variations and level of statistical dispersion agrees remarkably well with observation.

5.4. Initial twist

Finally, we turn to the possibility of twist in flux tubes *prior to their rise*. Present thinking holds that a dynamo operating at the base of the CZ generates a smooth layer of magnetic field which then fragments to form flux tubes [Cattaneo et al., 1990; Matthews et al., 1995]. A mean-field α-Ω dynamo generates flux with some magnetic helicity [Charbonneau and Gilman, 1998; Gilman and Charbonneau, 1999]. Present understanding of the fragmentation process, however, does not predict how this helicity might be converted to twist q_0 in the flux tubes which break free from the layer.

For the sake of a rough estimate we will consider the "twist" implied by the helicity in the dynamo layer itself, ignoring the fragmentation process entirely. An α-Ω dynamo produces magnetic fields and currents which are primarily torroidal. Magnitudes of these fields cannot be found from the standard linear treatments, however, their ratio can [Charbonneau and Gilman, 1998]. In particular an estimate of q_0 can be found,

$$q_0 \sim \frac{1}{2} \frac{\max{(\nabla \times \mathbf{B})_t}}{\max B_t} , \qquad (10)$$

where maxima over space are found for the torroidal components of each field. This quantity is quite sensitive to model parameters such as α and the dynamo number. Parameter variations give values of q_0 ranging from 3×10^{-6} rad/Mm to 3×10^{-2} rad/Mm, with a reasonable choice yielding $q_0 \simeq 4 \times 10^{-5}$ rad/Mm [Charbonneau and Gilman, 1998].

It is also possible that the poorly understood fragmentation process, which creates flux tube from the smooth dyanmo layer, introduces twist itself. It should be born in mind, however, that all of the mechansims mentioned in previous sections would add twist on top of q_0, present initially. The significant amount of twist contributed by the Σ-effect leaves little room in the observations for q_0.

Table 1. A quantitative summary of twist q in flux tubes

	q (rad/Mm)	source	Ref.
Observations			
AR	~ 0.01	Fig. 1	a
substructures	~ 0.2	—	b, c
Constraints on twist			
Integrity of tube	$q \gtrsim 0.1$	Eq. (3)	d, e
Kink stable		Eq. (4)	f
overshoot	$q \lesssim 0.5$		
always	$q \lesssim 0.1$		
Sources of twist			
Joy's Law	0.001	Eq. (9)	g
Diff'l rotation	0.002	—	—
Σ-effect	~ 0.02	Fig. 1	h
Created in dynamo	4×10^{-5}	Eq. (10)	i

[a] Pevtsov, Canfield and Metcalf 1995
[b] Pevtsov, Canfield and Metcalf 1994
[c] Leka et al. 1996
[d] Emonet and Moreno-Insertis 1998
[e] Fan, Zweibel and Lantz 1998
[f] Linton, Longcope and Fisher 1996
[g] Longcope and Klapper 1997
[h] Longcope, Fisher and Pevtsov 1998
[i] Charbonneau and Gilman 1998

6. SUMMARY AND DISCUSSION

Observations indicate that solar magnetic flux tubes consist of twisted magnetic field. The magnitude and sign of the twist are measured by the quantity $\alpha_{\rm pcm}$, which shows a latitudinal variation but with significant scatter. Table 1 lists the magnitude of observed twist along with theoretical predictions and constraints. Thin flux tube models show that large-scale sources, such as differential rotation or Joy's law, cannot explain the magnitude or latitudinal dependance of the observed twist. Turbulent buffeting of the rising tubes generates twist, through the Σ-effect, which does fit the observations, in magnitude, latitudinal dependance and degree of statistical dispersion.

A different category of comparison comes from application of two-dimensional MHD models to flux tubes. These predict minimum values of twist necessary for integrity or instability. Obsevations suggest that AR flux tubes remain in one piece during most of their rise from the base of the CZ. Almost none of the ARs observed, however, satisy the criterion for integrity (Equation [3]). This points to an inadequacy of the models or their interpretation. It has been suggested [Longcope et al., 1996; Emonet and Moreno-Insertis, 1998] that the requirement for tube integrity in a fully three dimensional MHD model would be different than for the perfectly straight two-dimensional models used to derive Equation (3). Numerical simulations presently underway will help to clarify this point.

The threshold for helical kinking also stands well below the observed twist of ARs. While proper motion studies provide evidence of kinking, only a portion of any AR appears kinked. Until models are developed for complexes of interlinked flux tubes, however, it is not easy to appreciate the significance of this. There is at least one example in which the axis of an entire AR flux tube was observed to be kinked [Pevtsov and Longcope, 1998], however, it is unclear if this is related to a helical kink instability.

Considerable uncertainty remains in the method of measuring the twist in an AR flux tube. The quantity $\alpha_{\rm pcm}$, developed by Pevtsov, Canfield and Metcalf [1995], has certain advantages over e.g. $\alpha_{\rm avg}$. Primary among these is the extensive dataset of ARs for which $\alpha_{\rm pcm}$ has been (laboriously) calculated. The technique itself should be studied in more detail to determine any biases it may introduce. More importantly, however, it must be determined how well any photospheric measurement can ascertain the twist which was present in the rising tube. In short, we must learn much more about the interface between CZ and the corona: the photosphere. Principally, does the photospheric twist reflect that in the rising tube, and does the photosphere influence the dynamics of the tube?

Acknowledgments. We wish to thank Drs. Charbonneau and Gilman for performing the calculations in Section 5.4, and for contributions to the discussion therein. This work was supported under NASA grant NAG5-5043

REFERENCES

Bao, S. and Zhang, H., Patterns of current helicity for solar cycle 22, *Astrophys. J.*, *496*, L43–L46, 1998.
Berger, M. A. and Field, G. B., The topological properties of magnetic helicity, *J. Fluid Mech.*, *147*, 133–148, 1984.
Böhm-Vitense, E., *Z. Astrophys.*, *46*, 108, 1958.
Caligari, P., Moreno-Insertis, F., and Schüssler, M., Emerging flux tubes in the solar convection zone. I. Asymmetry, tilt and emergence latitude, *Astrophys. J.*, *441*, 886–902, 1995.
Cattaneo, F., Chiueh, T., and Hughes, D., Buoyancy-driven instabilities and the nonlinear breakup of a sheared magnetic layer, *J. Fluid Mech.*, *219*, 1–23, 1990.
Charbonneau, P. and Gilman, P. 1998, Private communications.
Chou, D.-Y. and Fisher, G. H., Dynamics of anchored flux

tubes in the convection zone. I. Details of the model, *Astron. Astrophys.*, *341*, 533–548, 1989.

Choudhuri, A. R. and Gilman, P. A., The influence of the coriolis force on flux tubes rising through the solar convection zone, *Astrophys. J.*, *316*, 788–800, 1987.

Chui, A. Y. K. and Moffatt, H. K., The energy and helicity of knotted magnetic flux tubes, *Proc. Roy. Soc. Lond.*, *451*, 609–629, 1995.

DeLuca, E. E., Fisher, G. H., and Patten, B. M., The dynamics of magnetic flux rings, *Astrophys. J.*, *411*, 383–393, 1993.

D'Silva, S. and Choudhuri, A. R., A theoretical model for tilts of bipolar magnetic regions, *Astron. Astrophys.*, *272*, 621–633, 1993.

Emonet, T. and Moreno-Insertis, F., The physics of twisted magnetic flux tubes rising in a stratified medium: Two dimensional results, *Astrophys. J.*, *492*, 804, 1998.

Fan, Y., Fisher, G. H., and McClymont, A. N., Dynamics of emerging active region flux loops, *Astrophys. J.*, *436*, 907–928, 1994.

Fan, Y., Zweibel, E. G., Linton, M. G., and Fisher, G. H., The rise of kink unstable magnetic flux tubes in the solar convection zone, *Astrophys. J.*, *505*, L59–L64, 1998a.

Fan, Y.-H. and Fisher, G. H., Radiative heating and the buoyant rise of magnetic flux tubes in the solar interior, *Sol. Phys.*, *166*, 17–41, 1996.

Fan, Y.-H., Zweibel, E. G., and Lantz, S. R., Two-dimensional simulations of buoyantly rising, interacting magnetic flux tubes, *Astrophys. J.*, *493*, 480–493, 1998b.

Ferriz-Mas, A. and Schüssler, M., On the thin magnetic flux tube approximation, in *Physics of Magnetic Flux Ropes*, edited by Russel, C. T., Priest, E. R., and Lee, L. C., volume 58 of *Geophys. Monographs*, pp. 141–148, AGU 1990.

Ferriz-Mas, A., Schüssler, M., and Anton, V., Dynamics of magneitc flux concentrations: the second-order thin flux tube approximation, *Astron. Astrophys.*, *210*, 425–432, 1989.

Fisher, G. H., Fan, Y., and Howard, R. F., Comparisons between theory and observations of active region tilts, *Astrophys. J.*, *438*, 463–471, 1995.

Gilman, P. and Charbonneau, P., Creation of twist at the core-convection-zone interface, in *Magnetic Helicity in Space and Laboratory Plasmas*, edited by Brown, M. R., Canfield, R. C., and Pevtsov, R. C., Geophys. Monogr. Series, AGU, Washington, D.C. 1999, This volume.

Hood, A. W. and Priest, E. R., *Geophys. Astrophys. Fluid Dynamics*, *17*, 297, 1981.

Howard, R. F., Axial tilt angles of active regions, *Sol. Phys.*, *169*, 293–301, 1996.

Klapper, I. and Tabor, M., A new twist in the kinematics and elastic dynamics of thin filaments and ribbons, *J. Phys. A*, *27*(14), 4919–4924, 1994.

Krall, J., Chen, J., Santoro, R., Spicer, D. S., Zalesak, S. T., and Cargill, P. J., Simulation of buoyant flux ropes in a magnetized solar atmosphere, *Astrophys. J.*, *500*, 992–1002, 1998.

Leka, K. D., Canfield, R. C., McClymont, A. N., and Van Driel Gesztelyi, L., Evidence for current-carrying emerging flux, *Astrophys. J.*, *462*, 547–560, 1996.

Linton, M. G., Dahlburg, R. B., Fisher, G. H., and Longcope, D. W., Nonlinear evolution of kink unstable magnetic flux tubes and solar delta-spot active regions, *Astrophys. J.*, *507*, 404–416, 1998a.

Linton, M. G., Fisher, G. H., Dahlburg, R. B., Fan, Y.-H., and Longcope, D. W., Multi-mode kink instability as a mecahnism for delta-spot fomation, in *Proceedings of the 1998 COSPAR meeting* 1998b, in press.

Linton, M. G., Longcope, D. W., and Fisher, G. H., The kink instability of isolated, thin twisted flux tubes, *Astrophys. J.*, *469*, 954–963, 1996.

Longcope, D. W. and Fisher, G. H., The effect of convection zone turbulence on a rising flux tube, *Astrophys. J.*, *458*, 380–390, 1996.

Longcope, D. W., Fisher, G. H., and Arendt, S., The evolution and fragmentation of rising magnetic flux tubes, *Astrophys. J.*, *464*, 999–1011, 1996.

Longcope, D. W., Fisher, G. H., and Pevtsov, A. A., Flux tube twist resulting from helical turbulence: The Sigma-effect, *Astrophys. J.*, *507*, 417–432, 1998.

Longcope, D. W. and Klapper, I., Dynamics of thin twisted flux tubes, *Astrophys. J.*, *488*, 443–453, 1997.

Matthews, P., Hughes, D., and Proctor, M., Magnetic buoyancy, vorticity, and three-dimensional flux-tube formation, *Astrophys. J.*, *448*, 938–941, 1995.

Metcalf, T. R., Jiao, L., McClymont, A. N., Canfield, R. C., and Uitenbroek, H., Is the solar chromospheric magentic field force-free?, *Astrophys. J.*, *439*, 474–481, 1995.

Moffatt, H. K. and Ricca, R. L., Helicity and the Calugareanu invariant, *Proc. Roy Soc. Lond. A*, *439*, 411–429, 1992.

Moreno-Insertis, F., Rise times of horizontal magnetic flux tubes in the convection zone of the sun, *Astron. Astrophys.*, *122*, 241–250, 1983.

Parker, E. N., The formation of sunspots from the solar toroidal field, *Astrophys. J.*, *121*, 491–507, 1955a.

Parker, E. N., Hydromagnetic dynamo models, *Astrophys. J.*, *122*, 293–314, 1955b.

Parker, E. N., The generation of magnetic fields in astrophysical bodies: I. The dynamo equations, *Astrophys. J.*, *162*, 665–673, 1970.

Parker, E. N., The generation of magnetic fields in astrophysical bodies. X. Magnetic buoyancy and the solar dynamo, *Astrophys. J.*, *198*, 205–209, 1975.

Parker, E. N., *Cosmical Magnetic Fields, Their Origin and Their Activity*, Clarendon Press, Oxford 1979.

Pevtsov, A. A. and Canfield, R. C., Helicity of the photospheric magnetic field, in *Magnetic Helicity in Space and Laboratory Plasmas*, edited by Brown, M. R., Canfield, R. C., and Pevtsov, R. C., Geophys. Monogr. Series, AGU, Washington, D.C. 1999, This volume.

Pevtsov, A. A., Canfield, R. C., and Metcalf, T. R., Patterns of helicity in solar active regions, *Astrophys. J.*, *425*, L117–L119, 1994.

Pevtsov, A. A., Canfield, R. C., and Metcalf, T. R., Latitudinal variation of helicity of photospheric magnetic fields, *Astrophys. J.*, *440*, L109–L112, 1995.

Pevtsov, A. A. and Longcope, D. W., NOAA 7926: A kinked

Omega-loop?, *Astrophys. J.*, *508*, 908–915, 1998.

Priest, E. R., *Solar Magnetohydrodynamics*, volume 21, D. Reidel, Boston 1982.

Rust, D. M., Spawning and shedding helical magnetic fields in the solar atmosphere, *Geophys. Res. Lett.*, *21*, 241–244, 1994.

Schüssler, M., Magnetic buoyancy revisited: Analytical and numerical results for rising flux tubes, *Astron. Astrophys.*, *71*, 79–91, 1979.

Shafranov, V. D., *J. Nucl. Eng. II*, *5*, 86, 1957.

Spruit, H., A model of the solar convection zone, *Sol. Phys.*, *34*, 277–290, 1974.

Spruit, H. C., Motion of magnetic flux tubes in the solar convection zone and chromosphere, *Astron. Astrophys.*, *98*, 155–160, 1981.

Steenbeck, M. and Krause, F., The generation of stellar and planetary magnetic fields by turbulent dynamo action, *Z. Naturforsch*, *21a*, 1285–1296, 1966.

Tanaka, K., Studies on a very flare-active δ group: Peculiar δ spot evolution and inferred subsurface magnetic rope structure, *Sol. Phys.*, *136*, 133–149, 1991.

Tsinganos, K., Sunspots and the physics of magnetic flux tubes. X. On the hydrodynamic instability of buoyant fields, *Astrophys. J.*, *239*, 746–760, 1980.

Turner, J., *Buoyancy Effects in Fluids*, Cambridge University Press 1973.

Zhugzhda, Y. Y., Force-free thin flux tubes: Basic equations and stability, *Physics of Plasmas*, *3*(1), 10–21, 1996.

D. W. Longcope and A. A. Pevtsov, Dept. of Physics, Montana State University, Bozeman, MT 59717 (e-mail: dana@physics.montana.edu; pevtsov@physics.montana.edu)

G. H. Fisher and M. G. Linton, Space Sciences Laboratory, University of California Berkeley, CA 94720 (e-mail: gfisher@ssl.berkeley.edu; linton@ssl.berkeley.edu)

I. Klapper, Department of Mathematics, Montana State University, Bozeman, MT 59717 (e-mail: klapper@math.montana.edu)

Helicity of the Photospheric Magnetic Field

Alexei A. Pevtsov, Richard C. Canfield

Department of Physics, Montana State University, Bozeman

Observations of vector magnetic fields from various observatories have been used to study the helicity of the solar magnetic field, using different helicity proxies. Low resolution full disk longitudinal magnetograms have also been used to reconstruct the large-scale vector magnetic field and compute $h_c = B_z \cdot (\nabla \times B)_z$. All these studies show a hemispheric helicity rule – the photospheric magnetic field in the northern/southern hemisphere tends to have negative/positive current helicity. It appears, however, that the hemispheric asymmetry rule is weak. Carrington maps of h_c show that areas of both signs of current helicity are present in each hemisphere. In the Mees and Huairou data one sees several areas of abnormal helicity persisting for many rotations. We discuss the possible origin of the hemispheric helicity rule and conclude that neither differential rotation nor the Coriolis force can fully explain the observations for the following reasons. First, the hemispheric helicity rule is weak. Second, active regions exhibit persistent patterns of mixed helicity. Third, a correlation between magnetic field twist and the tilt of active regions is opposite in sign to what one should expect from the Coriolis force.

1. INTRODUCTION

Hale [1927] studied chromospheric H_α vortices around sunspots and found that they form counterclockwise spirals north of the equator and clockwise spirals south of it, independent of the solar cycle. Later *Richardson* [1941] confirmed this result on a larger dataset of more than 140 active regions. Although both *Hale* and *Richardson* explained their results in terms of solar atmospheric circulation, it is believed now that the chromospheric vortices reflect the hemispheric helicity rule for the magnetic field, discovered only recently, and not yet fully understood. The hemispheric helicity rule has been observed in the photospheric magnetic fields [*Seehafer*, 1990; *Pevtsov et al.*, 1995], interplanetary magnetic field [*Smith* 1999], chromospheric filaments [*Martin et al.*, 1994; *Rust*, 1999] and sheared coronal loops [*Rust and Kumar*, 1996; *Canfield and Pevtsov*, 1999]. It has alternatively been ascribed to either photospheric [*Zirker et al.*, 1997; *vanBallegooijen et al.*, 1998] or subphotospheric [*Rust and Kumar*, 1994; *Longcope et al.*, 1998] processes.

2. MEASURES OF HELICITY

It is widely believed that the solar magnetic field is generated by a dynamo operating at the base of the convection zone. The flux generated there is buoyant and rises to the surface as Ω-shaped loops whose ends are anchored in the convectively stable core. Where they intersect the photosphere, the Ω loops form solar active regions with areas of opposite polarity connected through the corona [*Babcock*, 1961]. The thin flux tube model has been used in several studies of Ω loops rising through the convective zone [*Fisher et al.*, 1995] and interacting with turbulent convection [*Longcope et al.*, 1998, 1999].

2.1. Thin Flux Tube

Consider a thin flux tube T defined by axial $B_a = \{0, 0, B_z(r)\}$ and meridional $B_m = \{0, B_\theta(r), 0\}$ components and $\mathbf{B}(x) = \nabla \times \mathbf{A}(x)$ in a domain $\mathcal{D}$ on whose surface $\mathbf{n} \cdot \mathbf{B} = 0$. Under these circumstances the magnetic helicity $\mathcal{H} = \int_\mathcal{D} \mathbf{A} \cdot \mathbf{B} dV = \int_T \mathbf{A}_a \cdot \mathbf{B}_a dV + 2 \int_T \mathbf{A}_m \cdot \mathbf{B}_m dV$, where first integral represents twist $\mathcal{T}$ and second represents writhe $\mathcal{W}$ [*Moffatt and Ricca*, 1992]. In absence of dissipative processes the magnetic helicity is a conserved quantity [*Field*, 1986].

Only the photospheric contribution of $\mathcal{T}$ can be measured using vector magnetograms. The writhe can be studied independently, using the tilt of active regions (see below).

2.2. Quantities Derived from Observations

Existing solar vector magnetographs measure three components of the magnetic field $\mathbf{B} = \{B_x, B_y, B_z\}$ at a single level in the solar atmosphere. Strictly speaking, neither the magnetic vector potential, nor the magnetic helicity can be computed from these data without additional assumptions, since the measurements are local, but these quantities are global.

2.2.1. Force-free field α. Consider a force free magnetic field $\nabla \times \mathbf{B} = \alpha \mathbf{B} = \mu \mathbf{J}$, where B - magnetic induction, μ - magnetic permeability in vacuum, J - electric current density. For a linear force-free field with $\alpha \equiv const$, the magnetic vector potential $\mathbf{A} = \alpha^{-1}\mathbf{B} + \nabla\Psi$, where Ψ is arbitrary scalar function and $\mathcal{H} = \int_\mathcal{D} (\alpha^{-1}\mathbf{B} + \nabla\Psi) \cdot \mathbf{B} dV = 2\mu\alpha^{-1} E_m$. Magnetic energy E_m can be computed using the virial theorem [*Priest*, 1984] $E_m = \mu^{-1} \int_S (xB_x - yB_y)B_z dxdy$. Values of α that best fit an entire active region (α_{best}) have been used by *Seehafer* [1990], *Pevtsov et al.* [1995] and *Hagyard et al.* [1999] to study the hemispheric helicity rule (Section 4), since $\mathcal{H}$ has the same sign as α.

For a non-linear force-free field $\alpha = \alpha(x, y)$ and $\alpha_z(x, y) = J_z/B_z$. Although α_z has no formal relation to $\mathcal{H}$, it may represent a current helicity density under the further assumption that the magnetic field and electric currents are collinear on a smaller scale.

2.2.2. Current helicity. Alternatively, one can compute the current helicity $\tilde{h}_c = \mathbf{B} \cdot (\nabla \times \mathbf{B})$. In general, only $(\nabla \times \mathbf{B})_z$ can be derived from a vector magnetogram; for a force-free field $(\nabla \times \mathbf{B})_x = (\nabla \times \mathbf{B})_y = (\nabla \times \mathbf{B})_z$. *Seehafer* [1990] showed that for a cylindrically symmetric infinite flux tube current helicity has the same sign and increases/decreases as the magnetic helicity does. For a linear force-free field $h_c = \alpha \cdot B^2 = \alpha^2 h_m$, where h_m is the magnetic helicity density. It is important to note, that although the term "current helicity" is widely used, the observed $h_c = \mu J_z B_z$ represents only the vertical component of the real current helicity density $\tilde{h}_c = \mu \mathbf{B} \cdot \mathbf{J} = \mu(B_x J_x + B_y J_y + B_z J_z)$. The disregard of the x and y terms may in principle change the sign of the current helicity density. Despite this limitation h_c averaged over a whole active region has been used to study the hemispheric helicity rule [*Abramenko et al.*, 1997; *Bao and Zhang*, 1998] and the large-scale pattern of helicity [*Pevtsov and Latushko*, 1995], Section 6.

3. OBSERVATIONAL ASPECTS

3.1. Vector Magnetographs

The results of studies that we review in this paper are based on observations from several vector magnetographs including the Haleakala Stokes Polarimeter (HSP) [*Mickey*, 1985], Advanced Stokes Polarimeter (ASP) [*Lites et al.*, 1993], Huairou Solar Station (HSS) vector magnetograph [*Bao and Zhang*, 1998] and Marshall Space Flight Center (MSFC) magnetograph [*Hagyard et al.*, 1985]. The differences between the instruments in observational technique and data reduction can affect helicity observations.

Both the HSP and the ASP derive magnetic field vectors from the full Stokes profiles of the two spectral lines Fe I $\lambda\lambda$ 6301.5 and 6102.5, using a nonlinear least-squares Unno profile-fitting routine [*Skumanich and Lites*, 1987]. It is important to note that the method makes a first-order correction for magneto-optical effects (Faraday rotation) and inadequate spatial resolution.

In contrast, the MSFC and HSS instruments take polarization measurements in a single fixed position of a spectral line profile (MSFC can make successive observations in several points across a spectral line). The value of $\mathbf{B}$ is then determined via comparison of the measured polarizations with theoretical ones, computed using a model of the solar atmosphere [*Semel et al.*, 1991]. The use of limited spectral information makes the MSFC and HSS magnetographs potentially susceptible to several known problems: magnetic saturation, under-resolution, and Faraday rotation. Recently, *Hagyard et al.* [1999] investigated the role of these factors on helicity computation using the MSFC magnetograph data. We are unaware of a similar study involving HSS data.

3.1.1. Magnetic saturation. Magnetic saturation is the result of nonlinear dependence of the field strength on the measured polarization. Thus, beginning at a certain field strength, polarization may not change, or even decrease with increasing field strength. Saturation alters both the total field strength and the (180° ambiguity resolved) direction of the transverse field and hence is important for helicity computation. However, *Hagyard et al.* [1999] concluded that careful choice of a threshold in the transverse field eliminates saturated pixels from α calculations. On the other hand, it is difficult, if not impossible, to correct for saturation in the computation of α_z and h_c.

3.1.2. Under-resolution. Solar photospheric magnetic fields are concentrated in small flux tubes, whose cross-sections are far below the spatial resolution limits of existing magnetographs [*Stenflo*, 1994]. Adequate spectral resolution and sampling allows one to estimate the relative contribution to the radiation from the magnetized and non-magnetized plasma. For a pixel of $\sim$ arc. sec diameter, the filling factor – the relative contribution from magnetized plasma – is 80–100% in sunspot umbrae, 45%–80% in penumbrae, $\approx$ 15% in plages [*Martinez et al.*, 1997] and $\leq$ 1% in the quiet Sun. In the MSFC and HSS magnetographs the information needed to determine the filling factor is lost. Lack of such information leads to underestimation of the field strength and may distort the local helicity pattern [*Stenflo*, 1994]. *Hagyard et al.* [1999] compared α from HSP and MSFC vector magnetograms for three active regions and found that they are in good agreement, perhaps because the value of the filling factor is large in sunspots, which dominate the determination of α.

3.1.3. Faraday rotation. *West and Hagyard* [1983] found that the Faraday rotation can change the azimuth of the transverse field by as much as 45° for measurements taken near the center of a spectral line or in the far wing. In a magnetic field of uniform polarity the azimuths of the transverse field will rotate in the same direction. Thus, Faraday rotation mimics $\nabla \times \mathbf{B}$ and hence may affect significantly the computation of helicity. *Hagyard et al.* [1999] concluded that the effect of Faraday rotation on the α computation is relatively unimportant for the MSFC dataset, since the measurements are usually taken far enough in the wing of a spectral line. On the other hand, they found a strong indication of the Faraday rotation in HSS data, at least in one case of a single active region they have analyzed. The HSS magnetograph measures transverse field in the center of a spectral line [*Bao and Zhang*, 1998], where Faraday rotation is strongest.

Table 1. Sign of α for 203 Active Regions

	Negative	Positive	Total
N hemisphere	58[a](62%)[b]	35 (38%)	93
S hemisphere	37 (34%)	73 (66%)	110
Total	95	108	203

[a] Absolute number of active regions
[b] Percent of active regions in the hemisphere

3.2. Full Disk Longitudinal Magnetograms

In Section 2 we have defined quantities, which one can compute using vector magnetograms. This approach, however, limits helicity studies to the solar active regions, since observations of the vector magnetic field for the full solar disk are unavailable at present time. To overcome this limitation *Pevtsov and Latushko* [1995] and *Latushko and Pevtsov* [1998] employed a reconstruction technique [*Grigoriev and Latushko*, 1992] to compute vector magnetic field from a set of longitudinal B_{long} magnetograms under assumption that the large-scale magnetic field evolves slowly and that all changes of B_{long} are the result of changing projection angle.

4. HELICITY OF ACTIVE REGION MAGNETIC FIELDS

4.1. Hemispheric Helicity Rule

Since the magnetic pressure in the photosphere is comparable with the gas pressure, the magnetic field in the photosphere is not force-free [*Priest*, 1984]. However, the force-free field assumption has been used widely to extrapolate the magnetic field from the photosphere up to the corona [*Semel et al.*, 1991]. Using α – a by-product of these calculations – *Seehafer* [1990] noticed a hemispheric asymmetry in the distribution of sign of α: negative in the northern hemisphere and positive in the southern. *Pevtsov et al.* [1995] studied 69 active regions and established the hemispheric helicity rule on $\sim$ 70%/30% level, although the extended HSP dataset [*Longcope et al.*, 1998] shows the tendency on $\sim$ 60%/40% level (Table 1). Figure 1 shows the latitudinal distribution of α_{best} for all 203 active regions of the *Longcope et al.* [1998] dataset.

Two other groups studied the current helicity density (vertical component only, see Section 2.2.2) of active regions by averaging h_c over whole region. *Abramenko et al.* [1997] examined 40 active regions and confirmed the hemispheric helicity rule on the same level as *Pevtsov et al.* [1995]. *Bao and Zhang* [1998] studied 421 active regions and found a stronger (80%/20%) tendency.

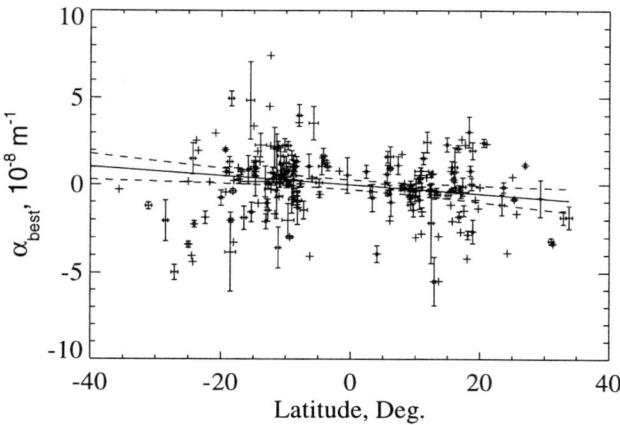

Figure 1. Values of α_{best} for 203 active regions. Error bars reflect the variation in α_{best} from independent measurements of the same AR. A linear fit to the data is shown by the solid line, dashed lines show 2σ errors of the fit. By permission [*Longcope et al.*, 1998]

Although the hemispheric helicity rule was established for solar cycle 22, observations show the same sign of the hemispheric asymmetry in solar cycles 21 [*Hagyard et al.*, 1999] and 23 [*Bao and Zhang*, 1998].

The current helicity of the large-scale magnetic field also reveals the same hemispheric asymmetry. Using MDI full disk magnetograms *Latushko and Pevtsov* [1998] reconstructed the vector magnetic field for one solar rotation (June-July 1996, Carrington rotation number 1910). Averaging radial B_r and toroidal B_λ components of the magnetic vector over all longitudes λ for each latitude φ, they computed the latitudinal profile of the current helicity $h_c = \frac{(\nabla \times B)_r}{B_r} B^2$, where $(\nabla \times B)_r = \frac{1}{r \sin \varphi}(\frac{\partial}{\partial \varphi}(\sin \varphi \cdot B_\lambda) - \frac{\partial B_\varphi}{\partial \lambda})$ and $\frac{\partial B_\varphi}{\partial \lambda} \equiv 0$, as the result of longitudinal averaging. The magnetic field was assumed to be force-free. Figure 2 shows the latitudinal profile of h_c from *Latushko and Pevtsov* [1998]. The large-scale magnetic field has predominantly negative/positive current helicity density in northern/southern hemisphere. It is important to note significant variations of h_c in low latitudes, where the sunspot activity is large, and much smaller h_c variations in high latitudes, where there is no sunspot activity. The current helicity increases toward the poles, a feature that can not be seen in active region data. The large variations of h_c between $\sim \pm 30°$ can be the result of the large-scale helicity areas described in Section 6.

4.2. Tilt-Twist Correlation

The relationship between twist and writhe can be used to discriminate between different possible origins of helicity. Consider helicity conservation in a flux tube with zero helicity. Twisting modifies both twist T and writhe W in the tube. However, the writhe will be opposite in sign to the twist. On the other hand, consider a highly twisted tube with non-zero helicity. It may "trade" its twist for writhe, conserving helicity. In a first case, T and W will be opposite in sign; in the second case they will have same sign. In the photosphere we can study the relationship between the internal twist of the magnetic field represented by α and the writhe represented by the "tilt" θ of a solar active region (Joy's law, [*Zirin*, 1988]), the angle between the N-S axis of the region and solar equator. It is believed that such tilt arises due to the Coriolis force acting on plasma flowing away from the apex of the rising Ω-loop [*Fisher et al.*, 1995]. As the result of helicity conservation, one should expect a strong anticorrelation between twist α and tilt, assuming that the Ω-loop was originally untwisted, and is in equilibrium.

For 99 active regions from the HSP dataset *Canfield and Pevtsov* [1998] computed both α and tilt per unit length $(-\theta/L)$, where L is the separation between sunspots of opposite polarity. To guide the analysis, *Canfield and Pevtsov* used a simple non-linear force-free field – a uniformly twisted cylindrical flux tube [*Priest*, 1984]. It is straightforward to show that for this field (which has no writhe), $\alpha = T/L$. Figure 3 shows a relationship between $-\theta/L$ and α. Clearly, the data show no reliable anticorrelation between twist and writhe, as one would expect if the Coriolis force had produced twist in originally untwisted flux tubes.

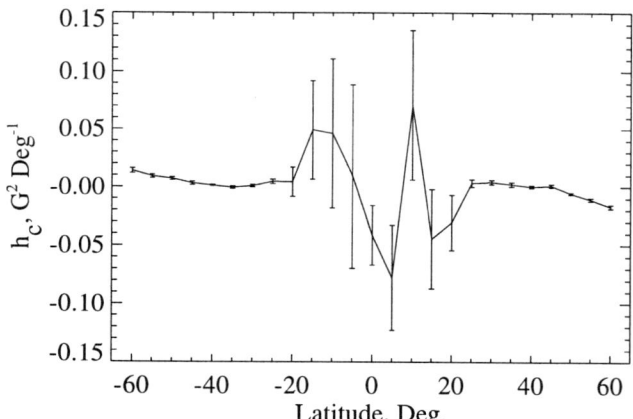

Figure 2. Variation with solar latitude of the current helicity density h_c of large-scale magnetic field computed using 184 arc. sec spatial averaging. Error bars refer to 1σ error.

5. α_z PATTERNS WITHIN ACTIVE REGIONS

In the previous section we presented observational results based on representation of each active region by a single parameter. Although this approach has proven to be useful in study of the hemispheric helicity rule, it ignores the small-scale patterns of oppositely directed currents which are known to exist inside active regions [*Gary et al*, 1987; *Pevtsov and Peregud*, 1990].

Pevtsov et al. [1994] described such patterns of α_z which they found to persist for several days, with a characteristic decay time $\tau \sim 27$ hours (see Section 2.2.1 for limitations of using α_z). It has been convincingly argued, albeit in a limited data set, that non-zero values of α_z are of sub-photospheric origin [*Leka et al.*, 1996]. On the other hand, *van Driel Gesztelyi et al.* [1997] argued that currents may also be induced by sunspot motions.

Patterns in α_z probably exist throughout the lifetime of active regions – from emergence through decay. Figure 4 shows α_z maps of the AR NOAA 7926 during its dissipation. The pattern of α_z evolved only gradually over four days of observation.

If photospheric α_z patterns originate in the convection zone, what can we learn about physical processes there? Numerical dynamo models [*Brummell et al.*, 1996] show that small scale flows with both signs of kinetic helicity develop in a dynamo region as a result of turbulent convection in a rotating coordinate system. These flows may generate magnetic fields with opposite twists, which one can observe in the photosphere as α_z pattern.

Longcope et al. [1996] studied the evolution of a cross-section of an isolated flux tube rising through the convection zone and found that the interaction between the flux tube and its surroundings lead to bifurcation. The circulation twists each tube of the pair in an opposite sense.

6. LARGE-SCALE HELICITY PATTERN

Several studies imply large-scale organization of fields and flows on scales larger than the size of a typical active region.

6.1. Clusters of Active Regions

Active regions tend to form in clusters or "nests" of activity that may last from three to six solar rotations [*Gaizauskas et al.*, 1983]. Perhaps such persistent magnetic activity reflects dynamo processes, and hence can appear in helicity data as well. *Canfield and*

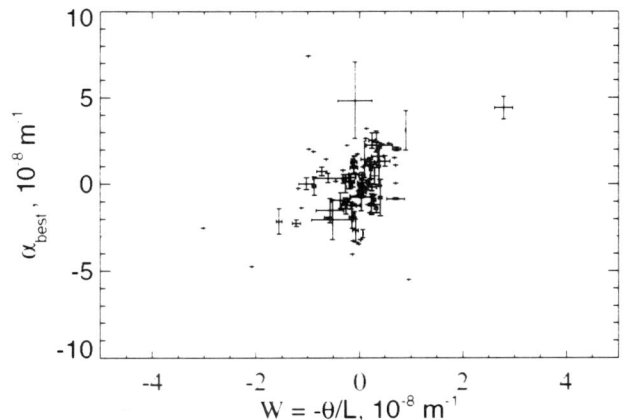

Figure 3. The dependence of observed active region α_{best} on the tilt θ per unit length L. Both solar hemispheres are included in this plot.

Pevtsov [1998] studied the distribution of active regions with longitude and found several abnormal areas which maintain the same sign of α for several successive solar rotations. The existence of nests of recurrent helicity has recently been confirmed by *Zhang and Bao* [1998].

The "helicity nests" may persist for up to 5 solar rotations, a few times longer than the theoretical estimates for the time a typical flux tube will rise from the bottom of the convection zone to the photosphere [*D'Silva and Howard*, 1994]. Hence, we speculate, the areas of abnormal α may reflect the helicity generated deep in the convection zone, not near the surface.

6.2. Large-Scale Helicity Areas

Latushko and Pevtsov [1998] studied the current helicity density h_c of the large-scale magnetic field, using longitudinal magnetograms from Wilcox Solar Observatory (WSO) and Michelson Doppler Imager (MDI, SOHO spacecraft). Figure 5 shows their results for three successive solar rotations. The authors found large-scale areas of both positive and negative chirality in both hemispheres. Some areas persisted for several solar rotations (Fig. 5, examples 1-3), while individual active regions came and went.

Coronal data from the Soft-X ray Telescope on the Yohkoh spacecraft seems to support such large-scale helicity areas, albeit indirectly. *Sandborgh et al.* [1998] used Yohkoh SXT data to identify boundaries of independent coronal flux systems and their chirality (using sigmoidal loops). Examining 10 solar rotations, they found several such flux systems with a typical size of 30-60° persisting for up to 5 solar rotations.

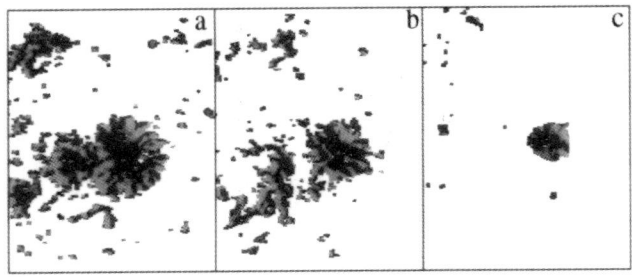

Figure 4. Local helicity (α_z) pattern of the dissipating sunspot of AR NOAA 7926 from Advanced Stokes Polarimeter (ASP) observations. Dark grey color indicates positive α_z, light grey indicates negative α. Pixels below noise level are shown in black (see Sec. 5).

Recently, giant convective cells were discovered [*Beck et al.*, 1998; *Hathaway et al.*, 1998]. The typical size of the convective cells is about 40°, and the individual cells persist for several solar rotations. Strikingly, the spatial and temporal scales of these features are very similar. We speculate that the large-scale helicity areas, coronal flux systems and giant convective cells may be simply different indications of same phenomenon.

7. DISCUSSION

Some researchers see the hemispheric helicity rule as an indication of sub-photospheric processes [*Rust and Kumar*, 1994; *Pevtsov et al.*, 1997]. Others see it as a result of large-scale photospheric shear motions [*Zirker et al.*, 1997; *Foukal*, 1997; *van Ballegooijen et al.*, 1998].

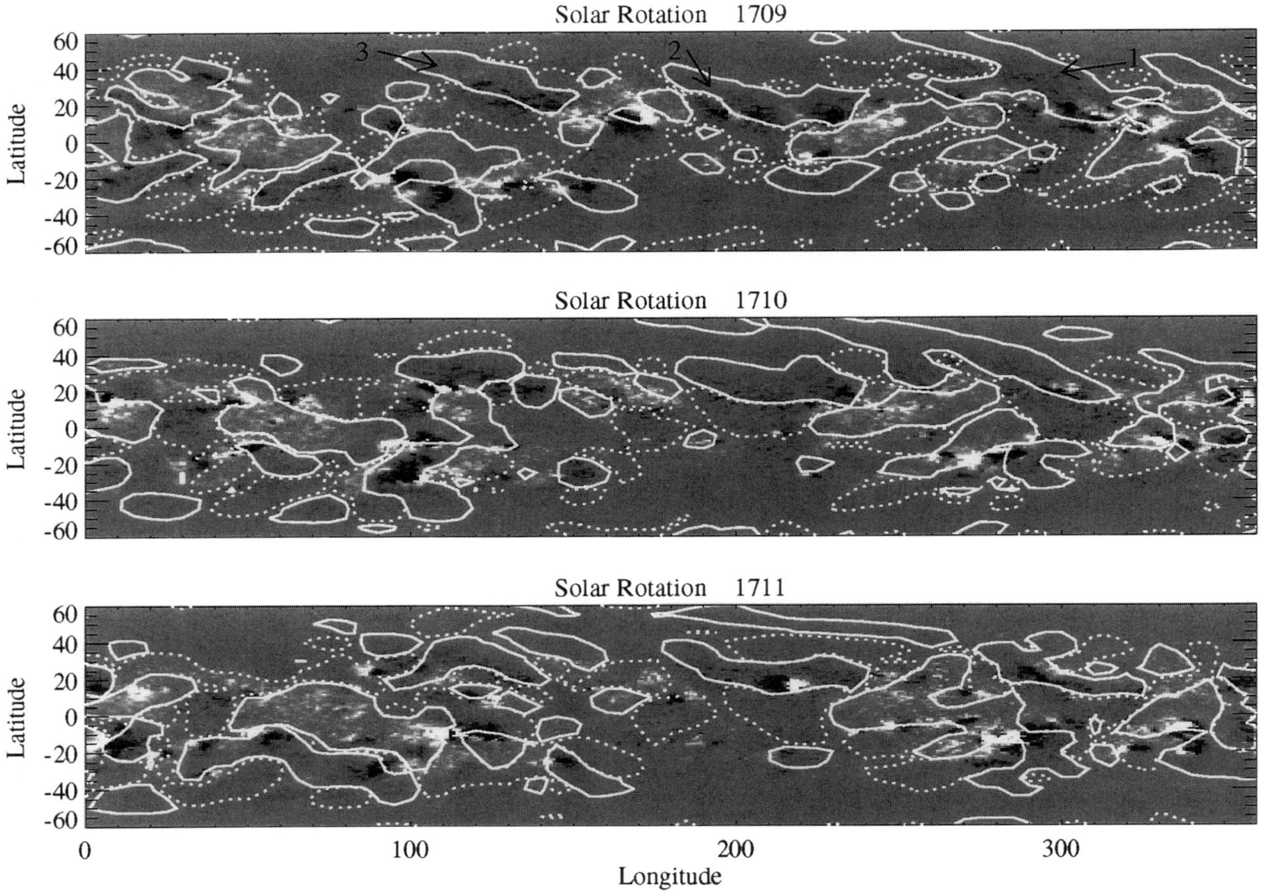

Figure 5. Carrington maps of $h_c = B_z \cdot (\nabla \times B)_z$ derived using daily full disk WSO magnetograms for the period of June – August 1981. Contours show h_c levels of $\pm$ 500 μT^2 degree^{-1} with solid contours showing positive and dashed - showing negative values. The underlying images are synoptic KPNO magnetograms. White indicates positive polarity (see Sec. 6). The numbered areas 1–3 show examples of the persisting h_c regions and their evolution.

What is the relative importance of each of these mechanisms on the Sun? Is there some signature of large-scale circulation in Fig. 5?

The orientation of bipolar active regions and coronal arcades, with one polarity situated closer to the solar equator than the other, allows differential rotation to shear their magnetic fields and produce S-shaped coronal loops in southern hemisphere and inverse-S shapes in the northern [*van Ballegooijen et al.*, 1998]. By its nature, differential rotation and Coriolis force should produce repeatedly the same sign of twist in the same hemisphere and hence, a strong hemispheric rule. Apparently, both chirality of active regions and current helicity of large-scale magnetic fields indicate that the rule is weak, *i.e.* there are areas of both signs of helicity in both hemispheres. We see the weakness of the hemispheric helicity rule as indirect evidence that neither differential rotation nor Coriolis force plays a dominant role in creation of twist. As well, the α_z pattern can not be understood in the framework of these two global mechanisms alone.

It appears that there is a certain disagreement between the Mees Solar Observatory HSP [*Pevtsov et al.*, 1995] and Huairou Solar Station [*Bao and Zhang*, 1998] datasets on the strength of the hemispheric helicity rule. The HSP has been used to study helicity via the α, the HSS employed current helicity h_c. Since for the force-free field $h_c = \alpha B^2$, both helicity proxies should have the same sign. Indeed, *Zhang and Bao* [1998] found that α and h_c correlate quite well with each other in their dataset. On the other hand, they found that the hemispheric helicity rule is much stronger in h_c data and weaker in α. We suspect that the presence of Faraday rotation in the HSS data may explain stronger hemispheric asymmetry in h_c (see Section 3.1.3 and *Hagyard et. al.* [1999]). It is also unclear how the neglect of x and y terms of the real current helicity may effect the h_c calculation (see Section 2.2.2).

The interaction between magnetic field rising through the convection zone and the turbulent convection – the Σ-effect, [*Longcope et al.*, 1998] – can explain many observed features: weakness of the hemispheric helicity rule, large scatter in Fig. 1, the typical value of α. It also can explain the α_z pattern. On the other hand, the observed tilt-twist correlation suggests that at least some amount of twist originates in the dynamo region.

Large-scale helicity areas open another challenge for helicity studies. If such areas do exist, do they reflect large-scale circulation in the dynamo region? What is the importance of such circulation on the dynamo? The study of large-scale helicity areas is in its infancy. Although the preliminary results are interesting, it would be premature to make serious conclusions based on them.

Acknowledgments. We are grateful to Dr. S.M. Latushko for providing us with the data for Figure 2. This research has been supported by NASA through SR&T grants NAG5-5043 and NAG8-1399 and the Yohkoh Soft X-Ray Telescope contract NAS8-40801. NSO/Kitt Peak data used here are produced cooperatively by NSF/NOAO, NASA/GSFC, and NOAA/SEL.

REFERENCES

Abramenko, V. I., Wang, T., and Yurchichin, V. B., Electric current helicity in 40 active regions in the maximum of solar cycle 22, *Solar Phys.*, *174*, 291–296, 1997.

Babcock, H., The topology of the Sun's magnetic field and the 22-year cycle, *Astrophys. J.*, *304*, 542–559, 1961.

Bao, S. and Zhang, H., Patterns of current helicity for the twenty-second solar cycle, *Astrophys. J.*, *496*, L43–L46, 1998.

Beck, J. G., Duvall, T. L., Scherrer, P. H., and Hoeksema, J. T., The detection of giant velocity cells on the Sun (abstract), *Eos Trans. AGU*, *79*(17), Spring Meet. Suppl., S281, 1998.

Brummell, N. H., Hurlburt, N. E., and Toomre, J., Turbulent compressible convection with rotation. I. Flow structure and evolution, *Astrophys. J.*, *473*, 494–513, 1996.

Canfield, R. C. and Pevtsov, A. A., Helicity of solar active-region magnetic fields, in *Synoptic Solar Physics, A.S.P Conf. Ser.*, edited by K.S. Balasubramaniam, J.W. Harvey and M. Rabin, *140*, pp. 131–143, A.S.P., 1998.

Canfield, R. C. and Pevtsov, A. A., Helicity and reconnection in the solar corona: observations, in *Magnetic Helicity in Space and Laboratory Plasmas*, edited by M.R. Brown, R.C. Canfield and A.A. Pevtsov, Geophys. Monogr. Ser., AGU, Washington, D.C., this volume, 1999.

D'Silva, S. and Howard, R. F., Sunspot rotation and the field strengths of subsurface flux tubes, *Solar Phys.*, *151*, 213–230, 1994.

Field, G., Magnetic helicity in astrophysics, in *Magnetospheric phenomena in astrophysics*, pp. 324–341, American Institute of Physics, New York, 1986.

Fisher, G. H., Fan, Y., and Howard, R. F., Comparisons between theory and observations of active region tilts, *Astrophys. J.*, *438*, 463–471, 1995.

Foukal, P., Chirality, helicity, and Joy's law, in *New Perspectives on Solar Prominences, Proceedings of a meeting held in Aussois, France (IAU Colloquium 167) 28 April -4 May 1997*, pp. 446, 1998.

Gaizauskas, V., Harvey, K. L., Harvey, J. W., and Zwaan, C., Large-scale patterns formed by solar active regions during the ascending phase of cycle 21, *Astrophys. J.*, *265*, 1056–1065, 1983.

Gary, G. A., Moore, R. L., Hagyard, M. J., and Haisch, B. M., Nonpotential features observed in the magnetic field of an active region, *Astrophys. J.*, *314*, 782–794, 1987.

Grigoryev, V. M. and Latushko, S. M., E-W motions of large-scale magnetic field structures of the Sun, *Solar Phys.*, *140*, 239–245, 1992.

Hagyard, M. J., Cumings, N. P., and West, E. A., The

MSFC vector magnetograph, in *Proceedings of Kunming Workshop on Solar Physics and Interplanetary Traveling Phenomena*, edited by C. De Jager and Chen Biao, pp. 204–204, Science Press, Beijing, China, 1985.

Hagyard, M. J., Pevtsov, A. A., and Canfield, R. C., Studies of solar magnetic helicity with the MSFC vector magnetograph, *Solar Phys.*, in preparation, 1999.

Hale, G. E., The fields of force in the atmosphere of the Sun, *Nature*, *119* (3002), 708–714, 1927.

Hathaway, D. H., Bogart, R. S., and Beck, J. G., A search for giant cells on the Sun (abstract), *Eos Trans. AGU*, *79*(17), Spring Meet. Suppl., S281, 1998.

Latushko, S. M. and Pevtsov, A. A., in preparation, *Astrophys. J.*, 1998.

Leka, K. D., Canfield, R. C., McClymont, A. N., and Van Driel-Gesztelyi, L., Evidence for current-carrying emerging flux, *Astrophys. J.*, *462*, 547–560, 1996.

Lites, B. W., Elmore, D. F., Seagraves, P., and Skumanich, A. P., Stokes profile analysis and vector magnetic fields. VI. Fine scale structure of a sunspot, *Astrophys. J.*, *418*, 928–942, 1993.

Longcope, D. W. and Fisher, G. H., The effect of convection zone turbulence on a rising flux tube, *Astrophys. J.*, *458*, 380–390, 1996.

Longcope, D. W., Fisher, G. H., and Pevtsov, A. A., Flux tube twist resulting from helical turbulence: the Σ-effect, *Astrophys. J.*, *507*, 417–432, 1998.

Longcope, D., Linton, M., Pevtsov, A., Fisher, G., and Klapper, I., Twisted flux tubes and how they get that way, in *Magnetic Helicity in Space and Laboratory Plasmas*, edited by M.R. Brown, R.C. Canfield and A.A. Pevtsov, Geophys. Monogr. Ser., AGU, Washington, D.C., this volume, 1999.

Martin, S. F., Bilimoria, R., and Tracadas, P. W., Magnetic field configurations basic to filament channels and filaments, in Rutten, C. J. and Schrjiver, C. J. (Eds.), *Solar Surface Magnetism*, pp. 303–338, Kluwer Academic Publishers, Dordrecht, 1994.

Martinez, P. V., Lites, B. W., and Skumanich, A. P., Active region magnetic fields. I. Plage fields *Astrophys. J.*, *474*, 810–842, 1997.

Mickey, D. L., The Haleakala stokes polarimeter, *Solar Phys.*, *97*, 223–238, 1985.

Moffatt, H. K. and Ricca, R. L., Helicity and the Călugăreanu invariant, *Proc. R. Soc. Lond. A*, *439*, 411–429, 1992.

Pevtsov, A. A., Canfield, R. C., and McClymont, A. N., On the subphotospheric origin of coronal electric currents, *Astrophys. J.*, *481*, 973–977, 1997.

Pevtsov, A. A., Canfield, R. C., and Metcalf, T. R., Patterns of helicity in solar active regions, *Astrophys. J.*, *425*, L117–L119, 1994.

Pevtsov, A. A., Canfield, R. C., and Metcalf, T. R., Latitudinal variation of helicity of photospheric magnetic fields, *Astrophys. J.*, *440*, L109–L112, 1995.

Pevtsov, A. A. and Latushko, S. M., Helicity of large scale photospheric magnetic fields (abstract), *Bull. Amer. Astron. Soc.*, *27*(2), 978, 1995.

Pevtsov, A. A. and Peregud, N. L., Electric currents in a unipolar sunspot, in *Physics of Magnetic Flux Ropes*, edited by C.T. Rassel, E.R. Priest and L.C. Lee, Geophys. Monogr. Ser., *58*, pp. 161–165, AGU, Washington, D.C., 1990.

Priest, E. R., *Solar magnetohydrodynamics*, Reidel, Dordrecht 1984.

Richardson, R. S., The nature of solar hydrogen vortices, *Astrophys. J.*, *93*, 24–28, 1941.

Rust, D. M., Magnetic helicity in solar filaments and coronal mass ejections, in *Magnetic Helicity in Space and Laboratory Plasmas*, edited by M.R. Brown, R.C. Canfield and A.A. Pevtsov, Geophys. Monogr. Ser., AGU, Washington, D.C., this volume, 1999.

Rust, D. M. and Kumar, A., Helical magnetic fields in filaments, *Solar Phys.*, *155*, 69–97, 1994.

Rust, D. M. and Kumar, A., Evidence for helically kinked magnetic flux ropes in solar eruptions, *Astrophys. J.*, *464*, L199–L202, 1996.

Sandborgh, S. C., Canfield, R. C., and Pevtsov, A. A., Chirality of large-scale flux systems in the solar corona (abstract), *Eos Trans. AGU*, *79*(17), Spring Meet. Suppl., S285, 1998.

Seehafer, N., Electric current helicity in the solar atmosphere, *Solar Phys.*, *125*, 219–232, 1990.

Semel, M., Mouradian, Z., Soru-Escaut, M., Maltby, P., Rees, D., Makita, M., and Sakurai, T., Active regions, sunspots and their magnetic fields, in *Solar Interior and Atmosphere*, edited by A. Cox, W. Livingston, and M. Matthews, pp. 844–889, The University of Arizona Press, Tucson, 1991.

Skumanich, A. and Lites, B. W., Stokes profile analysis and vector magnetic fields. I - Inversion of photospheric lines, *Astrophys. J.*, *322*, 473–482, 1987.

Smith, C. W., Solar-cycle, radial and latitudinal variations of magnetic helicity: IMF observations, in *Magnetic Helicity in Space and Laboratory Plasmas*, edited by M.R. Brown, R.C. Canfield and A.A. Pevtsov, Geophys. Monogr. Ser., AGU, Washington, D.C., this volume, 1999.

Stenflo, J. O., *Solar Magnetic Fields, Polarized Radiation Diagnostics*, Kluwer Academic Publishers, Dordrecht, 1994.

Van Ballegooijen, A. A., Cartledge, N. P., and Priest, E. R., Magnetic flux transport and the formation of filament channels on the Sun, *Astrophys. J.*, *501*, 866–881, 1998.

Van Driel-Gesztelyi, L., Csepura, G., Schmieder, B., Malherbe, J. M., and Metcalf, T., Evolution of a delta group in the photosphere and corona, *Solar Phys.*, *172*, 151–160, 1997.

West, E. A. and Hagyard, M. J., Interpretation of vector magnetograph data including magneto-optic effects. I - Azimuth angle of the transverse field, *Solar Phys.*, *88*, 51–64, 1983.

Zhang, H. and Bao, S., private communication, 1998.

Zirin, H., *Astrophysics of the Sun*, Cambridge Univ. Press, Cambridge, 1988.

Zirker, J. B., Martin, S. F., Harvey, K. H., and Gaizauskas, V., Global magnetic patterns of chirality, *Solar Phys.*, *175*, 45–58, 1997.

A. A. Pevtsov, R. C. Canfield, Department of Physics, Montana State University, P.O. Box 173840, Bozeman, MT 59717-3840. (e-mail: pevtsov@physics.montana.edu; canfield@physics.montana.edu)

Balance and Solar-Cycle Variations of Magnetic Helicity

Alexander Ruzmaikin

Jet Propulsion Laboratory, California Institute of Technology

Because magnetic helicity is conserved, the production and time evolution of helicity is due to its redistribution in space and in scale. Thus, the conservation of magnetic helicity implies that ejection of helicity into the solar wind is accompanied by production of an equal amount of helicity of opposite sign left at the Sun. The production of large-scale magnetic helicity by a mean-field dynamo is accompanied by a cascaded of helicity of opposite sign to the smaller scales. The helicity produced by the dynamo has opposite signs in the northern and southern solar hemispheres. A study of the evolution of magnetic helicity produced by mean-field dynamos indicates that the helicity of each hemisphere of the Sun oscillates about a mean with the period of the solar cycle (11 years). The magnetic helicity in a given hemisphere does not change sign from one 11 year period to the next. A rigidly rotating Sun creates a stationary helical configuration in the solar wind (the Parker spiral). Solar differential rotation produces an excess of helicity. It is suggested that this helicity excess is carried out of the Sun by solar mass ejections causing an overwinding of the Parker spiral, which varies with the solar cycle.

1. INTRODUCTION

The most striking property of magnetic helicity is its conservation. Being more strongly conserved than the magnetic energy, the helicity constrains possible equilibrium states [*Taylor*, 1974; *Moffatt* 1990]. It is a rugged invariant in MHD turbulence [*Matthaeus and Goldstein*, 1982]. A benefit of this robust nature of magnetic helicity is that the helicity can be calculated by many approximate methods or measured by many mirror-asymmetric proxies.

A problem with this robust nature of magnetic helicity is that helicity is not easy to create. The creation of magnetic helicity implies some dissipative processes and, with the large space scales involved, would require long characteristic times. However this does not preclude production and loss of magnetic helicity in space. Because the helicity has a sign it is easy to produce an equal amount of plus and minus helicity without violation of the conservation law. Then a separation of these opposite-signed helicities in space or in scales results in effective creation of magnetic helicity. For example, if a positive magnetic helicity is moved out of the Sun into the solar wind, the Sun is left with an equal amount of negative helicity.

The redistribution of magnetic helicity in and around the Sun poses an interesting problem of balance of the solar magnetic helicity: How much helicity is produced by the Sun in large and small scales, and how much of it is lost into the solar wind through the stationary flow and with mass ejections? At present, there is no clear understanding of all the processes involved in the redistribution of helicity so that only fragments of the balancing processes will be discussed in this chapter. Magnetic helicity redistribution is a key to understanding the long-term, in particular the solar-cycle variations, of the helicity.

The helicity balance can be checked with different types of observations. These include the coupling between the large-scale toroidal magnetic field, inferred from sunspots observations, and the poloidal field taken from photospheric magnetograms [*Seehafer*, 1990]; observations of the active region fields fitted to helical, force-free configurations [*Pevtsov et al.*, 1995; *Bao et al.*, 1998]; left or right-handedness of filaments and their overlying arcades [*Rust*, 1994; *Martin and McAllister*, 1996]; and the magnetic helicity in the solar wind [*Matthaeus and Goldstein*, 1982; *Bieber et al.*, 1987; *Smith and Bieber*, 1991].

Here, some basic mechanisms of helicity redistribution are discussed. A simple model of the helicity redistribution in the Sun due to the mean-field dynamo, is considered. The model indicates that the magnetic helicity of each hemisphere of the Sun oscillates about a mean with the half-period of the solar cycle (11 years), but that mean does not change sign from one 11 year period to the next. At any given time, the magnetic helicity in this model has different signs at large and small scales. The balance of total helicity produced by the dynamo is probably maintained by the distribution of magnetic helicity in scale sizes. It will also be illustrated how magnetic reconnection among magnetic loops observed to be rooted in the photosphere can introduce helicity into an erupting magnetic field accompanied with an encapture of helicity within the Sun. A discussion of the exchange of magnetic helicity between the Sun and the solar wind concludes the paper.

2. PRODUCTION OF MAGNETIC HELICITY BY A MEAN-FIELD DYNAMO

Consider first a mean axisymmetric magnetic field averaged over small solar scales and times, say over the supergranulation scales. The magnetic field can be split into the toroidal $B = B_\phi(t, r, \theta)$ and poloidal components

$$B_r = \frac{1}{r\sin\theta}\frac{\partial(\sin\theta A)}{\partial \theta}, \quad B_\theta = -\frac{1}{r}\frac{\partial(rA)}{\partial r}$$

defined by the vector potential $A = A_\phi(t, r, \theta)$. The magnetic helicity in one, say North, hemisphere of the convection zone is

$$H = 2\pi \int_{R_{cz}}^{R_\odot} \int_0^{\pi/2} ABr\sin\theta d\theta dr$$

where R_{cz} is the position of the bottom of the solar convection zone, and $R_\odot$ is the radius of the Sun. The sign of H and its change during the solar cycle immediately follow from symmetry considerations: Hale's law expresses the fact that the azimuthal component B has opposite signs in the northern and southern hemispheres and that the sign relation (+,-) changes into (-,+) in a half solar cycle, i.e. every 11 years [*Babcock*, 1961]. The radial component also has opposite signs in the two hemispheres, and the signs oscillate with the 11 year period. It follows that the vector potential A has the same sign in the North and South and this sign changes every 11 years. Hence the magnetic helicity has opposite signs in the northern and southern hemispheres of the Sun and this sign relation does not change from cycle to cycle.

We show now that the amplitude of magnetic helicity oscillates. The evolutionary equation for the magnetic helicity can easily be obtained from the mean field dynamo equations

$$\frac{\partial A}{\partial t} = \alpha B + \beta(\nabla^2 - \frac{1}{\sin^2\theta})A \quad (1)$$

$$\frac{\partial B}{\partial t} = (\nabla\Omega \times \nabla)_\phi r\sin\theta A + \beta(\nabla^2 - \frac{1}{\sin^2\theta})B \quad (2)$$

where α is proportional to the kinetic helicity of convective motions, β is the diffusivity which includes also a turbulent diffusivity, Ω is the angular velocity in the convective zone [*Krause and Rädler*, 1981; *Zeldovich et al.*, 1984]. (A relatively small α-term in the right side of Eq.(2) is omitted. Its inclusion would change $\int \alpha B^2 d^3r$ below into $\int \alpha \mathbf{B}^2 d^3r$.) Multiplying the first of these equations by B, the second by A, summing, integrating the resulting equations over a hemisphere and simplifying the effect of diffusion terms to a turbulent dissipation with a characteristic time τ_T we obtain

$$\frac{\partial H}{\partial t} = \int \alpha B^2 d^3r + \int (\nabla\Omega \times \nabla)_\phi r\sin\theta A^2 \frac{d^3r}{2} - \frac{H}{\tau_T} \quad (3)$$

where the integrals are taken over the convective zone in one hemisphere. This equation describes the long-term variations of the helicity produced by the mean-field dynamo. The quadratic dependence of the helicity on the magnetic field immediately tells us that the 22-year variation of the field results in a double-frequency, 11-year variation of the helicity [Ruzmaikin, 1996].

Further details depend on the signs and distributions of α and Ω. The distribution of $-\alpha$ in the standard mean-field theories is determined by the mean kinetic helicity of the convective flow (see for example Ch.11 in Zeldovich et al., 1984): A fluid element rising to the surface of the Sun expands due to the density decrease. The expansion produces a sideward velocity component of the element, $v_\perp$, and the element, subjected to the Coriolis force $-2\Omega v_\perp$, will acquire a spin, rot$\mathbf{v}$, directed opposite to $\boldsymbol{\Omega}$. The spin for a fluid element sinking in the convection zone will have the opposite sign but the product $\mathbf{v} \cdot$ rot$\mathbf{v}$, called "kinetic helicity", will have the same sign (negative in the northern hemisphere and positive in the southern hemisphere) for rising and sinking elements. Hence, α is positive in the northern and negative in the southern hemisphere. Note, however, that if α is determined by some other physical process its sign and distribution could be different.

The observational determination of α or the kinetic helicity in the solar convection zone is very important. An attempt to observationally determine α has been made by Patron et al., [1995] who used the Mount Wilson helioseismic data; however the results have not been confirmed yet with the SOHO/MDI data. If we accept that α is positive in the northern hemisphere, the contribution of the first term in Eq.(3), and hence the stationary helicity produced by the kinetic helicity, is positive.

The second term in Eq.(3) is determined by the differential rotation. The distribution of the rotation in the convection zone has recently been reconstructed from the helioseismic data [Kosovichev et al., 1997] so that we can evaluate the gradient of Ω directly from the reconstruction. The distribution is clearly separated into two parts: the near-equatorial one (below 30° in heliolatitude) in which the radial gradient of Ω is positive near the bottom of the convection zone (where the toroidal magnetic field is generated), and the part above 30° in which the radial gradient of Ω is negative near the bottom of the convection zone.

Thus the integrand in the second term on the right side of Eq.(3) has no definite sign. A simple evaluation of this integrand with Ω described above and a dipole-type poloidal field ($B_r \propto cos\theta, B_\theta \propto sin\theta, A \propto sin\theta$) leads to a positive value. However, the exact values of these integrals (and their signs) can only be calculated in specific dynamo models. The ratio of the first to the second integral is determined by the dimensionless parameter $(\alpha/|\nabla\Omega|h^2)(B/B_p)^2$, where h is the depth of the convection zone and $B_p \approx A/h$ is the poloidal component of the mean field. This parameter is expected to have a finite value, i.e. not extremely small or large, because the ratios $(\alpha/|\nabla\Omega|h^2)$, and $(B/B_p)^2$, in dynamo theories have the same order of magnitude.

Because of the conservation law, the mean magnetic helicity considered above can not be produced alone. An effective way to keep the total helicity conserved is to produce opposite-signed helicity in smaller scales. Such a process has been demonstrated in numerical simulations of a fully developed helical turbulence [Pouquet et al., 1978]. In these simulations the system was excited by injection of kinetic energy and kinetic helicity (which is not a conserved quantity) at some basic scale. A small amount of magnetic energy was also given initially. Then, as time increased, the magnetic energy grew at progressively larger scales. No magnetic helicity was given initially. However, scale-dependent magnetic helicity appeared as time increased. This helicity had the same sign as the kinetic helicity at the basic (the energy input) scale and the opposite sign at larger scales. To better understand this one can divide the magnetic field and its vector potential into large-scale (mean) $\mathbf{B}, \mathbf{A}$ and fluctuating $\mathbf{B}', \mathbf{A}'$ parts. Then the total magnetic helicity can be presented as $<\mathbf{AB}> + <\mathbf{A'B'}> \approx$ const.

Below we consider only the evolution of the mean, large-scale magnetic helicity. The role of the accompanying helicity produced by fluctuating fields in the Sun is not discussed. One can expect that the magnetic helicity associated with relatively small scale magnetic field has different signs in the two solar hemispheres and these are opposite to the signs of the mean helicity. It is tempting to identify the $<\mathbf{A'B'}>$ helicity with the helicity of the fields related to filaments. These have the signs of predominant chirality in the two solar hemispheres [Rust, 1994] which are opposite to the signs of the magnetic helicities associated with large-scale magnetic loops overlying the filaments [Martin and McAllister, 1996].

To study the mean magnetic helicity variations it is instructive to consider a model approximation to Eq.(3) for one (North) hemisphere [Ruzmaikin, 1996]

$$\frac{\partial H(t)}{\partial t} = 2\zeta E_B(t) - \frac{H(t)}{\tau_T} \quad (4)$$

where $E_B = 1/2 \int B^2 d^3r$ is the energy of the azimuthal magnetic field. The parameter ζ is close to α if the first integral in Eq.(3) is the larger and is determined mostly by Ω if the second integral is the larger.

Let us specify the solar cycle time dependence of the azimuthal field energy: $E_B = E_0 \cos^2 \omega t$, where $\omega = 2\pi/22$ years. Then Eq.(4) has an analytical solution

$$H(t) = C\exp(-\frac{t}{\tau_T}) + \zeta \tau_T E_0 (1 + \frac{\sin(2\omega t + \phi)}{\sqrt{1+4\omega^2 \tau_T^2}}) \quad (5)$$

where $\phi = \arctan(1/2\omega\tau_T)$ is a phase shift relative to the solar cycle. After a few diffusion times the first term vanishes, and the mean magnetic helicity will behave as a periodic function, with a period equal to the solar half-cycle period, oscillating around the mean value $\zeta \tau_T E_0$. The sign of the mean magnetic helicity is defined by the sign of ζ and, according to Eq.(5), does not change. The relative amplitude of the helicity oscillations and the phase shift are determined by $2\omega \tau_T = 4\pi \tau_T / 22 \text{years} \approx \tau_T/(2\text{years})$. The turbulent diffusivity in the solar convective zone has been estimated to be $\nu_T \approx 5 \times 10^{12}$ cm^2/sec [*Zeldovich et al.*, 1984] which gives an estimate $\tau_T = h^2/\nu_T \approx 2$ year with $h = 2 \times 10^{10}$ cm as the size of the convective zone. Hence $2\omega \tau_T \approx 1$ and the value of $H/(\zeta \tau_T E_0)$ oscillates between $1+1/\sqrt{2} \approx 1.7$ and $1-1/\sqrt{2} \approx 0.3$ with a phase shift (relative to sunspots) of about $1/8 \times$ 11-year, i.e. 1.4 years.

The simple theoretical consideration presented here apparently favors a positive sign for ζ. However analyses of the helical fields on the solar surface [*Seehafer*, 1990; *Pevtsov et al.*, 1995; *Bao et al.*, 1998] indicate that ζ is negative. The same negative (positive) helicity in the North (South) could already be inferred from the classical paper by *Babcock* [1961], see fig.3 of his paper. The simplified model discussed above does not take into account the effect of the solar surface. In fact, to be gauge invariant the mean helicity in this model has to be calculated under the assumption that this surface is a magnetic surface, i.e. $B_n = 0$ on it. This condition is not valid because we observe a normal component of the field on the solar surface.

To resolve this problem a full dynamo model–which includes the solar interior and exterior with realistic boundary conditions between them–has to be considered. The simplified model is also constrained to axisymmetric fields although the observed fields have very strong and often dominating non-axisymmetric components. Fortunately, we still can get some information about the magnetic helicity on the solar surface and in the space outside the Sun because the change in the solar helicity can be reduced to a helicity flux through its surface, see Section 4. The analysis of this flux shows that the helicity in the northern (southern) hemisphere is negative (positive), see *Berger and Ruzmaikin*, [1998] and Section 4 below.

3. ENCAPTURE OF MAGNETIC HELICITY AT THE SUN DUE TO EJECTIONS

The magnetic field structure plays an important role in the transfer of energy between the solar photosphere, the corona, and the solar wind. Dynamical changes in the field structure allow the release of stored magnetic energy during flares, filament eruptions, and coronal mass ejections. Magnetic structures are typically three-dimensional and expected to be topologically nontrivial. For example, Hα and x-ray observations provided by the Solar Maximum Mission and YOHKOH imply that twisted and braided magnetic flux tubes constitute a basic structural element of the solar atmosphere [*House and Berger*, 1987; *Shibata et al.*, 1992].

Interplanetary magnetic clouds also reveal helical magnetic fields [*Lepping et al.*, 1990]. It was pointed out that the sign of magnetic field twist (called "the chirality") of a cloud corresponds to the predominant chirality of filaments in the solar hemisphere in which that cloud originated [*Rust*, 1994]. The clouds' chirality agrees with that deduced from Hα observations [*Martin et al.*, 1994]: clouds with the left-handed magnetic configuration predominantly come from the northern hemisphere and right-handed ones come from the southern hemisphere [*Rust and Kumar*, 1994]. The problem is how the observed solar structures such as filaments and CMEs acquire the helical magnetic field. It has been proposed that helical magnetic configurations originate from three-dimensional magnetic reconnections of extended coronal loops which become CMEs as they moved out of the Sun [*Gosling*, 1990].

A coiled (writhed) magnetic flux tube can be produced by reconnections between magnetic field lines of two flux tubes that have emerged above the solar surface (Figure 1). The configuration above the surface, however, does not carry enough information to tell us whether or not the flux rope has magnetic helicity. Magnetic helicity is a non-local quantity –because the vector potential at a given point is determined by the magnetic field from all points. Conservation of magnetic helicity implies that if helical magnetic fields are produced by reconnections and then escape into the solar wind, an equal amount of magnetic helicity of opposite sign has to be encaptured at the Sun. The original loops can be helical or, as in Figure 1, non-helical.

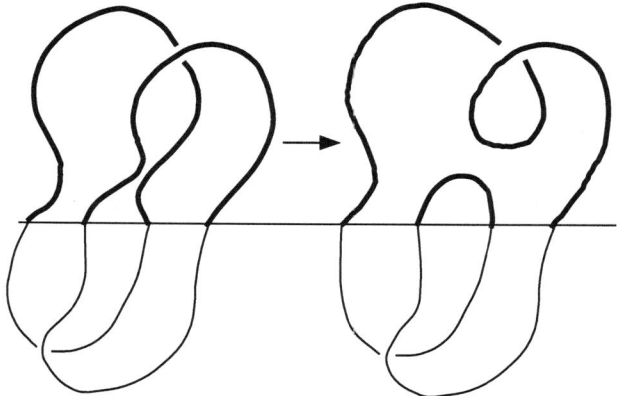

Figure 1. A sketch demonstrating how two unconnected flux tubes (left) can reconnect to create a coiled flux tube above the surface of the Sun (right). The parts of the flux tubes below the solar surface (shown by a horizontal line) are drawn with thin lines. The coil above the surface has a positive writhe, the coil mostly inside the Sun has a negative writhe so that the resulting magnetic helicity is zero.

4. FLUX OF MAGNETIC HELICITY THROUGH THE SOLAR SURFACE

With the exception of the previous section, we have discussed the helicity production inside the Sun. The balance of helicity, however, is strongly dependent on its flux through the solar surface and into the solar wind [*Berger and Ruzmaikin*, 1998].

To define the helicity transfer through the solar boundary consider first the helicity integrated over all space. In this case, no field lines have endpoints on boundaries; in effect all field lines eventually close upon themselves (or quasi-closed for ergodic fields) to form closed curves. If two bundles of closed field lines link each other, their linking (multiplied by the product of the fluxes) will add to the helicity. However, for subregions of space, field lines will indeed have endpoints on the boundaries between the regions. The definition of helicity requires some care because of this. Consider again the northern hemisphere of the Sun. Its volume, V_N, is bounded by a spherical boundary (the photosphere North of the equator) and a planar boundary (the equatorial slice through the interior). Neither of these boundaries are magnetic surfaces (on a magnetic surface the normal field component $B_n = 0$).

Note that the boundary flux, B_n, uniquely defines a potential field $\mathbf{B}_P$ within V_N. Potential fields have no currents (except for surface currents in the boundaries), and are the absolute minimum energy state of all fields with the same boundary flux. The magnetic helicity of the northern hemisphere becomes a unique and well defined topological measure when we set the helicity of this potential field to zero [*Berger and Field*, 1984; *Finn and Antonsen*, 1985]. Thus we first obtain the helicity $\int \mathbf{A}\mathbf{B} d^3 r$ for all space, then recalculate it substituting $\mathbf{B}_P$ for the field inside V_N: $\int_{outside} \mathbf{A}\mathbf{B} d^3 r + \int_{inside} \mathbf{A}_\mathbf{P} \mathbf{B}_\mathbf{P} d^3 r$. Subtract the second helicity from the first; the result, called "the relative helicity of the northern hemisphere, K_N", is independent of the fields outside V_N and is thus well defined. Technically, this procedure is required to guarantee gauge-invariance.

The time derivative of the relative helicity in the northern hemisphere includes Ohmic dissipation and boundary terms representing transfer of helicity from (to) the southern hemisphere and the corona. Strict limits can be placed on helicity dissipation [*Berger*, 1984], so we will neglect the dissipation effect. The boundary terms comprise integrals involving the normal component, B_n, and the plasma velocity $\mathbf{v}$ [*Berger and Field*, 1984]:

$$\frac{dK_N}{dt} = 2\int (\mathbf{A}_p \times \mathbf{E}) \cdot \mathbf{n} dS \qquad (6)$$
$$= 2\int (\mathbf{A}_p \cdot \mathbf{v}) \mathbf{B}_\mathbf{n} dS - 2\int (\mathbf{A}_p \cdot \mathbf{B}) v_n dS$$

where $\mathbf{E}$ is the electrical field and $\mathbf{A}_p$ is a vector potential derived from B_n

$$\mathbf{n} \cdot \nabla \times \mathbf{A}_p = B_n, \quad \nabla \mathbf{A}_p = 0, \quad \mathbf{A}_p \cdot \mathbf{n} = 0. \qquad (7)$$

Axial symmetry (where the magnetic field is independent of the coordinate φ) considerably simplifies the calculations. For the planar equatorial boundary with $\Phi(\rho)$ denoting the net flux within radius ρ, and $\hat{\varphi}$ the angular direction,

$$\mathbf{A}_p(\rho) = \frac{\Phi(\rho)}{2\pi\rho}\hat{\varphi}, \quad \Phi(\rho) = 2\pi \int_0^\rho B_z \rho d\rho. \qquad (8)$$

For the spherical part of the boundary with net flux $\Phi(\theta)$ between the North pole and co-latitude θ,

$$\mathbf{A}_p(\theta) = \frac{\Phi(\theta)}{2\pi R \sin\theta}\hat{\varphi}, \quad \Phi(\theta) = 2\pi \int_0^\theta B_r \sin\theta d\theta. \qquad (9)$$

Here we consider the simplest and important case of the pure rotational velocity: $\mathbf{v} = \mathbf{\Omega} \times \mathbf{r}$. With the assumption of axial symmetry we then have

$$\frac{dK_N}{dt} = 2R^2 \int_0^{\pi/2} B_r(\theta)\Phi(\theta)\Omega(R,\theta)\sin\theta d\theta$$
$$- 2\int_0^R B_z(\rho)\Phi(\rho)\Omega(\rho,\pi/2)\rho d\rho. \qquad (10)$$

The first integral describes the contribution from the rotation on the solar surface, the second integral describes the contribution of the equatorial rotation. The radial component of the solar magnetic field can be found from the synoptic maps employing spherical harmonics, however the B_z component is unobservable and can only be found from dynamo models of the internal field. Because no net helicity can be created by rotation, the helicity increase (decrease) inside the Sun has to be accompanied by the equal helicity decrease (increase) outside the Sun. To formally calculate the change of helicity outside the Sun, say to show that in the northern corona $dK_{NC} = -dK_N$, one has to reverse the sign of the normal in Eq.(6).

4.1. Helicity of a Rigidly Rotating Sun: the Parker Spiral

In the case of a rigidly rotating sun, $\Omega = \Omega_0 =$ constant and integrals in Eq.(10) can easily be calculated. The net magnetic flux Φ_0 through the northern photosphere equals the net flux through the equator.

The helicity change in the solar corona is especially interesting. For the northern part of the corona the first integral in Eqs.(6,10) with $\mathbf{n} = -\mathbf{r}/r$ gives $dK_{NC}/dt = -(\Omega_0/\pi) \int \Phi d\Phi = -\Omega_0 \Phi_0^2/(2\pi)$. In so far as K_{NC}/Φ_0^2 measures the twist, i.e. the number of turns per 2π radians, this means that the northern corona (northern solar wind) receives one negative twist per solar rotation. This twist is balanced by a positive twist inside the North solar interior, because $dK_N = -dK_{NC}$. However there is an additional contribution to the northern solar interior which comes from the South through the equatorial plane: $-(\Omega_0/\pi) \int \Phi d\Phi = -\Omega_0 \Phi_0^2/(2\pi)$, see the second integral in Eq.(10). This is because the equatorial plane rotates with the angular velocity Ω_0 so that field lines passing through this plane receive one twist per rotation. Thus the total change in the North interior (as well as in the South) is zero, $dK_N/dt = 0$. The helicity change in the northern corona is balanced by the opposite-signed change in the southern corona. The coronal (solar wind) helicity was first calculated and interpreted geometrically by *Bieber et al.*, [1987]. Because of the solar rotation, the large scale field lines have the shape of outward propagating spirals (the Parker spiral). This helicity is negative in the North and positive in the South.

Note that the solar wind is a well conducting fluid. If there were a vacuum outside the Sun, the electrical field **E** in Eq.(6) would be zero and thus there would be no flux of helicity from/into the Sun.

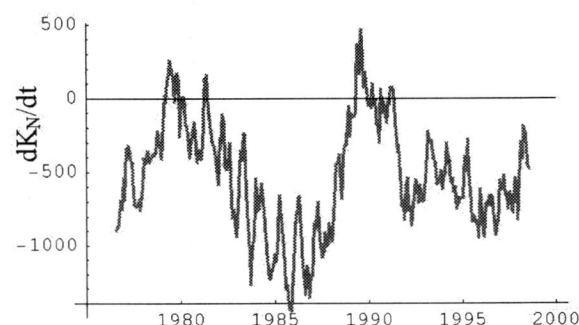

Figure 2. The relative helicity change (in units of $10^{40} \mathrm{Mx}^2/\mathrm{day}$) in the northern hemisphere [*Berger and Ruzmaikin*, [1998]].

4.2. Helicity of the Differentially Rotating Sun: Overwinding the Parker Spiral

Berger and Ruzmaikin [1998] calculated the flux of relative helicity for the real Sun using the differential rotation observed on the solar surface, the rotation in the equatorial plane estimated from the helioseismic data, and the photospheric magnetic field data obtained by the Wilcox Observatory during one solar cycle. The integrals in Eq.(6) were analyzed in terms of harmonic coefficients. In the second integral, it was assumed that B_z in the equatorial plane is uniform near the base of the convection zone. Figure 2 shows the evolution of helicity in the northern hemisphere calculated over 22-year period.

Here we consider a greatly simplified situation: a simple dipole field $B_r = b_1 \cos\theta$ on the surface and uniform at the solar equator, approximate the observed differential rotation by $\Omega(\theta) = \Omega_0(1 - a\cos^2\theta), a \approx 0.2$ and assume that the equator rotates uniformly at Ω_0. The magnetic flux through the northern hemisphere (equal to the flux through the equator) is $\Phi_N = \pi R^2 b_1$. Equation (10) gives now $dK_N/dt = -a\Omega_0 \Phi_N^2/(6\pi)$, i.e. a negative helicity is continuously pumping into the northern solar interior!

To avoid infinite accumulation of helicity, the Sun has to find a way to get rid of this excess. Because there are presumably no fluid flows across the solar equator, we conjecture that the Sun ejects the helicity through structures developed on its surface and propagating into the solar corona. In the other words, this excess of magnetic helicity is probably carried out by coronal mass ejections. The negative sign (in the northern hemisphere) of this extra helicity seems to be in agreement with the helicity of active region fields [*Pevtsov et al.*, 1995; *Bao et al.*, 1998] and the preferred left-skewed handedness of magnetic arcades associated with the coronal mass ejections [*Martin and McAllister*, 1996]. The actual amount of the excess can be evaluated from the helicity of magnetic clouds in the solar wind.

Because this excess of magnetic helicity is added to the stationary level carried by the Parker spiral in the interplanetary space, some overwinding of the spiral occurs. *Smith and Bieber* [1991] noted that interplanetary magnetic field at 1 AU is more tightly wound than would be expected for the Parker spiral. From spacecraft data taken near the ecliptic plane, they found that the winding angle of the interplanetary spiral near solar maximum is about 10° larger than at solar minimum. This variation results in part due to the cycle variation of the solar wind speed. This is because the one of basic sources of the solar wind is solar polar coronal holes. Coronal holes rotate almost rigidly [*Wagner*, 1976] and their size dramatically changes from solar minimum, when the centroidal location of the holes migrates to low latitudes, to solar maximum, when holes shrink toward the poles [*Wang et al.*, 1990]. Due to the asymmetry of the coronal holes this effect can possibly explain as well the north-south asymmetry in the winding angle, pointed out by these authors.

However an extra overwinding of about 1.5°±0.5° was observed and is apparently unrelated to the solar wind speed and solar rotation variations. An early suggestion that this overwinding arises from an escape of azimuthal magnetic field from the Sun [*Smith and Bieber*, 1991] has not been supported by further analyses; it was found instead that CMEs and shocks account for the excess overwinding [*Smith and Phillips*, 1991]. We suggest that these CMEs (perhaps along with some other, small-scale ejections) carry out the excess of magnetic helicity produced by the solar differential rotation. For the simple dipole approximation described above this excess can be evaluated as

$$\frac{dK_{difrot}}{dK_0} = \frac{a}{3} \approx 7\%.$$

where $dK_{difrot}(K_0)$ stands for the rate of change of helicity of the differentially (rigidly) rotating Sun. This simple estimate agrees within a factor of two with the observed relative spiral angle excess: $1.5°/45° \approx 3.3\%$. The solar cycle variation in the number of CMEs is in agreement with the cyclic variation of the extra overwinding of the Parker spiral.

5. CONCLUSIONS

This paper demonstrates how a relatively simple question of solar cycle variations of magnetic helicity runs into the difficult problem of helicity balance. The solution of this problem involves the study of many physical processes inside the Sun and in the solar wind. Among them are the dynamo, coronal mass ejections and overwinding of the spiral of the interplanetary magnetic field. The use of helicity allows us to look at these phenomena from a different point of view.

Acknowledgments. This research was conducted in part at the Jet Propulsion Laboratory, California Institute of Technology, under contract with the National Aeronautic and Space Administration, and was also supported in part by the NSF grant INT-9603415. I thank the referees for their comments.

REFERENCES

Babcock, H. W. The topology of the Sun's magnetic field and the 22-year cycle, *Astrophys. J.*, 572–587, 1961.

Bao, S. D., H. Q. Zhang, M. Zhang, and T. J. Wang, A study of helicity associated with flare occurrence in active regions, this Monograph, 1998.

Berger, M. A., and G. B. Field, The topological properties of magnetic helicity, *J. Fluid Mech.*, 147, 133–148, 1984.

Berger, M. A., Rigorous new limits on magnetic helicity dissipation in the solar corona, *Geophys. and Astrophys. Fluid Dyn.*, 30, 79, 1984.

Berger, M. A., and A. Ruzmaikin, Rate of Helicity Production by the Solar Rotation, *J. Geophys. Res.*, submitted, 1998.

Bieber, J. W., P. A. Evenson, and W. H. Matthaeus, Magnetic helicity of the Parker spiral, *Astrophys. J.*, 315, 700–705, 1987.

Finn, J. and T. M. Antonsen, Magnetic helicity: What is it and what is it good for, *Com. Plasma Phys. and Contr. Fusion*, 9, 111, 1985.

Gosling, J. T., Coronal mass ejection and magnetic ropes in interplanetary space, in *Physics of Magnetic Flux Ropes*, edited by C. Russell, E. Priest, and L. Lee, Geophys. Monog. 58, AGU, 343, 1990.

House, L. L. and M. A. Berger, The ejection of helical field structures through the outer corona, *Astrophys. J.*, 323, 406–413, 1987.

Kosovichev, A. G. et al (33 co-authors), Structure and rotation of the solar interior: initial results from the MDI medium-l program, *Solar Phys.*, 170, 43, 1997.

Krause, F., and K.-H. Rädler, *Mean-Field Dynamo and Mean-Field Magnetohydrodynamics*, Springer, Berlin, 1981.

Lepping, R. P., J. A. Jones, and L. Burlaga, Magnetic field structure of interplanetary magnetic clouds at 1 AU, *J. Geophys. Res.*, 95, 11,957, 1990.

Martin, S. F., R. Bilimoria, and P. W. Tracadas, in *Solar Surface Magnetism*, edited by R. J. Rutten and C. J. Schrijver, p.303, Springer-Verlag, New-York, 1994.

Martin, S., and McAllister, D., Predicting the sign of magnetic helicity in erupting filaments and coronal mass ejections, in *"Coronal Mass Ejections"*, *Geophys. Monograph 99*, eds. N. Crooker, JoAnn Joselyn, and J. Feynman, AGU, 1996.

Matthaeus, W. H., and M. L. Goldstein, Measurements of the rugged invariants of magnetohydrodynamic turbulence in the solar wind, *J. Geophys. Res., 87*, 6011–6028, 1982.

Moffatt, K., The energy spectrum of knots and links, *Nature, 347*, 367–369, 1990.

Patron, J., F. Hill, E. J. Rhodes, S. G. Korzennik, and A. Cacciani, Ring diagram analysis of Mt. Wilson data: Velocity field within the convection zone, *Astrophys. J., 455*, 746–757, 1995.

Pevtsov A. A., R. C. Canfield, and T. R. Metcalf, Latitudinal variation of helicity of photospheric magnetic fields, *Astrophys. J., 440*, L109–111, 1995; also this Monograph.

Pouquet, A., U. Frisch, and J. Léorat, Strong MHD turbulence and the nonlinear dynamo effect, *J. Fluid Mech., 77*, 321–354, 1976.

Rust, D. M., Spawning and shedding helical magnetic fields in the solar atmosphere, *Geophys. Res. Lett., 21*, 241–245, 1994.

Rust, D. M., and A. Kumar, Helical magnetic fields in filaments, *Solar Phys., 155*, 69–97, 1994.

Ruzmaikin, A., Redistribution of magnetic helicity at the Sun, *Geophys. Res. Lett. 23*, 2649–2652, 1996.

Seehafer, N. Electric current helicity in the solar atmosphere, *Solar Phys., 125*, 219–232, 1990.

Shibata, K., Ishido, Y., L. W. Acton, K. T. Strong et al., Observations of X-ray jets with the Yohkoh soft X-ray telescope, *Publ. Astron. Soc. Japan, 44*, L173–179, 1992.

Smith, C. W., and J. W. Bieber, Solar cycle variation of the interplanetary magnetic field spiral, *Astrophys. J., 370*, 435–441, 1991.

Smith, C. W., and J. L. Phillips, The role of CMEs and interplanetary shocks in IMF winding angle statistics, in *Solar Wind 8*, edited by D. Winterhalter, J. T. Gosling, S. Habbal, W. Kurth and M. Neugebauer, AIP Proc. 382, pp. 502–505, 1996.

Taylor, J. B., Relaxation of toroidal plasma and generation of reverse magnetic fields, *Phys. Rev. Lett., 33*, 1139, 1974.

Wagner W. J., Rotational characteristics of coronal holes, in *Basic Mechanisms of Solar Activity*, edited by V. Bumba and J. Kleczek, IAU, pp. 41–43, 1976.

Wang, Y.-M., N. R. Sheeley, and A. G. Nash, Latitudinal distribution of solar wind speed from magnetic observations of the Sun, *Nature, 347*, 439–444, 1990.

Zeldovich, Ya. B., A. Ruzmaikin, and D. D. Sokoloff, *Magnetic Fields in Astrophysics*, 365 pp., Gordon and Breach, New York, 1984.

A. Ruzmaikin, Jet Propulsion Laboratory, m/s 169-506, California Institute of Technology, 4800 Oak Grove Dr., Pasadena, CA 91109. (e-mail: aruzmaikin@jplsp.jpl.nasa.gov)

Magnetic Helicity and Relaxation: Theory

Paul M. Bellan

Applied Physics, California Institute of Technology

The magnetic helicity concept is considered for situations where open field lines intercept a flux-conserving bounding surface, i.e., situations corresponding to physically realistic problems such as the solar corona. Relative helicity rather than absolute helicity must be used when there are open field lines because absolute helicity is gauge-dependent and therefore ambiguous. Using conservation of relative helicity as a constraint, the relaxed state (self-organized state) for open field line situations is derived. It is shown that evolution of relaxed states typically involves discontinuous changes in the topology of magnetic field lines. Furthermore, individual field lines are not frozen into the bounding flux-conserving surface as a relaxed state evolves. The S-shapes often seen on the solar corona are shown to be consistent with open-field-line relaxed states.

1. INTRODUCTION

Magnetized plasmas generally have very complicated dynamics (e.g., instability, waves, reconnection, etc.). Often one is only interested in the outcome of the dynamics and not the dynamics itself. Relaxation theory [*Woltjer*, 1958; *Taylor*, 1974,1986] provides a simple and general method for determining the outcome of arbitrarily complicated dynamics in a low β plasma. The basis for this remarkable theory is the postulate that magnetic helicity, a measure of global topology, is nearly invariant regardless of dynamical details whereas magnetic energy tends to decay as a result of dynamics.

This paper is organized as follows: Section 2 defines the magnetic helicity K and shows how K quantifies flux linkage. Section 3 shows that K is inadequate for situations having open field lines (e.g., the solar corona and most other situations of practical interest) and introduces the relative helicity K_{rel}. Section 4 derives the conservation equation for K_{rel}. Section 5 derives relaxed states using conservation of K_{rel}, giving a theory of plasma self-organization. Section 6 discusses λ, the parameter which characterizes the relaxed state. Section 7 considers geometrical properties of relaxed states. Section 8 discusses the relation between relaxed states and the MHD energy principle.

2. MAGNETIC HELICITY, K

The magnetic helicity in a volume V is defined as

$$K = \int_V \mathbf{A} \cdot \mathbf{B} d^3 r \,. \tag{1}$$

Following *Moffatt* [1978] we consider two thin, untwisted, isolated, linked, closed flux tubes (labeled #1 and #2) as shown in Fig. 1; the magnetic field is assumed to vanish outside these flux tubes. We show that if V completely encloses the two flux tubes, then K provides a measure of their linkage.

Since $\mathbf{B} = 0$ outside the flux tubes, (1) involves only contributions from the respective interior volumes V_1 and V_2 of the two flux tubes, i.e.,

Magnetic Helicity in Space and Laboratory Plasmas
Geophysical Monograph 111
Copyright 1999 by the American Geophysical Union

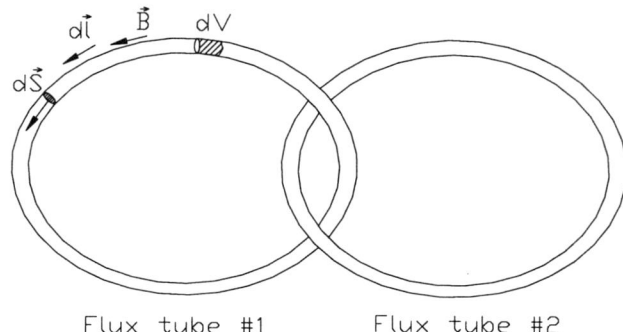

Figure 1. Two linked, untwisted, closed flux tubes.

$$K = K_1 + K_2 \tag{2}$$

where $K_j = \int_{V_j} \mathbf{A} \cdot \mathbf{B} d^3 r$. The differential element of volume for each flux tube is $d^3 r = d\mathbf{l} \cdot d\mathbf{s}$ where $d\mathbf{l}$ is an element of length along the flux tube and $d\mathbf{s}$ is the surface cross-section of the flux tube. For a thin, untwisted flux tube both $d\mathbf{l}$ and $d\mathbf{s}$ are parallel to $\mathbf{B}$ and so

$$\mathbf{A} \cdot \mathbf{B} d^3 r = \mathbf{A} \cdot \mathbf{B}\, d\mathbf{l} \cdot d\mathbf{s} = \mathbf{A} \cdot d\mathbf{l}\, \mathbf{B} \cdot d\mathbf{s} \tag{3}$$

so that

$$K_1 = \oint_{C_1} \int_{S_1} \mathbf{A} \cdot d\mathbf{l}\, \mathbf{B} \cdot d\mathbf{s}. \tag{4}$$

The magnetic flux through a surface S is $\Phi = \int_S \mathbf{B} \cdot d\mathbf{s}$ and, using Stokes' theorem, can also be expressed as $\Phi = \oint_C \mathbf{A} \cdot d\mathbf{l}$ where C is the contour following the perimeter of S. The flux $\Phi_1 = \int_{S_1} \mathbf{B} \cdot d\mathbf{s}$ is invariant along C_1 and so may be factored from the integral in (4) giving

$$K_1 = \Phi_1 \oint_{C_1} \mathbf{A} \cdot d\mathbf{l}. \tag{5}$$

Since C_1 follows the length of tube #1, C_1 links tube #2. Because $\mathbf{B} = 0$ outside both V_1 and V_2,

$$\oint_{C_1} \mathbf{A} \cdot d\mathbf{l} = \Phi_2 \tag{6}$$

and so $K_1 = \Phi_1 \Phi_2$. From symmetry, it is seen that $K_2 = \Phi_1 \Phi_2$ also. Hence, the helicity of two linked, thin, untwisted flux tubes is

$$K = 2\Phi_1 \Phi_2. \tag{7}$$

Since (6) depended on linkage of the flux tubes, K quantifies flux tube linkage (if the flux tubes had not been linked then K would have been zero).

More generally, K can also be shown [Berger and Field, 1984; Pfister and Gekelman, 1991] to measure the twist of an individual flux tube and also cross-overs of segments of untwisted flux tubes. These various forms of helicity are equivalent and can be transformed into each other via continuous topological deformations [Pfister and Gekelman, 1991].

3. RELATIVE HELICITY, K_{rel}

Isolated, interlinked, untwisted, closed flux tubes constitute a highly contrived arrangement and are unlikely to occur in nature. Instead, there is usually a flux-conserving boundary penetrated by field lines as shown in Fig. 2. In this case some field lines are *open* (i.e., leave the volume of interest). Both driven laboratory spheromaks and solar corona magnetic structures have open field lines.

K is ambiguous when there are open field lines because then K depends on gauge, a physically meaningless quantity [Berger and Field, 1984; Finn and Antonsen, 1985; Jensen and Chu, 1984]. To demonstrate this ambiguity, we make a gauge transformation so the vector potential becomes $\mathbf{A}' = \mathbf{A} + \nabla f$ where f is an arbitrary scalar function. The helicity in the new gauge is defined as K' and so helicity will be gauge-invariant only if $K' = K$.

Explicit evaluation shows

$$\begin{aligned} K' &= \int_V (\mathbf{A} + \nabla f) \cdot \mathbf{B} d^3 r \\ &= K + \int_V \nabla \cdot (f\mathbf{B}) d^3 r \\ &= K + \int_S f \mathbf{B} \cdot d\mathbf{s}\,. \end{aligned} \tag{8}$$

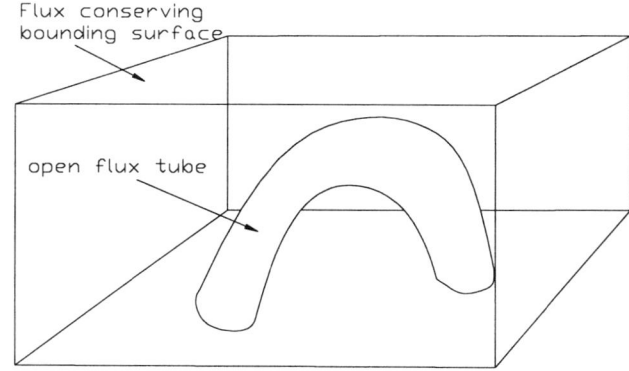

Figure 2. An open flux tube: field lines comprising the flux tube penetrate the flux-conserving bounding surface.

Since f is arbitrary, it is clear that in order to have $K' = K$ it is necessary to have $\mathbf{B}\cdot\hat{n}=0$ everywhere on S where $\hat{n}$ is the outward normal to the surface S. Equation (1) is gauge-dependent for open field line situations such as in Fig. 2 and therefore is physically meaningless.

Gauge ambiguity corresponds to ambiguity in counting flux linkages. If flux tubes leaving V have linkages, twist, writhe, or cross-overs external to V, these topological features are not taken into account by (1). Figure 3 shows this problem graphically. The two flux tubes shown in Fig. 3 are linked once inside V, but are also linked outside V. An integral over V cannot quantify how many times the flux tubes are linked.

This problem of linkage ambiguity (or equivalently gauge dependence) is resolved by defining a *relative helicity* which depends only on quantities defined inside V [*Berger and Field*,1984]. The relative helicity is gauge invariant and physically meaningful because it is independent of properties external to V.

The relative helicity is obtained by first defining a second volume V_b external to the volume of interest which we now call V_a. The volume V_b is defined so that the sum of the two volumes $V = V_a + V_b$ has no open field lines, i.e., $\mathbf{B}\cdot d\mathbf{s} =0$ on the surface of V but $\mathbf{B}\cdot d\mathbf{s} \neq 0$ on the interface surface between V_a and V_b. Often V_b is taken to be all space except V_a, but this is not necessary as long as $\mathbf{B}\cdot d\mathbf{s} =0$ on the surface bounding V. The surfaces bounding V_a, V_b, and V are denoted S_a, S_b, and S respectively, and the unit vector normal to S_a is $\hat{n}_a$, etc. We now construct a hypothetical reference magnetic field $\mathbf{B}_{ref}$ which:

1. equals $\mathbf{B}$ in V_b but differs in V_a,

2. has the same normal boundary conditions on S_a as does $\mathbf{B}$,

3. is easily calculated inside V_a.

The simplest magnetic field satisfying prescribed normal boundary conditions on a surface is the vacuum magnetic field $\mathbf{B}_{vac}$ (also called potential magnetic field). The vacuum field satisfies the current-free condition $\nabla \times \mathbf{B}_{vac} = 0$ and so can always be expressed as the gradient of a potential, i.e., $\mathbf{B}_{vac} = \nabla \chi_{vac}$. Because all magnetic fields must be divergence-free, this potential must satisfy Laplace's equation, i.e., $\nabla^2 \chi_{vac} = 0$. The properties of Laplace's equation indicate that, except for a constant, χ is completely determined by its normal derivative on the bounding surface (Neumann boundary condition); thus $\mathbf{B}_{vac}$ in a given volume is uniquely determined by $\hat{n} \cdot \mathbf{B}_{vac}$ on the volume's bounding sur-

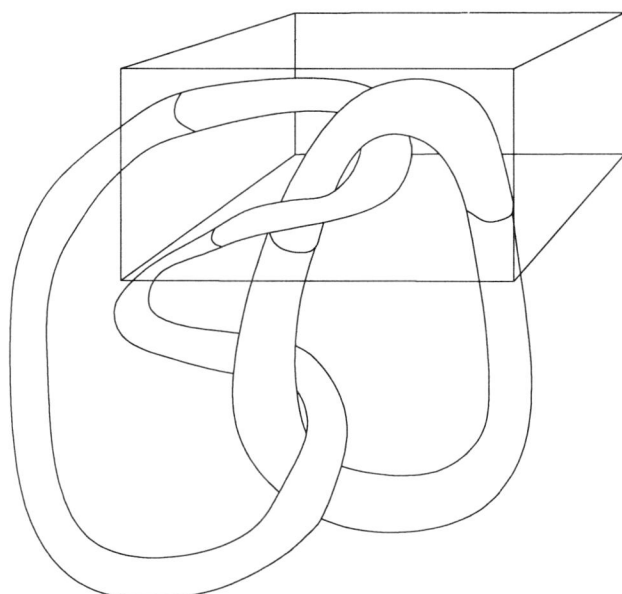

Figure 3. The two flux tubes have one linkage inside V (rectangular box) and one linkage outside V. The helicity K cannot account for the linkage outside V.

face. These simplifying features of vacuum magnetic fields are now exploited by constructing the reference magnetic field and its associated vector potential as

$$\mathbf{B}_{ref} = \begin{cases} \mathbf{B}_{vac} & \text{in } V_a \\ \mathbf{B} & \text{in } V_b \end{cases}$$

$$\mathbf{A}_{ref} = \begin{cases} \mathbf{A}_{vac} & \text{in } V_a \\ \mathbf{A}+\nabla h & \text{in } V_b \end{cases} \quad (9)$$

where $\nabla \times \mathbf{A}_{vac} = \mathbf{B}_{vac}$ in V_a, $\nabla \times \mathbf{A} = \mathbf{B}$ in V_b, $\hat{n}_a \cdot \mathbf{B} =\hat{n}_a \cdot \mathbf{B}_{vac}$ on S, and ∇h characterizes any allowed difference between $\mathbf{A}$ and $\mathbf{A}_{vac}$ in V_b. An integral over V will not suffer from gauge ambiguity or fail to count any linkages because, by assumption, V has no open field lines. Thus both $K_v = \int_V \mathbf{A} \cdot \mathbf{B} d^3r$ and $K_v^{ref} = \int_V \mathbf{A}_{ref}\cdot \mathbf{B}_{ref} d^3r$ are gauge invariant. To demonstrate gauge invariance we make separate gauge transformations for $\mathbf{A}$ and $\mathbf{A}_{ref}$ so that $\mathbf{A}' = \mathbf{A}+\nabla f$ and $\mathbf{A}'_{ref} = \mathbf{A}_{ref} + \nabla g$ and define the relative helicity as

$$K_{rel} = \int_V (\mathbf{A}+\nabla f) \cdot \mathbf{B} d^3r - \int_V (\mathbf{A}_{ref} + \nabla g) \cdot \mathbf{B}_{ref} d^3r. \quad (10)$$

Because each term is gauge invariant (i.e., independent of f and g respectively), K_{rel} is gauge invariant.

We now separate the volume integrals into integrals over V_a and V_b, so that

$$K_{rel} = \int_{V_a} (\mathbf{A}\cdot\mathbf{B} - \mathbf{A}_{ref}\cdot\mathbf{B}_{ref})\, d^3r$$
$$+ \int_{V_b} (\mathbf{A} - \mathbf{A}_{ref})\cdot\mathbf{B}\, d^3r. \quad (11)$$

The field in V_b is arbitrary except for the condition that it can only penetrate that portion of the surface of V_b which interfaces V_a. Thus, Eq.(11) becomes

$$\begin{aligned}K_{rel} &= \int_{V_a}(\mathbf{A}\cdot\mathbf{B}-\mathbf{A}_{vac}\cdot\mathbf{B}_{vac})\,d^3r \\ &+ \int_{V_b}\nabla h\cdot\mathbf{B}\,d^3r \\ &= \int_{V_a}(\mathbf{A}\cdot\mathbf{B}-\mathbf{A}_{vac}\cdot\mathbf{B}_{vac})\,d^3r \\ &- \int_{S_{int}} h\mathbf{B}\cdot\hat{n}_a\,ds \end{aligned} \quad (12)$$

where S_{int} is the interface between V_a and V_b.

Finn and Antonsen [1985] defined the relative helicity in slightly different form as

$$K_{rel}^{FA} = \int_{V_a} d^3r\,(\mathbf{A}+\mathbf{A}_{vac})\cdot(\mathbf{B}-\mathbf{B}_{vac}) \quad (13)$$

which can also be expressed as

$$K_{rel}^{FA} = \int_{V_a} d^3r\,(\mathbf{A}\cdot\mathbf{B}-\mathbf{A}_{vac}\cdot\mathbf{B}_{vac})$$
$$+ \int_{V_a}\nabla\cdot(\mathbf{A}\times\mathbf{A}_{vac})\,d^3r. \quad (14)$$

Equation (13) is equivalent to (10) since

$$\begin{aligned}\int_{V_a}\nabla\cdot(\mathbf{A}\times\mathbf{A}_{vac})\,d^3r &= -\int_{V_b}\nabla\cdot(\mathbf{A}\times\mathbf{A}_{vac})\,d^3r \\ &= -\int_{V_b}\nabla\cdot(\nabla h\times\mathbf{A}_{vac})\,d^3r \\ &= \int_{V_b}\nabla h\cdot\mathbf{B}_{vac}\,d^3r \\ &= \int_{S_{int}} h\mathbf{B}_{vac}\cdot\hat{n}_b\,ds \\ &= -\int_{S_{int}} h\mathbf{B}\cdot\hat{n}_a\,ds. \end{aligned} \quad (15)$$

In the first line we have used the condition that the tangential components of both $\mathbf{A}$ and $\mathbf{A}_{vac}$ must be gradients of potentials on the surface of V. The advantage of (13) is that it involves integration over V_a only and so does not require evaluation of h.

4. CONSERVATION EQUATION FOR RELATIVE HELICITY; INJECTION

The electric field can be decomposed into electrostatic and inductive components $\mathbf{E} = -\nabla\phi - \partial\mathbf{A}/\partial t$ so that

$$\partial\mathbf{A}/\partial t = -\mathbf{E} - \nabla\phi. \quad (16)$$

In order to clarify the algebra we define

$$\begin{aligned}\mathbf{E}_\pm &= \mathbf{E}\pm\mathbf{E}_{vac}, & \mathbf{B}_\pm &= \mathbf{B}\pm\mathbf{B}_{vac} \\ \mathbf{A}_\pm &= \mathbf{A}\pm\mathbf{A}_{vac}, & \phi_\pm &= \phi\pm\phi_{vac}; \end{aligned} \quad (17)$$

note that ϕ_{vac} is effectively a free parameter since it does not appear in (13). Following *Finn and Antonsen* [1985], we now calculate the time derivative of (13) obtaining

$$\begin{aligned}\frac{dK_{rel}}{dt} &= \int_{V_a} d^3r\left(\frac{\partial\mathbf{A}_+}{\partial t}\cdot\mathbf{B}_- + \mathbf{A}_+\cdot\frac{\partial\mathbf{B}_-}{\partial t}\right) \\ &= -\int_{V_a} d^3r\left[\begin{array}{c}(\mathbf{E}_++\nabla\phi_+)\cdot\mathbf{B}_- \\ +\mathbf{A}_+\cdot\nabla\times\mathbf{E}_-\end{array}\right] \\ &= -\int_{V_a} d^3r\,(\mathbf{E}_+\cdot\mathbf{B}_- + \mathbf{E}_-\cdot\nabla\times\mathbf{A}_+) \\ &\quad -\int_{S_a} d\mathbf{s}\cdot(\phi_+\mathbf{B}_- + \mathbf{E}_-\times\mathbf{A}_+) \\ &= -\int_{V_a} d^3r\,(\mathbf{E}_+\cdot\mathbf{B}_- + \mathbf{E}_-\cdot\mathbf{B}_+) \\ &= -2\int_{V_a} d^3r\,(\mathbf{E}\cdot\mathbf{B}-\mathbf{E}_{vac}\cdot\mathbf{B}_{vac}) \end{aligned} \quad (18)$$

where we have used $\mathbf{B}_-\cdot d\mathbf{s} = 0$ on S_a and have chosen $\nabla\phi_{vac}$ so that $\mathbf{E}_-\times d\mathbf{s} = 0$ on S_a.

If S_a is a flux conserver, then $\mathbf{B}\cdot\hat{n}$ must be constant in time on S_a. Since the vacuum field was defined to have the same normal component on S_a as the actual field, we must also have $\mathbf{B}_{vac}\cdot\hat{n}_a$ constant in time on S_a. Because $\mathbf{B}_{vac}$ inside V_a is completely determined by its normal component on S_a, the entire vacuum field inside V_a must therefore also be constant in time. Direct evaluation of the time derivative of $\mathbf{A}\cdot\mathbf{B}$ shows that

$$\frac{\partial}{\partial t}(\mathbf{A}\cdot\mathbf{B}) + \nabla\cdot\left(2\phi\mathbf{B}+\mathbf{A}\times\frac{\partial\mathbf{A}}{\partial t}\right) = -2\mathbf{E}\cdot\mathbf{B}. \quad (19)$$

Since we have shown that $\mathbf{A}_{vac}$ and $\mathbf{B}_{vac}$ are both constant in time, (19) reduces to

$$\nabla\cdot(\phi_{vac}\mathbf{B}_{vac}) = -\mathbf{E}_{vac}\cdot\mathbf{B}_{vac} \quad (20)$$

for a vacuum field bounded by a flux conserving surface. Thus (18) becomes

$$\frac{dK_{rel}}{dt} = -2\int_{V_a} d^3r \mathbf{E}\cdot\mathbf{B} - 2\int_{S_a} d\mathbf{s}\cdot(\phi_{vac}\mathbf{B}_{vac}). \quad (21)$$

Since $\mathbf{B}\cdot\hat{n}_a$ and $\mathbf{B}_{vac}\cdot\hat{n}_a$ are both constant in time on S_a, the tangential components of both $\partial\mathbf{A}/\partial t$ and $\partial\mathbf{A}_{vac}/\partial t$ must vanish on S_a. Because we chose ϕ_{vac} to make the tangential components of $\mathbf{E}$ and $\mathbf{E}_{vac}$ equal on S_a, the tangential components of $\nabla\phi$ and $\nabla\phi_{vac}$ must therefore also be equal on S_a. Thus ϕ and ϕ_{vac} differ at most by some constant c on S_a. The magnitude of c is of no consequence, because it does not affect the surface integral in (21), i.e., $\int_{S_a} d\mathbf{s}\cdot[(\phi-\phi_{vac})\mathbf{B}_{vac}] = c\int_{S_a} d\mathbf{s}\cdot\mathbf{B}_{vac} = 0$. The rate of change of relative helicity is therefore

$$\frac{dK_{rel}}{dt} = -2\int_{V_a} d^3r\eta\mathbf{J}\cdot\mathbf{B} - 2\int_{S_a} d\mathbf{s}\cdot(\phi\mathbf{B}) \quad (22)$$

where we have again used $\mathbf{B}\cdot\hat{n}_a = \mathbf{B}_{vac}\cdot\hat{n}_a$ on S_a and have invoked the MHD Ohm's law

$$\mathbf{E} + \mathbf{U}\times\mathbf{B} = \eta\mathbf{J} \quad (23)$$

inside V_a. If the plasma is a perfect conductor (i.e., $\eta=0$) and the bounding surface is an equipotential (i.e., ϕ =uniform on S_a), then the relative helicity is conserved inside V_a.

Now suppose there is just one open flux tube and let σ_1 and σ_2 be the two footpoint surfaces of this flux tube. Let ϕ_1, ϕ_2 be the respective electrostatic potentials of σ_1 and σ_2. In this case, (22) becomes

$$\frac{dK_{rel}}{dt} + 2\phi_1\int_{\sigma_1}\mathbf{B}\cdot d\mathbf{s} + 2\phi_2\int_{\sigma_2}\mathbf{B}\cdot d\mathbf{s} = -2\int_{V_a}\eta\mathbf{J}\cdot\mathbf{B}d^3r. \quad (24)$$

Since σ_1 and σ_2 are the two footpoints on S_a of the same flux tube, we must have

$$\int_{\sigma_1}\mathbf{B}\cdot d\mathbf{s} = -\int_{\sigma_2}\mathbf{B}\cdot d\mathbf{s}. \quad (25)$$

Defining $\psi(\sigma_1) = -\int_{\sigma_1}\mathbf{B}\cdot d\mathbf{s}$ as the flux into S_a through σ_1, we can express (24) as

$$\frac{dK_{rel}}{dt} = 2(\phi_1 - \phi_2)\psi(\sigma_1) - 2\int\eta\mathbf{J}\cdot\mathbf{B}d^3r. \quad (26)$$

The rate of helicity injection into the volume is therefore $2(\phi_1 - \phi_2)\psi(\sigma_1)$. In steady state, the two right-hand terms in (26) balance each other.

5. RELAXATION IN SYSTEMS WITH OPEN FIELD LINES

We now consider the long-term behavior of an arbitrarily complicated configuration inside a volume V_a which has open field lines. The dynamical evolution of this configuration involves turbulence, MHD instabilities, magnetic reconnection, waves, dissipation, etc.; the complexity of this dynamics would be difficult or impossible to analyze.

We impose only a few reasonable assumptions on this problem. First, it is assumed that the configuration is bounded by a flux-conserving surface S_a so that the open magnetic flux intercepted by S_a will be invariant. This invariance will be true for each differential element $d\mathbf{s}$ constituting S_a, i.e., $\mathbf{B}\cdot d\mathbf{s} = const.$ for each $d\mathbf{s}$. It is also allowed that S_a could be divided into electrically isolated flux-conserving segments having different electrostatic potentials [Jensen and Chu, 1984].

Second, it is assumed that $\beta = 2\mu_0 P/B^2 << 1$ so that hydrodynamic pressure may be neglected. Thus, any free energy must be magnetic.

Third, it is conjectured that when dissipation is added, the magnetic energy

$$W = \frac{1}{2\mu_0}\int_{V_a} B^2 d^3r \quad (27)$$

decays on a much faster time-scale than K_{rel} provided the characteristic scale length of the dissipative phenomena is much smaller than the scale length of the system [Taylor, 1974,1986]. Thus, in the limit that the scale length of the dissipative phenomena is microscopic, K_{rel} will be invariant on the time-scale of the magnetic energy decay. This conjecture is justified by the dimensional argument outlined in the next paragraph.

Magnetic energy scales as B^2L^3 whereas magnetic helicity scales as (flux)$^2 \sim B^2L^4$ (here L is the characteristic linear dimension). Suppose a certain dissipative process has a characteristic linear dimension L_1 and causes a reduction in magnetic energy $\Delta W_1 = \alpha B_1^2 L_1^3/\mu_0$ and associated reduction in magnetic helicity $\Delta K_1 = \gamma B_1^2 L_1^4$ where α and γ are dimensionless constants. If this process is re-scaled to a smaller linear dimension L_2 in a manner where the reduction in magnetic energy is kept the same, i.e., $B_1^2 L_1^3 = B_2^2 L_2^3$, then the associated reduction in magnetic helicity will be $\Delta K_2 = \gamma B_2^2 L_2^4 = \Delta K_1 \times L_2/L_1$. In the limit $L_2 \to 0$, negligible helicity will destroyed. Hence, if a certain amount of energy dissipation is caused by a set of intense microscopic events (e.g., microscopic reconnection events) negligible helicity will be destroyed compared to a spatially uniform dissipation of the same amount of energy. Detailed analytic models, numerical calculations, and experimental observations [*Barnes et al.*, 1986; *Ji et al.*, 1995] all support this conjecture. Geometrically complex phenomena involve smaller scale characteristic lengths than geometrically simple phenomena, and so helicity is expected to be conserved when there is complex microscopic activity.

Summarizing, the system is assumed to evolve in a spatially complex way such as to

1. conserve the relative helicity in V_a,
2. minimize the magnetic energy in V_a,
3. maintain invariant flux everywhere on S_a.

Condition 3 means that the variation of the component of $\mathbf{A}$ and $\mathbf{A}_{vac}$ tangential to the bounding surface must vanish, so that $\hat{n} \cdot \delta \mathbf{B} = \hat{n} \cdot \nabla \times \delta \mathbf{A} = 0$ and $\hat{n} \cdot \delta \mathbf{B}_{vac} = \hat{n} \cdot \nabla \times \delta \mathbf{A}_{vac} = 0$ on S_a.

Conditions 1-3 can be expressed as a variational problem, i.e., for prescribed boundary conditions, we minimize W subject to the constraint of conservation of K_{rel}. The constraint of helicity conservation is invoked by a Lagrange multiplier λ for the variation in K_{rel}. The variational problem is therefore

$$\delta W - \lambda \delta K_{rel} = 0 \quad (28)$$

or upon expansion,

$$\int_{V_a} \mathbf{B} \cdot (\nabla \times \delta \mathbf{A}) \, d^3 r$$
$$-\lambda \int_{V_a} [\delta \mathbf{A}_+ \cdot \mathbf{B}_- + \mathbf{A}_+ \cdot \nabla \times \delta \mathbf{A}_-] \, d^3 r = 0 \quad (29)$$

where $\delta \mathbf{A}_\pm = \delta \mathbf{A} \pm \delta \mathbf{A}_{vac}$ and $\delta \mathbf{A}, \delta \mathbf{A}_{vac}$ are arbitrary variations with vanishing tangential components on S_a. Note that some constant factors have been absorbed into λ. Because $\delta \mathbf{A}_{vac}$ is completely determined inside V_a by the tangential component of $\delta \mathbf{A}_{vac}$ on S_a, we realize that $\delta \mathbf{A}_{vac} = 0$ everywhere inside V_a. In other words, $\delta \mathbf{A}_{vac}$ must be zero everywhere because it is determined by $\mathbf{B} \cdot \hat{n}_a$, a quantity which cannot have a variation because S_a is flux-conserving. Setting to zero all terms involving $\delta \mathbf{A}_{vac}$ gives

$$\int_{V_a} \mathbf{B} \cdot (\nabla \times \delta \mathbf{A}) \, d^3 r$$
$$-\lambda \int_{V_a} [\delta \mathbf{A} \cdot \mathbf{B}_- + \mathbf{A}_+ \cdot \nabla \times \delta \mathbf{A}] \, d^3 r = 0. \quad (30)$$

After integration by parts on the two curl terms, (30) becomes

$$\int_{V_a} \delta \mathbf{A} \cdot (\nabla \times \mathbf{B}) \, d^3 r$$
$$-\lambda \int_{V_a} [\delta \mathbf{A} \cdot (\mathbf{B} - \mathbf{B}_{vac}) + \delta \mathbf{A} \cdot \nabla \times (\mathbf{A} + \mathbf{A}_{vac})] \, d^3 r = 0 \quad (31)$$

where all surface integrals vanish because the tangential component of $\delta \mathbf{A}$ vanishes on S_a. The terms involving $\mathbf{B}_{vac}$ cancel each other and the factor of 2 is absorbed into λ giving

$$\int_{V_a} \delta \mathbf{A} \cdot (\nabla \times \mathbf{B} - \lambda \mathbf{B}) \, d^3 r = 0. \quad (32)$$

Since $\delta \mathbf{A}$ is arbitrary within V_a, we must have

$$\nabla \times \mathbf{B} = \lambda \mathbf{B} \quad (33)$$

where $\mathbf{B}$ satisfies the normal boundary conditions prescribed on S_a. The solutions to (33) are called relaxed states and give the minimum energy configuration having the original inventory of relative helicity and satisfying the prescribed flux-conserving boundary conditions on S_a. *Any initial configuration will self-organize to the relaxed state after sufficient time.* Equation (33) has been shown to provide an excellent representation of the magnetic field profile observed in certain low β laboratory devices such as reversed field pinches [*Ortolani and Schnack*, 1993] and spheromaks [*Jarboe*, 1994]. Of equal importance (33) provides an excellent representation for interplanetary magnetic clouds [*Burlaga*,1988]. It also provides a compelling model for structures in the solar corona [*Heyvaerts and Priest*, 1984; *Browning*,1988; *Rust*, 1994; *Pevtsov and Canfield*,1997]. Equation (33) has been shown by *Dixon et al.*[1989] to apply also to the case of a free boundary subjected to external magnetic or plasma pressure forces.

6. DISCUSSION OF λ

λ has various interpretations and discussion of these interpretations provides useful insights into the properties of relaxed states [*Fernandez et al.*, 1989]. If $\lambda = 0$ then the configuration is current-free and so the magnetic field will be the vacuum field $\mathbf{B}_{vac} = \nabla \chi_{vac}$. In this case the relative helicity will be zero.

Suppose σ is some arbitrary surface either within V_a or on the bounding surface S_a of V_a. Integrating (33) over σ gives

$$\int_\sigma d\mathbf{s} \cdot \nabla \times \mathbf{B} = \int_\sigma d\mathbf{s} \cdot \lambda \mathbf{B}. \quad (34)$$

or
$$\mu_0 I(\sigma) = \lambda \psi(\sigma) \quad (35)$$

where $I(\sigma)$ is the current through σ and $\psi(\sigma)$ is the magnetic flux through σ. If σ is the cross-section an arbitrary flux tube, then λ is just the ratio of the axial current to the axial flux through the flux tube.

Now consider the relative magnetic energy, i.e., the increment in energy relative to the vacuum field energy

$$W_{rel} = \frac{1}{2\mu_0} \int_{V_a} \left(B^2 - B_{vac}^2\right) d^3r. \quad (36)$$

Writing $\mathbf{B}_{vac} = \nabla \chi_{vac}$, the relative magnetic energy can be evaluated as

$$\begin{aligned} W_{rel} &= \frac{1}{2\mu_0} \int_{V_a} (\mathbf{B} - \mathbf{B}_{vac}) \cdot (\mathbf{B} + \mathbf{B}_{vac}) d^3r \\ &= \frac{1}{2\mu_0} \left[\begin{array}{l} \int_{V_a} (\mathbf{B} - \mathbf{B}_{vac}) \cdot \mathbf{B} d^3r \\ + \int_{S_a} \chi_{vac} (\mathbf{B} - \mathbf{B}_{vac}) \cdot d\mathbf{s} \end{array} \right] \\ &= \frac{1}{2\mu_0} \int_{V_a} (\mathbf{B} - \mathbf{B}_{vac}) \cdot \nabla \times \mathbf{A} d^3r \quad (37) \end{aligned}$$

where the surface integral vanishes because by assumption $(\mathbf{B} - \mathbf{B}_{vac}) \cdot d\mathbf{s} = 0$ on S_a. Equation (33) can be integrated to give

$$\nabla \times \mathbf{A} = \lambda \mathbf{A} + \nabla f \quad (38)$$

where f is a gauge function. Inserting (38) into (37) gives

$$\begin{aligned} W_{rel} &= \frac{1}{2\mu_0} \int_{V_a} (\mathbf{B} - \mathbf{B}_{vac}) \cdot (\lambda \mathbf{A} + \nabla f) d^3r \\ &= \frac{\lambda}{2\mu_0} \int_{V_a} (\mathbf{B} - \mathbf{B}_{vac}) \cdot \mathbf{A} d^3r \quad (39) \end{aligned}$$

This provides another interpretation for λ, namely

$$\lambda = \frac{2\mu_0 W_{rel}}{\int_{V_a} (\mathbf{B} - \mathbf{B}_{vac}) \cdot \mathbf{A} d^3r} \quad (40)$$

so that λ is very nearly the ratio of the relative energy to the relative helicity (the denominator of (40) differs slightly from K_{rel}).

7. EVOLUTION OF THE RELAXED STATE

Because the relaxed state depends on both λ and on boundary conditions, it will change if either λ or the boundary conditions are varied. It is essential to realize that this change is *not* a dynamical evolution, because by assumption, the relaxed state is what results after dynamical behavior has finished; there is no motion in a relaxed state. Since the evolution is not dynamical, there is no reason to expect one relaxed state to evolve continuously into another as some parameter (e.g., λ) is adjusted. This situation is analogous to a thermodynamic phase transition; for example water freezes when cooled from 0^+ Celsius to 0^- Celsius. At each temperature the system is in thermodynamic equilibrium (analogous to the relaxed state), but for a slight change of an independent parameter, a discontinuous change of equilibrium can occur. There is actually complicated dynamical behavior associated with the discontinuous change, but consideration of this dynamics has been avoided by restricting the model to characterizing the system *after* dynamical behavior is over.

To study the properties of particular relaxed states, it is necessary to choose a coordinate system, prescribe V_a, and prescribe boundary conditions on S_a. Symmetry is also important; it is tempting to prescribe a great deal of symmetry in order to make analysis easier, but this can lead to misleading oversimplification because flux tubes can break up when there is asymmetry.

One also has to be careful about paradoxes that can occur if V_a is assumed to be infinite in one or more dimensions (e.g., see *Aly*, 1992 and *Laurence and Avellaneda*, 1993). Relaxation theory is based on the assumption that all dynamics has concluded. If a system were truly infinite, it would take forever for information to propagate throughout the system and allow a relaxed state to develop. As a consequence, the relaxed state concept makes physical sense only if the system dimensions do not exceed the distance information can travel in the time interval under consideration. Information about changes in magnetic topology travels at v_A the Alfvén velocity [*Bellan*, 1998; *Song and Lysak*, 1998] and so, for relaxation to be a physically meaningful model, the dimensions of V_a must be substantially less than $v_A t$ where t is the time interval under consideration.

For definiteness, we now briefly consider an example of relaxed states relevant to the solar corona. The solar surface is assumed flat and z denotes the altitude above this surface. The general solution to (33) can be written as

$$\mathbf{B} = \lambda \nabla \chi \times \nabla z + \nabla \times (\nabla \chi \times \nabla z) \quad (41)$$

where

$$\nabla^2 \chi + \lambda^2 \chi = 0; \quad (42)$$

this can be verified by direct substitution of (41) into (33). Using cylindrical geometry with $\exp(im\phi)$ dependence (42) becomes

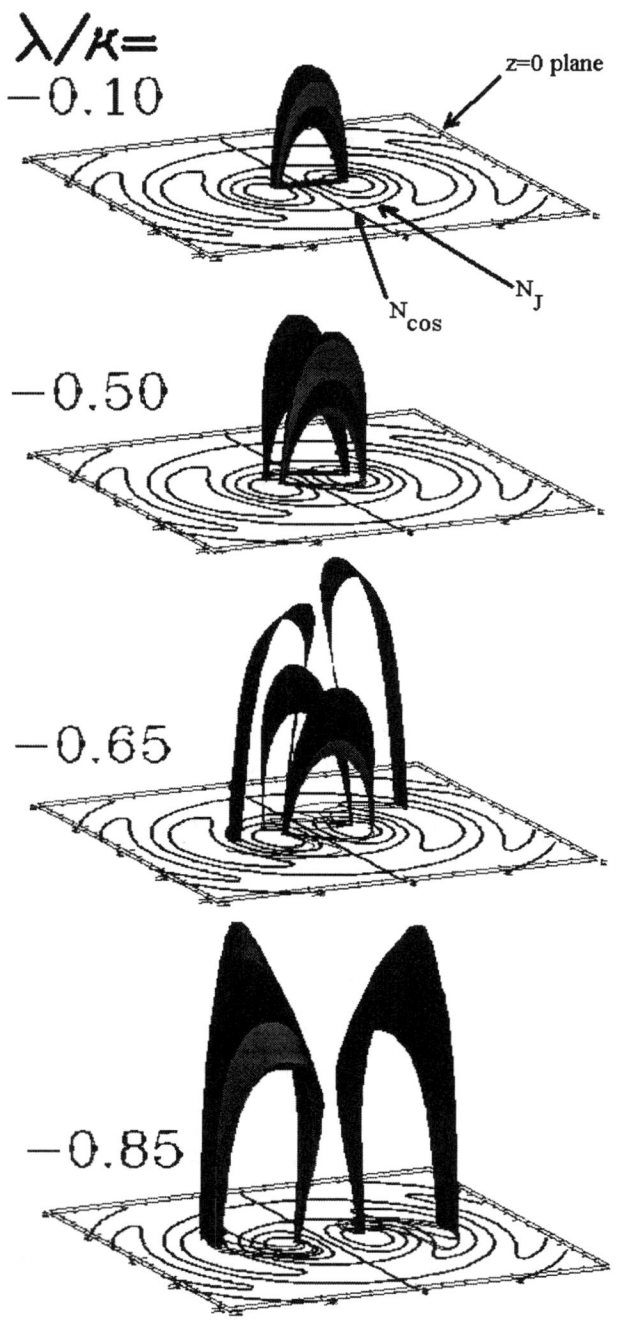

Figure 4. Flux tubes obtained by solving (47) using fields from (45) for a range of λ/κ. The neutral line N_J corresponding to first zero of $J_1(\kappa r)$ and the neutral line $N_{\cos}$ corresponding to the zero of the cosine are shown in top plot. Note discontinuous change in topology as λ varies and the bifurcation when $\lambda/\kappa = -0.65$ as field lines change from crossing $N_{\cos}$ to crossing N_J.

$$\frac{1}{r}\frac{\partial}{\partial r}r\frac{\partial \chi}{\partial r} + \frac{\partial^2 \chi}{\partial z^2} + \left(\lambda^2 - \frac{m^2}{r^2}\right)\chi = 0. \quad (43)$$

If we also assume χ has a radial dependence $\sim J_m(\kappa r)$, then (43) shows that the complete r, ϕ, z dependence of χ must be

$$\chi \sim J_m(\kappa r) \exp(\pm\sqrt{\kappa^2 - \lambda^2} z) \exp(im\phi). \quad (44)$$

Thus, equilibria can only exist if $\lambda^2 < \kappa^2$. From (41) and assuming a $\cos m\phi$ dependence for χ, the field components are

$$B_r = -\bar{\chi}\left[\begin{array}{c}\lambda r^{-1}m J_m(\kappa r)\sin m\phi \\ +k\kappa J'_m(\kappa r)\cos m\phi\end{array}\right]e^{-kz} \quad (45\text{a})$$

$$B_\phi = -\bar{\chi}\left[\begin{array}{c}\lambda\kappa J'_m(\kappa r)\cos m\phi - \\ mkr^{-1}J_m(\kappa r)\sin m\phi\end{array}\right]e^{-kz} \quad (45\text{b})$$

$$B_z = \bar{\chi}\kappa^2 J_m(\kappa r)e^{-kz}\cos m\phi \quad (45\text{c})$$

where $k = \sqrt{\kappa^2 - \lambda^2}$ and $\bar{\chi}$ is a constant.

The specific example of $m = 1$ provides a reasonable model for a solar prominence and also for the solar prominence simulation experiments underway at Caltech [*Bellan and Hansen*, 1998]. For these situations

$$\chi = \bar{\chi}J_1(\kappa r)\exp(-\sqrt{\kappa^2 - \lambda^2}z)\cos\phi \quad (46)$$

and the fields are obtained using $m = 1$ in (45). If $\lambda = 0$, the field is a potential field with the same normal boundary conditions as the actual field. The quantity κ is therefore determined by the radial behavior of the boundary condition in the $z = 0$ plane and in particular, κ^{-1} corresponds to the separation between footpoints.

Equation (45c) shows that B_z has the same functional dependence as χ; thus χ can be used as a proxy for B_z. Note that B_z is independent of λ on the ground plane $z = 0$ and so variation of λ does not change the flux in the ground plane. Consider a contour plot of B_z in the $z = 0$ plane; regions of positive and negative B_z are separated by neutral lines where $B_z = 0$. Equation (46) shows that these neutral lines correspond to the roots of $J_1(\kappa r)$ and the roots of $\cos\phi$; these neutral lines will be referred to as N_J and $N_{\cos}$. Figure 4 plots these contours in the $z = 0$ plane and indicates the neutral lines. The regions of positive B_z can be called "hills" and the regions of negative B_z can be called "valleys" of the function B_z.

The trajectory of a magnetic field line is obtained by numerically integrating

$$\frac{d\mathbf{r}}{d\tau} = \hat{B}. \quad (47)$$

If the integration starts from a hill in the $z = 0$ plane, then B_z is initially positive and so the field line trajectory will go upwards but is also deflected sideways because of finite B_r and B_ϕ. Because all components of **B** have the same exponential dependence with altitude for this single-mode situation, the field as well as the neutral lines are self-similar at all altitudes [*Bellan and Hansen*, 1998].

Consider a field line originating from the dominant hill of (46) on the side close to neutral line $N_{\cos}$. The field line trajectory will go up, be deflected to cross over $N_{\cos}$, and thus be over a valley so that it goes down to have its landing point (intersection with the $z = 0$ plane) in the dominant valley.

If λ is increased, the sideways deflection causes the field line to swerve to cross over the neutral line N_J. When this happens, the field line finds itself over a different valley from before. The field line heads down and lands in this different valley. Whether the field line crosses N_J or $N_{\cos}$ depends on its starting point and on λ. A small change in λ can therefore cause the field line trajectory to change discontinuously much like an airplane blown slightly off course may be forced to land in unfriendly territory and so have a very different flight path from the case with no side-winds. Figure 4 shows this bifurcation and discontinuous change in topology as $|\lambda|$ increases. In Fig. 4 two flux tubes and their bifurcations are plotted; the two flux tubes are most easily identified in the $\lambda/\kappa = -0.50$ plot (they are so close together in the $\lambda/\kappa = -0.10$ plot as to be indistinguishable). The foreground flux tube is obtained by plotting a set of adjacent field lines going from left to right where the starting point for these field lines is a small circle in the $z = 0$ plane. The ribbon surface between adjacent field lines is shaded and to distinguish ribbons, different shades are used for different ribbons. The background flux tube is formed by following a similar set of adjacent field lines, but integrating (47) backwards (for a more complete description see *Bellan and Hansen* [1998]). The starting points for the sets of field lines are the same in all plots; only λ is varied.

Thus, the topology of the system is a discontinuous function of λ and, in fact, if $\lambda^2 > \kappa^2$, relaxed states do not exist. It is obvious that there will have to be substantial magnetic reconnection to produce the discontinuous dependence on λ. Magnetic flux and magnetic energy are not conserved as λ is changed; however magnetic flux is conserved in the ground plane.

It is interesting to note that the dependence of field-line landing point on λ means that the two ends of the field lines are not frozen into the ground plane when λ is varied. Although flux is conserved in the ground plane, field lines are not frozen into this plane.

The relaxed state provides a reasonable explanation [*Bellan and Hansen*, 1998] for the $S-$shapes often seen [*Pevtsov and Canfield*, 1998] on the solar surface. Suppose λ is positive: When the field line trajectory is upwards ($B_z > 0$), then $(\nabla \times \mathbf{B})_z = \lambda B_z$ is positive, but when the trajectory is downward ($B_z < 0$) then $(\nabla \times \mathbf{B})_z = \lambda B_z$ is negative. Thus, the z component of the curl is positive for the upward portion and negative for the downward portion of the field line; the projection of this positive-then-negative curl on the solar surface gives an $S-$ shape.

8. RELAXED STATES, THE MHD ENERGY PRINCIPLE, AND STATIONARY STATES

The MHD energy principle [*Bernstein et al.*, 1958] provides a standardized technique for determining stability of an MHD equilibrium. Application of the energy principle involves calculating the change in system potential energy for a virtual displacement of system fluid elements subject to prescribed boundary conditions. The energy principle shows there are two general categories of instability, pressure-driven and current-driven. In a low β plasma the free energy associated with pressure is negligible and so only current-driven instabilities are significant. Current-driven instabilities are kink-like and involve the plasma coiling up to increase its self-inductance.

Using the energy principle it is straightforward to show that magnetic free energy is proportional to $|\nabla \lambda|$ for a helicity-conserving displacement. Since $\nabla \lambda = 0$ in a relaxed state, a relaxed state can be considered as the culmination of kink instability and so a relaxed state is inherently kink-stable. Thus MHD instabilities tend to smooth out λ; as *Fernandez et al.* [1989] noted, this can be considered as a flow from regions of large λ (large helicity) to regions of small λ.

This interpretation shows that λ plays a role somewhat analogous to the role of temperature in thermodynamics. Thermodynamic equilibria are states of maximum entropy for a given energy and are parameterized by the temperature. If two thermodynamic equilibrium states having different temperatures are connected, heat flows from the high temperature state to the low temperature state and in equilibrium the combined state has a uniform temperature. According to relaxation theory, magnetohydrodynamic equilibria are states of minimum energy for a given helicity; these equilibrium states are characterized by λ. If two magnetohydrodynamic equilibrium states having different λ's are connected, helicity flows from the high λ state to the low λ state and in equilibrium the combined state has uniform λ, i.e., is a Taylor state.

In more complicated thermodynamic problems incorporating physically separated heat sources and sinks, the stationary state is not isothermal, but has a temperature gradient from the heat source to the heat sink. In steady state (which is not the same as true thermodynamic equilibrium), the temperature gradient adjusts itself to give the required stationary heat flux $\Gamma = -D\nabla T$ between the source and sink. The heat diffusion coefficient D is determined by microscopic dynamics outside the realm of thermodynamics. For magnetohydrodynamic problems incorporating physically separated helicity sources and sinks (e.g., the solar dynamo as the source and infinite half-space as the sink), the analogous stationary state will not have uniform λ but instead will have a λ gradient from source to sink. There will be a helicity diffusion coefficient D_K depending on non-equilibrium microscopic processes and the λ gradient will adjust itself to produce the required stationary helicity flux $\Gamma_K = -D_K \nabla \lambda$ between source and sink. On scales much smaller than the distance between source and sink, λ can be considered as approximately uniform (in analogy to a local thermodynamic equilibrium). Thus, isothermal states and Taylor states are reasonable approximate descriptions of respective thermodynamic and magnetohydrodynamic stationary states having characteristic lengths much smaller than the distance between source and sink.

Acknowledgments. This work was supported by the US Department of Energy.

REFERENCES

Aly, J. J., Some Properties of finite energy constant α force-free magnetic fields in a half space *Solar Phys.* 138, 133, 1992.

Barnes, C. W., Fernandez, J. C., Henins, I., Hoida, H. W., Jarboe, T. R., Knox, S. O., Marklin, G. J., and McKenna, K. F., Experimental determination of the conservation of magnetic helicity from the balance between source and spheromak, *Phys. Fluids* 29, 3415, 1986.

Bellan, P. M. and Hansen, J. F., Laboratory Simulations of Solar Prominence Eruptions, *Phys. Plasmas* 5, 1991, 1998.

Bellan, P. M., Collisionless reconnection using Alfvén wave radiation resistance, to appear in *Phys. Plasmas*.

Berger, M. A. and Field, G. B., The topological properties of magnetic helicity, *J. Fluid Mech.* 147, 133, 1984.

Bernstein, I. B., Friedman, E. A., Kruskal, M. D., and Kulsrud, R. M., An energy principle for hydromagnetic stability problems, *Proc. Roy. Soc.* 244, 17, 1958.

Browning, P. K., Helicity injection and relaxation in a solar-coronal magnetic loop with a free surface, *J. Plasma Phys.* 40, 263, 1988.

Burlaga, L. F., Magnetic Clouds and Force-Free Fields with Constant α, *J. Geophys. Res.* 93, 7217, 1988.

Dixon, A. M., Berger, M. A., Browning, P. K., and Priest, E. R., A generalization of the Woltjer minimum energy principle, *Astron. Astrophys.* 225, 155, 1989.

Fernandez, J. C., Wright, B. L., Marklin, G. J., Platts, D. A., and Jarboe, T. R., The $m=1$ helicity source spheromak experiment, *Phys. Fluids B* 1, 1254, 1989.

Finn, J. M. and Antonsen, T. M., Jr., Magnetic Helicity: What it is and what is it good for?, *Comments Plasma Phys. Controlled Fusion* 9, 11, 1985.

Heyvaerts, J. and Priest, E. R., Coronal heating by reconnection in DC current systems. A theory based on Taylor's hypothesis, *Astron. Astrophys.* 137, 63, 1984.

Jarboe, T. M., Review of spheromak research, *Plasma Phys. Controlled Fusion* 36, 945, 1994.

Jensen, T. H. and Chu, M. S., Current drive and helicity injection, *Phys. Fluids* 27, 2881, 1984.

Ji, H., Prager, S., and Sarff, J., Conservation of magnetic helicity during plasma relaxation, *Phys. Rev. Letters* 74, 2945, 1995.

Laurence, P. and Avellaneda, M., Woltjer's variational principle, II: the case of unbounded domains, *Geophys. Astrophys. Fluid Dynamics* 69, 201, 1993.

Moffatt, H. K. *Magnetic Field Generation in Electrically Conducting Fluids* (Cambridge University Press, 1978), p.19.

Ortolani, S. and Schnack, D. D., *Magnetohydrodynamics of Plasma Relaxation*, World Scientific Publishing, 1993.

Pevtsov, A. A. and Canfield R. C., On the subphotospheric origin of coronal electric currents, *Ap. J.* 481, 973, 1997.

Pfister, H. and Gekelman, W., Demonstration of helicity conservation during magnetic reconnection using Christmas ribbons, *Am. J. Phys.* 59, 497, 1991.

Rust, D. M., Spawning and Shedding of Helical Magnetic Fields in the Solar Atmosphere, *Geophys. Res. Letters* 21, 241, 1994.

Song, Y., and Lysak, R. L., Magnetic Helicity and Solar Wind-Magnetosphere Interaction, this monograph.

Taylor, J. B., Relaxation of Toroidal Plasma and Generation of Reverse Magnetic Fields, *Phys. Rev. Lett.* 33, 1139, 1974.

Taylor, J. B., Relaxation and magnetic reconnection in plasmas, *Rev. Mod. Phys.* 58, 741, 1986.

Woltjer, L. W., A theorem on force-free magnetic fields, *Proc. Nat. Acad. Sci. (USA)* 44, 490, 1958.

P. M. Bellan, Applied Physics MS 128-95, California Institute of Technology, Pasadena CA 91125. (e-mail: pbellan@cco.caltech.edu)

Study of Magnetic Helicity and Relaxation Phenomena in Laboratory Plasmas

Masaaki Yamada

Plasma Physics Laboratory, Princeton University, James Forrestal Campus Princeton, New Jersey 08543

With recent growing interest in magnetic helicity in the solar physics community, the present paper reviews experimental results on evolution of magnetic helicity during the formation and relaxation of laboratory plasmas. Measurements made in spheromaks, reversed field pinch (RFP) and merging plasma experiments are reviewed. In laboratory plasmas boundaries are generally well defined thus it is straightforward to define a gauge invariant relative helicity. It was found by J.B. Taylor that in the formation process laboratory plasmas seek the minimum energy state, while keeping the relative magnetic helicity roughly constant. This state can be described as a force free state with an additional constraint $\mu = \mathbf{j}\cdot\mathbf{B}/B^2$ = constant throughout the plasma interior. Also in the sustained state of laboratory plasmas, it has been verified that plasmas repeatedly relax to the Taylor minimum energy states while conserving the relative magnetic helicity. It is seen that the confinement of these pinch plasmas is often significantly affected by the relaxation processes.

1. INTRODUCTION

Magnetic field generated by current in a plasma can effectivelyconfine high pressure plasmas by generating a stable toroidal pinch configuration. But this configuration is often susceptible to a relaxation phenomenon which deteriorates its plasma confinement feature. In order to quantitatively describe the characteristics of current carrying laboratory plasmas, magnetic helicity has been defined and studied as an important physics parameter [*Taylor* 1986]. Recently there has been a growing interest in magnetic helicity — a global quantity to express a "knottedness" of twisted magnetic field lines---in the solar physics community[*Rust*, 1994]. A primary objective of the present paper is to review experimental studies of evolution of magnetic helicity during the formation and relaxation processes of laboratory plasmas.

A toroidal pinch is one of the simplest systems for confining plasma by magnetic field and most utilized for confining hot plasmas for fusion research. In these systems a toroidal current is induced to heat and compress the plasma through the well known pinch effects. Tokamaks, reversed field (toroidal) pinch (RFP), and spheromak configurations belong to this category. The magnetic field produced by internal current provides an inward pinch force by generating a force balance with outward plasma pressure. Fig.1 presents schematics for these three configurations. While all these configuration generate self-pinching poloidal fields, toroidal fields are supplied differently; tokamak's toroidal field is very strong and is primarily created by external coils, while the toroidal field of RFP is created by combined effect of internal current and small external field and is much weaker than that of a tokamak. The spheromak does not have external toroidal field and its internal toroidal field is

Magnetic Helicity in Space and Laboratory Plasmas
Geophysical Monograph 111
Copyright 1999 by the American Geophysical Union

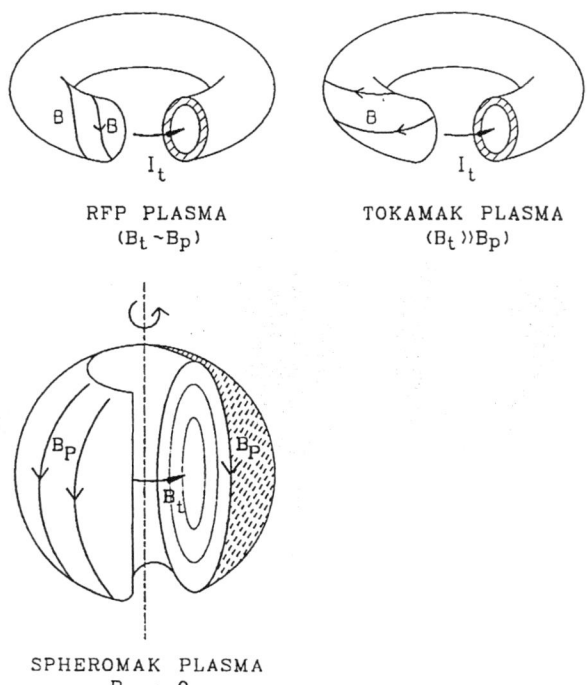

FIg.1. Schematic comparison of toroidal pinch configuration.

generated by the internal current. There is a remarkable feature common to all of these toroidal pinch discharges. It has been found that, after an initial highly turbulent state, the plasma settles into a more quiescent stable state in which the fluctuations are reduced.

To quantitatively describe the stable configuration of the twisted field lines in toroidal pinch plasmas, J.B. Taylor[1974] formulated a concept that during the relaxation of pinch plasma configuration the plasma seeks "a minimum energy state" while conserving magnetic helicity K. Here K is defined to monitor the linkage of magnetic field lines:

$$K = \int \mathbf{A} \cdot \mathbf{B}\, d^3 x,$$

where $\mathbf{A}$ is the vector potential for the magnetic field vector $\mathbf{B}$ and the integration is carried out inside the volume where the normal component of magnetic field is zero at the boundary ($\mathbf{B} \cdot \mathbf{n} = 0$). In laboratory experiments, MHD plasmas are created with variety of well defined boundary conditions which can be controlled externally.

During the formation of RFP's and spheromaks it was found that relaxation to the Taylor (minimum-energy, force-free)state was an important mechanism. In these pinch configurations, spatial profiles of $\mu = \mathbf{j} \cdot \mathbf{B}/B^2$ have been measured from internal probe data to verify the force-free minimum energy Taylor state which is characterized by μ = constant. Magnetic fluctuations with specific toroidal and poloidal mode numbers have been identified and found to be responsible for attaining the Taylor state. It has been observed that they often cause a conversion of poloidal to toroidal flux, or toroidal to poloidal flux, to occur.

In high temperature toroidal pinch discharges such as tokamaks, RFP's and spheromaks, "sawtooth"-type relaxation events were often observed and their primary mechanism was made clear by measuring their magnetic field and spatial electron temperature profiles[*Kadomtsev*, 1975]. If the plasma confinement is good at the plasma center, a deviation of the plasma configuration from the Taylor state is seen in most tokamak discharges. The good central confinement and large resistive loss at the plasma edge is considered to be responsible for this highly peaked current profile [$\mu \neq$ constant] , a deviation from Taylor state. This configuration can be usually maintained in stable manner in tokamak discharges.

When the magnetic configuration departs from the Taylor state in a RFP or a spheromak, it easily becomes unstable to low n (n = toroidal mode number) ideal MHD modes. Through these instabilities, magnetic reconnection takes place (relaxation) and a Taylor state is recovered [μ = constant]. It is found that the confinement time of these plasmas is significantly affected by this relaxation phenomenon.

Conservation of magnetic helicity has been verified during relaxation in many laboratory pinch plasmas in which the Taylor minimum energy states are identified after the relaxation events. Recently merging spheromaks have been utilized to investigate magnetic reconnection of magnetized toroidal plasmas or helical flux ropes [*Yamada et al.*,1990, *Ono et al.*, 1993]. Experiments with fully 3 dimensional magnetic reconnection are now possible in the MRX device [*Yamada et.al.*,1997] where evolution of magnetic helicity can be studied.

In this paper experimental findings on the relaxation processes of current carrying plasmas are reviewed to elucidate the fundamental physics mechanisms. The evolution of magnetic helicity of these pinch plasmas is studied. A special focus is put on spheromak characteristics since RFP properties will be described in another paper of this monograph [*Prager*]. In the end of this paper, a possible application of the laboratory results to the observed phenomena in the solar atmosphere is presented.

2. MAGNETIC HELICITY AND TAYLOR'S MINIMUM ENERGY STATE

In order to quantitatively describe the characteristics of current carrying laboratory plasmas, magnetic helicity has been defined and studied as an important physics parameter. Magnetic helicity represents a net twist, linkage or the "knottedness" of magnetic field, and defined by

$$K \equiv \int \mathbf{A} \cdot \mathbf{B} \, dV \qquad (1)$$

in a closed boundary. **B** is the magnetic field and **A** is its vector potential. If we express K by magnetic flux;

$$K = \int \mathbf{A} \cdot d\mathbf{l} \, B da = \int \psi \, d\Phi \qquad (2)$$

because magnetic flux can be expressed $B da = d\Phi$, $\int \mathbf{A} \cdot d\mathbf{l} = \Psi$ (where the integrals extend over a suitable boundary), we can write the magnetic helicity as the product of linked fluxes aas shown in Fig.2, where "da" represents differential cross section of the torus. For two simply connected loops of flux, ψ_o and Φ_o, the entire volume integration simply leads to [*Berger and Field*, 1984]

$$K = 2 \psi_o \Phi_o, \qquad (2a)$$

For laboratory plasmas, we often define the magnetic helicity after subtracting the helicity associated with vacuum magnetic fields thus making it gauge-invariant definition.

The gauge-invariant definition of the magnetic helicity in toroidal geometry is expressed by

$$K \equiv \int \mathbf{A} \cdot \mathbf{B} \, dV - \Phi_t(a)\psi_p(a) = \int [\mathbf{A} - \mathbf{A}_t(a)] \cdot \mathbf{B} \, dV \qquad (3)$$

where $\Phi_t(a)$ is the total toroidal flux, $\psi_p(a)$ is the external poloidal flux threading the central hole of the torus and $A_t(a) = \psi_p(a)/(2\pi R)$. Evolution of magnetic helicity becomes an issue whenever a plasma changes its topology thru magnetic reconnection or relaxation processes.

Based on his careful examination of RFP formation processes, J. B. Taylor proposed a conjecture that K should be conserved during plasma relaxation as the plasma rapidly approaches its minimum-energy state. The essential point of his hypothesis is that magnetic energy dissipates much faster than helicity during plasma relaxation inside of a boundary with $\delta \mathbf{B} \cdot \mathbf{n} = 0$.

With this assumption, Taylor theory predicts the minimum-energy state decribed by

$$\nabla \times \mathbf{B} = \mu \mathbf{B} \qquad (4)$$

with μ =constant. $\qquad (4a)$

The quantity μ (the Lagrange multiplier in the calculation) is a constant eigenvalue with units of (length)$^{-1}$. While μ is defined as a measure of the ratio of current density to magnetic field, $\mathbf{j} \cdot \mathbf{B}/B^2$ in the interior of the plasma, it is also equal to the ratio of total magnetic energy to helicity, $2W/K$. This finding has been very effective in describing many pinch plasma states in which internal current plays a major role in confining plasma.

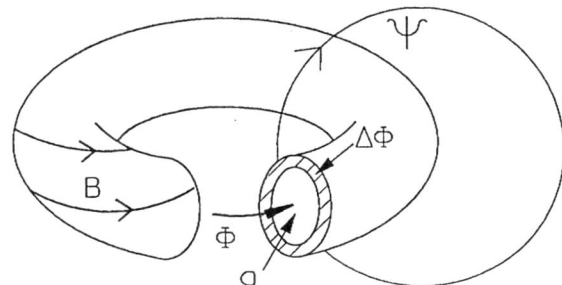

Fig. 2. Geometry of Magnetic helicity for toroidal plasma

3. FORMATION AND RELAXATION OF REVERSED FIELD PINCH

3.1 Formation of RFP Plasmas

The RFP is an axisymmetric toroidal pinch in which plasma is confined by a poloidal magnetic field Bp created by toroidal plasma current, and toroidal field B_t created by a poloidal plasma current and external coil currents. By carefully examining RFP discharges, Taylor postulated that the RFP configuration originates from a process of plasma relaxation in which plasma settles into a state of minimum energy with a constraint of constant helicity. If this state is described by simply by $\nabla \times \mathbf{B} = \mu \mathbf{B}$ with constant μ through out the large aspect ratio toroidal vacuum vessel, it would naturally lead to the well known Bessel function solution :

$$B_t(r) = B_{to} J_0(\mu r), \qquad (5a)$$

$$B_p(r) = B_{to} J_1(\mu r) \qquad (5b)$$

where B_{to} denotes toroidal field at the minor axis. This formulation is valid as long as the large aspect ratio toroidal plasma can be approximated by a cylinder of radius a. The relaxed force free states are independent of initial state and described by the dimensionless parameters, F and θ; F [$= B_t(a)/<B_t>$] is the ratio of toroidal field at the wall to the average toroidal field and F<0 implies a field reversal at the edge. θ [$= B_p(a)/<B_t>$] is called a pinch parameter defined as the ratio of poloidal field at the wall to the average toroidal field. In practice this force free state does not exactly describe RFP donfigurations since the assumption of μ= constant holds only in the inner part of the plasma and it goest down to zero at the edge. But the field profiles described by Eq.4(a,b) agree remarkably well with those with quiescent plasma obtained in RFP devices as shown in Fig. 3a. Fig. 3b shows points on the F-θ curve for several experiments together with a theoretical curve. In general, when the toroidal field reverses at the edge, F < 0, a more stable

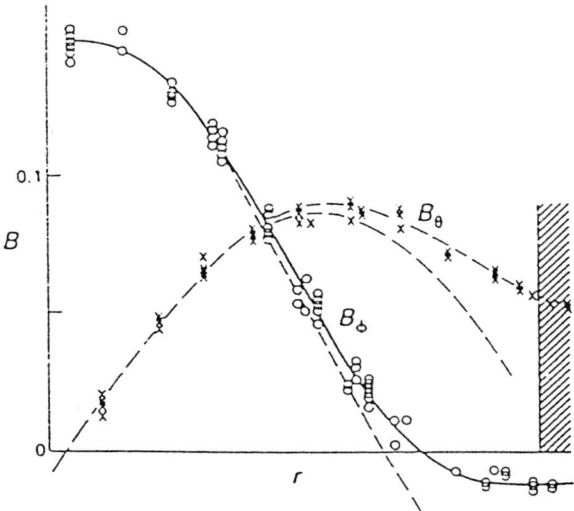

Fig. 3a. Experimental and theoretical magnetic profiles of RFP [*Bodin*, 1984].

plasma is obtained thus the configuration is maintained as "reversed field pinch" (RFP).

A tendency of RFP plasmas to relax into the Taylor state is more clearly demonstrated by the time evolution of F-q curves taken from discharges in the HBTX [*Bodin & Newton*, 1980] and ZT-40 devices [*DiMarco et al.*, 1983]. During fast pinch discharges in HBTX, the curve was temporarily forced away from the Taylor state but it quickly fell back into the relaxed state. When the discharge current rises slower, the curve always stays close to the Taylor state during the formation stage.

3.2. Sawtooth Relaxation in RFP plasmas

Once an RFP plasma configuration is established, the stable discharge can be maintained as long as the plasma current is sustained by inductive drive. Magnetic helicity is supplied by poloidal flux injection into the plasma. During the discharge the poloidal flux often becomes too excessive to maintain the Taylor state and a relaxation occurs as a sawtooth oscillation. If one draws the F-q diagram for the whole sawtooth process, the F-q curve gradually deviates from the Taylor state then quickly comes back to the relaxed state. During this relaxation process a conversion of flux from poloidal to toroidal flux has been seen many RFP discharges. It has been seen that the total helicity, K (= 2 $\Psi\Phi$) is kept roughly constant.

The magnetic helicity and energy evolution during the sawtooth cycle in RFP plasma was studied [*Ji et al.* '95] in the Madison Symmetric Torus (MST) RFP. The MST is a large RFP device (major radius = 1.50m, minor radius = 0.52m) with plasma current Ip up to 700kA. The plasma is surrounded by a 5cm-thick aluminum shell with one toroidal and one poloidal gap. The shell also acts as the vacuum vessel and a single-turn toroidal field coil.

Sawtooth oscillations consist of a fast crash phase and a slow recovery phase. The plasma rapidly relaxes towards its minimum energy state during the crash within a short time. This was also illustrated by drawing a F-θ curve. With lack of measurement of the exact magnetic field profiles, the μ profiles (and hence energy and helicity) were assumed from other measured quantities by employing equilibrium models of $\mu=\mu_0[1- (r/a)^\alpha]$. With this assumption the changes of K and W during a sawtooth crash was carefully monitored. Fig. 4 presents inventory of magnetic helicity, magnetic energy, poloidal flux and toroidal flux during a sawtooth oscillation. It was found that during the relaxation event the magnetic helicity decreases by 1.3-5.1 %, while the magnetic energy decreases by 4.0-10.5 %. Hence, the helicity conservation conjecture is modestly well-satisfied in that the helicity decay is less than the energy decay by a factor of 2-3.

This result that the decay ratio of energy to helicity is a factor of 2-3 instead of orders of magnitude, indicates that helicity conservation is only a rough approximation. The helicity change is larger than the simple MHD prediction. Determination of detailed mechanisms for possible anomalous helicity dissipation during relaxation awaits further investigations

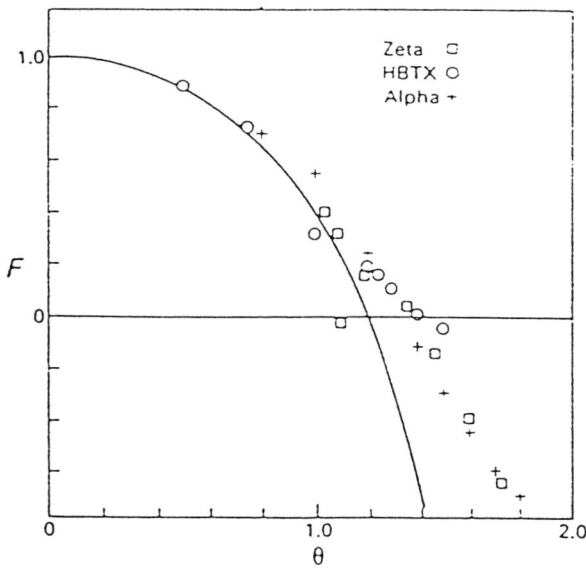

Fig. 3b. F-θ diagram. Data from HBTX, Alpha, ZETA[*Bodin and Newton*, 1984]

4. FORMATION AND RELAXATION OF SPHEROMAK PLASMAS

4.1 Spheromak Configuration

The spheromak confines a compact toroidal-shaped plasma in a spherical boundary. Poloidal fields are produced by plasma currents and external coils. Toroidal fields are produced entirely by plasma currents so that no coils link the plasma. The spheromak has many features in common with RFP including magnetic shear stabilization dynamics and confinement. The most notable difference between the RFP and the spheromak is that the spheromak does not have a reversal of the toroidal field at the outer edge of the plasma (B_t =0).

Spheromak is one type of toroidal pinch plasma which can be characterized by a Taylor minimum energy state. This configuration is generally enclosed by a spherical boundary or a conducting shell and the plasma is confined by the magnetic field produced by its own internal current. The magnetic field profile in this configuration can be straightfowardly derived from Eq. (3) with a spherical (can be oblate or prolate) boundary condition of $\mathbf{B} \cdot \mathbf{n}$ =0 .

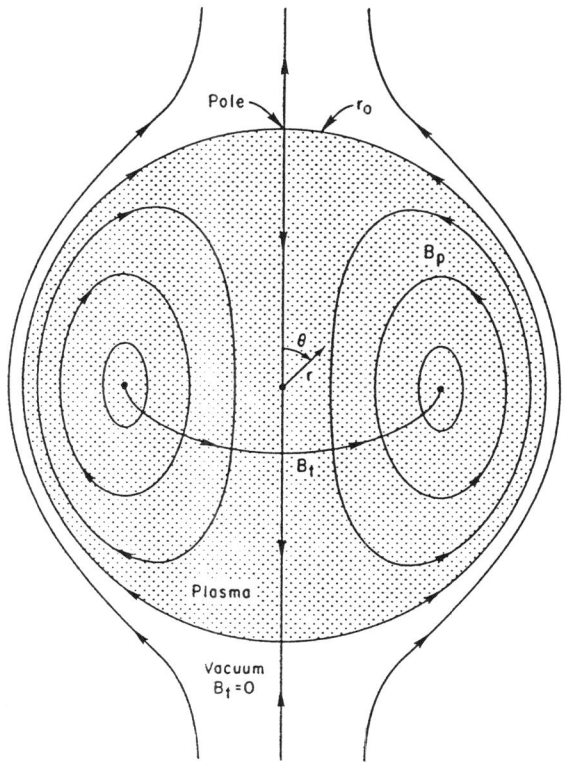

FIG. 5. Classic spheromak configuration.

The magnetic configuration of a spheromak can easily be derived by using a scaler function χ which satisfies Eq.(4),

$$\nabla^2 \chi + \mu^2 \chi = 0 \qquad (6)$$

Where ∇^2 is Laplacian operator. In a spherical container of radius a, the lowest eigenvalue is given by μa= 4.49, and the corresponding eigenfunction is obtained in spherical coordinates [ρ,θ,ϕ];

$$B_r = 2B_o [j_1(\mu\rho)/\mu\rho]\cos\theta, \qquad (6a)$$
$$B_\phi = B_o [j_1(\mu\rho)/\mu\rho]\sin\theta, \qquad (6b)$$
$$B_\theta = -B_o \, d/d\rho[j_1(\mu\rho)/\mu\rho]\sin\theta. \qquad (6c)$$

Where $j_1(x)$ is a spherical Bessel function defined by $j_1(x) = J_{3/2}(x)/x^{1/2}$ [*Rosenbluth and Bussac*,1979]. The flux contours of this configuration are shown in Fig. 5. The magnetic field has nested toroidal surfaces just like a toroidal pinch such as a tokamak or RFP, but the boundary is spherical instead of toroidal. The essential distinguishing feature of a spheromak is that the boundary is singly connected and there is no central conductor coils to create an external toroidal field. Therefore the toroidal

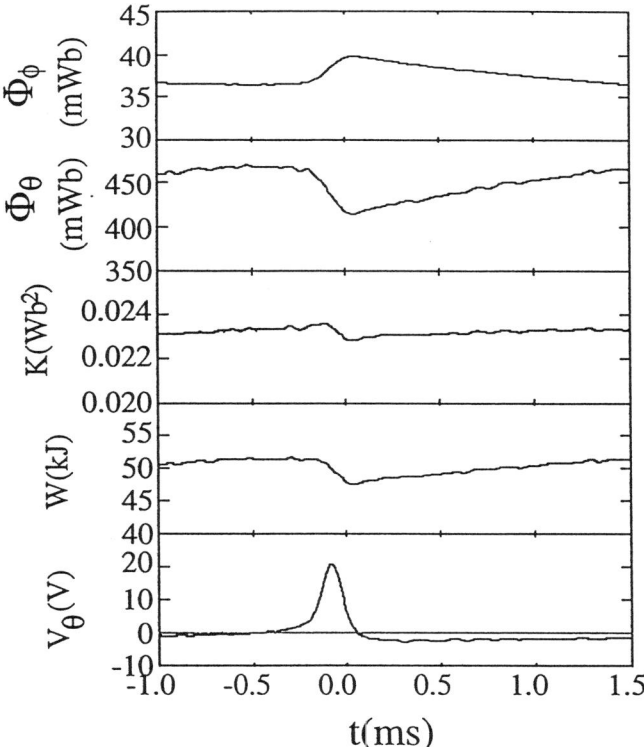

Fig. 4. Time evolution of toroidal and poloidal fluxes, the magnetic helicity K, magnetic energy W, one turn poloidal voltage V_θ during the sawtooth crash in RFP[*Ji etal*, 1995].

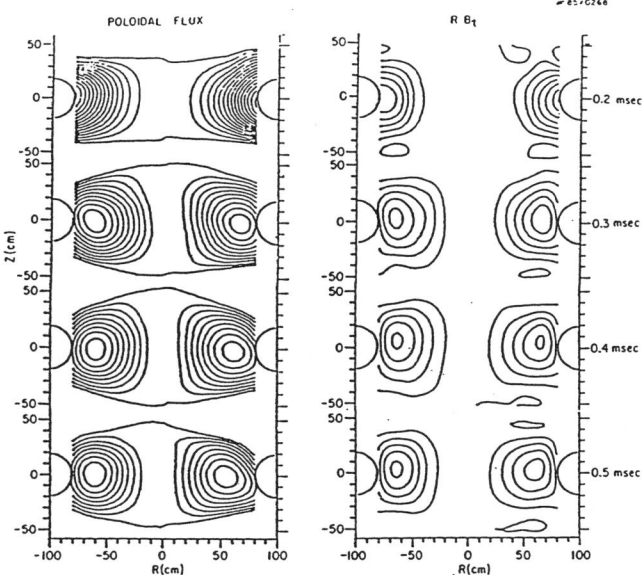

Fig. 6. Time evolution of poloidal flux and poloidal current contour plots taken 0.2, 0.3, 0.4, and 0.5 msec after the start of the S-1 discharge[*Yamada et al.*, 1985]. An excellent match of the poloidal flux ψ with the poloidal current RB_t implies the Taylor state in spheromak equilibrium. The separatrix is of the spheromak is shown by the outmost ψ contours.

field is zero everywhere on the boundary wall. It should be noted that the boundary does not have to be exactly spherical but can be oblate or prolate and even the cylindrical boundary can confine a spheromak as long as the boundary is singly connected. In an extreme case the force free minimum energy state in a doubly connected toroidal boundary, often used in describing a solar flare, can be regarded as one type of spermak.

4-2. Spheromak Formation Process; Governed by the Taylor Principle

There are many ways of forming spheromak plasmas; plasma gun formation [*Jarboe* et al. 1980], a mixture of linear and azimuthal discharges in the theta-pinch system [*Goldenbaum* et al., 1980], and inductive formation utilizing a flux core [*Yamada* et al., 1981, 1985]. In all formation schemes, both the poloidal and toroidal currents are induced either by electrodes or electromagnetic induction. Although each formation scheme is not elaborately described in this paper, there is a remarkable and important common feature in these formation schemes. After an initial highly turbulent state, the plasma settles into a more quiescent stable state in which the fluctuations are reduced. The forming plasma approaches the minimum-energy state theorized by Taylor and a spheromak configuration is created. The final plasma state is insensitive to initial conditions or details of the formation process so long as the geometry is not changed. Sometimes, an additional external field is supplied to support the final plasma equilibrium.

Let us take an example of S-1 Sphermak discharge[*Yamada*, 1985]. In the S-1 inductive formation scheme, the formation can be considered in terms of a slow injection of magnetic helicity into the plasma from a induction coil called a flux core[*Yamada* et al. 1981] which injects both toroidal and poloidal flux into the plasma. When this toroidal shaped solenoid is energized it induces poloidal and toroidal current around the ring. If the plasma is pushed inward toward the major axis by increasing (compressing) external field, the plasma pinches off from the flux core and spheromak is formed through reconnection and relaxation into a Taylor state. Figure 6 presents a time evolution of poloidal flux and poloidal current contour plots taken 0.2, 0.3, 0.4, and 0.5 msec after the start of the S-1 discharge. An excellent match of the poloidal flux ψ with the poloidal current RB_t implies the Taylor state in spheromak equilibrium.

In S-1 the electron temperature T_e ranges from 10 to 130 eV depending on plasma current and fill pressure. The electron density n_e reaches peak values of (0.2 to 1.5) $\times 10^{14}$ cm^{-3} during the formation phase and decays thereafter with a time constant of about 1 msec. The volume-averaged beta, β, is in the range of 5-10%.

During the spheromak formation large amplitude, globally coherent modes are observed [*Janos*, 1986] in coincidence with the flux conversion phenomena, suggesting that the modes play a vital role in the relaxation process by providing a means for flux conversion. These modes are low n-number, m = 1 modes where n and m are defined by the functional dependence exp[i(nφ + mθ)] of the fluctuations on toroidal angle φ and poloidal angle θ. During formation, peak amplitudes of the low n modes relative to the unperturbed field have been observed as high as 5- 20%.

4. 3 Verification of a Taylor State in Spheromaks

All magnetic configurations in a stable equilibrium have to satisfy the static force balance of $\mathbf{J} \times \mathbf{B} = \nabla p$ described by the well known Grad-Shafranov equation[*Wesson*, 1985]. Solving the axisymmetric Grad-Shafranov equation, it can be shown that

$$\mu = \mathbf{J} \cdot \mathbf{B}/B^2 = I'(\Psi) + p'(\Psi) I(\Psi)/B^2 \qquad (7)$$

where $I(\Psi) = RB_t$ is the poloidal current stream function; R is radial distance from the major axis and B_t is the local toroidal field, p(Ψ) is the plasma pressure, and the prime indicates differentiation with respect to Ψ. In order for μ to be a spatial constant, the second term must vanish: [*Hart* et al., 1986]. Since its numerator is a function of Ψ only, and the B^2 in the denominator is not in general a function of Ψ only, the second term will not even be constant on a flux surface. Therefore, it cannot be constant in the whole plasma volume as required by the Taylor state. Thus the constancy of Ψ requires that p'(Ψ) = 0. The equation remaining after dropping the p' term requires that I is a linear function of Ψ in order that I'(Ψ) is constant. The slope of I vs. Ψ is equal to the eigenvalue μ. For a given plasma shape, the eigenvalue is inversely proportional to the midplane separatrix radius.

Of fundamental importance to understanding the physics of spheromak equilibrium is the concept of the Taylor minimum energy state. The best way to determine how close this relaxed plasma is to the Taylor state is to measure the μ profile, where μ is defined by $\mu = \mathbf{J}\cdot\mathbf{B}/B^2$. If μ is constant in space, then the plasma is in the Taylor state. Profiles of the μ parameter were obtained by measuring the local B inside the plasma with magnetic probes[*Hart* et al, (1986)]. In order to deduce μ, the current density must be calculated. To measure the internal magnetic structure of the plasma, a one-dimensional probe array containing coils measuring all three components of the magnetic field was scanned. This facilitates taking the curl of the fields to calculate the current density. The calculation assumes the symmetry axis of the plasma is unperturbed. The profiles of μ(R) in the midplane are shown w.r.t. major radius R for two different times during the formation in Fig. 7a. Fig. 7b presents a two-dimensional contour plot of μ across the plasma cross section. At 0.2 msec, there is a deficit of current density near the symmetry axis because the driving electric fields are strongest near the plasma forming flux core. By 0.5 msec into the discharge, the current has diffused into the center to produce a flat μ profile. The variation of the t = 0.5 msec curve from the expected value is within the error bars of the measurement. Because of the limitations of the numerical differentiation, the error for μ is about 15 %. At early times, the μ profile peaks near the flux core and has a current hole near the symmetry axis. At 0.5 msec, the profile is basically flat throughout the plasma volume, with the gradients all occurring near the separatrix. This is in good agreement with the measured radius from the flux plot.

4.4 Helicity Injection and Sustainment of Spheromak Plasma by Coaxial Gun

The formation of spheromaks occurs on a wide time scale, generally longer than t_A ($t_{form} \geq 10^2 \tau_A$), where τ_A is the Alfven transit time; therefore, it is called a slow formation. As in the case of the RFP, spheromak formation can be considered in terms of an injection of

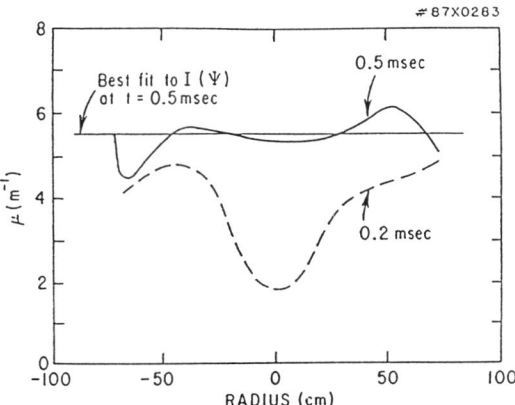

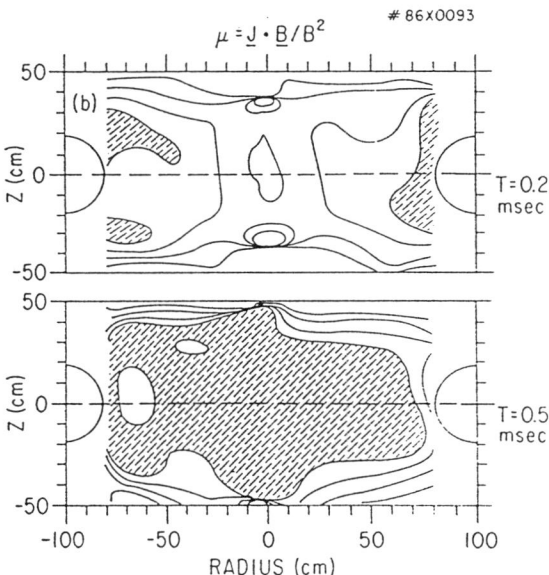

Fig. 7. 2-D contour plots, during (0.2 msec) and after (0.5 msec) formation, of $\mu = \mathbf{j}\cdot\mathbf{B}/B^2$. The shaded area is the region in which μ takes the nominal value of 5.5 m-1. Fig. 7b. μ profile versus major radius in the midplane[*Hart* et al., 1986].

magnetic helicity into the plasma from electrodes or induction coils. The CTX group at Los Alamos developed a magnetized, coaxial plasma gun to form and sustain spheromaks. The characteristics of the coaxial plasma gun plasma are well documented to optimize the efficiency of this formation[*Barnes* et al., 1986, *Wysocki* et al, 1988]. The gun consists of an inner electrode which is magnetized by an external coil as shown in Fig. 8 [*Jarboe* et al., 1980]. The initial poloidal flux links the inner to the outer electrode. Gas is puffed into the annular gap and high voltage is applied between the electrodes breaking down the gas. Current flows from the inner to the outer

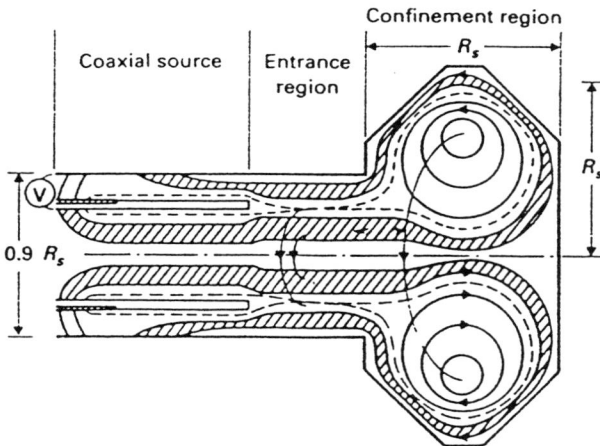

Fig. 8. Sustained configuration of CTX experiment [Jarboe et al. 1985]

electrode generating toroidal magnetic flux which encircles the inner electrode. If the toroidal field pressure force is sufficient then the plasma distends the gun flux and a spheromak is formed with the initial poloidal gun flux ψ_{gun} trapped in the plasma. During formation, helicity is injected at a rate $dK/dt = 2\psi_{gun}(d\Phi/dt)$. The voltage applied between the inner and outer electrode determines the rate at which toroidal flux is injected from the gun, $d\Phi/dt = V_{gun}$.

The helicity in the final spheromak has been compared to the injected helicity produced by a wide range of voltages on the poloidal- and toroidal field capacitor banks(or power supply). The global magnetic helicity in the plasma after formation was found to be proportional to the helicity injected by the gun, regardless of how that helicity was produced by the gun. Fig. 9 depicts the value of magnetic helicity measured in the plasma at 50 μsec after the initiation of the discharge versus the helcity injection from the gun based on the relationship shown above. This close relationship confirms conservation of magnetic helicity during the plasma formation.

Relaxation is also an essential requirement for this coaxial gun formation scheme. As the spheromak relaxes to the Taylor state, some of the injected toroidal flux is often converted to poloidal flux. The final plasma state is insensitive to the ratio of the poloidal to toroidal flux or to the details of the formation process so long as the geometry is not changed. Helicity balance between gun and spheromak has been verified in detail by the CTX group [Barnes et al., 1986, 1990]. The efficiency of this process depends on the relative sizes of the spheromak and the gun, helicity injector. It has been shown that energy coupling efficiency approaches unity if the gun and spheromak are of comparable size.

Spheromaks can be sustained near the Taylor state by continuously injecting helicity and energy through magnetic reconnection at a rate that balances helicity and energy dissipation. Steady-state spheromak sustainment (current drive) was attained by applying a sufficient voltage across the gun terminals while maintaining the linked gun flux[Jarboe et al., 1983; Browning, et al., 1992]. During the gun-formation process of spheromak, the gun flux must detach from the gun by magnetic reconnection in order to form the spheromak. In a similar way, spheromak sustainment requires continuous reconnection processes as the poloidal flux is replenished.

4.5 Plasma Relaxation During the Decay Phase Spheromak Plasmas

When a spheromak discharge is optimized its central temperature rises due to better confinement without radiation losses. With an enhanced pressure gradient the magnetic activity during the decay phase is observed to increase and the relaxation oscillations become pronounced. High T_e operation often allowed us to obtain a better picture of sawtooth-type relaxation events [Ono et al., 1988].

Figure 10a shows a time evolution of T_e profile measured in the S-1 spheromak by a multipoint Thomson scattering system. After the formation of the spheromak, the T_e profile gradually changes to a peaked profile up to 360 μsec. The maximum peak temperature (T_e = 80 eV) is obtained at 360 μsec when the peaking of T_e becomes maximum; afterwards, a large drop in T_e is observed and the T_e profile changes from peaked to broad. Simultaneously the measured poloidal and toroidal magnetic field profiles gradually become more peaked as

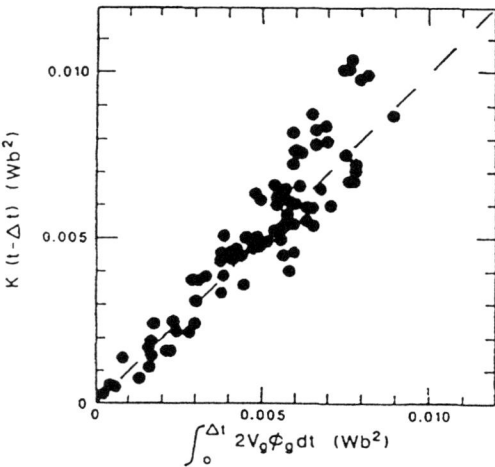

Fig. 9. Helicity conservation in CTX. Initial increase of helicity content of the plasma for δt= 50 μsec versus helicity injected from the gun[Barnes, et al., 1986].

time increases to 360 μsec, after which some redistribution restores a broad profile. The ratio of the poloidal flux Ψ to the toroidal flux Φ, relative flux inventory $\Psi'/\Psi - \Phi'/\Phi$, and q_0, the safety factor at the magnetic axis, are shown in Fig. 10b with respect to time during the relaxation period. The safety factor q represents the winding number of the field lines on the toroidal flux surface[*Wesson*, 1985] Initially, at about 300 μsec, Ψ/Φ is about 2.7 and the q value at the magnetic axis is 0.5, which are the expected values for the Taylor state. In the peaking phase, from 300 μsec to 360 μsec, Ψ/Φ increases to 4 and q_O decreases to 0.35 which indicates that the spheromak becomes poloidal-flux rich and therefore deviates from the Taylor state. The $\Psi'/\Psi - \Phi'/\Phi$ curve shows a preferential decay of Φ over Ψ until 360 μsec after which the relaxation occurs. Once the spheromak begins to relax from 360 μsec, Ψ/Φ drops to 3.0 and q_O value comes back to 0.5. The $\Psi'/\Psi - \Phi'/\Phi$ curve reverses its sign during relaxation from 360 μsec to 390 μsec, which suggests flux conversion from poloidal flux to toroidal flux ($\Delta\Psi < 0$, $\Delta\Phi > 0$). The flux conversion brings the spheromak from a non-Taylor state with excessive poloidal flux to the Taylor state[*Hart et al*, 1986].

According to the result of toroidal mode analysis using external pick-up coils, the n = 2, m = 1 mode starts growing just before the onset of the relaxation and becomes dominant during the relaxation. This suggests that the n = 2, m = 1 kink mode plays an important role in the flux conversion as conceived in an MHD simulation [*Knox et al.*, 1986].

5. EVOLUTION OF MAGNETIC FLUXES AND HELICITIES DURING PLASMA MERGING

Merging of two toroidal plasmas *[Katsurai and Yamada*, 1982] can provide fundamental data on the behavior of magnetic helicity during magnetic reconnection or topology changes of plasma configuration. In this set-up, the evolution of both global and local magnetic helicities can be studied to investigate helicity and dissipation during merging. Several experiments have been carried out in the past decade on this concept [*Ono et al*, 1993; *Yamada* et al., 1990, 1997; *Kornack et al*, 1998]. It is very important to know how three dimensional relaxation effects affect the helicity evolution. In some cases the plasma kinetic energy increases dramatically at the expense of magnetic energy and conversion of kinetic helicity to magnetic helicity and vice versa occur. In the TS-3 device at the University Tokyo, it was found that the helicity is conserved much better than the magnetic energy during plasma merging [*Ono et al*, 1993].

Let us consider how the magnetic fluxes and helicities evolve during plasma merging. We consider the co-

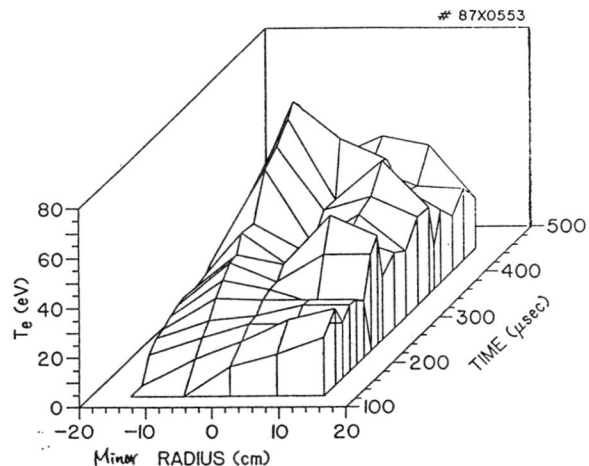

Fig. 10a. Time evolution Te (r) profile during sawtooth relaxation of S-1 Spheromak.

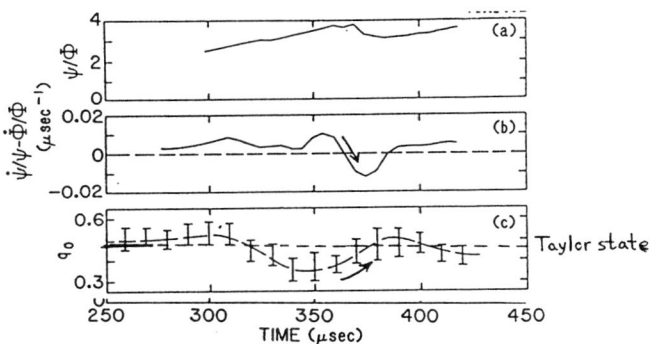

Fig. 10 b. Time evolution of Ψ/ϕ, $\Psi'/\Psi - \phi'/\phi$, and the q value* at the magnetic axis. Ψ/ϕ increases and q_O decreases to 0.35 which indicates that the spheromak becomes poloidal-flux rich and therefore deviates from the Taylor state. The curve shows a preferential decay of ϕ over Ψ until 360 μsec after which the relaxation occurs.

helicity merging case in which two toroidal plasmas with equal helicity K_o, poloidal flux ψ_o, and toroidal flux Φ_o, are merging to each other, as shown in Fig.11. Since it is expected that the toroidal fluxes and heliciites add while the poloidal fluxes do not, the magnetic fluxes after the merging(ψ_m, Φ_m) and helicity can be expressed utilizing the relatinship derived in Eq. (2a),

$$\Psi_m = \Psi_o, \tag{8a}$$
$$\Phi_m = 2\Phi_o, \tag{8b}$$
$$K_1 = 2c_o \psi_o \Phi_o, \tag{8c}$$
$$K_m = 2c_m \psi_m \Phi_m. \tag{8d}$$

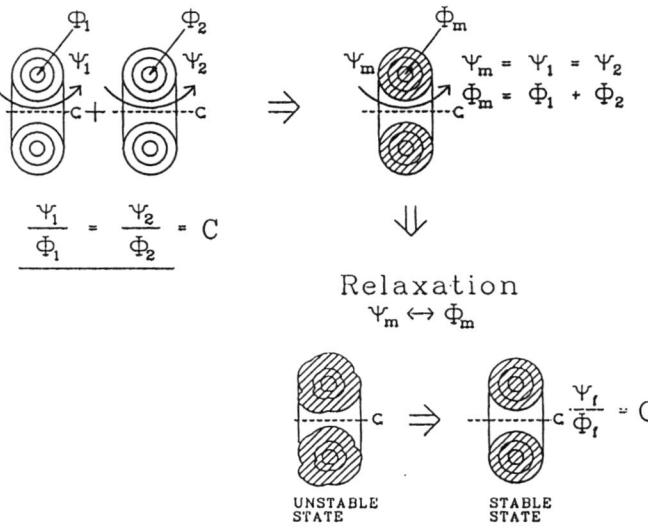

Fig. 11. Merging process of two identical spheromak plasmas

where (c_o & c_m are the numerical coefficient (~1) based on current profiles of the plasmas before and after merging) and both plasmas are assumed in the minimum energy state(ψ_o/Φ_o= const.= α) before merging. If the same current profile is recovered after merging($c_m = c_o$), the total magnetic helicity is expected to be conserved by the magnetic reconnection process as shown by

$$K_m = K_o + K_o = 2K_o \qquad (8e)$$

However, if Taylor's conjecture is applicable, the plasma should relax into the minimum energy state in which $\psi_m/\Phi_m = \psi_o/\Phi_o = \alpha$. Thus, a flux conversion has to occur to create a final stable state(ψ_f, Φ_f) by reducing toroidal flux from $\Phi_m = 2\Phi_o$ to $\Phi_f = \sqrt{2}\ \Phi_o$, and increasing poloidal flux from $\psi_m = \psi_o$ to $\psi_f = \sqrt{2}\ \psi_o$. Whether this relaxation process occurs or not is an important question. Although detailed study has not been carried out yet, recent data from TS-3 and MRX suggest that this relaxation process into a new Taylor state is occurring during co-helicity merging. As in other spheromak formation schemes, this process invokes flux conversion between the toroidal and poloidal fluxes during and after the merging of the plasmas.

6. IMPACT OF EXPERIMENTAL DATA ON SOLAR FLARE RESEARCH

What impact will laboratory research have on our understanding of phenomena where magnetic helicity is crucial? Solar flares are the best-known example of rapid conversion of magnetic energy into heat [*Shibata et al.*, 1995; *Tsuneta*, 1996]. The above mentioned laboratory experiments have provided important data for systems where magnetic helicity is stored in a force-free equilibrium configuration and then via slow adjustment of an external parameter (such as in the current in external coils), brought to a situation of ideal-MHD instability which forces a topology change. However, it is not clear whether the total magnetic helicity, without a well defined (conductor shell) boundary with $\mathbf{B}\cdot\mathbf{n} = 0$, would be conserved or not during evolution of solar flares and their reconnection process.

Recent theoretical work has focused on two-dimensional models of the evolution of force-free magnetic arcades [*Low*, 1987] whose field-line foot points are advected by flows in the solar photosphere. It gave an example of an evolution which produces a situation analogous to the antiparallel-toroidal-field double spheromak merging. Hence the merging or reconnection experiments bear directly on the physics of solar flares. One can measure how and at what rate magnetic helicity changes and magnetic energy is converted into heat.

Vector magnetograph measurements of the solar surface show that fields in active region generally exhibit positive helicity in the south hemisphere and negative helicity in the north. D. Rust[1994] considered that helical magnetic field structures are in a force-free state and interaction or merging of helical ropes can result in eruptive events and ejection of magnetic flux and energy. This event is analogous to the spheromak merging experiment mentioned in the previous section. It would be very important to find out how helicity evolves during the merging processes. A detailed study of evolution of the magnetic helicity and energy, of flux conversion processes, plasma flow and heating during the merging of two toroidal plasma could make an important impact on the analysis of solar flare interaction.

In general toroidal pinch plasmas can provide physics data for the role of helicity in energy conversion to directed flows, the role of turbulence in thermalizing directed-flow energies, and the role of turbulence in changing the reconnection region from the x-point-line of single two-dimensional (2-D) pictures to volume-filling processes. The spatial extent, intensity, and scale of departures from 2-D symmetry can be measured.

7. SUMMARY

It was found that relaxation to the Taylor (force-free, minimum energy) state was an important mechanism for the formation processes of RFP and spheromak plasmas [Taylor 1986]. There is a remarkable feature common to all of these toroidal pinch discharges. It was found that, after an initial highly turbulent state, the plasma settled into a more quiescent stable state in which the fluctuations were reduced. The concept of magnetic helicity has been proven to be very effectivein describing magnetic characteristics of these pasmas. Taylor's conjecture, conservation of magnetic helicity, $K \equiv \int \mathbf{A} \cdot \mathbf{B} dV$, has been verified during the relaxation of laboratory plasmas, in which the plasmas rapidly approach a minimum-energy state. There has been an evidence that the magnetic energy dissipates much faster than the helicity during plasma relaxation.

In some cases two-dimensional spatial profiles of $\mu = \mathbf{j} \cdot \mathbf{B}/B^2$, where $\nabla \times \mathbf{B} = \mu \mathbf{B}$ for a force-freestate, were deduced from internal probe data to verify the Taylor state (with μ = constant). Another important finding was the identification of flux conversion between the toroidal and poloidal fluxes of the spheromak during and after the formation of the plasma. The correlation between the MHD mode behavior and the Taylor relaxation process has been described.

In the high temperature plasmas, a "sawtooth-type" relaxation cycle was observed both in the magnetic configuration and electron temperature profile. It was found that the confinement characteristics of the spheromak were significantly affected by the relaxation phenomena.

When the magnetic configuration departs from the Taylor state in a RFP or a spheromak, it easily becomes unstable to low n (n = toroidal mode number) ideal MHD modes. Through these instabilities, magnetic reconnection takes place (relaxation) and a Taylor state is recovered. It is found that the confinement time of these plasmas is significantly affected by this relaxation phenomenon.

Conservation of magnetic helicity has been verified during relaxation in many laboratory pinch plasmas in which the Taylor minimum energy states are identified after the relaxation events. Recently merging spheromaks are utilized to investigate a magnetic reconnection of magnetized toroidal plasmas or helical flux ropes. Experiments with fully 3 dimensional magnetic reconnection are now possible in the MRX device where evolution of magnetic helicity can be studied.

In recent years, theoretical work for solar flares [*Parker*(1973), *Priest*(1984), *Low*(1992] has shown that a flux-conserving sequence of ideal-MHD equilibria must develop singular current sheets. Laboratory experiments have been investigating how theoretical singular current sheets are manifested in an often turbulent laboratory plasmas, as well as the relation between global forcing, the extent of current sheets, and the reconnection rate. To date an important finding in the current carrying laboratory plasma has been a conservation of magnetic helicity during its relaxation. A detailed experiment-theory-computation comparison will test our ability to describe nontrivial reconnection configurations with the same codes used to simulate MHD processes in magnetospheric and solar physics. Close collaborations of laboratory experiments with solar and magnetospheric research would bring a better understanding of essential mechanism of relaxation phenomena in plasmas.

Acknowledgments: The authors appreciate many useful and enlightening discussions with participants of this Chapman conference.

REFERENCES

Barnes, C.W. et al., "Experimental Determination of the Conservation of Magnetic Helicity From the Balance Between Source and Spheromak," *Phys. Fluids* 29, 3415 (1986).

Barnes, C.W. et al., "The Impedance and Energy Efficiency of a Coaxial Magnetized Plasma Source used for Spheromak Formation and Sustainment," *Phys. Fluids B* 8, 1871 (1990).

Berger, M. A. and G.B. Field, "Topological properties of magnetic helicity, *J. Fluid Mech.* 147, 133 (1984]

Bodin, H. A. B. and Newton, A. A., *Nucl. Fusion* 20 (1980) 1255.

Bodin, H.A.B., in *Proceedings of IAEA Int. Conf. Plasma Phys.* vol.1, 417 (1984)

Brown M. R., and A. Martin, ``Spheromak Experiment using Separate Guns for Formation and Sustainment'', *Fusion Technology* 30, 300 (1996).

Browning, P. K. et al., "Injection and Sustainment of Plasma in a Preexisting Toroidal Field using a Coaxial Helicity Source," *Phys. Rev. Lett.* 68, 1722 (1992).

DiMarco, J. N. Proc. Mirror Based Field Reversed Approaches to Nucl. fusion, vol.2, 681 (1983)

Goldenbaum, G. C. et al, *Phys. Rev. Lett.* 44 (1980) 393.

Hart, G. W., et al., Phys. Fluids 29 (1986) 1994.

Janos A. C., Phys. Fluids 29 (1986) 3342.

Jarboe, T.R. et al., "Motion of a Compact Toroid inside a Cylindrical Flux Conserver," *Phys. Rev. Lett.* 45, 1264 (1980).

Jarboe, T.R. et al., "Slow Formation and Sustainment of Spheromaks by a Coaxial Magnetized Plasma Source," *Phys. Rev. Lett.* 51, 39 (1983).

Ji H. et al. "Conservation of Magnetic Helicity during Plasma Relaxation,", *Phys. Rev. Lett.* 75, 2945 (1995).

Ji, H. et al, "Experimental test of the Sweet-Parker model of Magnetic Reconnection" *Phys. Rev. Lett.* 80, 3256 (1998).

Kadomtsev, B.B., "Disruptive Instability in Tokamaks," Sov. J. Plasma Phys. 1, 389 (1975).

Katsurai, M and M. Yamada, Nucl. Fusion 22 (1982) 1407.

Knox S. O., et al., ``Observations of Spheromak Equilibria which differ from the Minimum-Energy State and have Internal Kink Distortions'', *Phys. Rev. Lett.* , 56, 842 (1986).

Kornack T. W., et al., ``Experimental Observation of Correlated Magnetic Reconnection and Alfvenic Ion Jets'', *Phys. Rev. E* 58, R36 (1998).

Low, B.C. "Electric Current Sheet Formation in a Magnetic

Field Induced by Continuous Magnetic Footpoint Displacements," *Astrophys. J.* **323**, 358 (1987).

Low, B.C., "Formation of Electric-Current Sheets in the Magneto-Static Atmosphere," *Astron. Astrophys.* 253, 311 (1992).

Ono, Y et al, *Phys. Rev. Lett.* **61**, 2847 (1988

Ono, Y et al., "Experimental Investigation of Three-dimensional Magnetic Reconnection by use of Two Colliding Spheromaks," *Phys. Fluids* B 5, 3691 (1993)

Parker, E.N., *Cosmical Magnetic Fields* (Oxford: Claredon Press, 1979), Parker, E.N., *Astrophys. J.* 180, 247 (1973).

Prager S. In this Proceeding (1999)

Priest, E.R., "Solar Magnetohydrodynamics" (P. Reidel, Dordrecht, 1984), Chapter 10.

Rosenbluth, M. N. and Bussac, M. N. *Nucl. Fusion* 19, 489, (1979)

Rust, D.M., "Spawning and Shedding of Helical Magnetic Fields in the Solar Atmosphere," *Geophys. Res. Lett.* 21, 241(1994).

Shibata, K. et al., "Hot-plasma Ejections Associated with Compact-loop Solar Flares," *Ap. J.* **451**, L83 (1995).

Taylor, J.B. "Relaxation of Toroidal Plasma and Generation of Reverse Magnetic Fields," *Phys. Rev. Lett.* 33, 1139 (1974).

Taylor, J.B., "Relaxation and Magnetic Reconnection in Laboratory Plasmas," Rev. Mod. Phys. 28, 243 (1986).

Tsuneta, K. "Structure and Dynamics of Magnetic Reconnection in a Solar Flare," *Ap. J.* **456**, 840 (1996).

Wesson, J. *"Tokamaks"* Oxford Uni. Press, Oxford, (1987)

Wysocki, F.J. et al., "Evidence for a Pressure-Driven Instability in the CTX Spheromak," *Phys. Rev. Lett.* 61, 2457 (1988).

Yamada, M. et al., "Quasistatic Formation of the Spheromak Plasma Configuration," *Phys. Rev. Lett.* 46, 188 (1981)

Yamada, M., "S-1 Spheromak", *Nucl. Fusion* 25 (1985) 1327

Yamada, M et al., in *Plasma Physics and Controlled Nuclear Fusion Research, Proc. 10th Int. Conf.,* London, 1984 (IAEA, Vienna, 1985) Vol. 2, 535.

Yamada, M. et al., "Magnetic Reconnection of Plasma Toroids with Co- and Counter-Helicity," *Phys. Rev. Lett.* 65, 721 (1990);

Yamada, M. et al., "Identification of Y-shaped and O-shaped Diffusion Regions during Magnetic Reconnection in a Laboratory Plasma," *Phys. Rev. Letts.* 78, 3117 (1997).

Magnetic Helicity and Relaxation Phenomena in the Solar Corona

E.R. Priest

Mathematical and Computational Sciences Dept., St Andrews University, St Andrews KY16 9SS, Scotland.

The magnetic field of our Sun's corona is highly complex; the concept of magnetic helicity is useful for describing and understanding that complexity. Furthermore, the corona continually evolves in response to a variety of magnetic footpoint motions in the solar surface, and so magnetic relaxation is likely to play an important role in that evolution. Here I give a brief overview of the role of magnetic helicity and relaxation in the solar corona. Section 1 describes the corona and the magnetic helicity of the building blocks of coronal configurations. Section 2 discusses the way the coronal field evolves by relaxation and reconnection. Finally, Section 3 examines the role of magnetic helicity in a variety of coronal phenomena such as prominences, flares, jets and the heating of the corona.

1. THE CORONAL MAGNETIC FIELD

1.1. The Solar Corona

The corona has a three-part structure of magnetically open coronal holes, magnetically closed coronal loops and x-ray bright points (where reconnection is likely to be taking place). The Yohkoh satellite in particular has revealed the corona to be magnetohydrodynamic (Figure 1) with myriads of coronal loops continually interacting with one another.

In the magnetically open regions (coronal holes) magnetic helicity propagates away as Alfvén waves, whereas in magnetically closed regions (coronal loops) the magnetic helicity may build up in time. Indeed, prominences, which often form near the base of large-scale coronal arcades (coronal streamers) are large-scale twisted flux tubes where magnetic helicity is stored.

Active regions lying above sunspot groups show up as collections of coronal loops in soft x-rays. About 60 % of them have significant magnetic helicity which shows up as a so-called sigmoid structure [e.g. *Pevtsov et al*, 1995]. There is a large scatter in values of the helicity with a tendancy for S-shaped structures in the northern hemisphere having negative helicity and N-shaped in the southern hemisphere with positive helicity. This hemispheric pattern is also present in sunspot whirls and in the twist in prominences. Left-handed twist with $\alpha < 0$ and negative magnetic helicity (where $\alpha = \nabla \times \mathbf{B}/\mathbf{B}$ is the linear force-free parameter) dominates the northern hemisphere, while right-handed twist with $\alpha > 0$ and positive magnetic helicity is prevalent in the southern hemisphere.

In the corona the plasma velocity is generally much smaller than the Alfvén speed (V_A), which is typically $1000 \ km\,s^{-1}$ (e.g., for a typical coronal magnetic field B of 10 *Gauss* and density n of $10^{15}\,m^{-3}$). We therefore have a force balance

$$0 = -\nabla p + \rho \mathbf{g} + \mathbf{j} \times \mathbf{B} \quad (1)$$

between pressure gradients, gravity and magnetic forces. The ratio of the pressure gradient to the magnetic force is the plasma beta

$$\beta \simeq \frac{3.5nT}{10^{21}B^2}, \quad (2)$$

Magnetic Helicity in Space and Laboratory Plasmas
Geophysical Monograph 111
Copyright 1999 by the American Geophysical Union

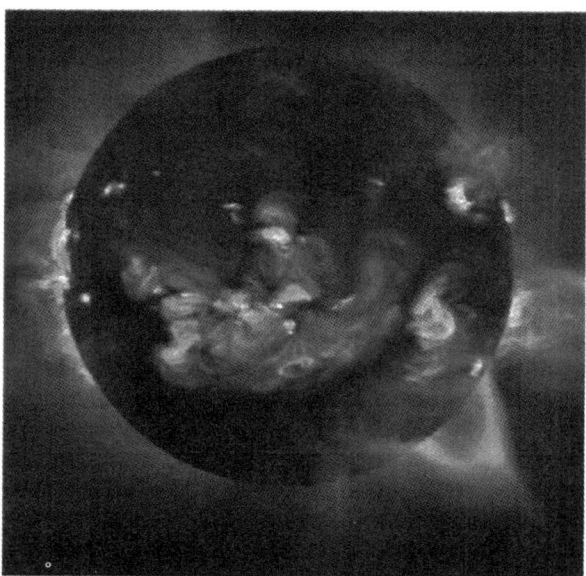

Figure 1. An image of the corona from the Soft X-ray Telescope on Yohkoh (courtesy S Tsuneta).

which is of order unity in strongly magnetic parts of the photosphere (where $n \approx 10^{23} m^{-3}, B \approx 1500 G, T \approx 6000 K$) and also in the high corona (where $n \approx 10^{14} m^{-3}, B \approx 2G, T \approx 10^6 K$). The ratio of the gravitational to the magnetic force is $\beta\delta$, where $\delta = \Lambda/h$ in terms of the height (h) of the structure and the scale-height (Λ), which is typically 100 Mm (for a temperature of $2 \times 10^6 K$).

Both pressure and gravitational forces are therefore important in the photosphere, in prominences and in the high corona, but in the low corona $\beta << 1, \beta\delta << 1$ and so the force-balance reduces to the force-free equation $0 = \mathbf{j} \times \mathbf{B}$, so that $\nabla \times \mathbf{B} = \alpha\mathbf{B}$, where α is in general constant along a magnetic field line.

The magnetic field does not thread the surface of the Sun uniformly but is concentrated to intense values of 1 kG or more at the edges of convection cells known as supergranules, which have a typical diameter of 30 Mm by comparison with the solar radius of 700 Mm. New magnetic flux continually emerges from below the solar surface, migrates to the cell boundaries, merges or reconnects. According to the *Converging Flux Model* [*Priest et al*, 1994], it is the reconnection of such flux at the edge of the cell that creates x-ray bright points and many x-ray jets (Figure 2b).

A surprise is that the flux of this so-called *magnetic carpet* (Figure 2a) is replaced very rapidly, every 40 hours [*Schrijver et al*, 1998]. During the accompanying reconnection, magnetic helicity (both self and mutual) is continually transferred from the emerging flux to the large-scale overlying corona. The corona is also in a state of small-scale turbulence, with typical nonthermal velocities of 10 - 30 $km\,s^{-1}$.

1.2. The Building Blocks of Coronal Magnetic Fields

The building blocks of the corona are coronal loops and coronal arcades, so how can we calculate their magnetic helicity? The relative magnetic helicity (a gauge-invariant quantity) in a volume V bounded by a surface S may be defined as

$$K = \int (\mathbf{A} + \mathbf{A}_0)(\mathbf{B} - \mathbf{B}_0) dV, \qquad (3)$$

where $\mathbf{B} = \nabla \times \mathbf{A}$, $\mathbf{B}_0 = \nabla \times \mathbf{A}_0$ is a potential field with the same normal component on S ($\mathbf{B}.\mathbf{n} = \mathbf{B}_0.\mathbf{n}$), such that $\nabla.\mathbf{A}_0 = 0$ and $\mathbf{A}.\mathbf{n} = 0$ on S. Then in ideal MHD it may be shown that the rate of change of magnetic helicity is

$$\frac{dK}{dt} = 2\int (\mathbf{B}.\mathbf{A}_0)(\mathbf{v}.\mathbf{n}) - (\mathbf{v}.\mathbf{A}_0)(\mathbf{B}.\mathbf{n}) dS. \qquad (4)$$

For N thin magnetic tubes of flux F_i undergoing translations and rotations (with angular speed ω_i), Berger (1988) showed that this expression reduces to

$$\frac{dK}{dt} = -\frac{1}{2\pi}\left(\sum_{i=1}^{N}\omega_i F_i^2 + \sum_{i=1}^{N}\sum_{j=1}^{N}\dot{\theta}_{ij}F_i F_j\right), \qquad (5)$$

where θ_{ij} is the inclination of the line joining the base of the ith and jth tubes. The first term on the right represents the effect of twisting and the second of braiding. This expression may be used to deduce the magnetic helicity of N tubes built up from zero as

$$K = \sum_{i=1}^{N} S_i + \sum_{i<j=1}^{N} M_{ij}, \qquad (6)$$

where $S_i = T_i F_i^2$ is the *self helicity* in terms of the *twist* T_i (the angle through which a tube is twisted in units of 2π) and $M_{ij} = 2L_{ij}F_i F_j$ is the *mutual helicity* in terms of the *linking number* L_{ij}.

Several examples of mutual and self helicity are illustrated in Figure 3. Suppose that two braided tubes of flux F_1 and F_2 are stretched between two planes such that the footpoints rotate through and angle θ. Then the mutual helicity is $K = \theta F_1 F_2/2\pi$. If instead two nearby tubes arch above the photospheric plane such that the angles subtended by the footpoints of one coronal loop at those of the other are θ_1 and θ_2 (Figure 3a), then the mutual helicity is

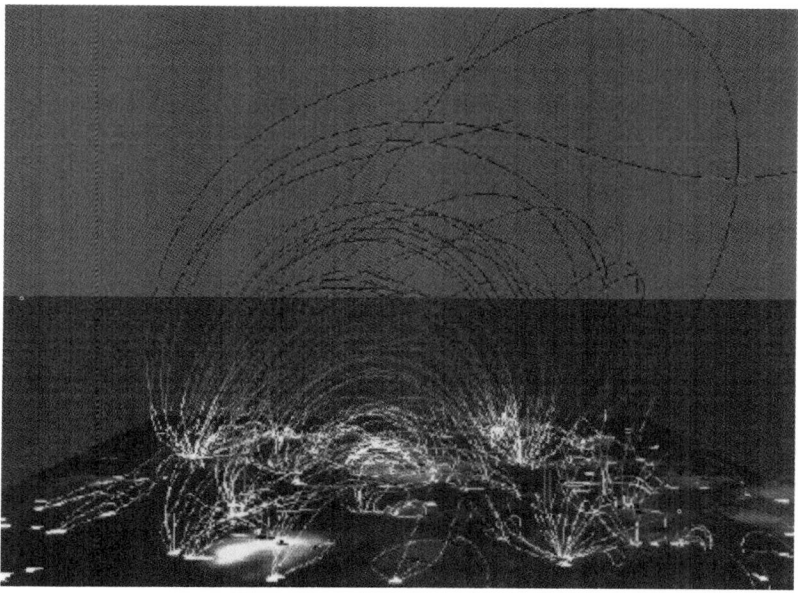

Figure 2. (a) The magnetic carpet showing potential magnetic field lines arching above the solar surface from a set of observed magnetic sources (courtesy K Schrijver). (b) The emergence of magnetic flux near the centre of a supergranule cell and its reconnection at the edge to create an x-ray bright point.

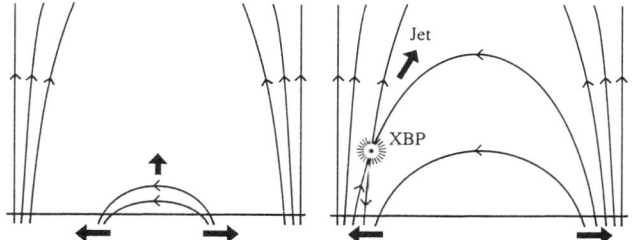

$$K = \frac{\theta_1 - \theta_2}{\pi} F_1 F_2. \quad (7)$$

Tubes that cross possess a mutual helicity

$$K = \frac{\theta_3 + \theta_4}{\pi} F_1 F_2, \quad (8)$$

where the angles (θ_3 and θ_4) that the footpoints of one loop subtend at the other are shown in Figure 3b.

A single flux tube of flux F and twist T (Figure 3c) has self-helicity

$$K = TF^2 \quad (9)$$

if the helicity is uniform or $K = 2\int_0^F T(f) f\, df$ if the twist varies with flux (f) from the axis. Finally, a coronal arcade of flux F and shear S per unit length (Figure 3d) has self-helicity

$$K = SF^2 \quad (10)$$

if the arcade is a shell of field lines, or $K = 2\int_0^F S(f) f\, df$ if the shear varies with flux (f) from the arcade axis.

However, the formula (10) applies to an infinitely long arcade with all the field lines going straight across from the source to the sink. More realistic arcades of finite length are more complex, since the magnetic helicity then depends on the separation and dimensions of the sources and on how force-free the field is, in such a way

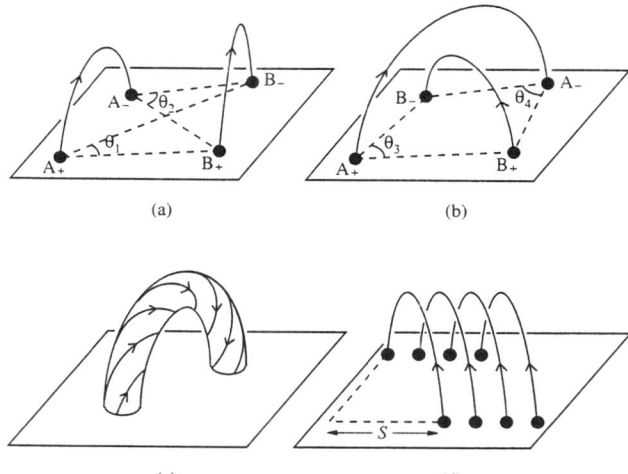

Figure 3. Building blocks of coronal fields: (a) nearby flux tubes; (b) crossing flux tubes; (c) a twisted flux tube; (d) a sheared arcade.

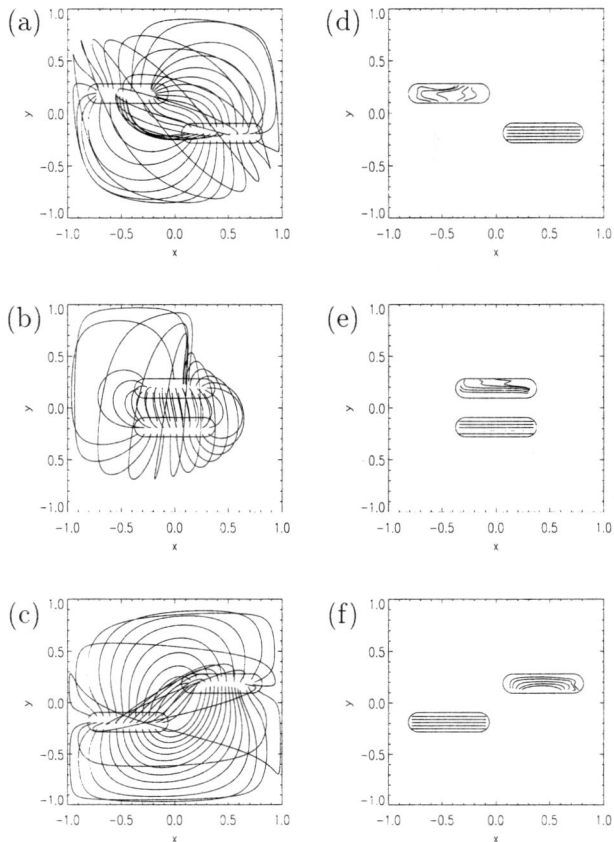

Figure 4. The magnetic field of a linear force-free field ($\alpha = 1$) with different source positions ((a)-(c)) and the corresponding footpoint mappings [*Parnell and Priest*, 1998].

that it increases with the separation or as the sources become shorter and fatter or as α increases [*Parnell and Priest*, 1998].

For example, for a simple potential arcade with symmetrically placed sources, straight rows of initial footpoints map to straight rows of final footpoints in the other source. When the sources are sheared with the field remaining potential, there is no magnetic helicity, but the structure of the field lines is not simple, with complex crossings in such a way that the net mutual helicity vanishes (similar to Figure 4c, which was obtained for a non-potential field). Also the mapping of the footpoints becomes distorted.

The corresponding results for a linear force-free field with $\alpha = 1$ (Figure 4) show that when the sources are symmetrically located (Figure 4b) the field lines twist in a clockwise direction. If the shear reinforces the twist, the field structure (Figure 4c) and footpoint mapping (Figure 4e) are simple, but if the shear competes with the twist the structure (Figure 4a) and mapping (Figure 4d) are complex and the magnetic helicity (K) is lower. K is found to increase from zero with α up to infinity at the resonant value associated with the finite computational box. Also, K increases nonlinearly with shear and with α.

2. THE EVOLUTION AND RELAXATION OF THE CORONA

2.1. Coronal Evolution

How does the corona evolve and relax? The footpoints of coronal field lines move due to a variety of photospheric motions. Granulation with a horizontal velocity (v) of $1\ km\,s^{-1}$ and a width (l) of $1\ Mm$ acts on a time-scale (τ_0) of $\lesssim 1000$ sec. The larger supergranule cells have $v \simeq 0.3\,km\,s^{-1}, l \simeq 30 Mm$ and $\tau_0 \approx 1$ day. Similarly, general active-region evolution and sunspot or pore motion has speeds of about $1\ km\,s^{-1}$ and $\tau_0 \approx 1$ day. Differential rotation at about $5\ m\,s^{-1}$ per supergranule or meridional flow of $20\ m\,s^{-1}$ act on much longer time-scales ($\tau_0 \sim 1$ year).

The response of the coronal magnetic field is to produce waves if τ_0 is smaller than the time ($\tau_A = L/V_A$) it takes an Alfvén wave to propagate along a coronal loop of length L. For $L = 100\ Mm$, say, and $V_A = 1000\ km\,s^{-1}$, the Alfvén time is $\tau_A = 100$ seconds. If however, $\tau_0 \gg \tau_A$ then the coronal magnetic field evolves through a series of force-free equilibria. The footpoint motions inject energy as a Poynting flux. They also inject magnetic helicity at a rate (Berger, 1998)

$$\frac{dK}{dt} = 2\int[(\mathbf{B}.\mathbf{A}_0)(\mathbf{v}.\mathbf{n}) - (\mathbf{v}.\mathbf{A}_0)(\mathbf{B}.\mathbf{n})]dS - 2\int\frac{\mathbf{j}.\mathbf{B}}{\sigma}dV, \quad (11)$$

where the first term on the right represents a advection of helicity through the boundary in emerging flux tubes and the second represents the effect of twisting and braiding. The third term represents magnetic helicity dissipation which is extremely slow, much slower than energy decay.

In the solar atmosphere, magnetic helicity cannot be easily destroyed. It is either injected as twisted flux emerging from below or it may be added or subtracted by twisting and braiding, or finally it may be ejected by eruptions of prominences and coronal mass ejections.

2.2. Coronal Relaxation

At the same time as coronal structures are trying to evolve through nonlinear force-free equilibria, they also tend to relax towards linear force-free states by three-dimensional magnetic reconnection. Relaxation tends not to destroy much magnetic helicity (when the magnetic Reynolds number is very large), but it can convert it from one kind to another, such as from mutual to self helicity.

The response of the corona depends on the values of the driving time (τ_0) and the relaxation time (τ_{relax}). If $\tau_A < \tau_{relax} < \tau_0$, there is a partial relaxation to a field with an energy between the linear and nonlinear force-free fields. Either this or the previous case seem often to be appropriate in active regions, where linear force-free models do not match the observed coronal structures well. If on the other hand, $\tau_A < \tau_{relax} \ll \tau_0$ the field relaxes easily to a linear force-free field.

What is the relaxation time in the corona? Perhaps it is the tearing time or the slow (Sweet-Parker) reconnection time of about $R_m^{1/2}\tau_A$. For typical magnetic Reynolds numbers of, say, 10^8 and Alfvén times (τ_A) of 100 sec, this would be about 10^6 sec $\sim$ 2 weeks, in which case relaxation in response to driving times of say a few days or less would be rare. Perhaps instead relaxation occurs on, say, 100 τ_A for fast reconnection, which would be about 10^4 sec $\simeq$ 3 hrs, so that relaxation or partial relaxation would be common. Perhaps also extra physical effects such as rippling modes due to temperature gradients or current filamentation could speed up the relaxation. At present we do not know.

A related question is: how braided and tangled do magnetic field lines become? How efficient is relaxation? An observational answer from the new TRACE satellite is that the magnetic field lines appear to run fairly parallel to one another with very little braiding. A theoretical answer from a numerical experiment on braiding by *Galsgaard and Nordlund* [1996] is that very little braiding persists, since reconnection tends to limit the twist of neighbouring field lines to at most one turn.

2.3. The Nature of 3D Magnetic Reconnection

Magnetic reconnection in three dimensions is quite different from in two dimensions and is only beginning

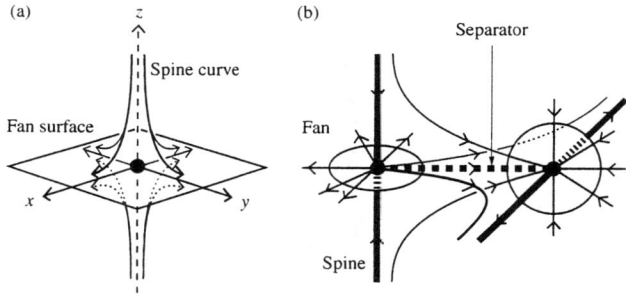

Figure 5. (a) The structure of an isolated null point, showing the spine and fan. (b) The intersction of the fans of two nulls to form a separator joining the one null to another.

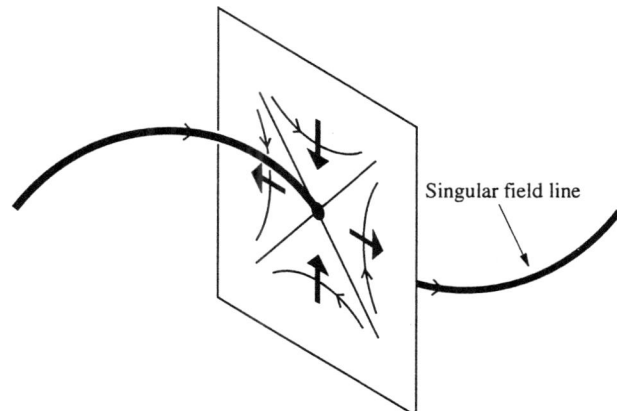

Figure 6. Singular field line reconnection.

to be explored. A generic null point in three dimensions possesses two classes of magnetic field lines that pass through the null point. A *spine* is an isolated field line that approaches (or recedes from) the null from two opposite directions, while a *fan* is a surface of field lines that recede from the null (or approach it). The intersection of the fans of two nulls is called a *separator* and is a special field line that links the nulls.

Three types of reconnection may occur near nulls [*Priest and Titov*, 1996]. In *spine reconnection* the current concentrates along the spine, whereas in *fan reconnection* it is maximized in the fan plane and in *separator reconnection* it is focussed along a separator (Figure 5).

Reconnection may also occur, however, in the absence of nulls by *singular field line reconnection* (Figure 6) when there is an electric field parallel to a particular magnetic field line (called a *singular field line*) and, in a plane perpendicular to the singular line, the magnetic field possesses an X-type topology with a hyperbolic flow [*Priest and Forbes*, 1989; *Hornig and Rastätter*, 1998].

Even though magnetic helicity is conserved to a high degree, small changes in magnetic helicity (K) are associated with reconnection, since [*Schindler et al*, 1988]

$$\frac{dK}{dt} = -2\int E_{\|} B\, dV \qquad (12)$$

when $E = 0$ on the bounding surface.

3. THE ROLE OF MAGNETIC HELICITY IN THE EVOLUTION AND DYNAMICS OF THE SOLAR CORONA

Magnetic helicity plays a key role in several coronal phenomena, such as prominences, solar flares and possibly coronal heating. The birth of a prominence may either be by the simple emergence of a large twisted flux tube through the photosphere [*Rust*, 1995] or it may be by the sum of many small flux emergences followed by reconnections that conserve the magnetic helicity to produce a large flux tube [*Priest et al*, 1996]. Once it has formed, the structure of a prominence is typically *dextral* in the northern hemisphere with negative magnetic helicity and *sinistral* in the southern hemisphere with positive helicity [*Martin et al*, 1994]. The eruption of a prominence is likely to occur when its magnetic helicity becomes too large for stability to be maintained. It represents a prime way that the Sun sheds magnetic helicity from closed magnetic regions.

The scenario for a large eruptive flare is as follows. First of all, the magnetic helicity builds up in a twisted flux tube and an overlying magnetic arcade. During the eruption mutual helicity between the tube and arcade is transferred by reconnection into the self-helicity of the erupting tube. Then, during the late stages of the eruption the tube itself reconnects, so that some helicity is ejected out into the solar wind as a closed plasmoid while some remains in the arcade that is left behind. During non-eruptive flares two flux systems reconnect with one another (either by emerging flux or interacting flux). The result is that a change in the mutual and self-helicities takes place while their sum is preserved to a high degree.

3.1. Coronal Heating

Magnetic helicity may also be implicated in the very existence of the solar corona, namely how it is heated. The heat flux that is required is about 300 Wm^{-2} in quiet regions or 5000 Wm^{-2} in active regions. The flux (vB^2/μ) of electromagnetic energy from the photosphere up into the corona is plentiful, namely about $10^4 Wm^{-2}$ for a typical footpoint speed (v) of $0.1 km\,s^{-1}$ and a magnetic field (B) of 100G. However, the question is: how efficient is the heating process? By what factor less than unity should this estimate be multiplied?

Heyvaerts and Priest [1984] suggested that the corona may be heated by relaxation as it evolves through a series of linear force-free states with the footpoint connections not preserved but the force-free constant α_0 determined from the evolution of magnetic helicity (K) Boundary motions cause the helicity to change in time according to (4). Conceptually, photospheric motions tend to build up energy in a nonlinear force-free field, which then goes unstable and relaxes to a linear force-free field by reconnection. Heyvaerts and Priest suggested that flares occur when the driving time is faster than the relaxation time, so that a nonlinear force-free field builds up and eventually goes unstable when the magnetic helicity is too great.

They applied their ideas to a model of a coronal arcade. The first step was to impose a footpoint displacement and calculate the resulting nonlinear force-free field and change in helicity. Then they supposed relaxation takes place and calculated the new linear force-free field with the same magnetic helicity. This enabled them to deduce the change in magnetic energy and therefore the heat that is liberated. The resulting heat flux is of the form

$$\frac{vB^2}{\mu}\left(\frac{L_B}{L_B+L_v}\right)\frac{\tau_{relax}}{\tau_{drive}}, \qquad (13)$$

which is of the form vB^2/μ (as expected) multiplied by two factors that are less than unity, the first involving the scales (L_B and L_v) for magnetic and velocity variations and the second the ratio of the relaxation time to the driving time. The optimum heating occurs when $L_v \approx L_B$ for motions that build up the nonlinear field on a time-scale of order the relaxation time. Negligible heating takes place when relaxation is rapid with $\tau_{relax} \ll \tau_{drive}$.

Several extensions to the basic theory have been constructed. *Vekstein et al* [1991] suggested that when $\tau_{drive} \approx \tau_{relax}$ there is *intermediate relaxation* to a state between the nonlinear field $\mathbf{B}^{nl}$ and the linear field $\mathbf{B}^l$. They calculated such an intermediate state from a phenomenological relaxation equation

$$\frac{\partial}{\partial t}(\mathbf{B}-\mathbf{B}^{nl}) = -\frac{(\mathbf{B}-\mathbf{B}^l)}{\tau_{relax}}, \qquad (14)$$

so that on times much shorter than τ_{relax}, $\mathbf{B} \approx \mathbf{B}^{nl}$, and on times much larger than τ_{relax}, $\mathbf{B} \approx \mathbf{B}^l$. They applied this theory to a variety of configurations including an arcade and a structure of close-packed flux tubes.

Vekstein et al [1993] suggested the corona is in a state of *partial relaxation*, in which closed magnetic fields are relaxed linear force-free states and the surrounding open field regions are essentially potential states. The open regions cannot support twist or shear, which would just propagate away to infinity. This overcomes one of the main disadvantages of linear force-free states, namely that they possess infinite energy and reversals when they occupy an infinite or semi-infinite domain. *Wolfson et al* [1994] constructed numerical solutions of such partially relaxed states and found that no solution exists when a critical helicity is exceeded, suggesting onset of an eruption.

Heyvaerts and Priest [1992] realised that the previous theory is incomplete since it depends on an unknown relaxation time, so they proposed a more self-consistent approach assuming that photospheric motions maintain the corona in a state with a turbulent viscosity (ν^*) and turbulent diffusivity (η^*). First of all, they calculate the global resistive MHD state driven by boundary motions and deduce the heat flux $(F_H(\nu^*, \eta^*))$. Secondly, they use cascade theories of turbulence to determine the values of ν^* and η^* that result from that F_H. This approach was applied to an arcade, a flux tube and to heating by wave motions. It gives reasonable heating and turbulence levels.

3.2. Heating of the Large-Scale Corona

In a letter that has just appeared in Nature, Priest et al (1998) build on previous work by *Kano and Tsuneta* [1996]. They realised that the temperature profile along a loop is highly sensitive to the nature of the heating. Turbulent magnetic dissipation in many small current sheets tends to deposit the heat uniformly along a loop and produce a profile in which $T^{\frac{7}{2}}$ is a quadratic function of distance (s) along the loop. Heat mainly near the summit, such as may be produced by long-wavelength standing waves, gives a $T^{\frac{7}{2}}(s)$ profile that has constant gradient in the legs of the loop and has strong curvature near the summit. Heating near the feet, such as produced by X-ray bright points, would give a steep rise near the feet and a flat profile near the summit.

They used the Japanese Yohkoh satellite to measure the temperature along the large loop shown in Figure 7 and compared with a series of models to deduce the likely form of the heating. Figure 8 shows the results. The observed temperature increases from about 1.5 MK near the feet up to 2.2 MK at the loop summit. If the heat is localised near the feet the best fit is rather poor (Figure 8a). If the heat is dumped at the summit the fit is better but still not very good (Figure 8b), but a uniform deposition of heat produces an excellent fit with the observations, exactly what turbulent relaxation would tend to give.

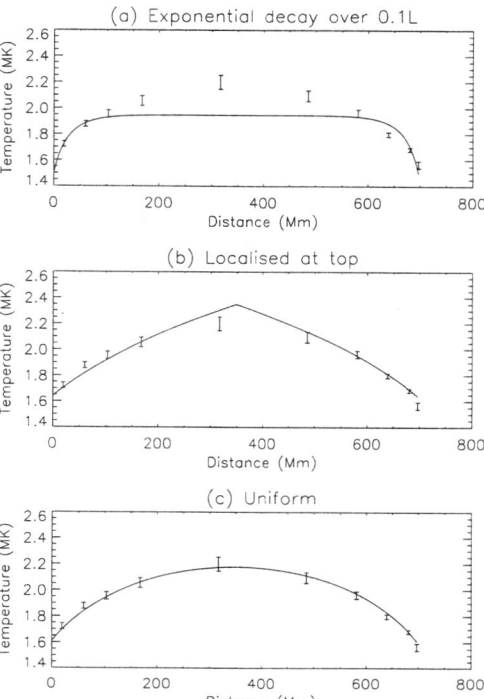

Figure 8. A comparison of the observations (vertical bars) and models (curves) of the temperature along the large-scale coronal loop of Figure 7.

4. CONCLUSION

The solar corona is a useful environment for exploring the role of magnetic helicity and relaxation in cosmical plasmas. It may play a key role in solar prominences, flares and coronal heating. In future, however, we need to develop the theory of three-dimensional reconnection, to determine the relaxation time in the corona and to observe magnetic structures in more detail. In particular, it is essential to pursue more sophisticated modelling and observation of the corona, so as to judge whether or not turbulent relaxation really is heating the corona.

Acknowledgments. I am most grateful to Loren Acton and Dick Canfield for enabling me to spend such an enjoyable summer in Montana and to Alex Pevtsov and Dick Canfield for organising such a splendid meeting.

REFERENCES

Berger, M.A. (1998), Magnetic helicity and filaments, in *New Perspectives on Solar Prominences*, edited by D. Webb, B. Schmeider and D. Rust, 1998.

Galsgaard, K. and Nordlund, A.,The heating and activity of the solar corona: I Boundary shearing of an initially ho-

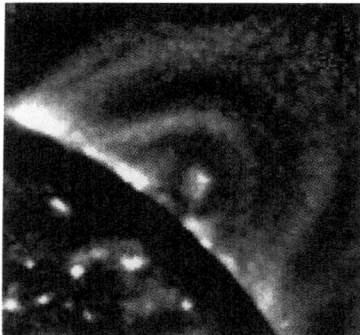

Figure 7. A large-scale loop observed by the Yohkoh Soft X-ray Telescope

mogeneous magnetic field, *J. Geophys. Res.*, *101*, 13445-13460, 1996.

Heyvaerts, J. and Priest, E.R., Coronal heating by reconnection in DC current systems - A theory based on Taylor's hypothesis, *Astron. and Astrophys.*, *137*, 63-78, 1984.

Heyvaerts, J. and Priest, E.R., A self-consistent turbulent model for solar coronal heating, *Astrophys. J.*, *390*, 297-308, 1992.

Hornig, G. and Rastätter, The magnetic structure of $B \neq 0$ reconnection, *Physica Scripta, T74*, 34-39, 1998.

Kano, R. and Tsuneta, S., Temperature distribution and energy scaling law of solar coronal loops obtained with Yohkoh, *Pub. Astron. Soc. Japan*, *48*, 535-543, 1996.

Martin, S.F., Bilimoria, R. and Tracadas, P.W., Magnetic field configurations basic to filament channels, in *Solar Surface Magnetism*, edited by R.J. Rutten and C.J. Schrijver, pp 303, Springer-Verlag, New York, 1994.

Parnell, C.E. and Priest, E.R., in preparation, 1998.

Pevtsov, A.A., Canfield, R.C. and Metcalf, T.R., Latitudinal variation of helicity in photospheric magnetic fields, *Astrophys. J.*, *440*, L109-L112, 1995.

Priest, E.R., Foley, C.R., Heyvaerts, J., Arber, T., Culhane, J.L. and Acton, L.W., The nature of the heating mechanism for the diffuse solar corona, *Nature*, 1997.

Priest, E.R. and Forbes, T.G., Steady magnetic reconnection in three dimensions, *Solar Phys.*, *119*, 211-214, 1989.

Priest, E.R., Parnell, C.E. and Martin, S.F., A converging flux model of an X-ray bright point and an associated cancelling magnetic feature, *Astrophys. J.*, *427*, 459-474, 1994.

Priest, E.R. and Titov, V.S., Reconnection at 3D nulls, *Phil. Trans. Roy. Soc. Lond.*, *355*, 2951-2992, 1996.

Priest, E.R., Van Ballegooijen, A.A. and MacKay, D.H., A model for dextral and sinistral prominences, *Astrophys. J.*, *460*, 530-543, 1996.

Rust, D. and Kumar, A., *Solar Phys.*, *155*, 69, 1995.

Schindler, K., Hesse, M. and Birn, J., General magnetic reconnection, parallel electric fields and helicity, *J. Geophys. Res.*, *93*, 5547-5557, 1988.

Schrijver, C.J., Title, A.M., Harvey, K.L., Sheeley, N.R., Wang, Y-M., van der Oord, G.H.J., Shine, R.A., Tarbell, T.D. and Hurlburt, N.F., Large-scale coronal heating by the dynamic small-scale magnetic field of the Sun, *Nature*, *487*, 424-425, 1998.

Vekstein, G.E., Priest, E.R. and Steele, C.D.C., Magnetic reconnection and energy release in the solar corona by Taylor relaxation, *Solar Phys.*, *131*, 297-318, 1991.

Vekstein, G.E., Priest, E.R. and Steele, C.D.C., On the problem of magnetic coronal heating by turbulent relaxation, *Astrophys. J.*, *417*, 781-789, 1993.

Wolfson, R., Vekstein, G.E. and Priest, E.R., Nonlinear evolution of the coronal magnetic field under reconnective relaxation, *Astrophys. J.*, *428*, 345-353, 1994.

E.R. Priest, Mathematical and Computational Sciences Dept.,St Andrews University,St Andrews KY16 9SS, Scotland. (e-mail: eric@dcs.st-and.ac.uk)

Magnetic Helicity and Stability in Solar Corona

K. Kusano

Institute for Nonlinear Sciences and Applied Mathematics, Hiroshima University, Higashi-Hiroshima, Japan

Solar coronal magnetic field plays a role as a channel through which the magnetic helicity is transported from the sun into the interplanetary space. If the coronal field forms the arcades, and if the magnetic Reynolds number is sufficiently large, the helicity transportation proceeds intermittently, even in the case that the helicity is constantly supplied from the sun. The intermittent transportation is a cyclic process which consists of the helicity storage, magnetic reconnection, and the plasmoid ejection. It is a result of the competition between the destabilization by the photospheric activity and the spontaneous stabilization caused by the magnetic arcade instability. The total helicity is well conserved even in the energy relaxation process, so that Taylor's hypothesis on the selective dissipation is applicable also to the solar corona. However, the numerical simulations indicate that the energy relaxation is even partial, and that the free energy cannot be exhausted, because the photospheric boundary gives a stronger constraint than Taylor's theory predicted. The coronal field could evolve around the marginally stable limit, which locates in the intermediate region between the stable and the unstable linear force free solutions. The results suggest that, even if there is no external trigger, the coronal plasma is able to generate intermittent activities like flares.

1. INTRODUCTION

It is widely believed that the solar coronal activities like flares are some intermittent liberation of the free energy stored in the coronal magnetic field. Since the solar corona is basically low β plasma, the force free field condition,

$$\nabla \times \boldsymbol{B} = \alpha \boldsymbol{B}, \qquad (1)$$

may be almost satisfied. It implies that, when the field has an excess energy compared to the potential field, there is a field aligned current generating the magnetic helicity,

$$H = \int \boldsymbol{A} \cdot \boldsymbol{B} dV, \qquad (2)$$

where the integration is over a whole domain under consideration.

In the solar corona, the magnetic helicity associated with the magnetic excess energy must be supplied by the photospheric activity. However, since the coronal field cannot possess an infinite capacity for the helicity, the injected helicity must be ejected into the interplanetary space. It means that the corona plays a role as a helicity channel, which connects between the sun and the interplanetary space. The main purpose of this paper is to understand the relationship between the magnetohydrodynamic (MHD) activities in the solar corona and the helicity transportation. In the following two sections, we will review the fundamental properties on the equilibrium and the stability of the linear force free field in the solar corona. Then, in Sec.4, the coronal dynamics in the helicity transportation process will be discussed based on the numerical simulations. Finally, the summaries and some important remarks will be presented in Sec.5.

2. LINEAR FORCE FREE FIELD

The principal theory related to the magnetic helicity is given by *Taylor* [1974, 1986]. Taylor proposed a hypothesis that the magnetic energy should be selectively dissipated faster than the magnetic helicity in the strongly magnetized plasma with small resistivity, and he predicted that the linear force free field, in which α is constant in space, might be self-organized. The applicability of this theory to torus plasmas (RFP and spheromak) is successfully confirmed by the experiments [*Ortolani and Schnack*, 1993] and also by the numerical simulations [*Kusano and Sato*, 1990]. However, the applicability to the coronal plasma is still an open question. In order to give an answer to that, let us first consider the basic feature of the linear force free field in the solar corona.

A major difference between the torus plasma and the coronal plasma is for the boundary condition. While in the torus plasma the boundary is almost identical to the flux surface, the corona contacts to two different types of boundaries. One is the photosphere, where the magnetic field is line-tied, and another is the interplanetary space continued infinitely. Hereafter, we consider the domain bounded by two horizontal planes $z = 0$, L_z, where z denotes the vertical coordinate. A plane at $z = 0$ corresponds to the photospheric boundary, where the vertical magnetic field is given by a sinusoidal function of the horizontal coordinate y,

$$B_z = B_0 \cos(ky), \tag{3}$$

and $k = 2\pi/L_y$. Another plane at $z = L_z$ forms the top boundary, where $B_z = 0$. Since the normal component of the magnetic field at the boundary is fixed in our model, the helicity (2) is a gauge invariant quantity, as long as the appropriate boundary condition for the vector potential is adopted [*Jensen and Chu*, 1984].

An infinite half space bounded just by the bottom plane can be treated as a subset of this geometry, since it corresponds to the limiting case where the aspect ratio $a \equiv L_z/L_y \to \infty$. Here, just for the simplicity, we restrict ourselves into the two dimensional system ($\partial/\partial x = 0$), and the effects of more complicated boundary conditions are left for the future works.

Let us consider the solution of the linear force free field in this geometry. According to *Yoshida and Giga* [1990], the general vector field can be described by

$$\boldsymbol{B} = \boldsymbol{B}_0 + \sum_i c_i \boldsymbol{b}_i.$$

Here, $\boldsymbol{B}_0$ is the potential field ($\nabla \times \boldsymbol{B}_0 = 0$) satisfying the boundary condition of the normal component ($\boldsymbol{B}_0 \cdot \boldsymbol{n} = \boldsymbol{B} \cdot \boldsymbol{n}$), and the vector $\boldsymbol{b}_i$ is the eigenfunction of curl operator,

$$\nabla \times \boldsymbol{b}_i = \lambda_i \boldsymbol{b}_i,$$

satisfying the uniform boundary condition ($\boldsymbol{b}_i \cdot \boldsymbol{n} = 0$), where $\boldsymbol{n}$ is the normal vector on the boundary. The eigenfunction $\boldsymbol{b}_i$ should be classified by whether the mutual helicity with the potential field, $I_i = \int \boldsymbol{a}_i \cdot \boldsymbol{B}_0 dV$, is vanished or not, where $\boldsymbol{a}_i$ is the vector potential for $\boldsymbol{b}_i$ [*Taylor*, 1986]. The eigenfunctions $\boldsymbol{b}_i$ for $I_i \neq 0$ and $\boldsymbol{b}_j$ for $I_j = 0$ are called the *coupled* function and the *decoupled* function, respectively.

The linear force free field should be classified into two solutions by the class of eigenfunctions composing the solution [*Taylor*, 1986; *Yoshida and Giga*, 1990]. The solution, which consists of the coupled functions only, is called the *coupled solution*, and the solution, which is given by the mixture of two sorts of eigenfunctions, is called the *mixed solution*. The former exists unless α is equal to the eigenvalue of the coupled function, and the latter can exist just when α is equal to the eigenvalue of the decoupled function.

Kusano et al. [1995] found that the linear force free field in the coronal geometry bifurcates into two solutions above when $a > \sqrt{5}/2$, and $H > H_0$, which is the helicity of the coupled solution for $\alpha = 2\pi/a$, as shown in Fig. 1. It means that, when the helicity exceeds H_0 the minimum energy state is switched from the coupled solution into the mixed solution. Furthermore, we should note that two solutions have different topology in the magnetic field each other. As shown in Fig. 2, while any field lines in the coupled solution connect to the photosphere, the mixed solution has an isolated magnetic island, which is detached from the photosphere. It implies that, when $H > H_0$, the self-organization into the minimum energy state must be obtained with magnetic reconnection.

Since the helicity is a monotonically increasing function of α, H_0 must decrease as a increases. In particular, $H_0 \to 0$, when $a \to \infty$. The minimum energy state in an infinite half space must be given by the potential field ($\alpha = 0$) for any helicity, as shown in Fig.1 (c). The linear force free field in a half space,

$$B_x = -B_0 \frac{\alpha}{k} \exp(-Kz) \sin ky, \quad (4)$$

$$B_y = B_0 \frac{K}{k} \exp(-Kz) \sin ky, \quad (5)$$

$$B_z = B_0 \exp(-Kz) \cos ky, \quad (6)$$

where $K^2 = k^2 - \alpha^2$ [*Heyvaerts and Priest*, 1984], is no more the minimum energy state when $\alpha > 0$, but it corresponds to the coupled solution in Fig.1 (c).

3. MAGNETIC ARCADE INSTABILITY

3.1. Linear Stability

The stability of the linear force free field is not guaranteed when α is larger than the lowest eigenvalue λ_1, since the Taylor's energy principle vanishes just the first energy variation [*Taylor*, 1986]. Actually, *Kusano and Nishikawa* [1996a] revealed that the coupled solution for $H > H_0$ is even linearly unstable. It is an instability breaking the symmetry in the series of magnetic ar-

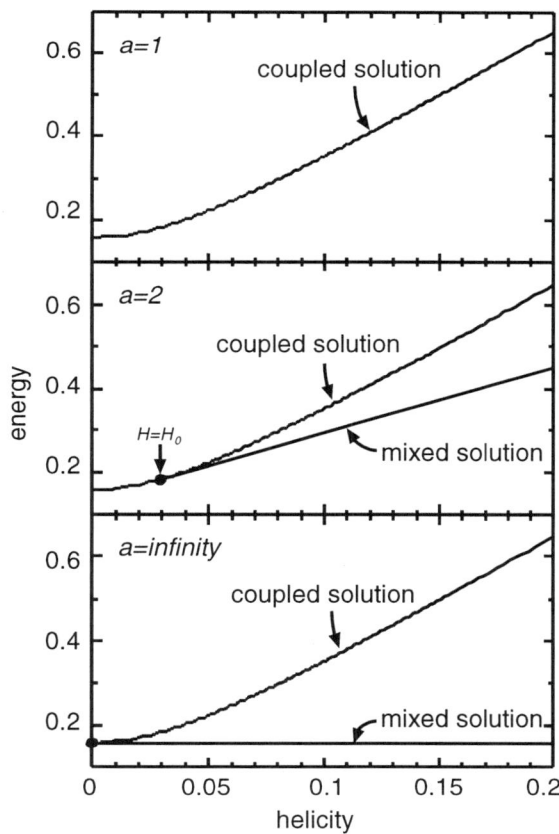

Figure 1. The relationship between the helicity and the magnetic energy of the linear force free field for different aspect ratio ($a = 1$, 2, and ∞).

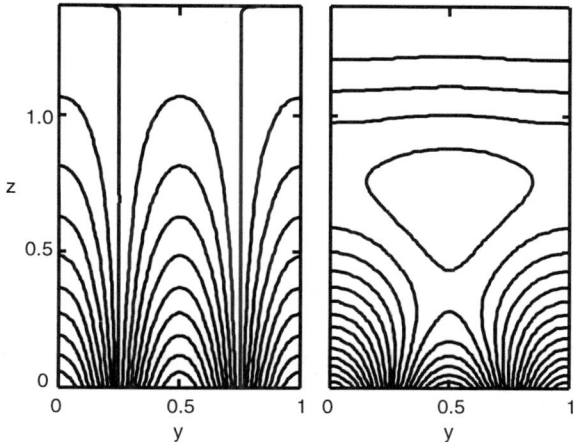

Figure 2. Contour plots of the magnetic flux for the coupled solution (*left*) and the mixed solution (*right*).

cades, and hereafter we will call it the *magnetic arcade instability*. *Kusano and Nishikawa* [1996b] numerically showed that the growth rate of this instability is proportional to $1/\sqrt{S}$, where S is the magnetic Reynolds number defined by the global scale L_y.

3.2. Energy Relaxation

Here, two questions arise for the nonlinear property of the magnetic arcade instability: How effectively can that release the free energy in the magnetic field, and whether is the magnetic helicity well conserved even in the energy relaxation process, as predicted by Taylor?

To achieve the answers to this question, *Kusano and Nishikawa* [1996b] carried out the numerical simulations, in those the initial state was given by the coupled solution plus the most unstable eigenfunction with small amplitude. The simulation indicated that, as a result of the nonlinear growth of the magnetic arcade instability, the current sheet is formed above the neutral line and magnetic reconnection happens. While reconnection releases the substantial energy, which is up to about 60 % of the energy difference between two branches of the linear force free field, the helicity is hardly changed. Therefore, we can conclude that the selective dissipation is an applicable principle also to the coronal plasma.

The mechanism of the selective dissipation is strongly related to the structure of the current sheet, in which the field lines reconnect. The simulations showed that the magnetic field parallel to the current channel at the reconnection point is extremely smaller than the field strength in the in-flow region. Therefore, the helicity dissipation rate $2\eta \boldsymbol{J} \cdot \boldsymbol{B}$ is minimized at the center of the current sheet, while the energy dissipation rate $\eta \boldsymbol{J} \cdot \boldsymbol{J}$ is maximized there, where η and $\boldsymbol{J}$ are the resistivity and the electric current density, respectively. The magnetic arcade instability drives the so-called *total* reconnection proposed by *Sato et al.* [1989].

However, the simulations showed also that the final states in the energy relaxation process do not coincide with the Taylor's minimum energy state, which should be described by the mixed solution. It might be due to the fact that, the current sheet cannot touch the photosphere, since the field lines just above the photosphere is hardly compressed horizontally because of the line-tied condition. Consequently, the magnetic flux below the low end of the current sheet is not subject to reconnection, and the free energy remains there. Hence, the boundary condition gives a constraint stronger than the Taylor's hypothesis, and hence the energy relaxation is halted in the intermediate state before the mixed solution is obtained.

Here, we should remark that the eigenfunction of the magnetic arcade instability has to have a longer wavelength than the characteristic scale on the photospheric boundary [*Kusano et al.*, 1995; *Kusano and Nishikawa*, 1996a]. It is well consistent with the fact that the single arcade system confined by the conductive box is not destabilized [*Biskamp and Welter*, 1989]. The reason is that, in contrast to the periodic condition, the conductive condition on the latelal wall inhibits the longer mode than the box size which is equivalent to the scale of the photospheric field in the single arcade system. On the other hand, in the case that the multiple arcades are included in the box, the instability may appear even in the conductive box, since the photospheric scale is smaller than the box size. This property may explain the variety of the arcade dynamics produced with different models [*Mikic et al.*, 1988; *Biskamp and Welter*, 1989; *Inhester et al.*, 1992; *Amari et al.*, 1996].

On the other hand, the simulation results [*Kusano et al.*, 1996] are well consistent with the recent X-ray observations of flares. Particularly, the generation of the hard X-ray hot spot [*Masuda et al.*, 1994] well agrees with the simulation in which reconnection generates an extremely hot region through the fast magnetosonic shock of the downward jet. Furthermore, the morphology of the magnetic field is also consistent with the soft X-ray images in flares [*Tsuneta*, 1996].

4. DYNAMIC PROCESS

4.1. Energy Accumulation

As explained in the previous section, once the magnetic arcade system becomes unstable, magnetic reconnection may release the free energy. However, according to the linear analyses [*Kusano and Nishikawa*, 1996a], when the helicity is larger than H_0, the coupled solution is always unstable. It means that, when the helicity is slowly supplied into the corona, the coronal field is destabilized before a substantial free energy is stored. Nevertheless, why and how the free energy is accumulated in the coronal field?

A solution of this paradox could be hidden in the fact that the dynamics is a competitive process between the accumulation and the relaxation of the free energy. Since the energy relaxation begins with magnetic reconnection, the system can stay on the unstable branch until the instability enables reconnection. Before then, some period $T = \gamma^{-1} \ln(A_R/A_0)$ has to been elapsed, where γ, A_0, and A_R are the growth rate, and the initial amplitude of the instability, and the amplitude required for reconnection, respectively. Whereas we can estimate γ and A_R from the linear analyses and/or the nonlinear calculation of the single instability event, A_0 should be determined in the dynamic process rather than as the initial condition. Hence, we have to self-consistently calculate the dynamics which is longer than a single instability event.

4.2. Numerical Model

Let us study the long term dynamics using the numerical model. The basic equations to be solved are given by the MHD equations as follows:

$$\frac{\partial \boldsymbol{V}}{\partial t} = -\boldsymbol{V} \cdot \nabla \boldsymbol{V} + \boldsymbol{J} \times \boldsymbol{B} + \nu \nabla^2 \boldsymbol{V}, \quad (7)$$

$$\frac{\partial \boldsymbol{B}}{\partial t} = \nabla \times \{\boldsymbol{V} \times \boldsymbol{B} - \eta \boldsymbol{J}\}, \quad (8)$$

$$\boldsymbol{J} = \nabla \times \boldsymbol{B}, \quad (9)$$

in which time t, space $\boldsymbol{x}$, velocity $\boldsymbol{V}$, magnetic flux density $\boldsymbol{B}$, electric current $\boldsymbol{J}$, plasma density ρ, viscosity ν, and electric resistivity η are normalized by $\tau_A \equiv L_y/V_A$, L_y, $V_A \equiv B_0/(\mu_0 \rho_0)^{1/2}$, B_0, $B_0/\mu_0 L_y$, ρ_0, L_y^2/τ_A, and $L_y^2 \mu_0/\tau_A$, respectively. Here, the plasma pressure and the gravity forces are omitted, and the viscosity ν is fixed to 1.0×10^{-4} in the normalized unit.

The simulation box is the two dimensional domain $(L_y \times L_z)$, and the lateral boundaries $(y = 0, L_y)$ satisfy the periodic condition. The velocity at the bottom and the top is given by $V_x = V_0 \sin(2ky)$, $V_y = V_z = 0$ at $z = 0$, and $V_x = V_y = V_z = 0$ at $z = L_z$, where V_0 is fixed to 5×10^{-3}. The photospheric motion continuously supplies the magnetic helicity into the coronal region with the constant rate,

$$\dot{H}_{in} = -\int_0^{L_y} A_x V_x B_z dy|_{z=0} = 2\pi V_0 B_0^2/k^2. \quad (10)$$

The initial state is composed of the potential field ($\nabla \times \boldsymbol{B} = 0$) plus a small perturbation, which is necessary as the seed to break the mirror symmetry for $y = L_y/2$.

The aspect ratio of the simulation box ($a = L_z/L_y$) is 10^2, which is much larger than the previous models [*Mikic et al., 1988; Kusano et al., 1996*]. The top boundary locates sufficiently far from the bottom, so that it does not influence any dynamics in the coronal region, even if the calculation continues as long as the helicity injection time-scale (L_y/V_0).

The numerical solver is composed of the finite difference scheme with two point spatial difference and the Runge-Kutta-Gill method. The grid number ($N_y \times N_z$) in the simulation box ($L_y \times L_z$) is 256×400, and the grid size is spatially modulated in order to accurately capture the thin current layers which appear in the reconnection site. The mesh size for the horizontal dimension Δy is minimized on the center of the magnetic arcades, where $\Delta y = 7.83 \times 10^{-4}$. The mesh size for the vertical dimension Δz is 1.25×10^{-2} for $z < (5/2)L_y$, and is continuously increased with the altitude in the higher region.

4.3. Results and Discussions

Three simulations (cases 1, 2, and 3) are performed with different resistivity ($\eta = 10^{-4}, 2 \times 10^{-4}$, and 10^{-3}, respectively). Here, for convenience' sake, we regard the lower part ($0 < z < L_c \equiv 2L_y$) and the higher part ($L_c < z < L_z$) as the corona and the interplanetary space, separately. First, let us consider the evolution of the helicity and the energy

$$H_c = \int_0^{L_y} dy \int_0^{L_c} dz \boldsymbol{A} \cdot \boldsymbol{B},$$

$$E_c = \int_0^{L_y} dy \int_0^{L_c} dz \boldsymbol{B} \cdot \boldsymbol{B}/2,$$

those are integrated just over the coronal region. As shown in Fig. 3, the trajectories in cases 1 and 2 are subject to an explosive energy relaxation after following the branch of the coupled solution. It is a result of magnetic reconnection driven by the magnetic arcade instability as seen also in the previous papers [*Mikic et al., 1988; Kusano et al., 1996*].

After the first big relaxation, however, the trajectories enter into each cyclic orbit, which consists of three phases: The first phase is the storage phase, in which the magnetic energy and the helicity are sup-

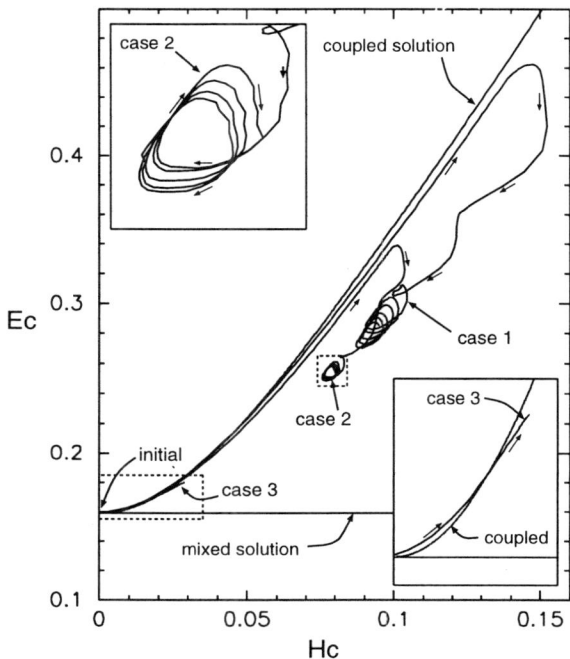

Figure 3. The trajectories of the simulation results as well as two branches of the linear force free field are plotted on the $H_c - E_c$ diagram. Subsets are the zoom-in views of the dotted squares, where the final states for cases 2 and 3 are included, respectively.

plied from the photosphere, while the magnetic arcade instability is slowly growing. The second phase is the energy relaxation phase, in which magnetic reconnection releases the energy. The final phase is the removal of the excess helicity. Reconnection in the second phase generates a flux tube (plasmoid), where the field lines have a helical structure, as shown in Fig. 4. The helical field lines clearly indicate that some amount of the helicity is contained. After the field line is disconnected from the photosphere, the plasmoid is ejected into the interplanetary space by the magnetic pressure, and removes a part of the helicity out of the coronal magnetic field.

It has been confirmed that the cyclic orbits in Fig. 3 are insensitive to the initial state of the calculation, so that we can call them a sort of the limit cycle. On the other hand, the trajectory of case 3, in contrast to the other cases, falls into the fixed point located just below the high branch of the linear force free field, and the steady state is obtained.

In Fig. 5, the averaged helicity and the oscillatory amplitude on the fixed point (case 1) or in the limit cycles (cases 2 and 3) are plotted as a function of the

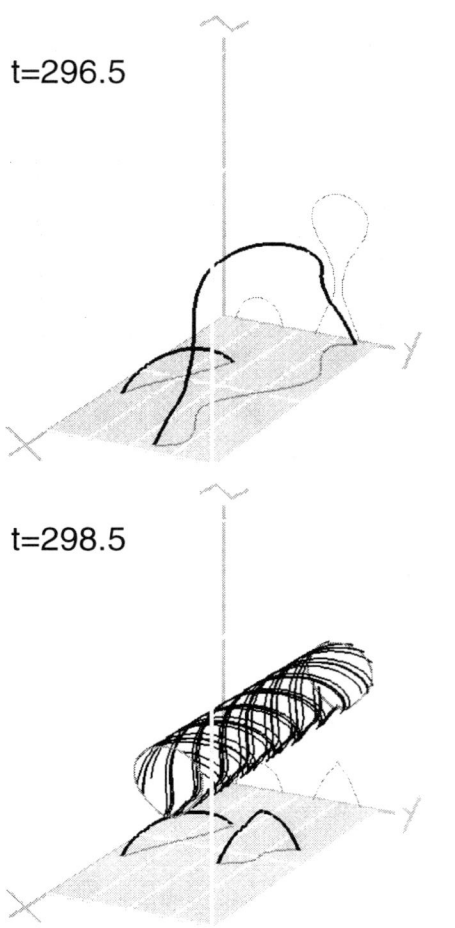

Figure 4. Three dimensional structure on the magnetic field lines before ($t = 296.5$) and after reconnection ($t = 298.5$) in case 1. The projections of the filed lines are also displayed on the vertical ($x = 0$) and the horizontal ($z = 0$) planes.

magnetic Reynolds number defined by the photospheric speed ($S_p = V_0 L_y/\eta$). Here, let us consider the equilibrium state, in which the helicity injection $\dot{H}_{in}$ is balanced with the resistive diffusion

$$\dot{H}_\eta = -2\eta \int_0^{L_y} dy \int_0^\infty dz \, \mathbf{J} \cdot \mathbf{B}. \quad (11)$$

When the magnetic field is given by (4) to (6), if the helicity balance condition,

$$\dot{H}_{in} + \dot{H}_\eta = 0, \quad (12)$$

is satisfied, we can derive

$$\alpha = \frac{2\pi S_p}{\sqrt{S_p^2 + (8\pi)^2}}. \quad (13)$$

The solid curve in Fig. 5 represents the helicity of the linear force free field for (13).

We can see that case 3 is located just near the steady state predicted by (13). However, as S_p increases, the deviation from the diffusive equilibrium is enhanced, and the increment of the coronal helicity is saturated. Furthermore, the oscillatory amplitude also increases with S_p. It indicates that the magnetic arcade instability effectively removes the excess helicity out of the corona. The results suggest that the coronal helicity must be extremely smaller compared to the resistive steady state, because the magnetic Reynolds number in the solar corona is much larger than the parameter adopted in the current model.

Another important result is that the limit cycle orbits seem to be distributed on a single curve, which lies in the intermediate region between the stable and the unstable branches of the linear force free field, as shown in Fig. 3. It suggests that the coronal magnetic field could evolve around a marginally stable region.

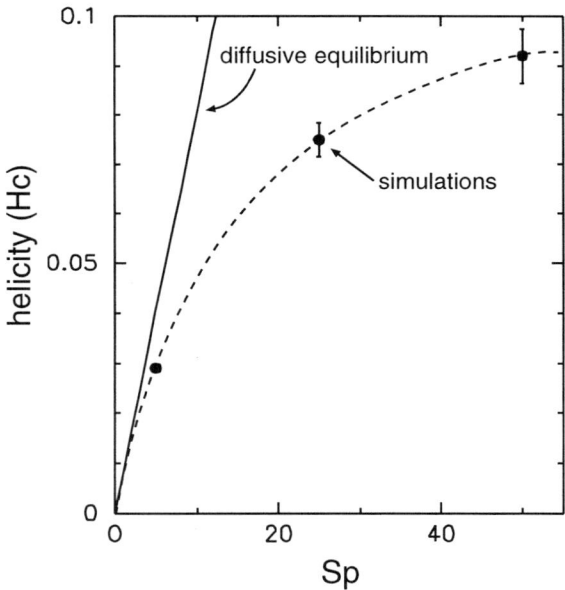

Figure 5. The solid circles and the vertical bars indicate the averaged helicity and the amplitude of the helicity oscillation for each limit cycle process of three simulations. The abscissa is $S_p = V_0 L_y/\eta$ and the solid curve is for the diffusive equilibrium derived from (13).

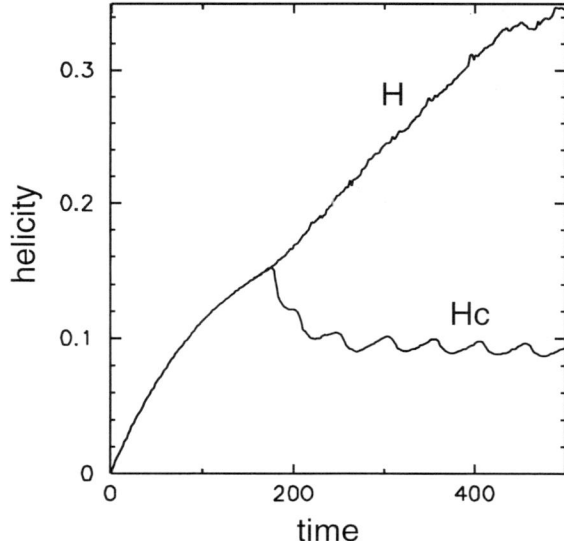

Figure 6. The evolution of the total magnetic helicity H and the magnetic helicity in the coronal region H_c for case 1.

5. CONCLUSION

The magnetic helicity is a key parameter to understand the coronal dynamics. In Fig. 6, the coronal helicity H_c as well as the total helicity H, which is integrated over the whole domain, are plotted as a function of time for case 1. Here, it is clearly seen that the total helicity monotonically increases without any oscillatory behavior, whereas the coronal helicity oscillates in saw-tooth shape. It means that the helicity is hardly dissipated even in the reconnection process, but the evolution of H_c is caused just by the transportation from the corona into the interplanetary space.

The helicity transportation is much similar to the dynamics of a dripping water. Even if the water is constantly supplied into the tap, the droplets appear intermittently and release the gravitational free energy. If we consider the physical analogy with the droplets, it can be understood that the helicity plays a role as the conserved quantity just like the water mass in the droplets, and that the magnetic tension and the magnetic pressure in the corona are counterparts of the surface tension and the gravity in the dripping process, respectively.

Before closing this chapter, we should mention that the numerical model is so much idealized compared to the solar corona. In particular, the magnetic Reynolds number is assumed to be several orders of magnitude less than the practical value because of the computational restriction. Furthermore, the boundary condition in the model might be much structured than the sun. Therefore, it is likely that the solar coronal dynamics is much more complicated than the simulation results. However, we can conclude here at least that the intermittent dynamics can be spontaneously generated in the helicity transportation through the solar corona.

REFERENCES

Amari, T., Luciani, J.F., Aly, J.J., & Tagger, M., Plasmoid formation in a single sheared arcade and application to coronal mass ejections, *Astronomy Astrophys.*, *306*, 913-923, 1996.

Biskamp, D. & Welter, H., Magnetic arcade evolution and instability, *Solar Phys.*, *120*, 49-77, 1989.

Heyvaerts, J., and Priest, E. R., Coronal heating by reconnection in DC current systems: A theory based on Taylor's hypothesis, *Astronomy Astrophys.*, *137*, 63-78, 1984.

Inhester, B., Birn, J., & Hesse, M., The evolution of line-tied coronal arcades including a converging footpoint motion, *Solar Phys.*, *138*, 257-281, 1992.

Jensen, T.H., and Chu, M., Current drive and helicity injection, *Phys. Fluids*, *27*, 2881-2885, 1984.

Kusano, K., and Sato, T., Simulation study of self-sustainment mechanism in reversed-field pinch configuration, *Nuclear Fusion*, *30*, 2075-2096, 1990.

Kusano, K., Suzuki, Y., and Nishikawa, K., A solar flare triggering mechanism based on the Woltjer-Taylor minimum energy principle, *Astrophys. J.*, *441*, 942-951, 1995.

Kusano, K., and Nishikawa, K., Bifurcation and stability of coronal magnetic arcades in a linear force-free field, *Astrophys. J.*, *461*, 415-423, 1996a.

Kusano, K., and Nishikawa, K., Magnetic reconnection in the solar atmosphere, in *Proc. of a Yohkoh Conference* edited by Bentley, R.D., and Mariska, J.T., pp.365-378, Astronomical Society of the Pacific, 1996b.

Kusano, K., Suzuki, Y., Fujie, K., Endo, Y., and Nishikawa, K., Solar flare as Taylor's relaxation, in *Proc. of a Yohkoh Conference* edited by Bentley, R.D., and Mariska, J.T., pp.280-285, Astronomical Society of the Pacific, 1996.

Masuda, S., Kosugi, T., Hara, H., Tsuneta, S., and Ogawara, Y., A loop-top hard X-ray source in a compact solar flare as evidence for magnetic reconnection, *Nature*, *371* No.6497, 495-497, 1994.

Mikic, Z., Barnes, D.C., and Schnack, D.D., Dynamical evolution of a solar coronal magnetic field arcade, *Astrophys. J.*, *328*, 830-847, 1988.

Ortolani, S., and Schnack, D. D., *Magnetohydrodynamics of Plasma Relaxation*, pp.88–93, World Scientific, Singapore, 1993.

Sato, T., Horiuchi, R., and Kusano, K., Global and local relaxation in magnetohydrodynamic plasma, *Phys. Fluids*, *B1*, 255-257, 1989.

Taylor, J. B., Relaxation of toroidal plasma and generation

of reverse magnetic fields, *Phys. Rev. Lett.*, *33*, 1139-1141, 1974.

Taylor, J. B., Relaxation and magnetic reconnection in plasmas, *Rev. Mod. Phys.*, *53*, 741-763, 1986.

Tsuneta, S., Structure and dynamics of magnetic reconnection in a solar flare, *Astrophys. J.*, *456*, 840-849, 1996.

Yoshida, Z., and Giga, Y., Remarks on spectra of operator rot, *Math. Z.*, *204*, 235–245, 1990.

K. Kusano, Institute for Nonlinear Sciences and Applied Mathematics, Hiroshima University, Higashi-Hiroshima 739-8526, Japan. (e-mail: kusano@sci.hiroshima-u.ac.jp)

The Evolution of Magnetic Helicity Under Reconnection

G. Hornig

Department of Physics and Astronomy, Ruhr-Universität Bochum, Germany

The evolution of magnetic helicity for magnetic reconnection is investigated on the basis of a general kinematic model of the reconnection process. The model allows for arbitrary 3-D reconnection in an almost ideal environment as given by the solar corona. An equation for the evolution of the magnetic helicity density is derived. In the most simple case, that is for reconnection with vanishing $\mathbf{E} \cdot \mathbf{B}$, this equation shows that the helicity density is frozen in a virtual fluid flow of stagnation type. Therefore, simple predictions about the redistribution of helicity in the reconnection process can be made. In the more general case of non-vanishing $\mathbf{E} \cdot \mathbf{B}$, the results still hold approximately for the case that the non-ideal reconnection region is small compared to the typical length scale of the magnetic structure. This is in accordance with previous results for the approximate conservation of helicity in a resistive plasma, but holds as well for reconnection processes where the non-idealness is not or not only a result of resistivity.

1. INTRODUCTION

The topology of magnetic fields is an important property of astrophysical and technical plasmas. It is crucial for the existence of equilibria, their energy content and their dynamics. In most astrophysical as well as many technical plasmas the evolution of the plasma and its magnetic field is almost free of large scale dissipation on dynamical time scales. In the framework magnetohydrodynamics this is expressed by the ideal form of Ohm's law, which excludes the dissipation of magnetic flux and conserves the magnetic topology. It also implies the vanishing of $\mathbf{E} \cdot \mathbf{B}$ which is the source term for magnetic helicity, $\mathbf{A} \cdot \mathbf{B}$. This quantity is a topological measure of the linkage of magnetic flux in the sense of an asymptotic linkage number of infinitesimal flux tubes [*Arnold*, 1986]. A balance equation for this quantity can be derived from Maxwell's equation,

$$\frac{\partial \mathbf{A} \cdot \mathbf{B}}{\partial t} + \nabla \cdot (\phi \mathbf{B} + \mathbf{E} \times \mathbf{A}) = -2\,\mathbf{E} \cdot \mathbf{B} \tag{1}$$

and this can be rewritten with the help of the ideal Ohm's law

$$\mathbf{E} + \mathbf{v} \times \mathbf{B} = 0, \tag{2}$$

into

$$\frac{\partial \mathbf{A} \cdot \mathbf{B}}{\partial t} + \nabla \cdot (\mathbf{v}\,\mathbf{A} \cdot \mathbf{B} - \mathbf{B}\,\mathbf{A} \cdot \mathbf{v}) = 0.$$

This leads to a conservation of the total helicity, that is the integral of $\mathbf{A} \cdot \mathbf{B}$ over a three-dimensional comoving volume bounded by a magnetic surface (a surface on which the normal component of $\mathbf{B}$ vanishes) is constant in time.

However, the absence of any kind of large scale dissipation or other non-idealness acting on the dynamical time scale of the system does not exclude the existence of non-ideal effects in strongly localized regions. This is

Magnetic Helicity in Space and Laboratory Plasmas
Geophysical Monograph 111
Copyright 1999 by the American Geophysical Union

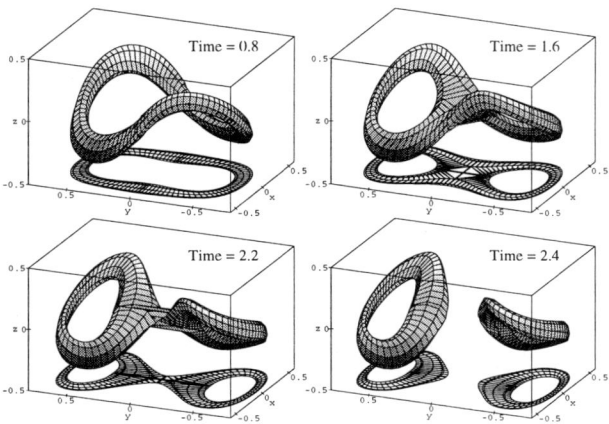

Figure 1. Example of a numerical simulation of magnetic reconnection. To visualize the process a closed flux surface of the magnetic field is chosen at an initial instant and followed in time.

observed for the process of magnetic reconnection where the self-organization of the plasma results in the formation of current sheets of decreasing thickness until eventually the length scale becomes small enough for non-ideal terms to become relevant. Which kind of non-idealness dominates the reconnection process is still a matter of debate (see for instance [*Biskamp*, 1997]). In most cases it is assumed to be an anomalous resistivity which is represented by a resistive term in Ohm's law. However, other terms such as the electron viscosity and inertial terms are also discussed in this context.

The possibility of reconnection and hence of a non-ideal evolution not only destroys the conservation of magnetic topology but also the simple argument which lead us to the conservation of magnetic helicity. However, first Taylor [*Taylor*, 1974] conjectured and later Berger proved for weakly resistive plasmas [*Berger*, 1984] that the total helicity, that is the helicity integrated over a volume bounded by a magnetic surface is still approximately conserved under certain conditions. It is the aim of this contribution to give a more detailed picture of the evolution of magnetic helicity during reconnection regardless of the special non-idealness which drives the process. Consider for instance a reconnection process as shown in Fig.1. Here the effect of reconnection on the magnetic field is demonstrated by choosing a closed magnetic surface (flux surface) at an initial instant and following it in time. The initial flux ring has a certain amount of total magnetic helicity and it is natural to ask for the helicity of the final stage. If we assume that this process is the effect of a resistive reconnection we can apply at least the result of Berger and state that the total helicity for a volume which incorporates the reconnection region is approximately conserved. But we have no information about how this helicity is distributed between the final flux rings. To shed more light on this and similar problems we first have to develop a model of the reconnection process which is as general as possible.

2. A KINEMATIC DESCRIPTION OF THE RECONNECTION PROCESS

To set up a general model of the reconnection process we assume that the reconnection takes place in a localized region of space which we assume to be embedded in an otherwise ideal plasma. This is the natural situation for reconnection taking place in a highly collisionless plasma where due to the formation of thin current sheets only locally non-ideal effects such as an anomalous resistivity enables reconnection.

We start with the most simple situation, i.e. a two-dimensional stationary reconnection as considered in the first models of reconnection by Dungey [*Dungey*, 1953], Sweet[*Sweet*, 1958], Parker [*Parker*, 1963], and others, a simple sketch of which is given in Fig. 2. A typical feature all of these examples have in common is that they are stationary and reconnection occurs at a magnetic null (the origin in Fig. 2).

We can define a transport velocity of magnetic field lines or magnetic flux respectively by

$$\mathbf{w} := \frac{\mathbf{E} \times \mathbf{B}}{B^2}. \qquad (3)$$

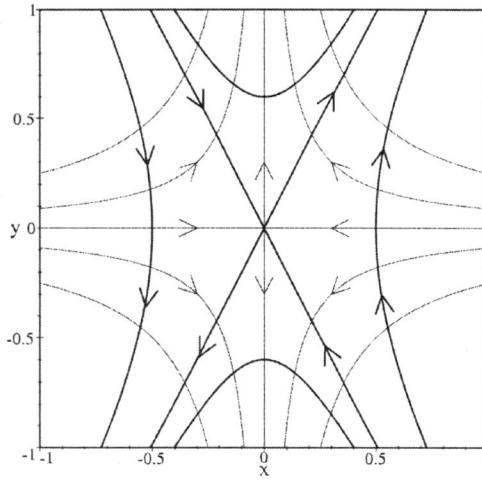

Figure 2. Topology of the magnetic field (thick) and the plasma velocity (thin lines) in two-dimensional reconnection.

This velocity field diverges at $\mathbf{B} = 0$ and is of X-type in the neighborhood of the null such that it transports the magnetic flux across the separatrices, i.e. the field lines which intersect at the null. It coincides with the plasma velocity in the ideal region outside the reconnection site, but it differs from it inside where some kind of non-idealness drives the process. For instance, time independence and two-dimensionality imply that the electric field is constant in the invariant direction ($\mathbf{E} = 1\mathbf{e}_z$) and if we assume that the magnetic field has a generic hyperbolic null ($\mathbf{B} = l\, y\mathbf{e}_x + k\, x\, \mathbf{e}_y$) this yields for the velocity

$$\mathbf{w} = \frac{-k\, x\, \mathbf{e}_x + l\, y\, \mathbf{e}_y}{l^2 y^2 + k^2 x^2} \quad (4)$$

which diverges proportional to $1/x$ along the inflow trajectory (x-axis). This divergence is not a peculiarity of the example chosen. In two dimensions every example of reconnection must have some kind of divergence in the transporting flow because it requires a null in the magnetic field and a finite electric field. Alternatively, if we assume that the velocity $\mathbf{w}$ is smooth, i.e. has a null point, then the time for the magnetic flux starting from some initial point x_0 on the X-axis to reach the null would be infinite and we would therefore not call this reconnection.

The existence and structure of the flow $\mathbf{w}$ can serve for a definition of reconnection. In view of the generalization to three dimensions, however, we will consider $\mathbf{w}$ as a flow in a four-dimensional space-time (e.g. the Minkowski space). Here the velocity has four components $W^{(4)} = (W^0, \mathbf{W})$. The superscript 0 denotes the time coordinate ct and a metric of the signature (1,-1,-1,-1) is used. The ideal Ohm's law (2) is equivalent to the covariant set of equations

$$W^\alpha F_{\alpha\beta} = 0 \quad \Leftrightarrow \quad \begin{cases} \mathbf{E} \cdot \mathbf{W} = 0 \\ W^0 \mathbf{E} + \mathbf{W} \times \mathbf{B} = 0 \end{cases} \quad (5)$$

where $F_{\alpha\beta}$ denotes the electromagnetic field tensor. The four-velocity $W^{(4)}$ is related to the usual velocity in space by

$$\mathbf{w} := \frac{\mathbf{W}}{W^0} = \frac{d\mathbf{X}/ds}{dX^0/ds}, \quad (6)$$

where $(X^0, \mathbf{X})$ are the coordinates of a point in space-time moving with the four-velocity $W^{(4)}$ and s is an arbitrary parameterization of its trajectory. The diverging velocity in space is now represented by a quotient of two quantities in space-time and this allows for a representation of the singularity of $\mathbf{w}$ by a null point of $W^{(4)}$. For instance, for the velocity defined by Eq. (4) we can choose $\mathbf{W} = -k\, x\, \mathbf{e}_x + l\, y\, \mathbf{e}_y$ and $W^0 = l^2 y^2 + k^2 x^2$. Note that we have not normalized $W^{(4)}$ to the eigen-time, because $W^{(4)}$ or $\mathbf{w}$ do not represent the flow of massive particles and hence can and in fact do exceed the velocity of light. Therefore, if we want to represent $\mathbf{w}$ by a smooth, i.e. at least differentiable, flow $W^{(4)}$ we cannot impose the condition $W^0 \geq 1$, but we can still postulate $W^0 \geq 0$. The way in which we represent $\mathbf{w}$ by $\mathbf{W}$ and W^0 is not unique, but for the condition that $W^{(4)}$ is smooth the existence and type of its null points are unique and this is the crucial point for our representation of reconnection.

The generic structure of eigenvalues of $\partial_\alpha W^\beta$ at the null is (0, -a, a, 0) in the rest frame of the singularity. This is due to the symplectic eigenvalue theorem [Abraham, 1978] which states that eigenvalues appear in pairs. With λ also $-\lambda$ and $\bar{\lambda}$ is an eigenvalue. Because one pair of eigenvalues is zero, the other is either real or purely imaginary. The latter case does not lead to reconnection. For the case of reconnection the positive and the negative eigenvalue result from the X-type structure of the flow in the space coordinates and the corresponding eigenvectors are tangent to the trajectories ending or starting at the null. The first zero eigenvalue is due to the stationarity of the system, and second due to the invariance of the flow in the third space direction. This means that $W^{(4)}$ vanishes for a surface in space-time spanned by the time axis and the space direction of the invariance. If the magnetic field has a non-hyperbolic null (e.g. $\mathbf{B} = y\mathbf{e}_x + x^3 \mathbf{e}_y$) all four eigenvalues of $W^{(4)}$ may vanish, but the topological structure of the null is the same, i.e. there are still two trajectories ending and two starting at every point of the null-surface. A sketch of the typical profile of W^0 and the structure of $\mathbf{W}$ is shown in Fig. 3.

Consider the derivation of the conservation of magnetic flux,

$$\mathbf{E} + \mathbf{w} \times \mathbf{B} = \nabla \Phi \quad (7)$$

$$\Rightarrow \quad \partial_t \mathbf{B} - \nabla \times (\mathbf{w} \times \mathbf{B}) = 0 \quad (8)$$

$$\Rightarrow \quad \int_C \mathbf{B} \cdot d\mathbf{a} = const., \quad (9)$$

from the magnetic flux conserving Ohm's law Eq. (7). Here the freedom to chose a transport velocity $\mathbf{w}$ different from the plasma velocity and to use a non-vanishing Φ reflects the existence of non-ideal evolutions (non-ideal terms in Ohm's law), which do not change the topology of the magnetic flux. The corresponding covariant set of equations (5) implies the conservation of electromagnetic flux [Hornig, 1997]:

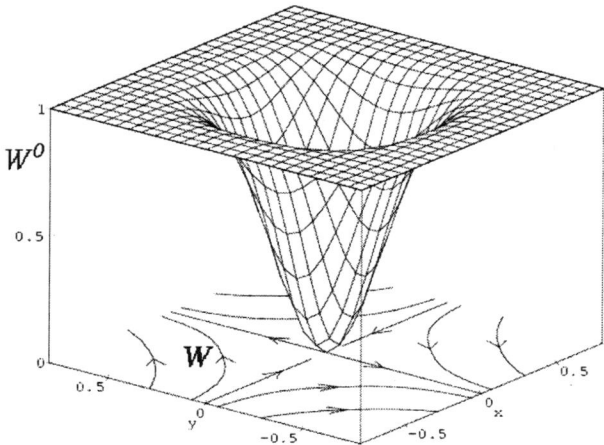

Figure 3. Local structure of $W^{(4)}$ near the reconnection line.

$$W^\nu F_{\nu\beta} = \partial_\beta \Phi \Leftrightarrow \begin{cases} \mathbf{E} \cdot \mathbf{W} = -\partial_0 \Phi \\ W^0 \mathbf{E} + \mathbf{W} \times \mathbf{B} = \nabla \Phi \end{cases} \quad (10)$$

$$\Leftrightarrow \epsilon^{\alpha\beta\gamma\delta} \partial_\alpha W^\nu F_{\nu\beta} = 0 \quad (11)$$

$$\Leftrightarrow \begin{cases} \partial_0(W^0 \mathbf{E} + \mathbf{W} \times \mathbf{B}) + \nabla(\mathbf{E} \cdot \mathbf{W}) = 0 \\ W^0 \partial_0 \mathbf{B} - \nabla \times (\mathbf{W} \times \mathbf{B}) - \nabla W^0 \times \mathbf{E} = 0 \end{cases}$$

$$\Leftrightarrow L_W \omega_F^2 = 0 \quad (12)$$

$$\Rightarrow \int_C F dA = \int_C \mathbf{B} \cdot d\mathbf{a} + \int_C \mathbf{E} \cdot d\mathbf{a_0} = const. \quad (13)$$

Here $L_W \omega_F^2 = 0$ is the Lie-derivative of the 2-form of the electromagnetic field with respect to the flow $W^{(4)}$. It is an equivalent formulation of the covariant equations in differential forms. Eq. (13) shows that in space-time the integration of the electromagnetic field tensor over a two-dimensional surface yields two terms. The first term on the right hand side is the well known magnetic flux through the surface C. It has three components corresponding to the three surface elements $dx \wedge dy$, $dy \wedge dz$ and $dz \wedge dx$. The second term is the corresponding contribution from $cdt \wedge dx$, $cdt \wedge dy$ and $cdt \wedge dz$. It is non-vanishing only if the surface has an extension along the time axis as shown in Figure 4. All together they form a Lorentz invariant measure of electromagnetic flux penetrating C and this quantity is constant for a comoving surface, that is a surface transported by the flow of $W^{(4)} = dX^{(4)}(s)/ds$ and parametrized no longer by t but by s. This can be proved with the help of the so called Lie-derivative theorem [Abraham et. al, 1988].

The second integral in Eq. (13) is important for reconnection due to the non-constant W^0. For example, if we use the $W^{(4)}$ from the above example (with $k=1, l=0$)

and start with a surface in the x-z-plane than this surface has initially only a magnetic component due to the vanishing component in the time direction as shown in Fig. 5a for $s = 0$. If it is transported with the velocity $W^{(4)} = dX^{(4)}/ds = (x^2, -x, 0, 0)$, or if integrated

$$X^0(s) = \frac{X^1(0)^2}{2}(1 - \exp(-2s))$$
$$X^1(s) = X^1(0) \exp(-s),$$

we see that as the parameter s increases the surface develops an increasing projection onto the ct-z-plane and hence an increasing electric part of the flux. Finally for $s \to \infty$ the projection onto the x-z-plane has vanished and so has the magnetic part of the flux. Hence the magnetic flux is completely converted to electric flux for $s \to \infty$.

This final stage ($s \to \infty$) of the surface is also the final stage of a surface starting from $-X^1(0)$. Moreover it is the initial stage (or final stage for $s \to -\infty$) of surfaces, which leave the z-axis in positive and negative y-directions. The evolution for two of these surfaces are shown in Fig. 5b. For $s \to \infty$ and $s \to -\infty$ respectively they are identical. For an observer it seems as if these surfaces are split and rejoined along the z-axis as shown in Fig. 6a.

The advantage of this representation of reconnection is that it has a natural generalization to finite reconnection processes, i.e. those processes which are not stationary but run for a finite time and which are not two-dimensional with an infinite reconnection line, but

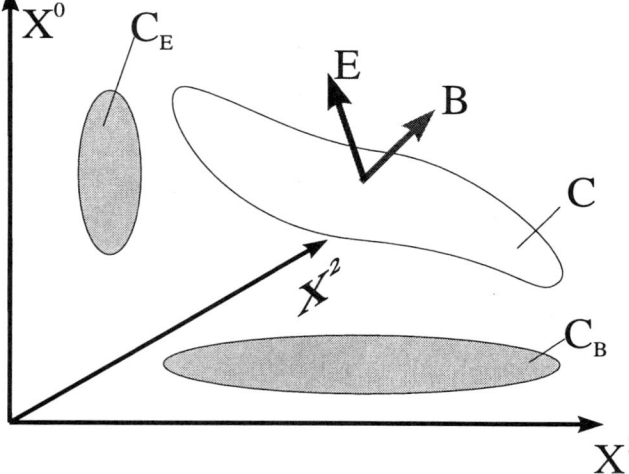

Figure 4. The domains of integration of the magnetic (C_B) and electric (C_E) part of the electromagnetic flux for a two-dimensional surface in $\mathbb{M}^4$.

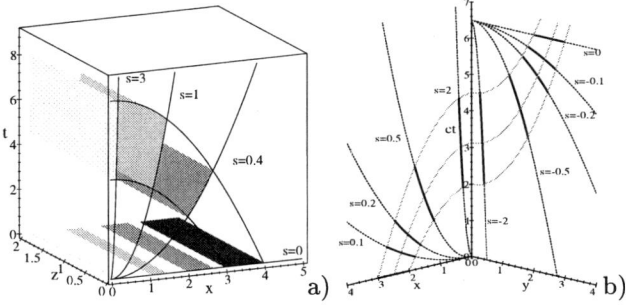

Figure 5. a) Evolution of a surface transported with $W^{(4)} = (x^2, -x, 0, 0)$ in space-time. At the bottom of the box the projection onto the x-z-plane is shown. b) Evolution of a surface starting in the x-z-plane for the flow $W^{(4)} = (x^2 + y^2, -x, y, 0)$. The z-dimension is suppressed and therefore the thick lines correspond to the projection of the surface onto the remaining coordinates. The thin lines are trajectories of $W^{(4)}$, the dotted lines are lines of constant s.

occur along a finite line. Also in general the reconnection line will be a curved line moving in space. Thus the generic scenario is that we have a bounded surface in space-time where $W^{(4)}$ vanishes and in which $\partial_\alpha W^\beta$ has the above mentioned structure of eigenvalues. The reconnection line is the intersection of this null-surface with an R^3-hyperplane as shown in Fig. 6b.

The magnetic flux reconnected in a finite reconnection process is given by the electromagnetic flux through the null-surface N,

$$\Phi_N := \int_N F_{\mu\nu} dx^\mu dx^\nu, \qquad (14)$$

This covariant scalar quantity is the amount of magnetic flux undergoing reconnection because the surface N is the limit surface of four different flows which map for $s \to \infty$, and $s \to -\infty$ respectively, surfaces lying in a R^3-hyperplane onto N as sketched in Fig. 7a. These four surfaces denoted by C_1, C_2, C_3, C_4 in Fig. 7a are mapped onto N by the electromagnetic flux conserving flow of $W^{(4)}$ and therefore,

$$\int_N F_{\mu\nu} dx^\mu dx^\nu = \int_{C_j} F_{\mu\nu} dx^\mu dx^\nu \quad j = 1, 2, 3, 4. \quad (15)$$

Moreover, because $C_1, ..C_4$ are all lying in a R^3-hyperplane we have

$$\int_N F_{\mu\nu} dx^\mu dx^\nu = \int_{C_j} \mathbf{B} \cdot d\mathbf{a} \quad j = 1, 2, 3, 4. \quad (16)$$

Therefore the magnetic flux undergoing reconnection is the same as the electromagnetic flux through the surface N.

The reader may have noticed that although we started with two-dimensional systems we used this invariance only at two points. First we used it to define $\mathbf{w}$ in Eq. (3), but this definition would also work for the more general case of $\mathbf{E} \cdot \mathbf{B} = 0$. The second time we used the invariance to show the existence of a null-line of $\mathbf{B}$, which resulted in a second zero eigenvalue of $\partial_\alpha W^\beta$. However, if there is no second zero eigenvalue for the points of the surface N it would degenerate to a line and hence the magnetic flux undergoing reconnection would vanish according to Eq. (16). Thus we do not need the invariance for this argument to hold, provided we only consider processes where a finite amount of magnetic flux is reconnected. This excludes processes such as those considered by *Priest and Titov*, [1996] where a change of topology occurs at a singular null because either they do not lead to a finite amount of reconnected flux or because they require a different type of null-point of $W^{(4)}$ and therefore differ significantly from the classical notion of reconnection. Hence we can weaken

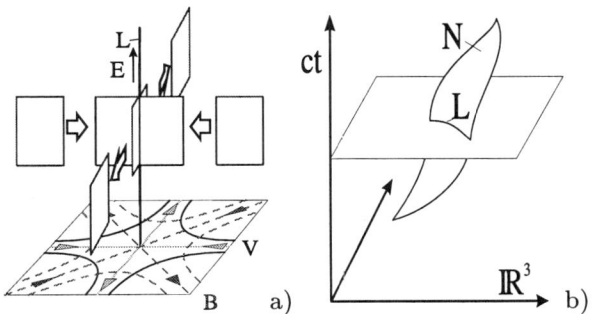

Figure 6. a) Evolution of the surfaces transported by $\mathbf{w} = \mathbf{W}/W^0$. b) The reconnection line is the intersection of the null-surface of $W^{(4)}$ with an R^3-hyperplane.

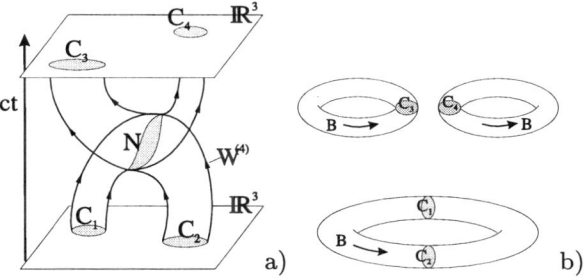

Figure 7. a) A set of four surfaces which for $s \to \infty$, and $s \to -\infty$ are mapped by the flow of $W^{(4)}$ onto the null-surface N. b) Example of the four surfaces $C_1, ..C_4$ of the previous figure for a simple reconnection of a single magnetic flux tube into two separated flux tubes.

our condition of two-dimensionality if we use $\mathbf{E}\cdot\mathbf{B} \equiv 0$ instead.

Even this restriction can be dropped if we use the freedom of specifying Φ in (11). A non-vanishing Φ allows for $\mathbf{E}\cdot\mathbf{B} \neq 0$ and thus allows us to consider reconnection for $\mathbf{B} \neq 0$ in the same way. Although $\mathbf{E}$ and $\mathbf{B}$ are no longer perpendicular we can still apply the results from our consideration of electromagnetic flux conserving flows as these are based on induction equation (11) in which Φ does not appear. However, we can no longer use Eq. (3) to determine $\mathbf{w}$. Thus whether or not a flow $W^{(4)}$ satisfying (10) has to have a null-set of the reconnection type and where this null-set is located is not as obvious as in the case of $\mathbf{E}\cdot\mathbf{B} \equiv 0$. This corresponds to the fact that in a region of non-vanishing magnetic field the site of reconnection is not determined by local properties of the field, but by the global evolution of magnetic flux, because there are no distinctive field lines. In this case questions about reconnection cannot be answered by a local analysis instead we have to check whether there exists a solution of $W^{(4)}$ satisfying Eq. (10) which has a null-set of the type mentioned above. Finally we propose the following general definition of reconnection.

Definition *A magnetic field shows reconnection if there exists no global solution of Eq. (11) with $W^0 > 0$ but a solution with $W^0 \geq 0$ locally and $W^{(4)} = 0$ for a two-dimensional surface where $\partial_\alpha W^\beta$ has a positive and negative eigenvalue.*

The restriction that there exists no global solution with $W^0 > 0$ was made for not artificially introducing reconnection where the evolution of electromagnetic field can be represented by a smooth deformation ($W^0 > 0$). For instance the stationary two-dimensional example of reconnection from above can be modified to serve as an example for $\mathbf{E}\cdot\mathbf{B} \neq 0$-reconnection by adding a constant B_z component,

$$\begin{aligned}
\mathbf{E} &= 1\,\mathbf{e}_z \\
\mathbf{B} &= l\,y\mathbf{e}_x + k\,x\mathbf{e}_y + B_z\mathbf{e}_z \\
W^0 &= (kx^2 + l\,y^2) \\
\mathbf{W} &= -x\,\mathbf{e}_x + y\,\mathbf{e}_y \\
\Phi &= B_z xy.
\end{aligned}$$

Another solution for the same $\mathbf{E}$ and $\mathbf{B}$-fields is given by

$$\begin{aligned}
W^0 &= 1 \\
\mathbf{W} &= 0 \\
\Phi &= z.
\end{aligned}$$

While the first solution shows reconnection according to our definition, the second one does not. To determine which solution is appropriate for a given situation we must match $\mathbf{W}$ to the true plasma velocity $\mathbf{V}$ outside the non-ideal reconnection region. This matching may also impose boundary conditions on Φ. Although this is only a rough sketch of the interesting topic of reconnection in non-vanishing magnetic fields, it is sufficient for the following investigation of magnetic helicity.

3. THE EVOLUTION OF HELICITY

Aside from the advantage of having a unified representation of $B = 0$- and $B \neq 0$-reconnection the electromagnetic flux conservation is also a very useful starting point for an investigation of magnetic helicity. The covariant equations (10) can be rewritten in terms of the vector potential [*Hornig*, 1997] as

$$\begin{cases} \partial_0(W^0 A^0) + \mathbf{W}\cdot\nabla A^0 - \mathbf{A}\partial_0\mathbf{W} = 0 \\ W^0\partial_0\mathbf{A} + \nabla(\mathbf{W}\cdot\mathbf{A}) - \mathbf{W}\times\nabla\times\mathbf{A} \\ \qquad\qquad - A^0\nabla W^0 = 0. \end{cases}$$

$$\Leftrightarrow L_W \omega_A^1 = 0, \tag{17}$$

provided the potential Φ has the form

$$\Phi = -(W^0 A^0 - \mathbf{W}\cdot\mathbf{A}) + \phi_c. \tag{18}$$

A non-covariant version of these equations has already been used by [*Holm*, 1986]. We can assume in the following that the constant ϕ_c is zero. Eq. (18) can be satisfied with the help of a gauge

$$\begin{cases} A^0 \to A^0 - \partial_0\Psi \\ \mathbf{A} \to \mathbf{A} + \nabla\Psi, \end{cases}$$

which allows us to write

$$W^0\partial_0\Psi = W^0 A^0 - \mathbf{W}\cdot(\mathbf{A}+\nabla\Psi) + \Phi \tag{19}$$

$$\Leftrightarrow \frac{d\Psi}{ds} = W^0 A^0 - \mathbf{W}\cdot\mathbf{A} + \Phi. \tag{20}$$

Therefore, given an arbitrary vector potential $(A^0, \mathbf{A})$ for $\mathbf{E}, \mathbf{B}$ and the scalar potential Φ, we can always integrate the gauge function Ψ such that the equations (17) hold. This is always possible for $W^0 > 0$, but it is also true for a flow with a null-set of the type we used for the definition of reconnection. Here we encounter the difficulty that integrating (20) along the trajectories of $W^{(4)}$, starting from a given time t_0 before the onset of reconnection, may result in some trajectories meeting at the null-surface N with different values of the gauge function Ψ (see Fig. 7a). However, we can use the freedom to define Ψ at t_0 to adjust the initial values of Ψ

such that they end with the same value at N. Thus even if W^0 vanishes there exists a gauge such that (17) holds.

We are now in a position to derive the transport equation for the helicity, which is in space-time a four-vector,

$$H^\nu = A_\mu \mathcal{F}^{\nu\mu} \Leftrightarrow \begin{cases} H^0 = \mathbf{A}\cdot\mathbf{B} \\ \mathbf{H} = A^0\mathbf{B} + \mathbf{E}\times\mathbf{A} \end{cases} \quad (21)$$

$$\Leftrightarrow \omega_H^3 = \omega_A^1 \wedge \omega_F^2, \quad (22)$$

where H^0 is the usual magnetic helicity and $\mathbf{H}$ denotes the helicity current, while ω_H^3 is the corresponding differential 3-form of helicity. Eq. (17) and (11) yield

$$\begin{cases} W^0\partial_0 H^0 + \nabla\cdot(\mathbf{W}H^0) - \mathbf{H}\cdot\nabla W^0 = 0 \\ \partial_0(W^0\mathbf{H}) - \nabla\times(\mathbf{W}\times\mathbf{H}) + \mathbf{W}\nabla\cdot\mathbf{H} \\ \qquad\qquad\qquad\qquad\qquad\qquad - H^0\partial_0\mathbf{W} = 0 \end{cases}$$

$$\Leftrightarrow L_W \omega_H^3 = 0 \quad (23)$$

The last equation states that these transport equations are equivalent to the helicity Lie-transported as a three-form, which in turn implies that the helicity integrated over a three-dimensional volume comoving with the flow $W^{(4)}$ is constant (see [Hornig, 1997]). Although this integral is an elegant conservation law and reveals the beauty of the underlying mathematical structure, it is not very useful for our investigation of reconnection. This is due to the fact that a volume which is initially in an R^3-hyperplane of the space-time becomes distorted by the non-constant W^0 component of $W^{(4)}$ so that it develops a finite extension in the time direction. This makes it extremely difficult to interpret this conservation law. However, the equation for the helicity density (H^0) in Eq. (23) already gives important informations about the evolution of helicity. This equation shows in particular that the evolution of magnetic helicity can be interpreted as a scalar transported by the flow of $W^{(4)}$ and, moreover, that the source term of helicity is confined to the reconnection region where $\nabla W^0 \neq 0$ (see the following sections for the details). This result is independent of the non-ideal term which drives the reconnection process. One can prove that this equation is equivalent to (1) using (10) and (18) to rewrite $\mathbf{H}$ as

$$W^0\mathbf{H} = \mathbf{W}H^0 - 2\Phi\mathbf{B} + \nabla\times\Phi\mathbf{A}. \quad (24)$$

3.1. Reconnection with $\mathbf{E}\cdot\mathbf{B} = 0$

For the classical two-dimensional models of reconnection $\mathbf{E}\cdot\mathbf{B}$ vanishes everywhere and in many other cases $\mathbf{E}\cdot\mathbf{B} = 0$ might still be a good approximation. For these cases we can assume without loss of generality that $\Phi \equiv 0$ because Φ has to vanish at the reconnection site according to (18) where $W^{(4)}$ vanishes and it can be subsumed in $\mathbf{W}$ where $\mathbf{B} \neq 0$. For these processes we can state that for every flux tube which undergoes reconnection there exists a well defined counterpart it becomes connected to. Although it is tempting to always imagine reconnection as an interaction of pairs of flux tubes it is not the general case as we will see below. It is possible for $\mathbf{E}\cdot\mathbf{B} = 0$-reconnection because we can represent the process not only by a smooth $W^{(4)}$-flow for the electromagnetic flux conservation, but also (dividing Eq. (10) by W^0) by a diverging stagnation flow $\mathbf{w}$ for the ideal Ohm's law, which except for the singularity of $\mathbf{w}$ conserves magnetic field lines and flux surfaces.

We can derive the evolution of helicity for this case either from (1) or (23)

$$\partial_0 H^0 + \nabla\cdot\left(H^0\,\mathbf{W}/W^0\right) = 0 \quad (25)$$

which means that the helicity is convected in the stagnation flow of $\mathbf{w}$. Note that $H^0\mathbf{W}/W^0$ does not diverge because $\mathbf{B}$ vanishes at the reconnection line. This equation shows that the helicity density is transported by the same flow which transports the magnetic field. Hence there is no production of helicity during the process nor is there any exchange of helicity across the separatrices of the flow $\mathbf{W}$, i.e. no helicity goes through the cross sections along which the magnetic flux is cut and reconnected (denoted in Fig. 7b as C_1, C_2 before, and C_3, C_4 after the reconnection). Therefore the question posed in the introduction regarding the helicity content of the final flux tubes for the process shown in Fig. 1 (also shown schematically in Fig. 7b) can be answered because this was actually a process of vanishing $\mathbf{E}\cdot\mathbf{B}$. (The tubes cross each other in the second frame of the figure, which requires a null of the magnetic field and hence $\mathbf{E}\cdot\mathbf{B} = 0$. For comparison see Fig. 8 for a process with non-vanishing magnetic field.) For such processes the helicity of the final flux tubes is precisely given by the helicity content of those parts of the initial flux tubes of which they are formed and which are bounded by the cross sections C_1, C_2. This means in Fig. 1 that the left flux tube has the helicity of the left half of the initial tube and the right one of the right half of the initial flux tube. On first sight this appears to be unphysical, since we can redistribute the helicity density in the initial flux tube arbitrarily by using the freedom of gauge. On the other hand, the total helicity of each of the final flux tubes is well defined. The solution of this apparent contradiction is that the freedom of gauge for the initial flux tube is restricted by Eq. (20) in a way which fixes the helicity content of each half tube.

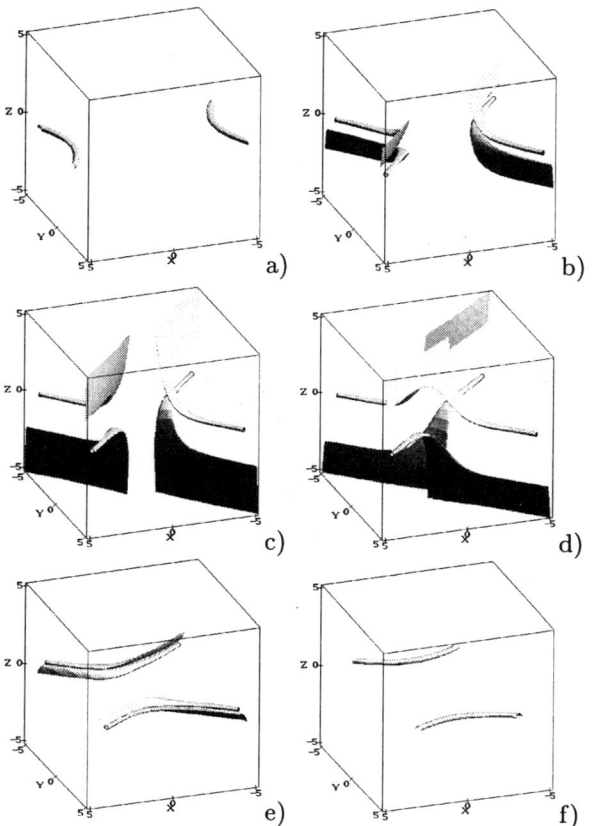

Figure 8. Evolution of a pair of flux tubes undergoing a $\mathbf{B} \neq 0$-reconnection.

Because the condition (20) involves the flow $W^{(4)}$, the determination of the gauge requires the knowledge of the reconnecting flow. Without this knowledge there are no general statements possible on the distribution of the total helicity of the initial flux tube to the final ones. For example the flux tube might get twisted on its way to the reconnection site, which can induce twist of opposite polarity into the final flux tubes. Thus one should keep in mind that even if the helicity density of a certain region vanishes identically at an initial moment, a suitable flow can twist flux tubes and reconnect them such that pairs of flux tubes of opposite total helicity are produced.

3.2. Reconnection with $\mathbf{E} \cdot \mathbf{B} \neq 0$

The existence of a component of the magnetic field along the reconnection line, however small it is, should be considered as the generic case for reconnection. Unfortunately this considerably complicates the situation. Firstly there is no flow, either diverging or smooth, in which the magnetic flux is frozen-in. Thus there is no well defined velocity of magnetic flux or field lines within the non-ideal region. However, we can follow flux tubes which are connected to the ideal surrounding the reconnection site where the magnetic flux is frozen-in. With the help of this method Fig. 8 gives us an impression of a reconnection event for $\mathbf{B} \neq 0$. Two flux tubes rooted in the ideal external region approach the reconnection site. We see that they do not stay connected as they enter the non-ideal regions, instead they split in four different flux tubes which, as time goes on, flip around each other. Each tube eventually approach its counterpart before leaving the non-ideal region. Due to the symmetry of the example shown in Fig. 8 the two pairs of flux tubes merge eventually as they leave the non-ideal region, but if there is a substantial production of magnetic helicity, this is not necessarily the case. In contrast to the $\mathbf{E} \cdot \mathbf{B} = 0$ case $\mathbf{E} \cdot \mathbf{B} \neq 0$-reconnection (so called magnetic flipping) is not a process where only two flux tubes are involved. Here the magnetic flux undergoing reconnection is opened up to the ambient magnetic field.

Also in contrast to the case of $\mathbf{E} \cdot \mathbf{B} = 0$-reconnection the evolution of the helicity density now has a non-vanishing source term $\mathbf{H} \cdot \nabla W^0$, see Eq. (23). This source term is confined to the region where W^0 is non-constant. This is only a part of the non-ideal region, where W^0 increases from $W^0 = 0$ along the reconnection line to $W^0 = 1$ (see Fig. 3). The source of helicity vanishes in the ideal region, where $W^0 \equiv 1$. For examples of helicity producing reconnection see [*Hornig and Rastätter, 1997a/b*]. Note that although the magnetic flux opens to the ambient field during reconnection which results in a component of the helicity current proportional to $\mathbf{B}$ in the non-ideal region, the helicity current leaving the non-ideal region is still given only by $H^0 \mathbf{W}/W^0$. Thus there is no exchange of helicity with the ambient magnetic field.

The question of to what extent the total helicity is preserved during the relaxation of plasmas due to reconnection, can be answered by comparing the characteristic time scale of the generation of helicity with the characteristic time scale of dissipation by estimating roughly,

$$\tau_{hel} = \left| \frac{\int_V \mathbf{A} \cdot \mathbf{B} d^3x}{\int_V \mathbf{E} \cdot \mathbf{B} d^3x} \right| = \frac{B\,V\,L}{E\,V_{rec}},$$

$$\tau_{diss} = \left| \frac{\int_V B^2/(8\pi) d^3x}{\int_V \mathbf{E} \cdot \mathbf{J} d^3x} \right| = \frac{B\,V\,d}{E\,V_{rec}}.$$

Here V is the volume containing the magnetic flux with

characteristic length scale L, V_{rec} is the volume of the reconnection region and d its width perpendicular to the current sheet. Thus the characteristic value of **J** is B/d and of **A** is LB. For collisionless plasmas the width d of the current sheet is very small compared to the overall length scale L of the magnetic field and therefore the total helicity is conserved on a longer time scale than the energy for all mechanism which lead to a localized reconnection. The effect of decreasing helicity production for vanishing width of the reconnection region is shown in Figure 9 for the helicity production normalized to the reconnected flux for various inverse width $\sigma = 1/d$. Although it is a somewhat different approach, the result is in agreement with the result of [Berger, 1984] derived for weakly resistive plasmas.

4. CONCLUSIONS

The evolution of magnetic helicity during reconnection is determined to a high degree by the structure of reconnection itself and does not depend on the particular non-ideal term which drives the process. This is expressed by Eq. (23), which shows that the evolution of helicity in reconnection is given by the same flow $W^{(4)}$ which determines the evolution of electromagnetic flux. It also shows that in general the generation of helicity is confined to a small region near the reconnection site where the transport velocity of the magnetic flux tends to infinity in such a way that the time component of the corresponding four-vector has to decrease to zero. Due to the small size of this non-ideal region with respect to the volume of the magnetic flux involved, the production of magnetic helicity during the process itself will be very small for astrophysical applications. However, even for vanishing production of magnetic helicity during the actual process, reconnection can separate parts of a flux tube containing magnetic helicity of opposite sign, or vice versa, join flux tubes of opposite helicity and produce or destroy helicity in this way.

Acknowledgment. This work was supported by the *Volkswagen-Foundation*. The author gratefully acknowledges the valuable comments of the referees.

REFERENCES

Abraham, R. and J.E. Marsden, *Foundations of Mechanics*, 170 pp., Addison-Wesley, Reading Mass., 1978.
Abraham, R., J.E. Marsden and T. Ratiu, *Manifolds Tensor Analysis and Applications*, Applied Mathematical Sciences 75, Springer, New York, 1988.
Arnold, V.I., The asymptotic Hopf invariant and its application, Sel. Math. Sov., 5(4), 327, 1986.
Berger, M.A., Rigorous new limits on magnetic helicity dissipation in the solar corona, Geophys. Astrophys. Fluid Dynamics, *30*, 79, 1984.
Biskamp, D., Collisional and Collisionless Magnetic Reconnection, Phys. Plasmas, *4*, 1964, 1997.
Dungey, J.W., Phil. Mag., Series 7, *44*, 725, 1953.
Holm, D.D., Hamilton dynamics of a charged fluid including electro- and magnetohydrodynamics, Phys. Lett. A, *114*(3), 137, 1986.
Hornig, G., The covariant transport of electromagnetic fields and its relation to magnetohydrodynamics, Phys. Plasmas, *4*(3), 646, 1997.
Hornig, G., and L. Rastätter, The role of Helicity in the Reconnection Process, Adv. Space Res., *19*, 1789, 1997a.
Hornig, G., and L. Rastätter, The magnetic structure of $B \neq 0$-reconnection, Physica Scripta, *T 74*, 34, 1997b.
Parker, E.N., The solar flare phenomenon and the theory of reconnection and annihilation of magnetic fields, Astrophys. J. Suppl. Ser. *8*, 177, 1963.
Priest, E.R., and V.S. Titov, Magnetic reconnection at three-dimensional null points, Phil. Trans. Roy. Soc. London (A), *354*, 2952, 1996.
Sweet, P.A., The production of high energy particles in solar flares, Nuovo Cimento Supp. *8*, Series X, 188, 1958.
Taylor, J.B., Relaxation of toroidal plasma and generation of reverse magnetic fields, Physical Review Letters, *33*(19), 1139, 1974.

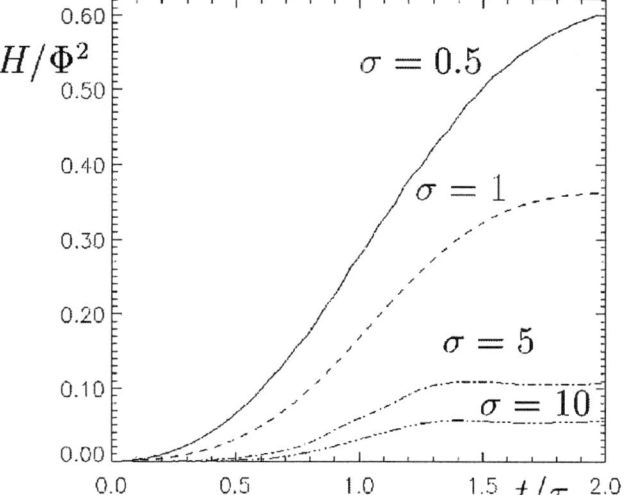

Figure 9. Production of helicity normalized to the total reconnected flux for different values of the inverse width σ [*Hornig and Rastätter*, 1997a]

G. Hornig, Topologische Fluiddynamik NB 7/30, Fakultät für Physik und Astronomie, Ruhr-Universität, 44780 Bochum, Germany (email: gh@tp4.ruhr-uni-bochum.de)

Helicity, Reconnection, and Dynamo Effects

Hantao Ji

Princeton Plasma Physics Laboratory, Princeton University Princeton, New Jersey

The inter-relationships between magnetic helicity, magnetic reconnection, and dynamo effects are discussed. In laboratory experiments, where two plasmas are driven to merge, the helicity content of each plasma strongly affects the reconnection rate as well as the shape of the diffusion region. Conversely, magnetic reconnection events also strongly affect the global helicity, resulting in efficient helicity cancellation (but not dissipation) during counter-helicity reconnection and a finite helicity increase or decrease (but less efficiently than dissipation of magnetic energy) during co-helicity reconnection. Close relationships also exist between magnetic helicity and dynamo effects. The turbulent electromotive force along the mean magnetic field (α-effect), due to either electrostatic turbulence or the electron diamagnetic effect, transports mean-field helicity across space without dissipation. This has been supported by direct measurements of helicity flux in a laboratory plasma. When the dynamo effect is driven by electromagnetic turbulence, helicity in the turbulent field is converted to mean-field helicity. In all cases, however, dynamo processes conserve total helicity except for a small battery effect, consistent with the observation that the helicity is approximately conserved during magnetic relaxation.

1. INTRODUCTION

Magnetic helicity, a measure of the "knottedness" and the "twistedness" of magnetic field [*Woltjer*, 1958], is closely related to field line topology [*Berger and Field*, 1984]. It is defined by

$$K = \int \mathbf{A} \cdot \mathbf{B} dV \qquad (1)$$

where **A** is the vector potential of the magnetic field **B** and the integration is over a volume V. The magnetic helicity is an invariant within a flux tube in a perfectly conducting plasma. *Taylor* [1974] conjectured that in a "slightly" resistive plasma the *total* helicity is well-conserved during plasma relaxation in which the magnetic energy decays toward a minimum-energy state. This well-known hypothesis has been successful [*Taylor*, 1986] in explaining magnetic structures in laboratory toroidal plasmas, such as the reversed-field-pinch (RFP), spheromak, and multipinch. Recently, there has been growing awareness that the global magnetic helicity contained in flux tubes also plays an important role in solar flare evolution [*Berger*, 1984; *Rust*, 1994].

As an elementary process in resistive plasmas, magnetic reconnection [*Vasyliunas*, 1975; *Biskamp*, 1993] has long been regarded as a key mechanism in deciding the dynamics of solar flares [e.g., *Parker*, 1979], magnetospheric substorms [e.g., *Akasofu*, 1972], and relaxation phenomena in laboratory plasmas [e.g., *Taylor*, 1974]. Although this is a localized process, it of-

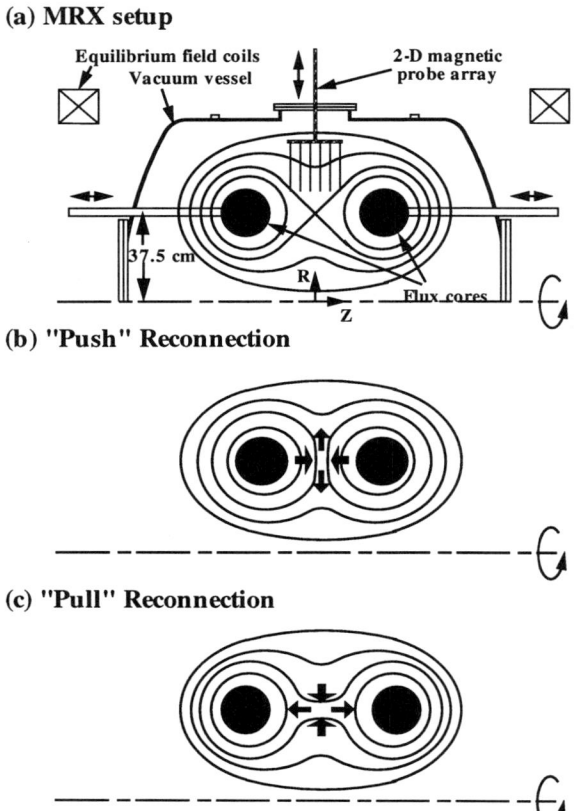

Figure 1. Experimental setup in MRX (a) and illustration of "push" (b) and "pull" (c) reconnection.

ten causes fundamental changes in macroscopic magnetic field topology through cutting and rejoining of field lines. Therefore, an inherent relationship between global helicity and local reconnection events must exist.

Dynamo effects also have been another focal point of research in electrically conductive fluids or plasmas attempting to explain the observed solar and planetary magnetic fields. In particular, generation of an electromotive force (EMF) along a mean field by turbulence, or the well-known α-effect [*Parker*, 1955], is an essential process in amplifying large-scale magnetic fields [e.g., *Proctor and Gilbert*, 1994]. These dynamo effects drive parallel current which twists up the field lines thus creating magnetic helicity on large scales. Therefore, the helicity also must be closely related to dynamo effects.

In the following sections, we discuss the relationship between magnetic helicity and magnetic reconnection with emphasis on results from recent laboratory experiments, followed by discussions on the relationships between magnetic helicity and dynamo effects in a plasma.

2. RELATIONSHIP BETWEEN HELICITY AND RECONNECTION

2.1. "Push" and "Pull" Reconnection

Consider the situation where two plasmas with parallel toroidal current interact with each other. Reconnection of the poloidal fields results in the formation of three regions in the system: private regions associated with each plasma and a public region surrounding both plasmas, as illustrated in Fig. 1 (a). This configuration has been created in a laboratory experiment, MRX (Magnetic Reconnection Experiment) [*Yamada et al.*, 1997], where two internal coils are inserted in each private region (denoted by the closed circles in Fig. 1). After the plasma is formed, the private flux can be further increased to generate "push" reconnection as shown in Fig. 1(b), where the field lines are reconnected from the private regions to public region. On the other hand, when the private flux is decreased the poloidal flux is "pulled" back from the public region to the private regions by reconnection, as shown in Fig. 1(c). Both types of reconnection can occur in nature and can be induced in MRX by changing operational procedures. (Two examples of natural pull reconnection are solar flare and magnetotail reconnection.)

2.2. Effects of Helicity Content on Reconnection

The most common description of magnetic field line reconnection is shown in Fig. 2(a), on which the two-dimensional theories have been based [*Vasyliunas*, 1975; *Biskamp*, 1993]. However, magnetic field lines have three vector components. The third component (toroidal field B_T) decides not only the helicity content of each plasma but also the local reconnection angle, as seen in the three-dimensional pictures of Fig. 2(b). When B_T vanishes in both plasmas [null-helicity case, Fig. 2(i)],

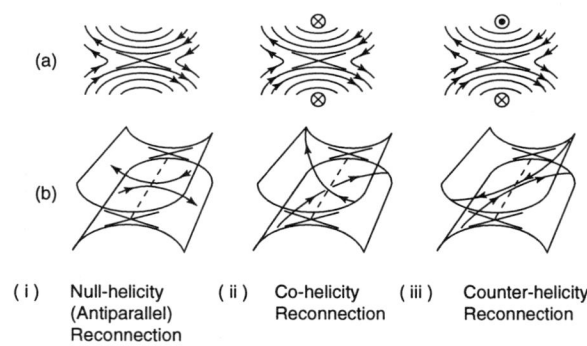

Figure 2. (a) 2-D and (b) 3-D schematic views of magnetic reconnection for three cases: (i) null-helicity (ii) co-helicity and (iii) counter-helicity.

conventional 2-D reconnection is applicable. In the presence of a third component, (1) the field lines reconnect at an angle when uni-directional toroidal fields exist [co-helicity case, Fig. 2(ii)] or (2) they reconnect with anti-parallel geometry when the toroidal fields are oppositely directed [counter-helicity case, Fig. 2(iii)]. Note that the reconnecting field lines are anti-parallel for both null-helicity and counter-helicity cases.

In MRX, the effect of merging angle on reconnection has been studied extensively [*Yamada et al.*, 1997]. Figure 3 shows examples of time evolution of the poloidal flux contours obtained by a 90-channel 2-D magnetic probe array [Fig. 1(a)] during pull reconnection. Two different shapes of diffusion regions are found, depending on the third components of the reconnecting magnetic fields. Other operational conditions are held constant for each discharge. When no magnetic reconnection is induced, a typical X-shape separatrix region is observed as seen at $t = 260$ μs in both Fig. 3(a) and 3(b). As poloidal flux is driven toward the diffusion region, a neutral sheet is formed. Without the third component (null-helicity reconnection), a thin double-Y shaped diffusion region is clearly identified [Fig. 3(a)]. In the presence of an appreciable third component (co-helicity reconnection), an O-shaped sheet current appears [Fig. 3(b)].

The existence of a magnetic island in the co-helicity case indicates a much broader current sheet than the counter-helicity case. Figure 4 presents the radial profiles of poloidal field B_Z, toroidal field B_T, toroidal current density j_T, and pitch of field lines for co-helicity and null-helicity reconnection. In the null-helicity case, B_T is almost zero resulting in an abrupt transition of

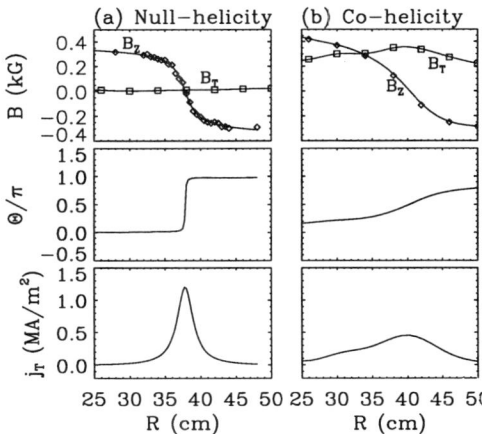

Figure 4. Radial profiles of measured B_Z, B_T, field line angle (Θ), and j_T at $Z = 0$ cm and $t = 290$ μs in the (a) null-helicity and (b) co-helicity cases.

the pitch of field lines (Θ) at the reconnection point, while in the co-helicity case, B_T is order B_Z resulting in a gradual change of Θ over R. In the co-helicity case, the j_T profile is broad with width of order 10 cm. In the null-helicity case, one observes a steepening of magnetic field slope at the diffusion region and therefore a sharp neutral sheet current. The thickness of this current sheet is seen to be as narrow as 1 cm, which is found to be roughly equal to the ion gyro-radius, ρ_i, defined using the ion temperature at the center and B_Z at the knee.

The existence of a sizable toroidal field (co-helicity) also results in a much slower reconnection rate than in the null-helicity case. Similar observations also have been made in an earlier experiment, where two spheromaks were driven to merge [*Yamada et al.*, 1990; *Ono et al.*, 1993]. A detailed quantitative analysis [*Ji et al.*, 1998; *Ji et al.*, 1999] in MRX shows that the slower reconnection rate in the co-helicity case is due to the combined effect of lower anomalous resistivity, lower compressibility, and larger pressure difference between the down-stream and up-stream regions of reconnection. It should be emphasized that the existence of a sizable toroidal field alters the detailed dynamics of the diffusion region since the current flows essentially in the perpendicular direction in the null-helicity or counter-helicity case but in the parallel direction in the the co-helicity case.

2.3. Effects of Reconnection on Helicity Conservation

Although magnetic reconnection is a localized process, it often causes topological changes in macroscopic configurations, affecting the globally defined magnetic

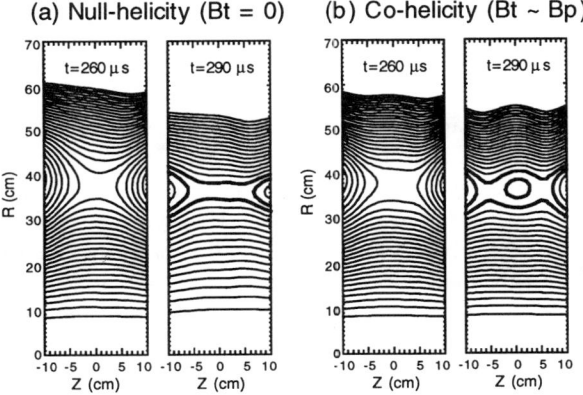

Figure 3. Time evolution of poloidal flux measured by internal magnetic probes By $t = 290$ μs, a double-Y shaped and O-shaped diffusion regions are formed in the (a) null-helicity case and (b) co-helicity case, respectively.

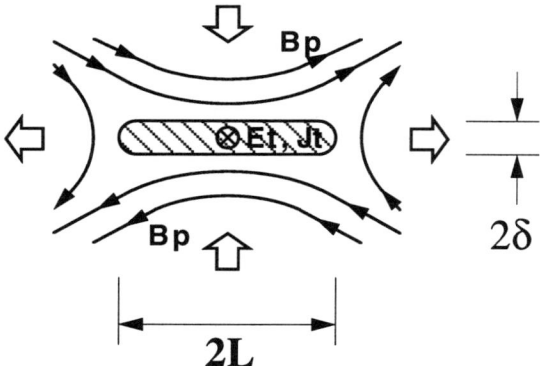

Figure 5. Illustration of a Sweet-Parker type of magnetic reconnection.

helicity. Some properties of helicity conservation during magnetic reconnection have been discussed [*Wright and Berger*, 1991; *Pfister and Gekelman*, 1991]. In this section, a detailed account of the effects of reconnection on helicity conservation and dissipation of magnetic energy is given.

2.3.1. Time evolution of helicity and energy. The time rate of change of helicity defined by Eq.(1) can be described by

$$\frac{dK}{dt} = -2\int \mathbf{E}\cdot\mathbf{B}\, dV - \int (2\phi\mathbf{B} + \mathbf{A}\times\frac{\partial\mathbf{A}}{\partial t})\cdot d\mathbf{S} \quad (2)$$

where ϕ is the electrostatic potential and $\mathbf{S}$ is the surface surrounding integrated volume. The first term on the right hand side (RHS) represents the volume helicity rate of change while the second and third terms are helicity flux through the surface via inductive and electrostatic means, respectively.

A similar equation can be derived for the time evolution of the volume integrated magnetic energy, $W = \int B^2/2\mu_0 dV$,

$$\frac{dW}{dt} = -\int \mathbf{E}\cdot\mathbf{j}\, dV - \int (\mathbf{E}\times\mathbf{B})\cdot d\mathbf{S}, \quad (3)$$

where the first term on the RHS represents the energy dissipation rate and the second term the Poynting flux.

What is of interest here is how much helicity and energy change occurs *within* the volume of integration during the reconnection process. Therefore, only the volume-dissipative terms $\mathbf{E}\cdot\mathbf{B}$ and $\mathbf{E}\cdot\mathbf{j}$ but not the surface terms (which represent helicity or energy flux) in Eq.(2) and Eq.(3) need to be examined.

2.3.2. Dissipation terms during reconnection. Now consider a Sweet-Parker type of reconnection [*Sweet*, 1958; *Parker*, 1957] as illustrated in Fig. 5, where a rectangular diffusion region forms with width 2δ and length $2L$. Recently, this type of reconnection has been verified experimentally in MRX, where some other effects, including an enhanced resistivity over the Spitzer value, are taken into account [*Ji et al.*, 1998; *Ji et al.*, 1999]. (On the other hand, an alternative model by *Petschek* [1964], which is based on a much smaller diffusion region and standing shock waves, has not yet been confirmed experimentally.) As is typical for magnetic reconnection, the region outside of the diffusion region can be treated by ideal MHD, where $\mathbf{E} + \mathbf{V}\times\mathbf{B} = 0$ holds. Therefore,

$$\mathbf{E}\cdot\mathbf{B} = E_T B_T + E_P B_P = 0$$
$$\mathbf{E}\cdot\mathbf{j} = (\mathbf{j}\times\mathbf{B})\cdot\mathbf{V},$$

leading to no changes in helicity ($E_T B_T$ balances with $-E_P B_P$) and no magnetic energy dissipation (except for an exchange with mechanical energy).

The situation changes when the diffusion region is considered. By definition, the reconnecting field $B_P \approx 0$ and the Ohm's law can be approximated by $\mathbf{E} + \mathbf{V}\times\mathbf{B} = \eta^*\mathbf{j}$, where η^* is an effective resistivity which includes all non-ideal MHD effects. Because both E_T and the non-reconnecting B_T are unchanged from outside the diffusion region, we have

$$\mathbf{E}\cdot\mathbf{B} \approx E_T B_T (= \eta^* j_T B_T) \quad (4)$$
$$\mathbf{E}\cdot\mathbf{j} = (\mathbf{j}\times\mathbf{B})\cdot\mathbf{V} + \eta^* j^2, \quad (5)$$

leading to possibly non-zero $\mathbf{E}\cdot\mathbf{B}$ (a net change in helicity) and a net energy dissipation. We shall discuss these terms in the following sections for both counter- and co-helicity reconnection.

2.3.3. Helicity neutralization during counter-helicity reconnection. Consider the case of counter-helicity reconnection shown in Fig. 2(iii), where the third component B_T changes sign across the current sheet. In this case, B_T is also reconnected as a result of field line diffusion. Therefore, $B_T \approx 0$ in the diffusion region, resulting in no helicity dissipation [Eq.(4)]. However, the original helicity contained in each plasma has been lost as a result of reconnection, and this process can be described as helicity neutralization or cancellation. The end product of the counter-helicity reconnection is a toroidal configuration (called a Field Reversed Configuration, or FRC) consisting of only toroidal current (or poloidal field) with no toroidal field, or no helicity. This physical process has been demonstrated experimentally [*Yamada et al.*, 1990; *Ono et al.*, 1993] as illustrated in Fig. 6.

One unique feature of an FRC is that a high β (~ 1) plasma heated by reconnection is confined by the perpendicularly flowing current, most of which is carried by ions. The observed stability of such a configuration cannot be explained by Taylor's theory since it contains no apparent magnetic helicity. Recently, there have been attempts [*Steinhaurer and Ishida*, 1997; *Hegna*, 1998] to generalize relaxation theories, using electron helicity to describe parallel current and ion helicity to describe perpendicular current or plasma flow, in order to minimize the total energy including kinetic and thermal energies. Then, FRC plasmas can be classified as a minimum energy state with zero electron helicity (or magnetic helicity) but finite ion helicity (or kinetic helicity) [*Steinhaurer and Ishida*, 1997].

2.3.4. Helicity change during co-helicity reconnection. As illustrated in Fig. 2(ii), the third component B_T will be non-zero during co-helicity reconnection, introducing a non-zero helicity change. With respect to $\mathbf{B}_T$, however, the direction of $\mathbf{E}_T$ can be parallel or anti-parallel, depending on push or pull reconnection, as discussed previously. For the case of positive helicity as shown in Fig. 7, $\mathbf{E}_T$ is parallel to $\mathbf{B}_T$ during pull reconnection resulting in a negative $-\mathbf{E}\cdot\mathbf{B}$ or decrease in helicity, while $\mathbf{E}_T$ is anti-parallel to $\mathbf{B}_T$ during push reconnection resulting in a positive $-\mathbf{E}\cdot\mathbf{B}$ or increase in helicity.

An intuitive picture of helicity change during co-helicity reconnection is given as follows. In the case of co-helicity reconnection, only the poloidal field B_P is diffused and reconnected in the diffusion region. The poloidal field lines slip from up-stream to down-stream as soon as it enters the diffusion region, with a speed on the order of c. Meanwhile, the toroidal field B_T is *not* diffused and still frozen with plasma, which moves with a speed on the order of the Alfvén speed ($V_A \ll c$). This slippage of B_P relative to B_T gives rise a change in linkage (or helicity) with toroidal flux contained in the diffusion region. Following this argument, two more examples for helicity change can be given as shown in

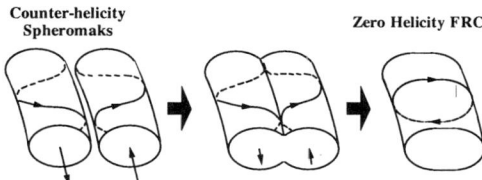

Figure 6. Helicity neutralization during a counter-helicity merging of spheromaks, resulting in a Field Reversed Configuration (FRC) with no magnetic helicity but finite ion kinetic helicity.

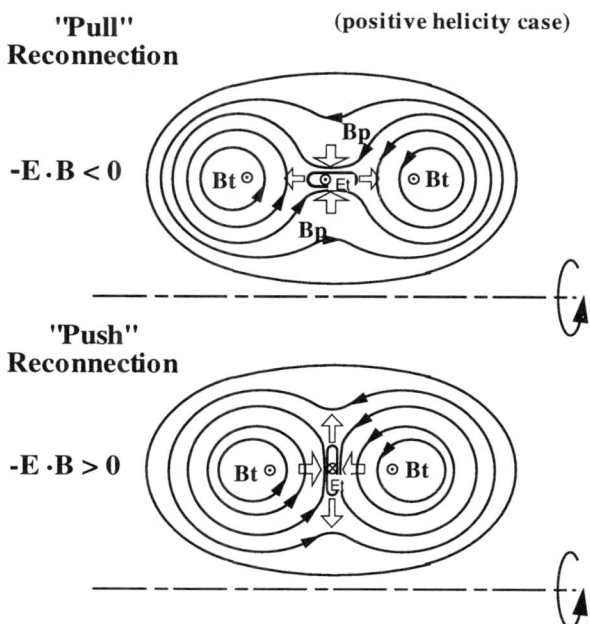

Figure 7. Illustration of co-helicity reconnection in the positive helicity case: "pull" ("push") reconnection leads to a decrease (an increase) in helicity.

Fig. 8, which are often seen in sawtooth reconnection in tokamaks [*Nagayama et al.*, 1991] and relaxation events in reversed field pinches (RFP's) [*Schnack et al.*, 1985] if one regards the "toroidal field" as the parallel component along a resonant line and the "poloidal field" as the perpendicular reconnecting component.

2.3.5. Relative rate of change in helicity and energy. A quantitative account of helicity change always needs to be compared with energy dissipation since only their relative difference has physical meaning. Volume integration of $\mathbf{E}\cdot\mathbf{B}$ and $\mathbf{E}\cdot\mathbf{j}$ over the diffusion region,

$$\frac{dK}{dt} = -2\,E_T B_T\,2\delta\,2L\,2\pi R$$
$$\frac{dW}{dt} = -E_T j_T\,2\delta\,2L\,2\pi R,$$

gives changes in helicity and energy per reconnected poloidal flux Ψ,

$$\left|\frac{dK}{d\Psi}\right| = 8\delta L B_T = 2\Phi_{\mathrm{DR}} \tag{6}$$

$$\frac{dW}{d\Psi} = -\frac{4LB_P}{\mu_0}, \tag{7}$$

where $2\pi R$ is the total length of diffusion region, Φ_{DR} is toroidal flux contained with the diffusion region. (To derive Eq.(6) and Eq.(7), the relations $d\Psi/dt = 2\pi R E_T$

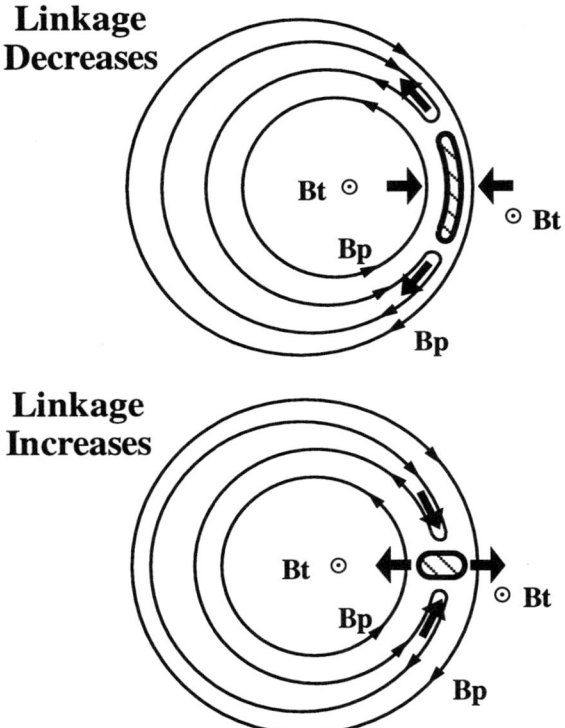

Figure 8. Two more examples for helicity change during co-helicity reconnection.

and $j_T \delta = B_P/\mu_0$ have been used.) Since generally W and K are related by $W/K \sim 1/(\mu_0 L)$, the ratio of the rates of change is given by

$$\left| \frac{W}{K} \frac{dK}{dW} \right| \sim 2 \frac{\delta}{L} \frac{|B_T|}{B_P}. \qquad (8)$$

We note that not only the current sheet thickness but also the reconnecting angle $\theta = 2\tan^{-1}(B_P/|B_T|)$ determine the relative rate of change.

An alternative form of Eq.(8) can be written as

$$\left| \frac{W}{K} \frac{dK}{dW} \right| \sim 2 \frac{\lambda_{\text{global}}}{\lambda_{\text{DR}}},$$

where $\lambda \equiv \mu_0 \mathbf{j} \cdot \mathbf{B}/B^2$, a parameter often used in the equation $\nabla \times \mathbf{B} = \lambda \mathbf{B}$ to describe the force free state [*Taylor*, 1974]. When the plasmas are close to their relaxed states, then $\mu_0 W/K \sim 1/L \sim \lambda_{\text{global}}$. Similarly, λ can also be defined in the diffusion region as $\lambda_{\text{DR}} \equiv \mu_0 j_T / B_T$, although it is not clear why the diffusion region has to be in a relaxed state also. But if this is so as suggested by *Biskamp* [1993], then the relative ratio of change should only be decided by $2\delta/L$ since $\lambda_{\text{DR}} \sim 1/\delta$ and $\lambda_{\text{global}} \sim 1/L$.

2.3.6. How well is helicity conserved during reconnection? Based on Eq.(8), the relative ratio of change of helicity to energy due to co-helicity reconnection can be estimated. For pull co-helicity reconnection in MRX, as shown in Fig. 3(b), $\delta/L \sim 1/3$ and $B_T/B_P \sim 1$ [*Ji et al.*, 1999] so that $|(W/K)(dK/dW)| \sim 2/3$ if the island structure is ignored. Taking into account the presence of the island structure would likely decrease the estimated ratio.

A second case of interest is the RFP plasmas from where the idea of helicity conservation originates. However, the current sheet thickness δ is not measured in RFP's during a reconnection event. One way to estimate δ is to relate it to the relative parallel drift parameter defined as $v_{\text{drift}}/v_{\text{th},e}$ ($v_{\text{drift}} \equiv j/en$ and $v_{\text{th},e}$ is the electron thermal velocity) by

$$\delta = \frac{B_{\text{rec}}}{\mu_0 j_\parallel} = \frac{B_{\text{rec}}}{\mu_0 e n v_{\text{th},e}} \left(\frac{v_{\text{drift}}}{v_{\text{th},e}} \right)^{-1}, \qquad (9)$$

where B_{rec} is the reconnecting field, which is typically the radial field B_r in the RFP. The typical drift parameter can be estimated to be on the order of 0.2-0.3 based on the observation of current carrying fast electrons [*Stoneking et al.*, 1994] or the measured $j_\parallel$, n, and T_e [*Ji et al.*, 1994]. Using typical parameters [*Ji, Prager, and Sarff*, 1995] in MST (Madison Symmetrical Torus) plasmas as we shall mention in the next section, i.e., $B_T = 2$ kG, $n = 1 \times 10^{19}/\text{m}^3$, $T_e = 100$ eV, and the plasma radius $a = 0.5$ m, we have $\delta/a \sim (1.5\text{-}2.5) \times 10^{-3}$. Using observed $B_T/B_r \approx 100$ and $\mu_0 W/K \approx 1/(0.73a)$, the relative rate of change of helicity to energy $|(W/K)(dK/dW)| \sim 0.4\text{-}0.7$, which is consistent with the observed ~ 0.4 during a relaxation event [*Ji, Prager, and Sarff*, 1995]. This estimate suggests that the helicity conservation is only marginally satisfied in the RFP plasmas.

The last interesting situation is how well helicity is conserved during a solar flare, where reconnection is considered to play an essential role. Again, the current sheet thickness δ is undetermined observationally. Using Eq.(9) with an assumed drift parameter of 0.2 and the typical parameters ($B_T = 500$ G, $n = 10^{15}/\text{m}^3$, $T_e = 100$ eV, $L = 10,000$ km), we have $\delta \sim 300$ m and $\delta/L \sim 3 \times 10^{-5}$. If we choose $B_T/B_P \sim 1$ as one might consider if the plasma is close to a relaxed state, an estimated relative rate of change $|(W/K)(dK/dW)| \sim 6 \times 10^{-5}$ is obtained. Because any adjustment of plasma parameters is unlikely to change this relative rate to a number close to unity, one may conclude that the helicity is indeed conserved relative to the energy change in the solar corona. It is noted that the classical Sweet-Parker model gives an even smaller δ (< 1 m) and thus a smaller helicity change.

3. RELATIONSHIP BETWEEN HELICITY AND DYNAMO EFFECTS

The dynamo effect, or, generation of magnetic field by motions in an electrically-conducting medium, is another focal point of understanding solar magnetic activities [*Parker*, 1979]. In this section, close relations between dynamo effects and magnetic helicity are discussed.

3.1. MHD Dynamo and Diamagnetic Dynamo

A widely used scheme to discuss dynamo effects arising from MHD turbulence has been based on the mean-field electrodynamics [*Krause and Rädler*, 1980], where every quantity x is divided into a mean part $\bar{x} \equiv <x>$, averaged over ensembles or space, and a turbulent part $\tilde{x}$: $x = \bar{x} + \tilde{x}$. Therefore, the mean MHD Ohm's law can be written as,

$$\overline{\mathbf{E}} + \overline{\mathbf{v}} \times \overline{\mathbf{B}} + <\tilde{\mathbf{v}} \times \tilde{\mathbf{B}}> = \eta \bar{\mathbf{j}},$$

where the last term on the left-hand side is the mean electromotive force (EMF) $\mathcal{E}$ arising from turbulence, and it can be expressed approximately as

$$\mathcal{E} = \alpha \overline{\mathbf{B}} - \beta \nabla \times \overline{\mathbf{B}}. \quad (10)$$

Here α and β are determined by turbulence, often called the α-effect and the β-effect [*Parker*, 1955]. It can be seen easily that the α-effect, which has been regarded as an essential process for a working dynamo, represents an electromotive force generated by turbulence in the direction *along* the mean magnetic field.

In order to include other possible dynamo effects in a plasma, the same process can be repeated for the generalized Ohm's law (ignoring the electron inertial term) [*Spitzer*, 1962]

$$\mathbf{E} + \mathbf{v} \times \mathbf{B} - \frac{\mathbf{j} \times \mathbf{B}}{en} + \frac{\nabla P_e}{en} = \eta \mathbf{j}, \quad (11)$$

where n is the electron density and P_e the electron pressure. Then the mean EMF in a turbulent plasma becomes

$$\mathcal{E} = <\tilde{\mathbf{v}} \times \tilde{\mathbf{B}}> - <\tilde{\mathbf{j}} \times \tilde{\mathbf{B}}>/e\bar{n}, \quad (12)$$

where the second term is often called the Hall term and we have neglected $<\tilde{n}\nabla \tilde{P}_e>/e\bar{n}^2$ (a battery effect, see the next section). Since $\mathbf{v} = (m_i \mathbf{v}_i + m_e \mathbf{v}_e)/(m_i + m_e) \approx \mathbf{v}_i$ and $\mathbf{j} = en(\mathbf{v}_i - \mathbf{v}_e)$, Eq.(12) can be rewritten as

$$\mathcal{E} = <(\tilde{\mathbf{v}} - \tilde{\mathbf{j}}/e\bar{n}) \times \tilde{\mathbf{B}}> \approx <\tilde{\mathbf{v}}_e \times \tilde{\mathbf{B}}>, \quad (13)$$

where $\mathbf{v}_i$ ($\mathbf{v}_e$) is the ion (electron) flow velocity. We note that the appearance of $\mathbf{v}_e$ only is consistent with the Ohm's law being a force balance of *electrons*.

The parallel component of $\mathcal{E}$, or the α-effect, along the mean field are of interest. Therefore, only the perpendicular turbulent flow and magnetic field are relevant, i.e.,

$$\mathcal{E}_{\|} = <\tilde{\mathbf{v}}_e \times \tilde{\mathbf{B}}>_{\|} = <\tilde{\mathbf{v}}_{e\perp} \times \tilde{\mathbf{B}}_{\perp}>. \quad (14)$$

An alternative form of the parallel Ohm's law can be derived by substituting the perpendicular component of Eq.(11),

$$\tilde{\mathbf{v}}_{e\perp} \approx \tilde{\mathbf{v}}_{\perp} - \frac{\tilde{\mathbf{j}}_{\perp}}{e\bar{n}} \approx \frac{\tilde{\mathbf{E}}_{\perp} \times \overline{\mathbf{B}}}{\overline{B}^2} + \frac{\nabla_{\perp} \tilde{P}_e \times \overline{\mathbf{B}}}{e\bar{n}\overline{B}^2}, \quad (15)$$

into Eq.(14) to yield

$$\mathcal{E}_{\|} = <\tilde{\mathbf{E}}_{\perp} \cdot \tilde{\mathbf{b}}_{\perp}> + <\nabla_{\perp} \tilde{P}_e \cdot \tilde{\mathbf{b}}_{\perp}>/e\bar{n} \quad (16)$$

where $\mathbf{b} \equiv \mathbf{B}/B$.

We identify two possible α-effects [*Ji et al.*, 1995, 1996] in Eq.(16). The first term $<\tilde{\mathbf{E}}_{\perp} \cdot \tilde{\mathbf{b}}_{\perp}>$, represents the contribution to $\tilde{\mathbf{v}}_{e\perp}$ from the turbulent $\tilde{\mathbf{E}}_{\perp} \times \overline{\mathbf{B}}/\overline{B}^2$ drift which is a MHD (single fluid) effect (MHD dynamo), while the second term, $<\nabla_{\perp} \tilde{P}_e \cdot \tilde{\mathbf{b}}_{\perp}>/e\bar{n}$, is the contribution from the turbulent electron diamagnetic drift $\nabla_{\perp} \tilde{P}_e \times \overline{\mathbf{B}}/\overline{B}^2$ which is an electron fluid effect in the two-fluid framework (diamagnetic dynamo). We emphasize here that only the MHD dynamo effect has been studied in most dynamo theories and simulations, while both MHD and diamagnetic dynamo effects have been detected in RFP plasmas [*Ji et al.*, 1994, 1995, 1996] (also see Fig. 9). MHD dynamo effects also have been measured [*al Karkhy et al.*, 1993] in spheromak plasmas [*Jarboe*, 1994].

3.2. Helicity Conservation During Dynamo Action

Both the MHD and diamagnetic dynamos drive a parallel current twisting up the field lines, which can be translated into creation of magnetic helicity. Then one question may arise: can magnetic helicity be generated by dynamo action without any constraints? The answer is no: the total helicity must be conserved, except for a battery effect, as discussed below.

By using the generalized Ohm's law, the rate of change of helicity can be rewritten as

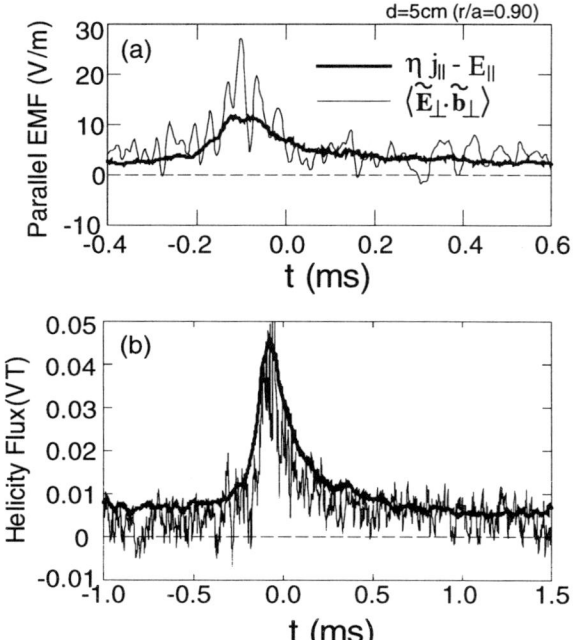

Figure 9. Measured (a) parallel EMF (α-effect) due to electrostatic turbulence, $< \widetilde{\mathbf{E}}_\perp \cdot \widetilde{\mathbf{b}}_\perp >$, and (b) helicity flux (dotted line), $< \widetilde{\phi}\widetilde{\mathbf{B}} >$ in a laboratory plasma. The solid line in (b) is the prediction from the helicity balance equation.

$$\frac{dK}{dt} = -2\int \eta \mathbf{j} \cdot \mathbf{B} dV - 2\int \frac{\nabla P_e \cdot \mathbf{B}}{en} dV - \int (2\phi \mathbf{B} + \mathbf{A} \times \frac{\partial \mathbf{A}}{\partial t}) \cdot d\mathbf{S}. \quad (17)$$

The first term on the RHS is not a dynamo effect but a resistive effect, which vanishes with zero resistivity. A finite resistivity introduces not only the usual resistive helicity decay but also a helicity increase or decrease during magnetic reconnection, a phenomenon which happens only in a resistive plasma (see Section 2). The last term on the RHS is surface integration, which transports helicity across space while conserving the total helicity. Indeed, the dynamo effect can originate from this surface term, as shall be seen in the next section.

The second term on the RHS of Eq.(17) can be rewritten as

$$\int \frac{\nabla P_e \cdot \mathbf{B}}{en} dV = \int \frac{T_e}{e} \mathbf{B} \cdot d\mathbf{S} + \int \frac{T_e}{en} \nabla n \cdot \mathbf{B} dV \quad (18)$$

where the first term is a surface term that does not change the total helicity while the second term does in certain conditions. Obviously, one such condition is a finite density gradient along the field line. However, this condition is not enough to change the total helicity. The integral of the second term in Eq.(18) is equivalent to

$$\frac{T_e}{en} \nabla n \cdot \mathbf{B} = \nabla \cdot \left(\frac{T_e}{e} \ln n \mathbf{B}\right) - \ln n \nabla \frac{T_e}{e} \cdot \mathbf{B}, \quad (19)$$

where the first term leads to a surface term with no effects on the total helicity. From the second term, it can be seen that a finite temperature gradient is required for a finite change in the total helicity. Therefore, both finite gradients in density and electron temperature (of course also in electron pressure) along the field line are necessary conditions to change the total helicity. However, we note that such parallel gradients, especially $\nabla_\parallel T_e$, are very small owing to fast electron flow along the field lines. Such effects, often called the battery effect [*Parker*, 1979], provide only a seed for magnetic field to grow in a dynamo process and, of course, it can be accompanied by a small but finite magnetic helicity.

In summary, dynamo effects conserve the total helicity except for a small battery effect. This conclusion is consistent with the observation that the helicity is approximately conserved during magnetic relaxation [*Ji, Prager, and Sarff*, 1995]. In the following section, the battery effect will be ignored for simplicity.

3.3. Helicity in Mean and Turbulent Fields

Dynamo action amplifies a seed magnetic field into a large-scale, mean field and maintains it against resistive decay. Magnetic helicity associated with the mean fields must also be generated and maintained by dynamo effects. According to the last section, however, the helicity in the total (mean plus turbulent) field cannot be created (except for a small battery effect). Therefore, only two possibilities exist for the mean-field helicity: either it is transported across space or it is separated from the helicity associated with turbulent fields. We shall see that both mechanisms are possible depending on the nature of the dynamo process.

We begin with the rate of change of the helicity in the mean field, $K_m = \int \overline{\mathbf{A}} \cdot \overline{\mathbf{B}} dV$, and the helicity in the turbulent field, $K_t = \int < \widetilde{\mathbf{A}} \cdot \widetilde{\mathbf{B}} > dV$:

$$\frac{dK_m}{dt} = -2\int \overline{\mathbf{E}} \cdot \overline{\mathbf{B}} dV - \int (2\bar{\phi}\overline{\mathbf{B}} + \overline{\mathbf{A}} \times \frac{\partial \overline{\mathbf{A}}}{\partial t}) \cdot d\mathbf{S} \quad (20)$$

$$\frac{dK_t}{dt} = -2\int < \widetilde{\mathbf{E}} \cdot \widetilde{\mathbf{B}} > dV - \int (2 < \widetilde{\phi}\widetilde{\mathbf{B}} > + < \widetilde{\mathbf{A}} \times \frac{\partial \widetilde{\mathbf{A}}}{\partial t} >) \cdot d\mathbf{S}, \quad (21)$$

and their sum, $K_m + K_t$, is the total mean helicity, $\overline{K}$. By using an alternative form of the generalized Ohm's law, $\mathbf{E} + \mathbf{v}_e \times \mathbf{B} + \boldsymbol{\nabla} P_e/en = \eta \mathbf{j}$, we have

$$\overline{\mathbf{E}} + \overline{\mathbf{v}}_e \times \overline{\mathbf{B}} + \frac{\boldsymbol{\nabla}\overline{P_e}}{en} + \boldsymbol{\mathcal{E}} = \eta \overline{\mathbf{j}} \quad (22)$$

$$\widetilde{\mathbf{E}} + \widetilde{\mathbf{v}}_e \times \overline{\mathbf{B}} + \overline{\mathbf{v}}_e \times \widetilde{\mathbf{B}} + \widetilde{\mathbf{v}}_e \times \widetilde{\mathbf{B}} - \boldsymbol{\mathcal{E}} + \frac{\boldsymbol{\nabla}\widetilde{P_e}}{en} = \eta \widetilde{\mathbf{j}}, \quad (23)$$

where $\boldsymbol{\mathcal{E}} = <\widetilde{\mathbf{v}}_e \times \widetilde{\mathbf{B}}>$. Substituting these two equations into Eqs.(20) and (21), after some algebra, we obtain

$$\frac{dK_m}{dt} = -2\int \eta \overline{\mathbf{j}} \cdot \overline{\mathbf{B}} dV + 2\int \boldsymbol{\mathcal{E}} \cdot \overline{\mathbf{B}} dV$$
$$- \int (2\overline{\phi}\overline{\mathbf{B}} - 2\frac{\overline{P_e}\overline{\mathbf{B}}}{en} + \overline{\mathbf{A}} \times \frac{\partial \overline{\mathbf{A}}}{\partial t}) \cdot d\mathbf{S} \quad (24)$$

$$\frac{dK_t}{dt} = -2\int \eta <\widetilde{\mathbf{j}} \cdot \widetilde{\mathbf{B}}> dV - 2\int \boldsymbol{\mathcal{E}} \cdot \overline{\mathbf{B}} dV$$
$$- \int <2\widetilde{\phi}\widetilde{\mathbf{B}} - 2\frac{\widetilde{P_e}\widetilde{\mathbf{B}}}{en} + \widetilde{\mathbf{A}} \times \frac{\partial \widetilde{\mathbf{A}}}{\partial t}> \cdot d\mathbf{S}. \quad (25)$$

The α-effect appears as the second terms in the RHS of these equations but with opposite signs. It might be concluded that the dynamo effects would generate the same amount of helicity but with opposite signs in the mean field and the turbulent field [*Seehafer*, 1996]. However, it may not be the case depending on types of dynamo effects.

Following Eq.(16), the α-effect is

$$\boldsymbol{\mathcal{E}} \cdot \overline{\mathbf{B}} = -<\boldsymbol{\nabla}_\perp \widetilde{\phi} \cdot \widetilde{\mathbf{B}}_\perp> - <\frac{\partial \widetilde{\mathbf{A}}_\perp}{\partial t} \cdot \widetilde{\mathbf{B}}_\perp>$$
$$+ \frac{<\boldsymbol{\nabla}_\perp \widetilde{P_e} \cdot \widetilde{\mathbf{B}}_\perp>}{en}, \quad (26)$$

where the three terms correspond to effects due to electrostatic, electromagnetic, and electron diamagnetic turbulence, respectively. Substituting Eq.(26) into Eqs.(24) and (25),

$$\frac{dK_m}{dt} = -2\int (\eta \overline{\mathbf{j}} \cdot \overline{\mathbf{B}} + <\frac{\partial \widetilde{\mathbf{A}}_\perp}{\partial t} \cdot \widetilde{\mathbf{B}}_\perp>) dV$$
$$- \int (2\overline{\phi}\overline{\mathbf{B}} - 2\frac{\overline{P_e}\overline{\mathbf{B}}}{en} + \overline{\mathbf{A}} \times \frac{\partial \overline{\mathbf{A}}}{\partial t}$$
$$+ 2<\widetilde{\phi}\widetilde{\mathbf{B}} - \frac{\widetilde{P_e}\widetilde{\mathbf{B}}}{en}>) \cdot d\mathbf{S} \quad (27)$$

$$\frac{dK_t}{dt} = -2\int (\eta <\widetilde{\mathbf{j}} \cdot \widetilde{\mathbf{B}}> - <\frac{\partial \widetilde{\mathbf{A}}_\perp}{\partial t} \cdot \widetilde{\mathbf{B}}_\perp>) dV$$
$$- \int <\widetilde{\mathbf{A}} \times \frac{\partial \widetilde{\mathbf{A}}}{\partial t}> \cdot d\mathbf{S}. \quad (28)$$

are obtained. In the case of electromagnetic turbulence, i.e., $\widetilde{\mathbf{v}}_e$ is driven by an inductive electric field, the dynamo effect generates the same amount of helicity both in the mean and turbulent fields but with opposite signs, as seen from the second terms of the above equations. Techniques often used in the laboratory to drive currents in a plasma by an incident electromagnetic wave fall into this category. The wave helicity is converted to the mean-field helicity by interaction between the wave and the background plasma. (We note that the last term in Eq.(28) represents an electromagnetic wave propagating across space without interacting with the mean field.) In the case of electrostatic or electron diamagnetic turbulence, i.e., $\widetilde{\mathbf{v}}_e$ is driven by electrostatic field or perpendicular electron pressure, the dynamo effect does not affect the turbulent helicity but merely transports the mean-field helicity across space, as seen from the surface terms in Eq.(27).

Therefore, it is crucial to know the type of turbulence which generates the dynamo effect in a turbulent plasma in order to assess the role of dynamo effects on magnetic helicity, even though the total helicity is always conserved. In the case of a laboratory plasma (the MST RFP), direct measurements indicated that the turbulence is predominantly electrostatic, thus causing helicity transport in the mean field with no effects on the turbulent field. Figure 9 shows such an example of measured helicity flux caused by the electrostatic turbulence [*Ji, Prager, and Sarff*, 1995] together with the measured α-effect [*Ji et al.*, 1994].

4. CONCLUSIONS

Magnetic helicity is closely related to magnetic reconnection and dynamo effects, both of which involve changes in magnetic field topology. Recent laboratory experiments have provided opportunities to test, verify, and discover the relationships between them as summarized below.

When two plasmas are driven to merge, the helicity content of each plasma determines the reconnecting angle which strongly affects the reconnection rate and the shape of the diffusion region. Conversely, magnetic reconnection events also strongly affect the global helicity, resulting in efficient helicity cancellation (but not dissipation) during counter-helicity reconnection and a finite helicity increase or decrease (but less efficiently than dissipation of magnetic energy) during co-helicity reconnection.

The turbulent electromotive force along the mean magnetic field (α-effect), due to either electrostatic turbulence or the electron diamagnetic effect, have been measured in an RFP plasma. These dynamo effects transport mean-field helicity across space without dissipation, as seen in direct measurements of helicity flux. When the dynamo effect is driven by electromagnetic turbulence, helicity in the turbulent field is converted to mean-field helicity. In all cases, however, dynamo processes conserve total helicity except for a small battery effect, consistent with the observation that the helicity is approximately conserved during magnetic relaxation.

Acknowledgments. The author is grateful to Drs. M. Yamada, R. Kulsrud, S. Hsu, and S. Prager for their collaborations.

REFERENCES

Akasofu, S.-I., *Physics of Magnetospheric Substorms*, p.12, D. Reidel, Norwood, Mass., 1977.

al Karkhy, A., P.K. Browning, G. Cunningham, S.J. Gee, M.G. Rusbridge, Obeservations of the Magnetohydrodynamic Dynamo Effect in a Spheromak Plasma, *Phys. Rev. Lett., 70*, 1814, 1993.

Berger, M.A., Rigorous new limits on magnetic helicity dissipation in the solar corona, *Geophys. Astrophys. Fluid Dyn., 30*, 79, 1984.

Berger, M.A., and G.B. Field, The topological properties of magnetic helicity, *J. Fluid Mech., 147*, 133, 1984.

Biskamp, D., Magnetic reconnection, *Phys. Rep., 237*, 179, 1994.

Biskamp, D., Current sheet profiles in two-dimensional magnetohydrodynamics, *Phys. Fluids B, 5*, 3893, 1993.

Hegna, C.C., Self-consistent mean-field forces in turbulent plasmas: Current and momentum relaxation, *Phys. Plasmas, 5*, 2257, 1998.

Jarboe, T.R., Review of Spheromak Research, *Plasma Phys. Control. Fusion, 36*, 945, 1994.

Ji, H., A.F. Almagri, S.C. Prager, J.S.Sarff, Time-resolved observation of discrete and continuous MHD dynamo in the reversed-field pinch edge, *Phys. Rev. Lett., 73*, 668, 1994.

Ji, H., S.C. Prager, J.S.Sarff, Conservation of magnetic helicity during plasma relaxation, *Phys. Rev. Lett., 74*, 2945, 1995.

Ji, H., Y. Yagi, K. Hattori, A.F. Almagri, S.C. Prager, Y. Hirano, J.S. Sarff, T. Shimada, Y. Maejima, and K. Hayase, Effect of Collisionality and Diamagnetism on the Plasma Dynamo, *Phys. Rev. Lett., 75*, 1085, 1995.

Ji, H., S.C. Prager, A.F. Almagri, J.S. Sarff, Y. Yagi, Y. Hirano, K. Hattori, and H. Toyama, Measurement of the dynamo effect in a plasma, *Phys. Plasmas, 3*, 1935, 1996.

Ji, H., M. Yamada, S. Hsu, R. Kulsrud, Experimental test of the Sweet-Parker model of magnetic reconnection, *Phys. Rev. Lett., 80*, 3256, 1998.

Ji, H., M. Yamada, S. Hsu, R. Kulsrud, T. Carter, and S. Zaharia, Magnetic reconnection with Sweet-Parker characteristics in two-dimensional laboratory plasmas, to be published in *Phys. Plasmas*, 1999.

Krause, F., and K.-H. Rädler, *Mean-Field Magnetohydrodynamics and Dynamo Theory*, Akademie-Verlag, Berlin, 1980.

Nagayama, Y., K.M. McGuire, M. Bitter, A. Cavallo, E.D. Fredrickson, K.W. Hill, H. Hsuan, A. Janos, W. Park, Analysis of sawtooth oscillations using simultaneous measurement of electron-cyclotron emission imaging and X-ray tomography on TFTR, *Phys. Rev. Lett., 67*, 3527, 1991.

Ono Y., A. Morita, M. Katsurai, M. Yamada, Experimental investigation of 3-dimensional magnetic reconnection by use of 2 colliding spheromaks, *Phys. Fluids B, 5*, 3691, 1993.

Parker, E.N., Hydromagnetic dynamo models, *Astrophys. J., , 121*, 293-314, 1955.

Parker, E.N., *J. Geophys. Res., 62*, 509, 1957.

Parker, E.N. *Cosmical Magnetic Fields*, 841 pp., Clarendon Press, Oxford, 1979.

Petschek, H.E., Magnetic field annihilation, NASA Spec. Pub. SP-50, 425, 1964.

Pfister, H., and W. Gekelman, Demonstration of helicity conservation during magnetic reconnection using Christmas ribbons, *Am. J. Phys., 59*, 497, 1991.

Proctor, M.R.E., and A.D. Gilbert (Eds.), *Lectures on Solar and Planetary Dynamos*, 375 pp., Cambridge University Press, New York, 1994.

Rust, D.M., Spawning and shedding helical magnetic fields in the solar atmosphere, *Geophys. Res. Lett., 21*, 241, 1994.

Schnack, D.D., E.J. Caramana, R.A. Nebel, Three-dimensional magnetohydrodynamic studies of the reversed field pinch, *Phys. Fluids, 28*, 321, 1985.

Seehafer, N., Nature of the α effect in magnetohydrodynamics, *Phys. Rev. E, 53*, 1283, 1996.

Spitzer, L. Jr., *Physics of Fully Ionized Gases* (2nd Revised Edition), Interscience Publishers, New York, 1962.

Steinhauer, L.C., and A. Ishida, Relaxation of a two-specie magnetofluid, *Phys. Rev. Lett.,* it 79, 3423, 1997; Relaxation of a two-species magnetofluid and application to finite-beta flowing plasmas, *Phys. Plasmas, 5*, 2609, 1998.

Stoneking, M.R., S.A. Hokin, S.C. Prager, G. Fiksel, H. Ji, D.J. Den Hartog, Particle transport due to magnetic fluctuations, *Phys. Rev. Lett., 73*, 549, 1994.

Sweet, P.A., and B. Lehnert (Eds.), *Electromagnetic Phenomena in Cosmical Physics*, 123pp., Cambridge University Press, New York, 1958.

Taylor, J.B., Relaxation of toroidal plasma and generation of reverse magnetic fields, *Phys. Rev. Lett., 33*, 1139, 1974.

Taylor, J.B., Relaxation and Magnetic Reconnection in Plasmas, *Rev. Mod. Phys., 58*, 741, 1986.

Vasyliunas, V.M., Theoretical models of magnetic line merging, *Rev. Geophys. Space Phys., 13*, 303, 1975.

Woltjer, L., A theorem on force-free magnetic fields, *Proc. Natl. Acad. Sci. USA, 44*, 489, 1958.

Wright, A.N., and M.A. Berger, A physical description of magnetic helicity evolution in the presence of reconnection lines, *J. Plasma Phys., 46*, 179, 1991.

Yamada, M., Y. Ono, A. Hayakawa, M. Katsurai, Magnetic Reconnection of Plasma Toroids with Cohelicity and Counterhelicity, *Phys. Rev. Lett., 65*, 721, 1990.

Yamada, M., H. Ji, S. Hsu, T. Carter, R. Kulsrud, Y. Ono, F. Perkins, Identification of Y-Shaped and O-Shaped Diffusion Regions during Magnetic Reconnection in a Laboratory Plasma, *Phys. Rev. Lett.*, *78*, 3117, 1997.

Yamada, M., H. Ji, S. Hsu, T. Carter, R. Kulsrud, N. Bretz, F. Jobes, Y. Ono, F. Perkins, Study of Driven Magnetic Reconnection in a Laboratory Plasma, *Phys. Plasmas 4*, 1936, 1997.

H. Ji, Princeton Plasma Physics Laboratory, Princeton University, P.O. Box 451, Princeton, NJ 08543. (e-mail: hji@pppl.gov)

Measurements of Helicity and Reconnection in Electron MHD Plasmas

R. L. Stenzel, J. M. Urrutia, and M. C. Griskey

Department of Physics and Astronomy, University of California, Los Angeles

Laboratory experiments on time-varying magnetic fields in Electron MHD plasmas are reviewed. Helicity is a fundamental property in EMHD. It is observed that three-dimensional fields usually assume the topology of vortices or flux ropes. The electromagnetic perturbations are force-free, frozen into the electron fluid, and propagate in the whistler mode along a uniform background magnetic field. Small-amplitude vortices do not interact when propagating through each other. Reflection of a vortex at a conducting boundary reverses its helicity, hence does not conserve helicity, but does conserve energy. Large amplitude vortices can create magnetic null points. Three-dimensional spiral null points in magnetic fields and fluid velocities are observed and the expansion matrix is measured. Three-dimensional reconnection can approach the whistler speed.

1. INTRODUCTION

The helicity of magnetic fields plays a fundamental role in many fields of physics such as cosmology [*Cornwall*, 1997], in astrophysics [*Kumar and Rust*, 1996], dynamos [*Boozer*, 1993], flares [*Ruzmaikin*, 1996], the solar wind [*Goldstein et al*, 1995], magnetic reconnection [*Pevtsov et al*, 1996]. Helicity has also been studied in laboratory plasmas, e.g., spheromaks [*Ono*, 1995; *Yamada et al*, 1997], reverse field pinches [*Ji et al*, 1996], tokamaks [*Ohkawa*, 1989], and in whistler wavepackets [*Isichenko and Marnachev*, 1987; *Urrutia et al*, 1995].

In the present work we consider the helicity of magnetic fields in the parameter regime of electron MHD [*Kingsep et al*, 1990], where the magnetic field interacts only with the electrons. The ions form a stationary background fluid. This situation arises on spatial scales smaller than an ion Larmor radius and on time scales shorter than an ion cyclotron period. The ions can also become unmagnetized by collisions. Typical examples include the vicinity of magnetic null points, fast instabilities such as electron tearing modes, and photospheric plasmas. In contrast to single-fluid MHD, space charge electric fields and electron Hall currents are important in EMHD, whose governing equations are discussed in Sec. 3.1. Transient fields are transported by whistlers rather than Alfvén waves. Reconnection involves magnetic energy transfer to electrons rather than ions.

After a brief description of the laboratory experiment and the measurement techniques the basic properties of whistler vortices will be reviewed. Examples of helicity injection, transport of helicity, and helicity reversal of propagating vortices will be presented. Finally, magnetic null points and reconnection processes of whistler vortices whose magnetic field exceeds that of the ambient field will be described.

2. EXPERIMENTAL ARRANGEMENT

The experiments are performed in a large laboratory plasma device sketched schematically in Figure 1. A 1 m diam × 2.5 m long plasma column of density $n_e \simeq 10^{12}$ cm^{-3}, electron temperature $kT_e \simeq 2$ eV, Argon gas pressure $p \simeq 3 \times 10^{-4}$ Torr, is produced in a

Magnetic Helicity in Space and Laboratory Plasmas
Geophysical Monograph 111
Copyright 1999 by the American Geophysical Union

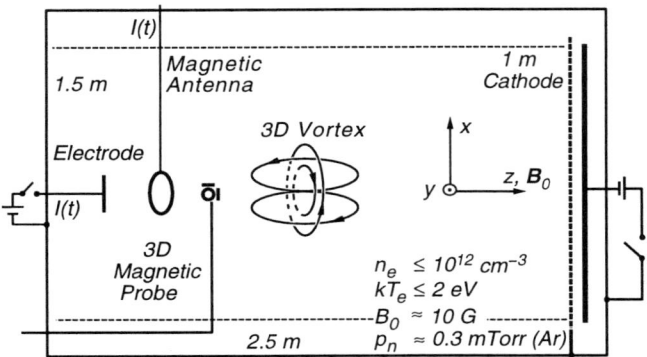

Figure 1. Experimental setup and basic parameters.

uniform axial magnetic field $B_0 \simeq 10$ G with a pulsed dc discharge (50 V, 600 A, $t_{pulse} \simeq 5$ ms, $t_{rep} \simeq 1$ s) with a large oxide-coated cathode. In the quiescent, uniform, current-free afterglow plasma pulsed currents are induced with magnetic loop antennas or drawn to biased electrodes. The time-varying magnetic fields associated with the plasma currents are measured with a triple magnetic probe, recording three orthogonal vector components versus time at a given position. By repeating the highly reproducible discharges and moving the probe to many positions in a three-dimensional volume, the vector field $\mathbf{B}(\mathbf{r},t)$ is obtained with high resolution ($\Delta r \simeq 1$ cm, $\Delta t \simeq 10$ ns). This allows us to calculate at any instant of time the current density $\mathbf{J}(\mathbf{r},t) = \nabla \times \mathbf{B}/\mu_0$ without making any assumptions about field symmetries or using $\nabla \cdot \mathbf{B} = 0$. The plasma parameters are obtained from a small Langmuir probe which is also movable in three dimensions.

3. EXPERIMENTAL RESULTS

3.1. Properties of Linear Whistler Vortices

When a current is injected from a positively biased electrode it flows through the plasma in the form of a spiral as shown in Figure 2. The helical current flow can be thought of a superposition of the field aligned current and an electron Hall current. The latter is produced by a radial electric field which is due to the collection of electrons at the electrode. Note that for $\mathbf{B}_0 > 0$ the Hall current $\mathbf{J}_{Hall} = \mathbf{B}_0 \times ne\mathbf{E}/B_0^2$ produces a right-handed helix and for $\mathbf{B}_0 < 0$ a left-handed helix. The front of the current system propagates at the whistler speed along $\mathbf{B}_0$. Since the current is closed ($\nabla \cdot \mathbf{J} = 0$), the current density lines at the front return as outer helices to the negative return electrode. The length of the current tube is determined by the applied pulse length and propagation speed. For short pulses a current vortex is formed which detaches from the electrodes and propa-

gates through the plasma. It exhibits knotted current density lines [*Urrutia et al*, 1995]. Figure 3 shows an experimental verification of a spheromak-like vortex in the perturbed magnetic field excited by a short current pulse. The magnetic/current vortex can be viewed as a whistler wave packet consisting of a single oscillation. Note that the vortex is not spherical because the propagation along $\mathbf{B}_0$ is faster than oblique to $\mathbf{B}_0$. Inside the vortex there are trefoil knots corresponding to the first torus knots in the magnetic field lines [*Urrutia et al*, 1995]. By induction secondary vortices are generated ahead and behind the main vortex which can also be interpreted by the dispersion of whistlers. Nested spheromaks can arise in unbounded plasmas.

Propagating vortices can also be excited with magnetic antennas which induce either the field-aligned current component or the Hall current component or couple to the corresponding magnetic fields. For example, a simple loop antenna with dipole moment along the uniform dc magnetic field induces Hall currents or produces an axial magnetic field perturbation. But, unlike an isotropic conductor, the induced current is not driven directly by the inductive electric field, it is a Hall current due to a radial space charge field associated with the radial $\mathbf{E}_\theta \times \mathbf{B}_0$ drift of electrons. Space charge and inductive electric fields have been obtained separately [*Rousculp et al*, 1994]. The incompressible electrons stream along the dc field which produces current/field linkage similar to the case of electrode excitation. Helicity is a fundamental property of EMHD pulses which can be easily explained as follows:

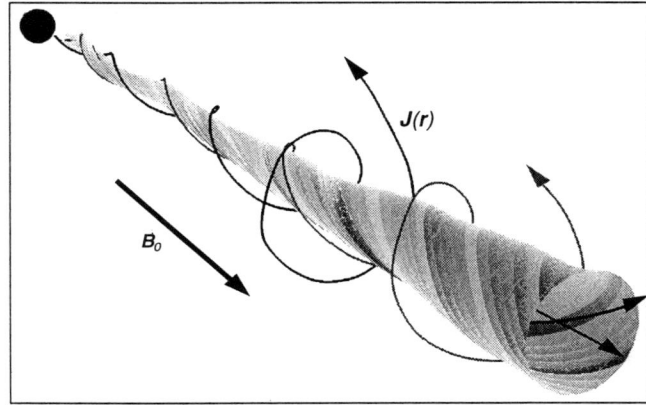

Figure 2. Measured current density lines, $\mathbf{J}(\mathbf{r}, t = const)$, and surface of a current tube ($I = const$) for a pulsed current from an electrode in a magnetoplasma. The flux-rope topology arises from the superposition of an electron Hall current and the field-aligned current. In Electron MHD the current front propagates at the speed of a whistler wave.

The penetration of the applied magnetic field into a plasma is theoretically described by Faraday's law and Ohm's law which for an ideal uniform plasma dominated by the Hall effect yields $\partial \mathbf{B}/\partial t = \nabla \times (\mathbf{v} \times \mathbf{B})$. Here, $\mathbf{v} = -\mathbf{J}/ne = -\nabla \times \mathbf{B}/ne\mu_0$ is the electron fluid velocity and $\mathbf{B} = \mathbf{B}(\mathbf{r},t) + \mathbf{B}_0$ the total magnetic field. Displacement currents are negligible compared to conduction currents, $J_{dis}/J_{cond} \simeq (\omega/\omega_p)^2 \simeq 10^{-7}$. Uniformity implies absence of pressure gradients. Fourier analysis of the equation yields the dispersion of low-frequency whistlers, $\omega \cong \omega_c(kc/\omega_p)^2$. For small field perturbations, $B(\mathbf{r},t) \ll B_0$, propagating with wave velocity $\partial z/\partial t = \pm v_\parallel$ along the dc magnetic field, the linearized solution of the differential equation yields $\mathbf{J}/ne = \pm v_\parallel \mathbf{B}(\mathbf{r},t)/B_0$. The perturbed field has a positive (negative) self-helicity density $\mathbf{J} \cdot \mathbf{B}(\mathbf{r},t)$ for propagation along (opposite to) the dc magnetic field. The same holds for the magnetic self-helicity density $\mathbf{A}(\mathbf{r},t) \cdot \mathbf{B}(\mathbf{r},t)$ and the kinetic helicity density $\mathbf{v} \cdot \boldsymbol{\omega} \propto \mathbf{J} \cdot (\nabla \times \mathbf{J})$. The unique property of EMHD vortices that their helicity densities depend on propagation direction is the result of the Hall effect, $\mathbf{E} = \mathbf{J} \times \mathbf{B}/ne$. The latter also shows that the electromagnetic fields are force-free provided that $\nabla p = 0$, and frozen into the electron fluid. When electron inertia is included the generalized vorticity $\boldsymbol{\Omega} = \nabla \times \mathbf{v} - (e/m)\mathbf{B}$ is frozen into a collisionless electron fluid (Avinash and Taylor, 1991).

Measurements have been performed on the relaxation of energy and helicity in propagating vortices. In the presence of weak collisional damping both quantities decay at the same rate [*Urrutia et al*, 1995]. The generalized vorticity has been obtained experimentally and its helicity also decays proportional to the magnetic energy [*Urrutia and Stenzel*, 1996]. There is no reason to expect Taylor's conjecture of preferred helicity conservation to hold [*Taylor*, 1974].

3.2. Helicity Injection and Directionality of Propagation

When a current pulse is applied to a loop antenna whose symmetry axis is along $\mathbf{B}_0$ two vortices are excited, one propagating along $\mathbf{B}_0$ with positive self-helicity density, the other one opposite to $\mathbf{B}_0$ with negative self-helicity density. Thus, the total self-helicity is zero, consistent with the fact that the loop does not inject net helicity. This also holds for the mutual magnetic helicity, $\int \mathbf{A}_0 \cdot \mathbf{B}(\mathbf{r},t) \, dV = 0$, where $\nabla \times \mathbf{A}_0 = \mathbf{B}_0$. There is no mutual current helicity, $\int \mathbf{B}_0 \cdot \mathbf{J}(\mathbf{r},t) \, dV = 0$.

Two vortices of opposite self-helicities are also excited by a torus antenna, which excites the toroidal magnetic field, $\mathbf{B}_\theta$, of the vortices. While the self-helicity vanishes, the mutual helicity does not vanish since the applied toroidal field links with the uniform field so as

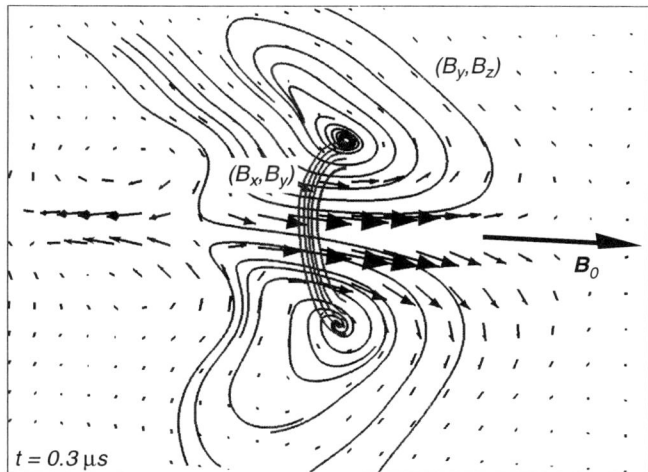

Figure 3. Vector field of the perturbed magnetic field $\mathbf{B}(\mathbf{r}.t)$ and selected field lines in orthogonal planes. The field topology forms a three-dimensional vortex which propagates in the whistler mode along $\mathbf{B}_0$.

to inject net helicity. Both vortices have the same $\mathbf{B}_\theta$, hence carry the same mutual helicity, as expected from helicity conservation.

A loop placed on the axis of a torus can be used to apply helical fields to a plasma, i.e., to inject helicity. The sign of the applied helicity depends on the relative direction of loop and torus currents. When positive helicity is injected a vortex with positive self-helicity is excited which travels along $\mathbf{B}_0$, for negative helicity the propagation direction reverses. Thus, helicity conservation implies directional radiation of EMHD fields. This new concept has been tested in a computer simulation [*Rousculp and Stenzel*, 1997] and verified in recent experiments. Figure 4 shows a snapshot of magnetic field components in two orthogonal planes demonstrating that a loop-torus antenna radiates predominantly one vortex into one hemisphere with the same sign of helicity as applied. The completely symmetric loop-torus antenna exhibits a directivity of 20dB (power ratio $P_{left}/P_{right} = 100$). The antenna directionality also holds when receiving vortices. Transmission between two antennas is unidirectional, where the direction can be selected by the antenna helicity or the direction of $\mathbf{B}_0$. These interesting helicity properties may eventually lead to useful antenna applications.

3.3. Collisions and Reflections of Vortices

We first consider the collision of two EMHD vortices propagating on the same field line $\mathbf{B}_0$ against each other. Since they carry opposite helicities either their toroidal or axial field components must cancel during the collision. The interaction has some similarity with

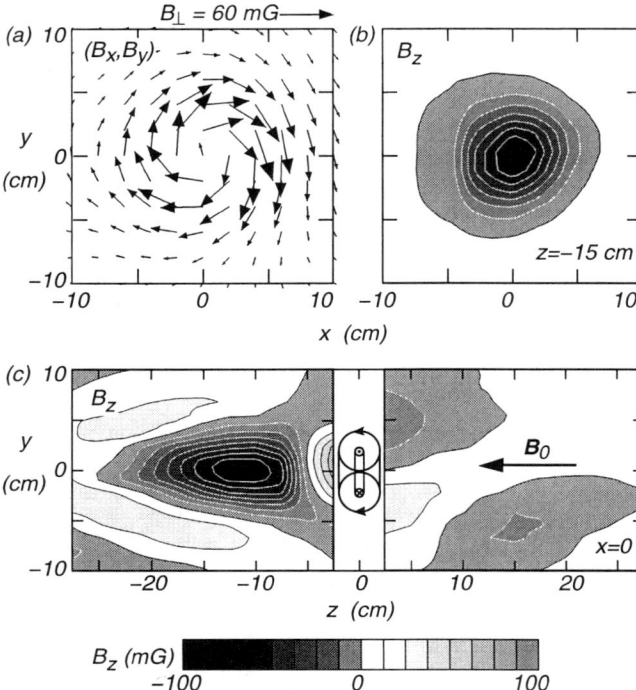

A second experiment with surprising results on helicity conservation is the reflection of a vortex from a conducting boundary. The vortex is excited with a torus antenna and propagates in the direction opposite to $\mathbf{B}_0$ against a conducting plate large compared to the vortex and with surface normal along $\mathbf{B}_0$. Figure 6a shows magnetic field components for both the incident and reflected vortex, demonstrating that the magnetic self-helicity has been reversed as expected from the change of propagation direction. Also the mutual helicity, hence total magnetic helicity, is reversed since it is the toroidal field component which reverses direction upon reflection. The axial (poloidal or dipolar) field component does not change sign upon reflection. The boundary conditions imply that the normal magnetic

Figure 4. Demonstration that helicity injection from a loop-torus antenna results in the directional radiation of a whistler vortex. Snapshot of (a) the perturbed toroidal magnetic field component $B_\theta(x,y)$ excited by the torus and (b) the axial field $B_z(x,y)$ due to the linked loop, forming a vortex of positive helicity. (c) Axial field component in the central $y-z$ plane showing that the injection of positive helicity produces only one vortex propagating along $\mathbf{B}_0$.

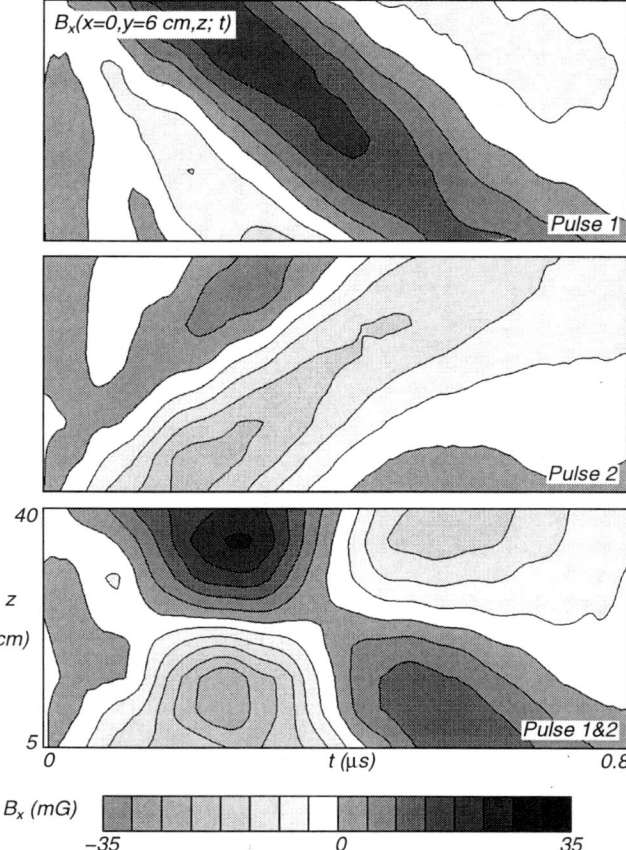

the merging of MHD spheromaks [*Ono*, 1995; *Yamada et al*, 1997] so that questions arise about driven reconnection/annihilation of opposing magnetic field components during the collision. Of course, there are fundamental differences such as the presence of a dc magnetic field, no coupling between field and ions, the absence of flux-conserving boundaries for stabilizing the spheromak motion, etc. Figure 5 shows experimental results of the head-on collision between two identical vortices with opposite toroidal magnetic fields. The toroidal field component B_θ is displayed in a position-time diagram when (i) each vortex propagates separately and (ii) both vortices propagate simultaneously. In the latter case the fields are found to be a linear superposition of the individual pulses. This implies that there is no nonlinear interaction between two force-free vortices as they propagate through one another, that the opposing fields annihilate and recreate at the whistler speed without energy transfer to particles, and that zero helicity is conserved at all times. Thus, reconnection is not an issue in the merging of EMHD vortices.

Figure 5. Collision of two vortices of opposite helicity excited by antennas at $z=0$ and $z=50$ cm and propagating along $\mathbf{B}_0$ against one another. Contours of the field components $B_\theta(z,t)$ of each individual pulse separately (top 2 frames) and both pulses simultaneously (bottom frame). The latter case is a linear superposition of the former, indicating that there is no interaction between two force-free vortices. Zero total helicity is conserved at all times.

field and the tangential electric field vanish at the conductor. Near the plate the magnetic field lines are predominantly radial due to induced toroidal currents in the conductor. The plate draws no axial current from the plasma since $B_\theta \simeq 0$ at the surface. During the pulse reflection the axial magnetic flux near the plate grows and decays which causes the toroidal inductive electric field to reverse sign. This implies a sign reversal of the radial Hall current and axial current, which explains the reversal of the toroidal magnetic field component. Thus, during the reflection the linkage between toroidal and poloidal field lines is reversed by a temporary annihilation of the toroidal field component. As in the collision process, no energy is transferred to particles. However, in contrast to the collision process, the magnetic helicity is *not* conserved. In contrast to the loop-torus antenna the wall cannot inject helicity since it carries only a toroidal current. No fields are transmitted through the plate. Before the reflection the net helicity was zero, after the reflection it is positive. This is an interesting example of a violation of magnetic helicity conservation. Figure 6b,c show that also the current helicity $\mathbf{J} \cdot \mathbf{B}(\mathbf{r},t)$ and the electron fluid helicity $\mathbf{v} \cdot \boldsymbol{\omega}$ reverse sign upon reflection, i.e., are *neither* conserved quantities. However, the angular fluid momentum ($\propto J_\theta$) is conserved, the linear momentum ($\propto J_z$) change is taken up by the plate, and energy is conserved except for weak collisional damping.

3.4. EMHD Fields with Magnetic Null Points

Penetration of magnetic fields, described by $\partial \mathbf{B}/\partial t = \nabla \times (\mathbf{v} \times \mathbf{B})$, appears inherently nonlinear since $\mathbf{v} \times \mathbf{B} \propto (\nabla \times \mathbf{B}) \times \mathbf{B}$. However, in the presence of a uniform background magnetic field $\mathbf{B}_0$ the vortex solutions showed that $(\nabla \times \mathbf{B}) \parallel \mathbf{B}(\mathbf{r},t)$, i.e., the nonlinearity is weak due to the cross product. This is experimentally verified by propagating pulses with large self-magnetic fields ($B(\mathbf{r},t) < B_0$) and observing little difference in propagation speed, topology and amplitude distribution, provided the plasma parameters are not modified [*Urrutia and Stenzel*, 1996]. An interesting consequence of such robust, force-free fields is the existence of stable, elongated current sheets in EMHD, an example of which is shown in Figure 7. Contours of the axial current density J_z are shown in a transverse $x - y$ plane and an axial $y - z$ plane for a current sheet generated by drawing electrons to a 1 cm × 30 cm electrode located at $z = 0$. In addition to the axial current component there is a linked Hall current such that the topology of $\mathbf{J}$ and its associated field $\mathbf{B}(\mathbf{r},t)$ is that of a flux rope of elliptical cross section. On EMHD time scales the current density can exceed the Langmuir limit ($v_{drift} > v_{thermal}$) without producing any tearing of the current sheet as

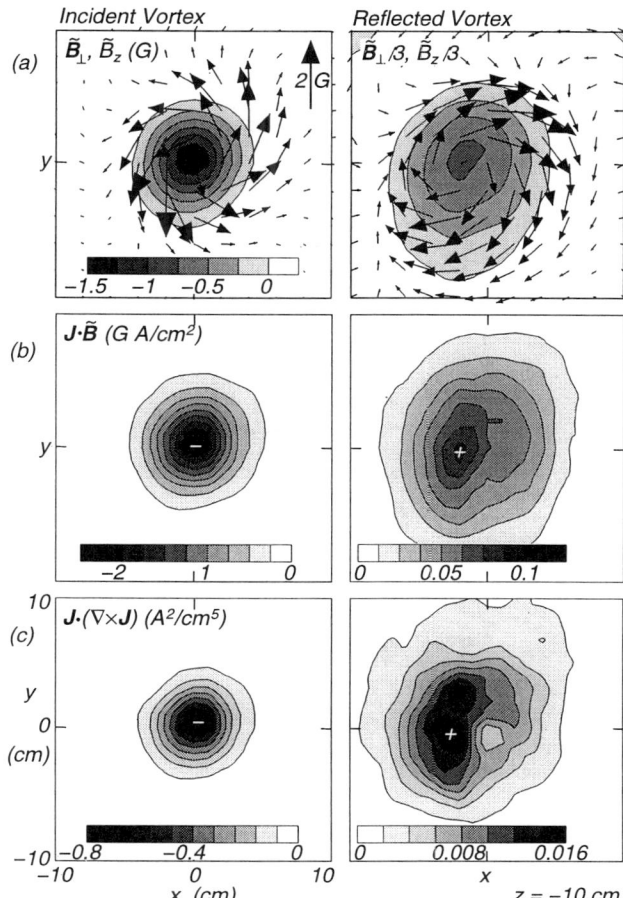

Figure 6. Helicity reversal of an EMHD vortex reflected by a conducting boundary with surface normal along $\mathbf{B}_0$. (a) Vector field (B_x, B_y) and contours $B_z(\mathbf{r},t)$. Note that upon reflection only B_θ reverses sign, implying that also the mutual, hence total magnetic helicity change sign upon reflection. (b) Contours of constant $\mathbf{J} \cdot \mathbf{B}(\mathbf{r},t)$. (c) Contours of constant fluid helicity $\mathbf{v} \cdot \boldsymbol{\omega}$ which also changes sign upon reflection.

seen in computer simulations [*Drake et al*, 1994]. However, the large currents/fields cannot be maintained for long time scales due to the formation of density depletions and double layers near the electrode [*Urrutia and Stenzel*, 1997].

For self-fields $B(\mathbf{r},t) > B_0$ the linearized vortex solution is not appropriate and the problem becomes clearly nonlinear. Experiments have been performed on the penetration/propagation of fields into plasmas with $\beta = nkT_e/(B_0^2/2\mu_0) > 1$ where the plasma-generated magnetic fields can exceed the dc magnetic field. A current pulse (150 A, 6 μs) is applied to a shielded loop antenna (12 cm diam, four turns) arranged so as to produce a strong dipole field opposite to $\mathbf{B}_0$. Figure 8

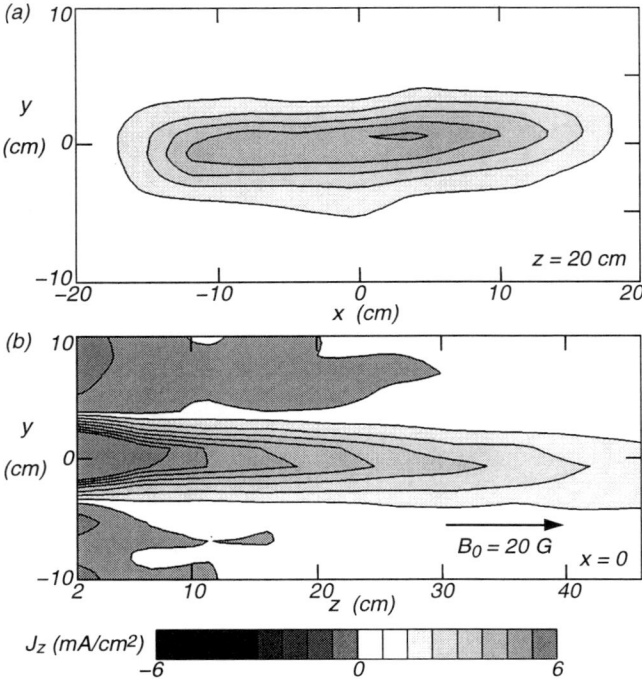

Figure 7. Observation of a stable thin long EMHD current sheet (half widths $\Delta y \simeq c/\omega_p$, $\Delta x \simeq 30\Delta y$). Contours of axial current density in orthogonal planes, (a) $J_z(x, y, z = 25$ cm$)$ and (b) $J_z(x = 0, y, z)$. There are also Hall currents J_x such that the current/field topology is that of a flux rope with ellipsoidal cross section. No whistler instabilities are observed even for large current densities ($v_{drift} \simeq v_{thermal}$).

shows a snapshot of magnetic field lines traced through the two cusp-type null points created on axis where the dipole field cancels $\mathbf{B}_0$. The field lines lie on a 3D separatrix surface which divides "closed" dipolar field lines from "open" field lines similar to an idealized planetary magnetosphere in a uniform interplanetary field [*Parks*,1991]. The field lines enter/leave the null points along a "spine" and a "fan" [*Parnell et al*, 1996]. The objective is to study the penetration and relaxation of the dipole field inside the high beta plasma. For comparison, the field topology has also been measured in vacuum.

Figure 9 displays the dynamics of the null points, i.e., its axial distance from the loop vs time. In vacuum the null point location is simply determined by the waveform of the loop current. In the plasma, during turn-on, the induced shielding currents delay the penetration of the null point, which is eventually achieved during quasi-steady state ($\partial I/\partial t \simeq 0$), while at turn-off the plasma prevents the rapid loss of the dipole field. After the end of the applied current pulse ($t \geq 8$ μs) the null point is entirely maintained by plasma currents. Since the transient EMHD fields/currents propagate in the whistler mode the null point is dragged a long distance away from the loop. The propagation is highly nonlinear since the whistler dispersion depends on the magnetic field which is now space and time-dependent. On the average, the axial propagation speed decreases/increases when a large wave field opposes/adds to the ambient field. In the vicinity of the

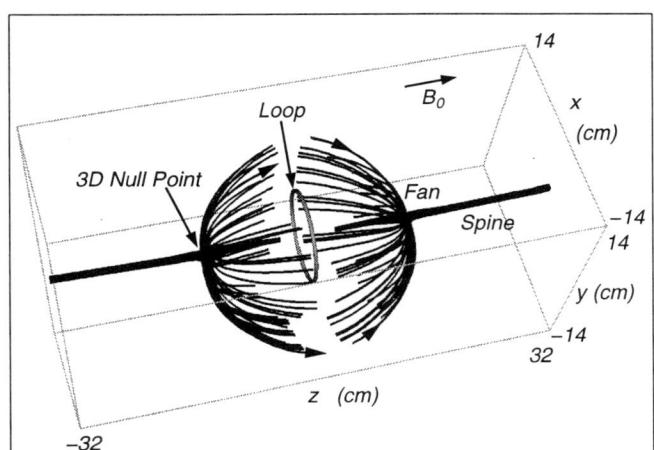

Figure 8. Selected field lines through the cusp magnetic null points created by the dipolar field of a current-carrying loop antenna and an opposing uniform background field $\mathbf{B}_0$. The separatrix consists of an axial spine through the 3D null points and a shell-like fan separating closed from open field lines. The penetration and relaxation of the dipole field inside a high beta plasma is studied.

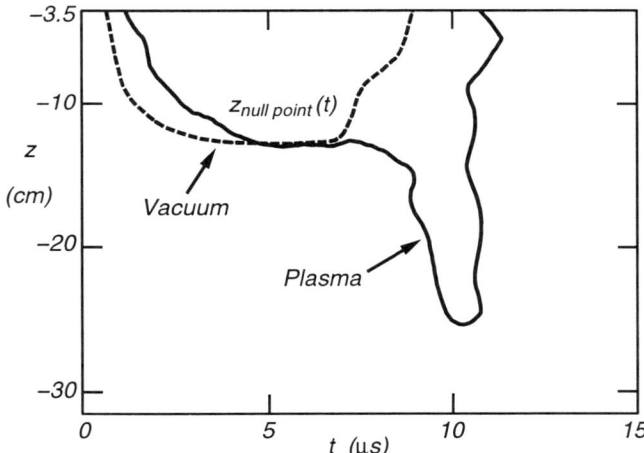

Figure 9. Axial location of the 3D null point in time. In vacuum the trajectory reflects the applied current waveform. In plasma the transient EMHD currents prevent the rapid penetration of the field at turn-on, and maintain the null points well after the end of the loop current. The perturbed field with null point propagates in the whistler mode away from the loop at $z = 0$.

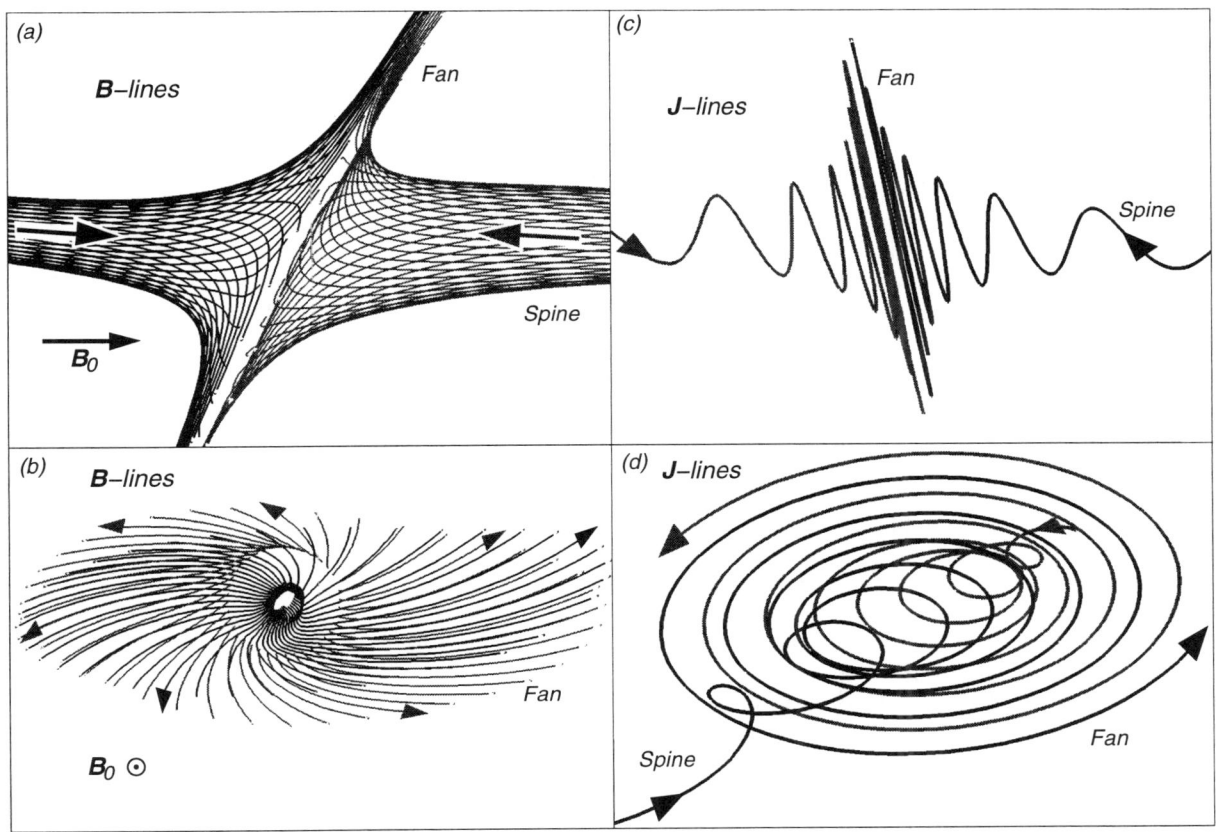

Figure 10. Measured topology of 3D null points after turn-off of the loop current. Magnetic field lines viewed normal to (a) the spine and (b) the fan. The expansion matrix identifies the configuration as that of a improper non-potential spiral null. Current density lines (c,d) also exhibit a spiral null whose location lies within the separatrix. Note that the helicity density changes sign across the null points in both the magnetic field and fluid flow.

null point the EMHD approximation breaks down, first by large Larmor radius effects ($r_{ce} \geq B/|\nabla B|$), then collisional ($\nu_{ei} \geq \omega_c$) and inertial effects ($\omega \geq \omega_c$).

Magnetic reconnection must take place since the flux inside and outside the separatrix changes on the whistler time scale. Unlike in 2D reconnection geometries, no current sheets are formed near a 3D magnetic null point but instead there are distributed field-aligned and Hall currents on either side of the separatrix. These currents define the configuration of the magnetic null point, which is typically observed to be an improper non-potential spiral null, an example of which is shown in Figure 10a,b. The two views of the field lines transverse to (Figure 10a) and along (Figure 10b) the spine show that the fan is twisted and elliptical due to currents in the direction of the spine, and the normal of the fan is inclined with respect to the spine due to currents across the spine. The field expansion matrix $\mathbf{M}$, defined by the expansion $\mathbf{B} = \mathbf{M} \cdot \mathbf{r}$, has been determined experimentally. The helicity density changes sign across the fan.

It is interesting to note that during the decay a second pair of null points is created because the plasma carries away two dipole-like fields to either side of the loop antenna. Equally interesting is the observation of 3D null points in the current density or electron fluid velocity, $\mathbf{J} = -ne\mathbf{v}$. The null point of $\mathbf{J}$ lies inside the separatrix on axis slightly away from $\mathbf{B} = 0$. Figure 10c,d show $\mathbf{J}$-lines near the $\mathbf{J}$-null point along and across the spine which also identify it as a spiral null. Vorticity $\boldsymbol{\omega} = \nabla \times \mathbf{v}$ flows both along and across the spine. The null point in $\mathbf{v}$ is a consequence of the helicity properties of EMHD fields: The current helicity density $\mathbf{J} \cdot \mathbf{B}(\mathbf{r}, t)$ or cross-helicity density $\mathbf{v} \cdot \mathbf{B}(\mathbf{r}, t)$ changes sign when a whistler wave propagates in opposite directions along the net magnetic field which, in the presence of a null point, reverses direction.

A further manifestation of the $\mathbf{v} \times \mathbf{B}$ nonlinearity is a "rectification" of oscillating magnetic fields, which produces dc magnetic fields and harmonics [*Stenzel and Urrutia*, 1998].

4. SUMMARY AND CONCLUSIONS

Basic laboratory experiments have shown that helicity is a fundamental property of fields and currents in EMHD plasmas. Our main findings are that transient EMHD fields form vortex topologies, propagate in the whistler mode, have a unique sign of helicity depending on propagation direction, conserve helicity except for a sign change upon reflection, and become nonlinear when the wave field exceeds the background field.

The physics of EMHD is rarely considered in space physics, presumably because there are inadequate observational tools, i.e., no data. For solar magnetic fields the spatial and temporal resolution for field measurements near null points would have to be improved by many orders of magnitude to observe EMHD processes. Satellite measurements lack 3D spatial information which is essential to separate temporal from spatial variations. Nevertheless, it is obvious that EMHD physics enters all reconnection problems with magnetic neutral points, lines or sheets. In 2D reconnection, it has been the tradition to distinguish an outer ideal MHD region from an inner diffusion region [*Vasyliunas*, 1975]. However, the latter is dominated by EMHD physics on scale lengths where the ions are unmagnetized while the electrons are still magnetized ($r_{ce} < |B/\nabla B| < r_{ci}$). In this regime the magnetic field is decoupled from the ions, reconnects at the whistler speed, but exchanges no energy with electrons provided Ohm's law is given by the Hall effect.

Acknowledgments. The authors gratefully acknowledge support for this work by the National Science Foundation under grant PHY-9713240.

REFERENCES

Avinash, K, and J. B. Taylor, Relaxed states with plasma flows, *Comments Plasma Phys. Controlled Fusion*, *14*, 127, 1991.

Boozer, A. H., Magnetic helicity and dynamos, *Phys. Fluids B*, *5*, 2271, 1993.

Cornwall, J. M., Speculations on primordial magnetic helicity, *Phys. Rev. D*, *56*, 6146, 1997.

Drake, J. F., R. G. Kleva, and M. E. Mandt, Structure of thin current layers: Implication for magnetic reconnection, *Phys. Rev. Lett.*, *73*, 1251, 1994.

Goldstein, M. L., D. A. Roberts, and W. H. Matthaeus, Magnetohydrodynamic turbulence in the solar wind, in *Annual review of astronomy and astrophysics*, *33*, edited by G. Burbidge and A. Sandage, pp. 283-325, Annual Reviews, Palo Alto, CA, 1995.

Isichenko, M. B., and A. M. Marnachev, Nonlinear wave solutions of electron MHD in a uniform plasma, *Sov. Phys. JETP 66*, 702, 1987.

Ji, H., S. C. Prager, A. F. Almagri, J. S. Sarff, Y. Yagi, Y. Hirano, K. Hattori, and H. Toyama, Measurement of the dynamo effect in a plasma, *Phys. Plasmas*, *3*, 1935, 1996.

Kingsep, A. S., K. V. Chukbar, and V. V. Yankov, Electron magnetohydrodynamics, in *Reviews of Plasma Physics*, *16*, edited by B. B. Kadomtsev, pp. 243-291, Consultants Bureau, New York, 1990.

Kumar, A., and D. M. Rust, Interplanetary magnetic clouds, helicity conservation, and current-core flux-ropes, *J. Geophys. Res.*, *101*, 15667, 1996.

Ohkawa, T., Plasma current drive by injection of photons with helicity, *Comments on Plasma Phys. Contr. Fusion*, *12*, 165, 1989.

Ono, Y., Slow formation of field-reversed configuration by use of two merging spheromaks, *Fusion Technology*, *27*, 369, 1995.

Parks, G. K., Physics of Space Plasmas, Addison-Wesley Publ. Co., Redwood City, CA, 1991.

Parnell, C. E., J. M. Smith, T. Neukirch, and E. R. Priest, The structure of three-dimensional magnetic null points, *Phys. Plasmas*, *3*, 759, 1996.

Pevtsov, A. A., R. C. Canfield, and H. Zirin, Reconnection and helicity in a solar flare, *Astrophys. Journal*, *473*, 533, 1996.

Rousculp, C. L., R. L. Stenzel, and J. M. Urrutia, Inductive and space charge electric fields in a whistler wave packet, *Phys. Rev. Lett.*, *72*, 1658, 1994.

Rousculp, C. L., and R. L. Stenzel, Helicity injection by knotted antennas into electron magnetohydrodynamical plasmas, *Phys. Rev. Lett.*, *79*, 837, 1997.

Ruzmaikin, A. A., Redistribution of magnetic helicity at the Sun, *Geophys. Res. Lett.*, *23*, 2649, 1996.

Stenzel, R. L., and J. M. Urrutia, Generation of dc magnetic fields by rectifying nonlinear whistlers, *Phys. Rev. Lett.*, *81*, 2064, 1998.

Taylor, J. B., Relaxation of toroidal plasma and generation of reverse magnetic fields, *Phys. Rev. Lett.*, *33*, 1139, 1974.

Urrutia, J. M., R. L. Stenzel, and C. L. Rousculp, Pulsed currents carried by whistlers. III: Magnetic fields and currents excited by an electrode, *Phys. Plasmas*, *2*, 1100, 1995.

Urrutia, J. M., and R. L. Stenzel, Pulsed currents carried by whistlers. VI: Nonlinear effects, *Phys. Plasmas*, *3*, 2589, 1996.

Urrutia, J. M., and R. L. Stenzel, Pulsed currents carried by whistlers. IX: In-situ measurements of currents disrupted by plasma erosion, *Phys. Plasmas*, *4*, 36, 1997.

Vasyliunas, V. M., Theoretical models of magnetic field line merging, 1, *Rev. Geophys. Space Phys.*, *13*, 303, 1975.

Yamada, M., H. Ji, S. Hsu, T. Carter, R. Kulsrud, Y. Ono, and F. Perkins, Identification of Y-shaped and O-shaped diffusion regions during magnetic reconnection in a laboratory plasma, *Phys. Rev. Lett.*, *78*, 3117, 1997.

R. L. Stenzel, J. M. Urrutia, and M. C. Griskey, Department of Physics, University of California, Los Angeles, 90095-1547. (e-mail: stenzel@physics.ucla.edu; urrutia@ucla.edu; griskey@physics.ucla.edu

The Role of Helicity in Magnetic Reconnection: 3D Numerical Simulations

Spiro K. Antiochos and C. Richard DeVore

Naval Research Laboratory, Washington, D. C.

We demonstrate that conservation of global helicity plays only a minor role in determining the nature and consequences of magnetic reconnection in the solar atmosphere. First, we show that observations of the solar coronal magnetic field are in direct conflict with Taylor's theory. Next, we present results from three-dimensional MHD simulations of the shearing of bipolar and multi-polar coronal magnetic fields by photospheric footpoint motions, and discuss the implications of these results for Taylor's theory and for models of solar activity. The key conclusion of this work is that significant magnetic reconnection occurs only at very specific locations and, hence, the Sun's magnetic field cannot relax completely down to the minimum energy state predicted by conservation of global helicity.

1. INTRODUCTION

Magnetic reconnection has long been invoked as the physical mechanism underlying much of solar activity. For example, reconnection is believed to be the process driving many of the observed dynamic solar events ranging from spicules to the largest and most energetic manifestations of solar activity, coronal mass ejections (CME) and eruptive flares. In spite of the long and intensive study of reconnection in the solar atmosphere, the process is still not well understood, especially in three dimensions. One of the main difficulties in developing a comprehensive understanding is that reconnection may take on different forms depending on the details of the physical situation. Consequently, any theory that can provide some unifying insight into the nature of reconnection would be of great benefit to understanding many aspects of solar activity. This is the compelling motivation behind studies of magnetic helicity. Since magnetic helicity is believed to be conserved during re-

connection in general, the hope is that helicity conservation may allow one to determine the final state of a reconnecting system without having to calculate the detailed dynamics of the evolution. Helicity conservation may also be able to provide some valuable information on the dynamics. In this paper we argue, however, that helicity plays a negligible role in determining the evolution of reconnecting magnetic fields in the Sun's corona. It should be emphasized that by the term "helicity", we refer in this paper solely to the global relative helicity [*e.g., Berger*, 1985], which defines a single invariant. We are not referring to the helicity density which defines an infinite set of possible invariants. Only the global helicity is believed to be conserved during reconnection.

The basic theory for using helicity conservation to determine the evolution of magneto-plasmas has been developed by Taylor [1986]. For Taylor's theory to be applicable to the solar corona, three key statements must be true. First, the helicity (global) is conserved during reconnection. Our numerical simulations agree well with this statement — the higher the magnetic Reynolds number of the simulation, the better the agreement. Second, helicity is the only topological quantity that is generally conserved during reconnection. We believe that this assumption is also true, but our sim-

Magnetic Helicity in Space and Laboratory Plasmas
Geophysical Monograph 111
This paper not subject to U.S. copyright
Published in 1999 by the American Geophysical Union

ulations cannot test it, because they all begin with a potential field in which a simple shear or twist flow is imposed on the photospheric boundary. There are no knots or disconnected flux in the coronal field, and no braiding motions or higher-order topologies produced by the boundary flows. Since the complete topology of our fields is contained in the helicity density, it is unlikely that there are any global topological invariants other than helicity available to be conserved.

It appears, therefore, that the first two requirements for Taylor's theory are valid for our simulations, and probably for the corona as well. The final requirement is that *complete reconnection* occurs, *i.e.*, the reconnection continues until the magnetic energy achieves its lowest possible state. Note that this statement does not say anything about helicity, it is actually a model for reconnection. Unfortunately, this statement is completely wrong for our simulations and, we believe, also for the Sun.

The physical reason for the failure of complete reconnection in the corona is that it requires the formation of numerous current sheets, or sheet-like current structures. But we, and others, have found from both 2.5D and 3D simulations that due to photospheric line-tying, current sheets do not form easily in the corona [e.g., *Mikic, Schnack, and Van Hoven*, 1989; *Dahlburg, Antiochos, and Zang*, 1991; *Karpen, Antiochos, and DeVore*, 1990]. It is instructive to note that the Taylor theory is closely related to Parker's nonequilibrium theory for coronal heating [*Parker*, 1972; 1979]. The nonequilibrium theory also proposes that in a 3D system, current sheets will form spontaneously throughout the coronal volume. But, there have been numerous simulations testing nonequilibrium [*e.g., Van Ballegooijen*, 1985; *Mikic, Schnack, and Van Hoven*, 1989; *Dahlburg, Antiochos, and Zang*, 1991], and to our knowledge, no simulation produces these current sheets. This does not mean that current sheets cannot form or that reconnection does not occur in the corona. Many simulations find that current sheets readily form at magnetic separatrices [*e.g., Karpen, Antiochos, and DeVore*, 1995; 1996; 1998], and intense current concentrations do form at those locations where the photospheric motions produce exponentially growing gradients in footpoint displacements, in particular, at stagnation points of the flow [*e.g., Van Ballegooijen*, 1986; *Mikic, Schnack, and Van Hoven*, 1989; *Strauss*, 1993; *Antiochos and Dahlburg*, 1997]. But since reconnection occurs only at these very specific locations, it is far from complete, and Taylor's theory cannot be used to determine either the final state of the field or its evolution. We assert, therefore, that while the global helicity is conserved, it plays little role in determining the corona's dynamics and evolution.

This conclusion is also evident from observations. The Taylor theory would predict that the coronal field evolves towards a linear force-free field. For an infinite system like the corona, the only linear force-free field with finite energy is the field which is current-free in any finite volume [*Berger*, 1985]. Therefore if the theory held, the coronal field would evolve *via* reconnection to the potential field, in which case there would be no need for CMEs or eruptive flares. It may be argued that the Taylor theory should not be applied to the corona as a whole, since the helicity is not uniquely defined for an infinite system. But, in fact, the Taylor prediction for an infinite system is completely sensible. If reconnection could proceed freely, indeed it would be energetically favorable for the field to transfer all its shear and twist to the outermost field lines that extend toward infinity, such as the field lines at the poles. By transferring all the shear/twist to the longest field lines, the field conserves its helicity, but brings its energy down to the potential field value. The only problem with this type of evolution for the solar corona is that it is never observed.

One could argue, however, that a Taylor process may occur in some small portion of the corona, such as an active region, in which case the field should evolve to a linear force-free state inside this bounded domain. But this prediction also disagrees with observations. The canonical result from vector magnetograms and from H_α observations is that the field is strongly sheared near photospheric polarity-reversal lines ("neutral" lines), and unsheared or weakly sheared far from these lines [*e.g., Gary et al.*, 1987; *Falconer et al.*, 1997]. (By shear we mean that the field lines appear to be greatly stretched out along the reversal line.) We show below that such a shear distribution can explain the formation of prominences/filaments, which lends strong support to the observations. But this observed localization of the shear is *not* compatible with a linear force-free field.

In order to demonstrate this point, consider a simple analytic model for the field. Take the active region to consist of a 2.5D linear force-free field arcade:

$$\vec{B} = \nabla \times (A(y,z)\hat{x}) + B_x(y,z)\hat{x}. \quad (1)$$

Since this field must satisfy, $\nabla \times \vec{B} = \lambda \vec{B}$, where λ is a constant, we find that $B_x = \lambda A$, and the force-free equation reduces to the usual Helmholtz form, $\nabla^2 A + \lambda^2 A = 0$. One possible solution is:

$$A = \cos(ky)\exp(-\ell z), \quad (2)$$

where the wavenumbers k, ℓ, and λ are related by, $\lambda^2 = k^2 - \ell^2$. We have chosen the form of the flux function in Equation (2) so that it corresponds to a bipolar arcade with a photospheric polarity reversal line at $y = 0$, and a width $ky = \pi$ (this periodic solution actually corresponds to an infinite set of arcades.)

If the wavelengths in the vertical and horizontal direction are equal, $\ell = k$, then $\lambda = 0$, and the solution reduces to the potential field. However, if the vertical wavelength becomes larger than the horizontal one $\ell < k$ (we expect the force-free field to inflate upward), then the solution corresponds to a field with finite shear, $B_x \neq 0$. Assuming that our bipolar arcade is at disk center, then the observed shear of the field at the photosphere would be given by the angle, $\theta = \arctan(B_x/B_y)$. If the field is potential then $B_x = 0$, which implies that $\theta = 0$, and the field lines are perpendicular to the polarity reversal line (the x axis). For the nonpotential case we find from Equations (1) and (2) that $B_y = dA/dz = -\ell A$. Hence, $\theta = -\arctan(\lambda/\ell)$. The shear is constant throughout the region rather than being localized near the polarity-reversal line. Although this result has been derived for only one family of solutions, it seems likely to hold true in general. A linear force-free field must has a constant ratio of electric current magnitude to magnetic field magnitude, and hence must have shear everywhere. But a broad shear distribution is in total disagreement with numerous observations of the solar field [e.g., Gary et al., 1987; Schmieder et al., 1996].

We conclude, therefore, that complete reconnection does not occur even in small regions of the corona, and that helicity conservation is of limited usefulness for determining the structure and evolution of the coronal field. We verify this conclusion with large-scale 3D numerical simulations in the following sections. The goal of our simulations is to understand the formation and eruption of solar prominences and the accompanying CME, but as will be demonstrated below, the simulations also address the issues of the role of helicity conservation in magnetic reconnection and the applicability of the Taylor theory to the corona.

2. SIMULATIONS OF BIPOLAR FIELDS

The first simulation concerns the formation of prominences. Solar prominences or filaments consist of huge masses of cool ($\sim 10^4$ K), dense ($\sim 10^{11}$ cm^{-3}) material apparently floating high up in the hot ($\sim 10^6$ K), tenuous ($\sim 10^{-9}$ cm^{-3}) corona [e.g., Priest, 1989]. Prominences reach heights of over 10^5 km, which is approximately three orders of magnitude greater than the gravitational scale height of the cool material. Hence, the most basic question concerning prominences is the origin of their gravitational support. It must be due to the magnetic field; the field lines in the corona must have hammock-like geometry so that high-density plasma can be supported stably in the hammock [Priest, 1989].

A characteristic feature of all prominences is that they form over photospheric polarity-reversal lines which exhibit strong shear. Since many prominences are also observed to be very long compared to their width or height, 2.5D models for their magnetic structure (a magnetic arcade) have usually been considered. Both numerical simulations and analytic theory showed, however, that 2.5D models of a sheared bipolar arcade cannot produce field lines with the necessary dips to support prominence material [Klimchuk, 1990; Amari et al., 1991]. This led many to consider more complicated topologies involving multi-polar systems or topologies with flux disconnected from the photosphere, the so-called flux ropes [e.g., Priest and Forbes, 1990; van Ballegooijen and Martens, 1990].

We have shown, however, that the lack of dipped field lines is only an artifact of assuming translational symmetry, and that a sheared 3D bipolar field readily develops the correct geometry to support prominences [Antiochos, Dahlburg, and Klimchuk, 1994; Antiochos, 1995]. Our previous results were based on a 3D static equilibrium code that computed the force-free field in the corona given the connectivity of the field lines at the photosphere. Here we present results from recent fully time-dependent 3D simulations of photospheric shearing of a bipolar field. Since we include the dynamics, these simulations also address the issues of current-sheet formation, reconnection, and eruption.

The code uses a highly-optimized parallel version of our 3D flux-corrected transport algorithms to solve the ideal MHD equations in a finite-volume representation. The code is thoroughly documented and available on the WEB under the auspices of NASA's HPCC program (see http://www.lcp.nrl.navy.mil/hpcc-ess/). The computational domain consists of the rectangular box, $-20 \leq x \leq 20$, $-4 \leq y \leq 4$, $0 \leq z \leq 8$. We use a fixed, but very large non-uniform Cartesian mesh of $462 \times 150 \times 150$ points. The initial magnetic field is that due to a point dipole located at (0,0,-2) and oriented along the y-axis, so that the polarity reversal line at the photospheric plane (z = 0) corresponds to the x-axis. As boundary conditions, we impose line-tying

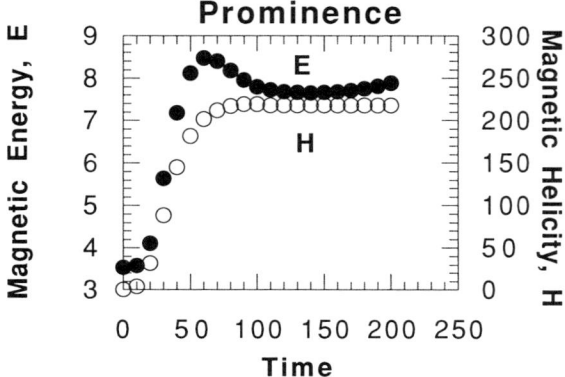

Figure 1. The total magnetic energy and magnetic helicity (relative) of a bipolar field that is sheared for 100 Alfven crossing times and then allowed to relax for another 100 Alfven times.

with an assumed shear flow and no flow-through conditions at the bottom, and zero gradient conditions on all quantities at the sides and top. The initial plasma density is uniform, and we neglect the effects of both plasma pressure and gravity in this simulation, corresponding to a zero beta approximation. Note, however, that the plasma is fully compressible and all Alfven waves are included in the calculation,

We shear the field by imposing a flow at the photosphere that is localized near the polarity reversal line. The shear vanishes for $|y| > 1$, and for $|y| \leq 1$ it has the form:

$$V_x = (8\pi/\tau)\sin(\pi t/\tau)\sin(\pi y), \qquad (3)$$

where the time scale for achieving the maximum shear $\tau = 100$. Even though we performed the simulations on the latest architecture massively-parallel machines, it is still not possible to use observed solar values for the shear properties. Our shear extends over roughly half the width of the strong field region on the photosphere, wider than is observed, and the average shearing velocity is approximately 10% the Alfven speed in the strong field region, rather than the 1% typical of the Sun. But even with these limitations, dipped field lines form readily in the corona.

Plate 1 shows the magnetic configuration halfway through the shearing, at $t = \tau/2$. It is evident that the strongly sheared field lines have dipped central portions. The dips form as a result of the balance of forces between the increased magnetic pressure of the low-lying sheared field lines and the increased tension of the unsheared overlying field. Since the unsheared flux is strongest at the center of the system, the downward tension force is strongest there, producing a local minimum in the height of the sheared flux. Also shown in the Plate is the half-maximum iso-surface of electric current magnitude. As expected, the current is concentrated where the gradient of the shear is largest, in the boundary between the sheared and unsheared field.

The field of Plate 1 reproduces all the basic observed features of prominences; hence, we conclude that the magnetic structure of solar prominences and filaments is simply that of a sheared 3D field. It is tempting to conjecture that continued shearing of this field eventually leads to eruption. This is the basic hypothesis of the tether-cutting model, which proposes that reconnection of the sheared field either with itself or with the unsheared flux destroys the force balance between sheared and unsheared flux, thereby allowing the field to erupt outward explosively [*Sturrock*, 1989; *Moore and Roumeliotis*, 1992]. Note that the tether-cutting model is physically similar to the Taylor theory since both hypothesize that reconnection transfers shear from inner to outer field lines.

Our simulation, however, shows no evidence for either tether-cutting or a Taylor process. We continued the shearing up to $t = 100$, twice as far as shown in Plate 1. We then set the photospheric velocity to zero, and let the system relax for another 100 Alfven times. The total magnetic energy and helicity are shown in Figure 1. The system appears to achieve a stable equilibrium with negligible loss of either energy or helicity. A key point is that the system appears stable, even though some "reconnection" does occur. By $t = 80$ the imposed boundary shear is so extreme that even with our large grid we cannot resolve it numerically, which produces "current-sheets" in the corona, in particular, the current structure seen in Plate 1. As a result, reconnection (or perhaps more appropriately diffusion) occurs, and helical field lines begin to appear at this time. However, the appearance of helical lines is confined to the regions of strongest shear, and does not propagate outward as would be necessary for tether-cutting or for a Taylor process. We conclude, therefore, that our simulation rules out both tether-cutting and the Taylor theory as viable models for the corona.

3. SIMULATIONS OF MULTI-POLAR FIELDS

There are two fundamental reasons for the lack of eruption in the simulation described above. First, line-tying inhibits magnetic reconnection in a topologically smooth field such as a simple bipole. Second, eruption of the low-lying sheared flux requires the overlying unsheared field to open as well, but the Aly-Sturrock limit

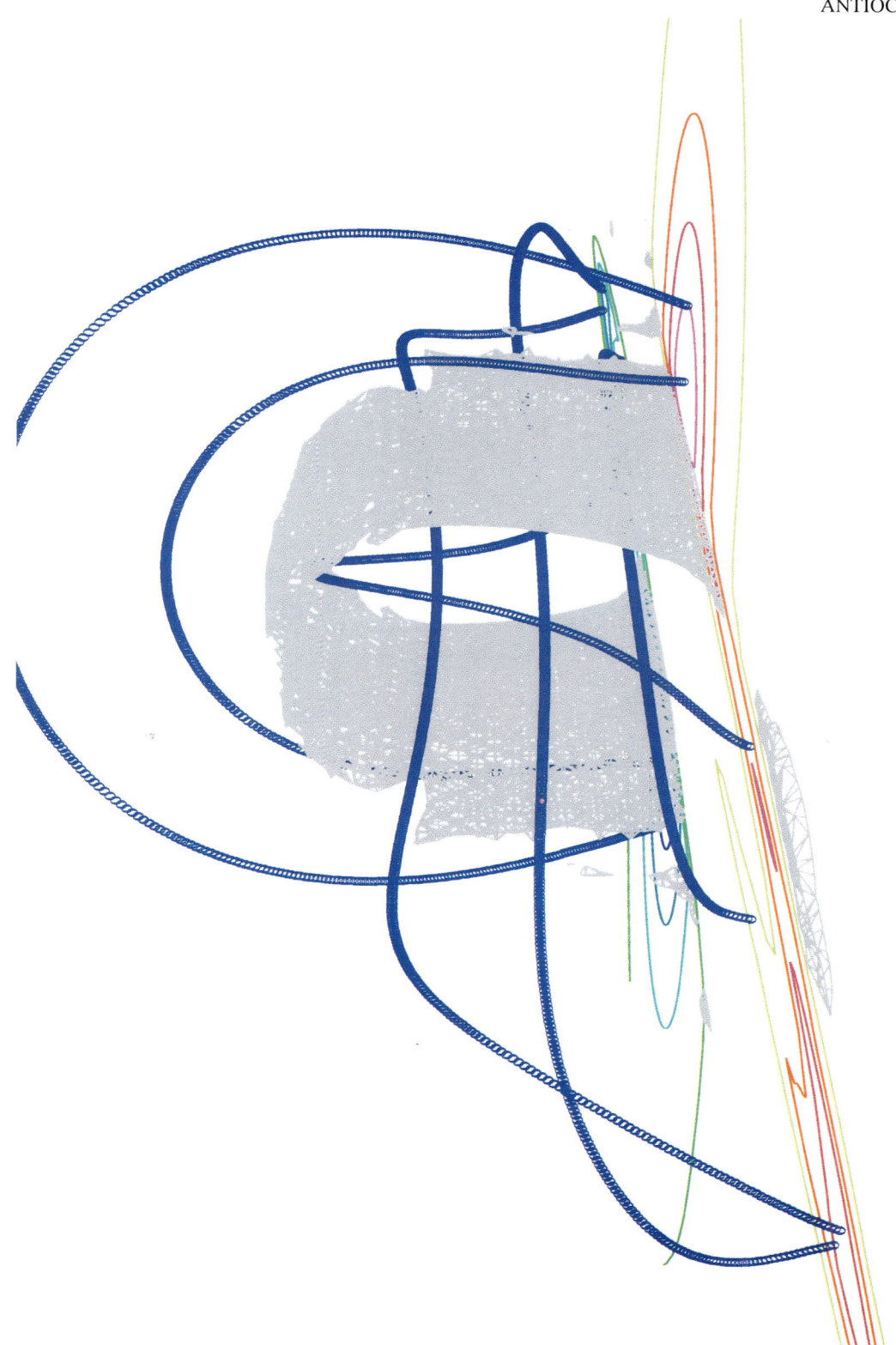

Plate 1. Structure of a bipolar magnetic field that has been sheared by footpoint motions at the photosphere, bottom plane in the Plate. Contours of normal-magnetic-field magnitude are plotted on the bottom plane. Four field lines with footpoints in the shear region and two field lines with footpoints outside this region are shown. Also plotted is the iso-surface of electric current magnitude at half-maximum.

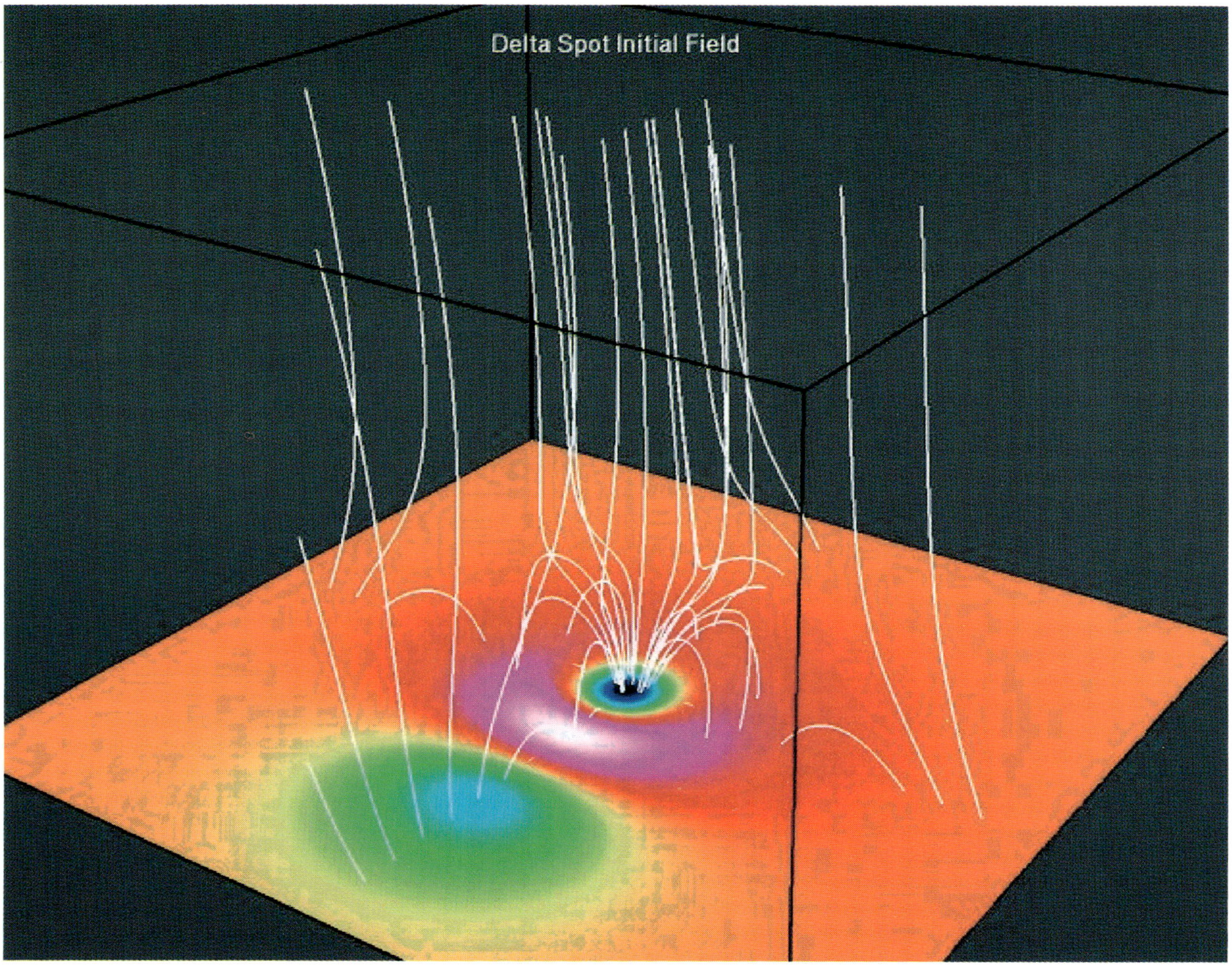

Plate 2. Evolution of a delta-sunspot magnetic field that is sheared by footpoint motions. **2a.** The initial current-free magnetic configuration. **2b.** The field lines after a shear of $\sim 2\pi$. **2c.** The field lines after a shear of $\sim 4\pi$.

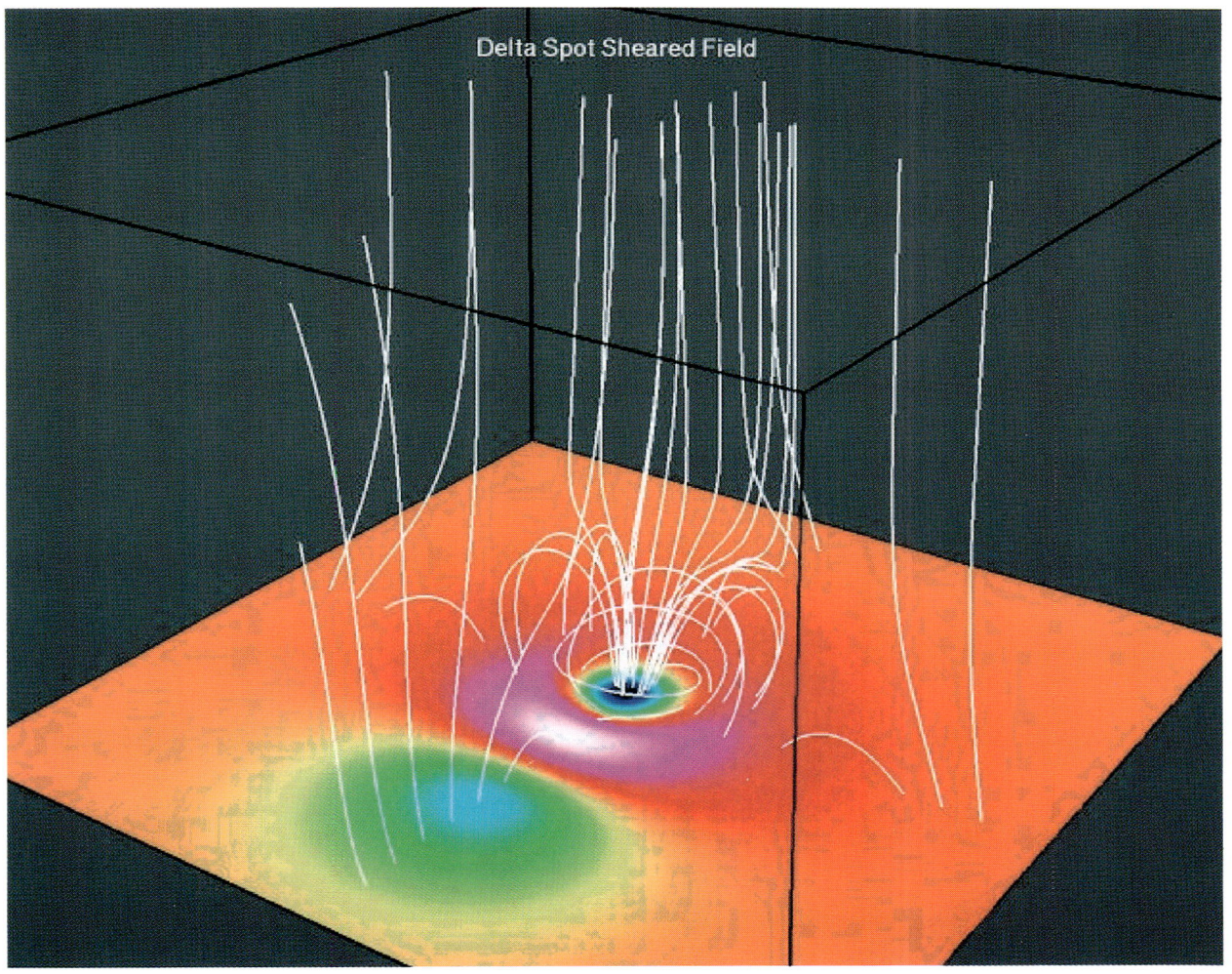

Plate 2b

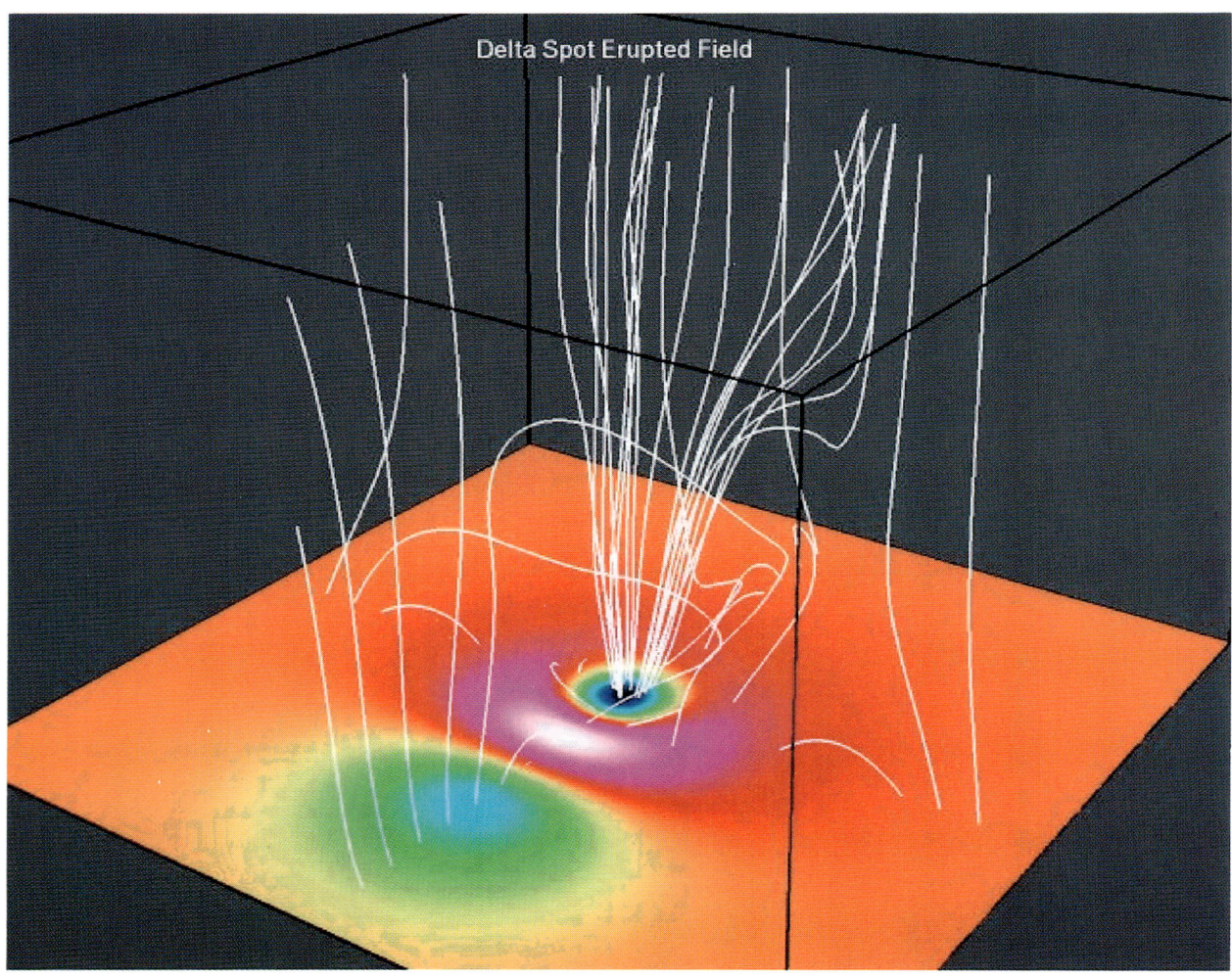

Plate 2c

implies that no closed configuration can have sufficient energy to reach this open state [Aly, 1984; 1991; Sturrock, 1991].

We have argued that a multi-polar magnetic topology overcomes both these problems, and have proposed a "breakout" model for prominence eruptions and coronal mass ejections [Antiochos, 1998; Antiochos, DeVore, and Klimchuk, 1999]. Line-tying does not inhibit current-sheet formation at the separatrix surfaces between flux systems, and these current sheets can lead to sustained reconnection at separator lines [Karpen, Antiochos, and DeVore, 1995; 1996; 1998]. Furthermore, a multi-flux topology makes it possible to transfer the unsheared overlying flux to neighboring flux systems, thereby allowing the sheared field to erupt outward while keeping the unsheared flux closed. This allows the system to erupt explosively while still satisfying the Aly-Sturrock energy limit [Antiochos, DeVore, and Klimchuk, 1999].

We show below results from our first 3D simulation of the breakout model. The simulation domain in this case consists of the region: $-3 \leq x \leq 3$, $-3 \leq y \leq 3$, $0 \leq z \leq 3$, with a fixed, non-uniform grid of $166 \times 166 \times 86$ points. The initial magnetic field is that due to three point dipoles: one located at (0, 25, -50), with magnitude unity, and pointing in the $+y$ direction; another located at (0, 1, -1), with magnitude 10, and pointing in the $-y$ direction; and the third at (0, 0, -0.5), with magnitude 10 and pointing in the $-z$ direction. The initial density and pressure were chosen so that the average plasma beta near the base is less than 0.1. (This simulation did not use the zero-beta approximation.)

Plate 2a. shows the initial potential field of the simulation. The magnetic topology consists of four flux systems due to four distinct polarity regions at the photosphere. There are two toroidal separatrix surfaces that define the boundaries of these flux systems, and their intersection in the corona defines a separator line, along which rapid reconnection can occur [Antiochos, 1998]. This configuration corresponds to a so-called delta-sunspot region. We impose similar boundary conditions as in the previous simulation, and apply a shearing motion localized near the circular polarity-reversal line of the delta-spot located at the center of the bottom plane. The shear is such that it produces a rotation of $\sim 2\pi$ over a time interval of 100 Alfven times.

Plate 2b. shows the effect of this shear. All field lines shown in Plates 2 and 2b are traced from exactly the same set of positions at the photosphere. It is evident that field lines with footpoints near the polarity reversal line have been twisted through almost a full

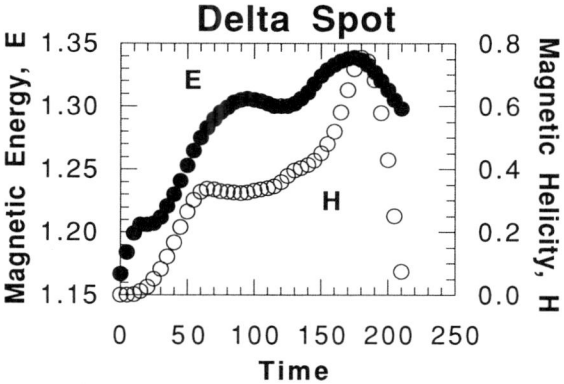

Figure 2. Total energy and helicity of a delta-sunspot field that undergoes 2 shearing phases, each of duration 100 Alfven crossing times.

rotation, (this sheared flux system corresponds to the sheared prominence bipole of Plate 1). A careful examination of the field lines in 2b reveals that some of the unsheared delta-spot flux overlying the sheared flux has become open, i.e., it extends to the top boundary of the simulation box rather than closing down over the sheared field. This transfer occurs as a result of reconnection between the delta-spot flux and the neighboring open flux system.

The effect of another 2π rotation is shown in Plate 2c. Almost all the delta spot flux in now open. We show in Figure 2 the total magnetic energy and helicity of the system, equivalent to Figure 1. Near the end of the second shearing phase there is clearly a burst of impulsive energy release and helicity ejection through the top of the system. The helicity within the simulation volume decreases much faster than the energy, because of the eruption. Therefore, the evolution of our simulated corona is the exact opposite of a Taylor process!

Of course, this claim is somewhat overstated because we expect that, in fact, the helicity of the whole system (including the erupted field) stays constant, whereas the magnetic energy of the whole system decreases. It is interesting to note, however, that since solar telescopes observe only a finite coronal volume, eruptive opening of the magnetic field in that volume implies that the observed helicity decreases to zero, but the observed magnetic energy asymptotes to the open field value, which is generally well above the potential field value. In this sense, an anti-Taylor process is an appropriate approximation for the evolution of coronal regions observed during an eruption.

A key result of our 3D delta-spot simulation is that that even though a great deal of reconnection occurs, it is far from complete. The reconnection is confined to the separator line between the sheared and unsheared regions. Consequently, the field never approaches a linear force-free field. Instead it evolves toward an open state in which the currents are concentrated in a thin sheet. This also is opposite to what is expected for a Taylor process.

In summary, we conclude that while helicity is conserved and may well be the only topological quantity that is generally conserved during reconnection, the actual amount of reconnection in the Sun's corona is determined by the detailed magnetic topology of the particular region. Consequently, the global helicity by itself yields little information on coronal evolution.

Acknowledgments. This work was supported in part by NASA and ONR.

REFERENCES

Aly, J. J., On some properties of force-free magnetic fields in infinite regions of space, *Astrophys. J., 283,* 349, 1984.

Aly, J. J., How much energy can be stored in a three-dimensional force-free magnetic field?, *Astrophys. J., 374,* L61, 1991.

Amari, T., et al., The creation of the magnetic environment for prominence formation in a coronal arcade, *Astron. Astrophys., 241,* 604, 1991.

Antiochos, S. K., Dahlburg, R. B., and Klimchuk, J. A., The magnetic field of solar prominences, *Astrophys. J., 420,* L41, 1994.

Antiochos, S.K., Solar Drivers of Space Weather, *Astron. Soc. Pacific Conf. Series, 95,* 1, 1996.

Antiochos, S. K. and Dahlburg, R. B., The implications of 3D for solar MHD modelling, *Solar Phys., 174,* 5, 1997.

Antiochos, S.K., The magnetic topology of solar eruptions, *Astrophys. J., 502,* L181, 1998.

Antiochos, S. K., DeVore, C. R., and Klimchuk, J. A., A model for solar coronal mass ejections, *Astrophys. J., 510,* in press, 1999.

Berger, M. A., Structure and stability of constant-α force-free fields, *Astrophys. J. Suppl., 59,* 433, 1985.

Dahlburg, R. B., Antiochos, S. K., and Zang, T. A., Dynamics of solar coronal magnetic fields, *Astrophys. J., 383,* 420, 1991.

Falconer, D. A., et al., Neutral-Line Magnetic Shear and Enhanced Coronal Heating in Solar Active Regions, *Astrophys. J., 482,* 519, 1997.

Gary, G. A., Moore, R. L., Hagyard, M. J., Haisch, B. M., Nonpotential features observed in the magnetic field of an active region, *Astrophys. J., 314,* 782, 1987.

Karpen, J. T., Antiochos, S. K., and DeVore, C. R., On the formation of current sheets in the solar corona, *Astrophys. J., 356,* L67, 1990.

Karpen, J. T., Antiochos, S. K., and DeVore, C. R., The role of magnetic reconnection in chromospheric eruptions, *Astrophys. J., 450,* 422, 1995.

Karpen, J. T., Antiochos, S. K., and DeVore, C. R., Reconnection driven current filamentation in solar arcades, *Astrophys. J., 460,* L73, 1996.

Karpen, J. T., Antiochos, S. K., and DeVore, C. R., Dynamic responses to magnetic reconnection of solar arcades, *Astrophys. J., 495,* 491, 1998.

Klimchuk, J. A., Shear-induced inflation of coronal magnetic fields, *Astrophys. J., 354,* 745, 1990.

Mikić, Z., Schnack, D. D., and Van Hoven, G., Creation of current filaments in the solar corona, *Astrophys. J., 338,* 1148, 1989.

Moore, R. L. and Roumeliotis, G., Triggering of eruptive flares: destabilization of the preflare magnetic field configuration, in *Eruptive Solar Flares,* edited by Z. Svestka, B. V. Jackson, and M. E. Machado, p. 69, Springer, Berlin, 1992.

Parker, E. N., Topological dissipation and the small-scale fields in turbulent gases, *Astrophys. J., , 174,* 499, 1972.

Parker, E. N., Cosmical Magnetic Fields, Clarendon Press, Oxford, U.K., 1979.

Priest, E. R., Dynamics and structure of quiescent solar prominences, Kluwer, Dordrecht, 1989.

Priest, E. R. and Forbes, T. G., Magnetic field evolution during prominence eruptions and two-ribbon flares, *Solar Phys., 126,* 319, 1990.

Schmieder, B., Demoulin, P., Aulanier, G., and Golub, L. Differential magnetic field shear in an active region, *Astrophys. J., 467,,* 881, 1996.

Strauss, H., Fast three dimensional driven reconnection, *Geophys. Res. Lett., 20,* 325, 1993.

Sturrock, P. A., The role of eruption in solar flares, *Solar Phys., 121,* 387, 1989.

Sturrock, P. A., Maximum energy of semi-infinite magnetic field configurations, *Astrophys. J., 380,* 655, 1991.

Taylor, J. B., Relaxation and magnetic reconnection in plasmas, in *Rev. Mod. Phys., 58,* 741, 1986.

van Ballegooijen, A. A., Electric currents in the solar corona and the existence of magnetostatic equilibrium, *Astrophys. J., 298,* 421, 1985.

van Ballegooijen, A. A., Cascade of magnetic energy as a mechanism of coronal heating, *Astrophys. J., 311,* 1001, 1986.

van Ballegooijen, A. A. and Martens, P. C. H., Magnetic fields in quiescent prominences, *Astrophys. J., 361,* 283, 1990.

S. K. Antiochos, Code 7675, Naval Research Lab, Washington, DC 20375-5352. (e-mail: antiochos@nrl.navy.mil

C. R. DeVore, Code 6440, Naval Research Lab, Washington, DC 20375. (e-mail: devore@nrl.navy.mil)

Helicity and Reconnection in the Solar Corona: Observations

Richard C. Canfield and Alexei A. Pevtsov

Department of Physics, Montana State University, Bozeman

Solar coronal magnetic fields are twisted on all observed spatial scales, though studies have emphasized active regions and their complexes. The overall twist of active region magnetic fields inferred from coronal images exhibits a hemispheric chirality preference. This twist has been shown to be consistent statistically with vector magnetic field observations in the photosphere, and is arguably of sub-photospheric origin. The manner in which active regions are observed to connect with one another through the corona implies that it is favorable, from the point of view of current closure and energetics, for active regions to reconnect with other of the same chirality. Transport of twist through reconnection from one coronal flux system to another has been observed in flares, jets, and large-scale eruptions, presumably associated with the shedding of helicity generated by the solar dynamo. Finally, X-ray observations show that structures that are significantly twisted are closely associated with coronal mass ejections.

1. INTRODUCTION

The striking coronal images from the the Soft X-ray Telescope [Tsuneta et al., 1991] on the Yohkoh mission [Ogawara et al., 1991] call attention to the topology of coronal magnetic fields. Figure 1 shows a striking equatorial array of S-shaped coronal structures within and between several bright active regions. We use the term sigmoidal [Rust and Kumar, 1996] to identify such structures of S or inverse-S form. Everywhere in the corona except within the negligible volumes of current sheets, the thermal/magnetic energy ratio β is small. Hence magnetic terms dominate over gravitational and pressure gradient terms in force balance, and the magnetic induction **B** satisfies the force-free field equation $\nabla \times \mathbf{B} = \alpha \mathbf{B}$. Sigmoidal structure can be quantified in this approximation, using α as a parameter.

Magnetic Helicity in Space and Laboratory Plasmas
Geophysical Monograph 111
Copyright 1999 by the American Geophysical Union

Models of force-free fields can be compared to measurements of sigmoidal structures to infer values of α, as a measure of twist. Although force-free fields can be modeled quite generally from photospheric magnetogram data using sophisticated computational methods [Jiao et al., 1997; McClymont et al., 1997], these techniques are too demanding computationally to apply to large datasets. Two approximate models have been developed for the latter purpose.

Rust and Kumar [1996] modeled sigmoidal structures, in the force-free approximation, as helically kinked flux ropes, shown in Figure 2. They measured the amplitude of the twist of the field lines by the ratio of the major radius R to the axial length of one turn of the helix L, which they parameterized by $k = 2\pi/L$. Values of both R and L can be derived from observations, taking care to avoid projection effects. In this model, α is related simply to R and the constant $x_0 = 2.4$, the first zero of the Bessel function of order 0, by $\alpha R = x_0$.

Pevtsov, Canfield and McClymont [1997] modeled the projection of three-dimensional constant-α force-free fields of a simple bipole, as shown in Figure 3. They

Figure 1. Yohkoh SXT X-ray image of the solar corona, showing magnetic flux systems and S-shaped sigmoidal structures within them. Such structures trace isolated magnetic field lines because thermal conduction along field lines is so much greater than that across it.

Table 1. Distribution of coronal sigmoids by hemisphere

	Forward S	Inverse S	Total
Rust and Kumar [1996]			
Northern	4	24	28
Southern	40	12	52
Cross-equator	12	11	23
Total	56	47	103
Pevtsov and Canfield [1998]			
Northern	32	47	79
Southern	70	33	103
Total	102	80	182

used a two-dimensional force-free-field model to derive the relationship between the crossing angle γ that the central sigmoidal field lines make to the line joining the conjugate polarities, separated by distance L. In this model $\alpha = (\pi/L) \sin \gamma$.

Neither of these models [Rust and Kumar, 1996; Pevtsov et al., 1997] is sufficiently robust to be useful for quantitative studies; the former assumes that sigmoidal structures can accurately be represented in terms of Bessel functions of order $m = 0$ and 1, and the latter uses a two-dimensional function which is just an approximation to the projection of three-dimensional structures. Appropriate improvements would enable a quantitative understanding of the relationship between currents measured in the photosphere and those present the corona, as well as values of α associated with magnetic stability.

2. HEMISPHERIC CHIRALITY PREFERENCE

The lowest-order quantity that can be derived from the sigmoidal structures is the chirality, or handedness, of the helical structures that give rise to them. Rust and Kumar [1996] and Pevtsov and Canfield [1998] used Yohkoh SXT coronal X-ray images to identify sigmoidal structures. The results of these two studies are compared in Table 1. When divided along simple hemispheric lines, the two studies are consistent: 60 – 70% of such brightenings in the Northern hemisphere are of inverse-S form, and in the Southern hemisphere, of forward-S form. It is important for subsequent interpretation to note that the preference is weak; not all of the sigmoids in each hemisphere have the same chirality. It is also useful to think of a simple memory device: those shaped like the letter S occur in the South, and those shaped like an inverse-S, which look like the letter N, in the North. Figure 3 shows that S shapes correspond to positive α values.

How is this hemispheric chirality rule related to that found in the photospheric magnetic fields of active regions [Pevtsov and Canfield, 1999]? Pevtsov, Canfield, and McClymont [1997] used a dataset of regions that had been observed both in the corona and in the photosphere. The coronal twist α_c was determined from the crossing angle γ. The photospheric twist α_p was determined from photospheric vector magnetograms. Figure 4 shows the measurements. The data set consists of only 44 active regions, since only they showed clear sigmoids, i.e. the angle γ was significantly non-zero; that is why there are no data points for small α_c values. The result is that such regions have the same sign of α in the photosphere and the corona in 39 of the 44 cases (90%), and the opposite sign in only 5 cases (10%). The reason for those of opposite sign is obvious when the images are examined in detail; both S and inverse-S structures are sometimes observed in a single active region.

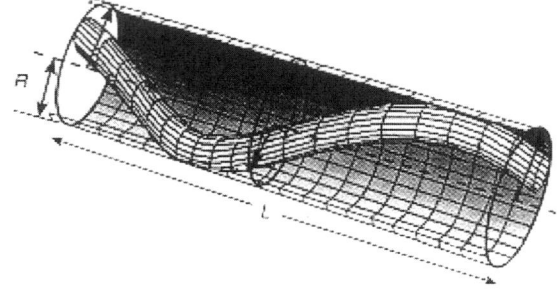

Figure 2. Model of a kinked ($m = 1$ mode) flux rope used to quantify the sigmoidal shape of coronal structures. By permission [Rust and Kumar, 1996].

Hα filaments in active regions [Rust, 1999] also show a hemispheric chirality rule, which is about as weak as that for coronal sigmoids in active regions. Ambiguities in the interpretation of the prominences images prevent a fully unambiguous inference of their chirality. If prominence material collects in concave-upward portions of coronal field lines, as is commonly believed, then prominences show the same hemispheric chirality rule as coronal loops do [Rust, 1999]. It would be very surprising if it were otherwise, given the largely force-free and therefore highly nonlocal nature of the magnetic field of the corona.

Twist has been studied in the corona on spatial scales larger than active regions, using Yohkoh SXT coronal images [Sandborgh et al., 1998] such as that in Figure 1. On the basis of the connections that can be seen in those images, [Sandborgh et al., 1998] determined the boundaries (separatrices) between flux systems. The sizes of these flux systems range up to tens of degrees in latitude and longitude; their lifetimes range up to five solar rotations. Some of these flux systems contain large-scale coronal loops of S or inverse-S shape that allow us to distinguish the chirality of their magnetic fields. Typically, the flux systems maintain their chirality throughout their lifetimes. They obey the same weak hemispheric chirality rule as active regions.

Systematic flows are present in the convection zone on variety of spatial scales [Kosovichev, 1999]. The same statement applies to turbulence. At the largest scales, there is now evidence for slow, long-lived cellular flows in the Sun's convection zone with typical diameters of tens of degrees and lifetimes of one or more rotations, both in magnetic fields [Ambroz, 1992] and the flows themselves [Beck et al., 1998; Hathaway et al., 1998]. This interpretation is supported by numerical simulations of convection and its interaction with magnetic flux [Brummell et al., 1996], which show large-scale coherent structures spanning the full vertical extent of the computational domain, involving multiple density scale heights (analogous to the full depth of the convection

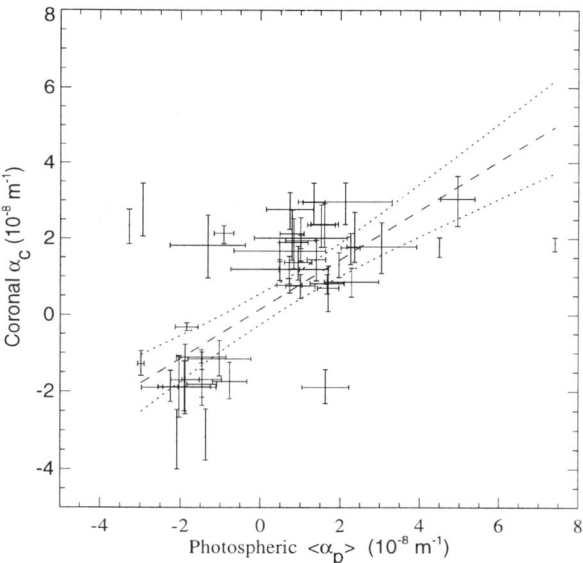

Figure 4. Observed force-free field parameter α from photospheric vector magnetograms (α_p) and from the geometry of coronal loops (α_c) for 44 active regions. Error bars show one standard deviations in both parameters. Error bars for α_c correspond to $10°$ standard deviation in the measured angle γ. Active regions without error bars in α_p are represented by only one magnetogram. The best linear fit (long dashed line) and 2σ error band (short dashed lines) are shown. By permission [Pevtsov et al., 1997].

zone). The Sandborgh et al [1998] study of flux systems show that there is significant twist on comparable scales in space and time. At active region scales, there is evidence of turbulence in the twist of photospheric magnetic fields [Pevtsov and Canfield, 1999; Longcope, 1999], though the relationship between still smaller surface flows and those in the underlying convection zone is not as clear. In view of the presence of a hemispheric asymmetry, and the weakness of the hemispheric rule at and above active region scales, we expect future researchers to look at on the relationship between coronal structures and sub-photospheric flows on all these scales.

3. ROLE OF CURRENTS IN MAGNETIC RECONNECTION

As bipolar active regions rise into the corona from beneath the photosphere, they initally are isolated [Babcock, 1961]. The fields of regions on opposite sides of the solar equator are anti-parallel [Hale et al., 1919], and therefore we would expect them to reconnect, forming loops across the equator. Figure 5 shows that this does not always happen. In the upper figure, there are

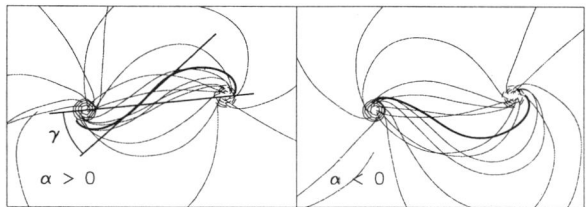

Figure 3. Model bipolar linear force-free field lines for positive and negative values of α. The heavy solid lines highlight low-lying sigmoidal field lines.

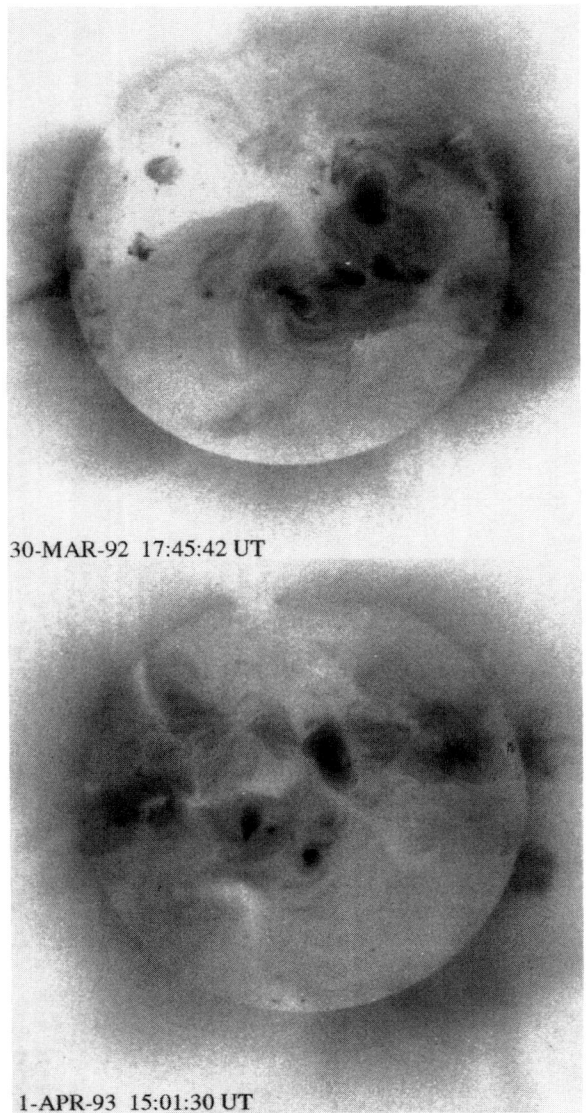

Figure 5. Yohkoh coronal X-ray images (negative) contrasting regions which are well connected across the equator (top) with others which are poorly connected (bottom).

plasmas [Ji, 1999]. In the latter, reconnection proceeds more rapidly in the counter-helicity case, and this is attributed to the fact that the anti-parallel components of **B** at the current sheet separating the current systems are greater. In the case of the solar observation, the anti-parallel components are not much different in the co-helicity and counter-helicity cases, since the fields are always rather close to potential. Hence, the importance of the field orientations at the current sheet is not clear. Moreover, in the laboratory reconnection experiments, the currents do not come in through the boundary, but close within the plasma, and current cancellation can take place.

There is evidence from observations that solar coronal current systems are of sub-photospheric origin, and do not close within the corona. Current conservation arguments have been advanced that take this into account [Canfield et al., 1996]. The argument is illustrated in Figure 6. Consider the photospheric footpoints of the pair of active regions as shown in Figure 6a. The pair on the left is the co-helicity case, the pair on the right, counter-helicity. Figure 6b shows that reconnection of the co-helicity bipoles along the paths indicated does not require a change of the currents at the footpoints, whereas the counter-helicity case, where the question marks appear, does require a change. This is a problem if the currents are imposed from below and continuity is required through the photosphere.

Energetics arguments have been advanced to explain reconnection in solar flares, and they, too, may fit

obvious connections between regions in opposite hemispheres. In the lower figure, there are no such connections. Canfield, Pevtsov, and McClymont [1996] studied the chirality of these active regions, and 25 other pairs, using photospheric vector magnetograms. They found a clear tendency for regions to connect preferentially with others of the same chirality (co-helicity)

There is an interesting difference between the explanations advanced for this observation and that advanced to explain reconnection in laboratory spheromak

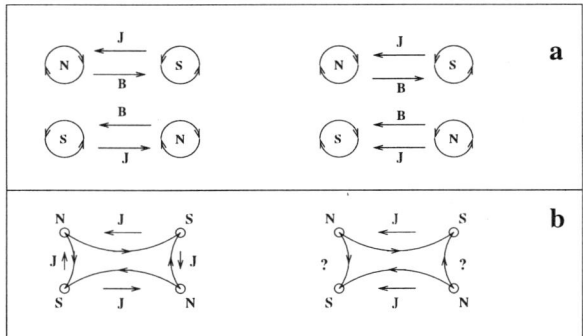

Figure 6. Diagram showing current closure argument for active regions in the co-helicity (left) and counter-helicity (right) cases. (a) Schematic representation, looking down on the photosphere, of the photospheric footpoints of two current-carrying flux tubes on opposite sides of the solar equator, before reconnection. Circles with arrows show the direction of the azimuthal component of the field. (b) Topology of the magnetic field lines and currents in the post-reconnection loops, viewed from above.

the trans-equatorial reconnection observation [Melrose, 1997]. Figure 7 shows a schematic diagram of current-carrying loops before (light) and after (dark) reconnection. The case shown is one of several modeled, but does not have quite the same geometry as the trans-equatorial observation. The general requirement of the model is that the reconnecting structures have the same flux and current at the footpoints before and after reconnection. Hence, the model allows reconnection only between loops of the same chirality – the co-helicity case. Melrose [1997] shows that co-helicity reconnections are preferred on energetics grounds.

Neither of the existing explanations are sufficiently complete or relevant to be considered adequate explanations of the observations. To our knowledge, it has not yet been shown theoretically how electric current systems present in the convection zone, with its high thermal/magnetic energy ratio β, thread through the photosphere and pass into the low-β corona. Hence, the assumption of conservation of both current and flux in the transition through the photosphere to the corona is not compelling. As well, in the spheromak experiments, the flux ropes are not rooted in their current generators; we are not aware of a demonstration in the laboratory of how currents imposed at the boundary affect the reconnection of current-carrying magnetic flux.

4. TWIST AND RECONNECTION IN PRE-ERUPTIVE STRUCTURES

Many solar observations imply that twist is found in magnetic structures that show a tendency to erupt.

Figure 7. Schematic diagram showing loops before and after reconnection [Melrose, 1997]. Initially loops 1 and 2 (lightly shaded) connect footpoints 1+, 1- and 2+, 2-, respectively. The final loops (darkly shaded) connect footpoints 1+, 2- and 2+, 1- respectively. By permission [Melrose, 1997].

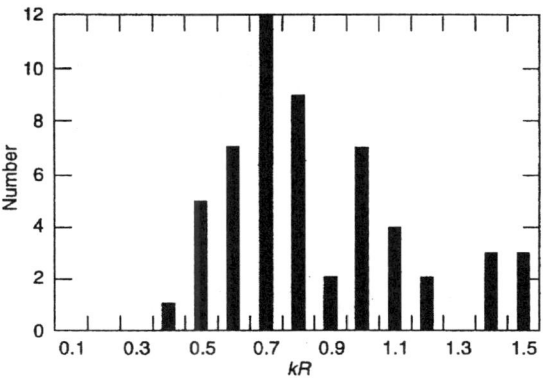

Figure 8. Distribution of sigmoid brightenings as a function of $kR = 2\pi R/L$. By permission [Rust and Kumar, 1996].

In an observational study of 28 solar prominences that showed helical patterns, Vrsnak, Ruzdjak, and Rompolt [1991] found that eruptive prominences show higher observed twist than quiet ones. Furthermore, no observed prominence had a value of twist greater than 2π. Rust and Kumar [1996] studied large coronal transient brightenings of the type known to be associated with filament eruptions and coronal mass ejections. These brightenings are predominantly sigmoidal. Figure 8 shows that if their twist is measured as discussed in Section 1, the distribution drops off abruptly below $kR = 0.6$, the threshold for the $m = 1$ kink mode [Rust and Kumar, 1996].

Twist may build up in a given flux system through shear motions [van Ballegooijen, 1999] and/or reconnection [van Ballegooijen and Martens, 1990]. However, it is the propagation of twist from one flux system in the photosphere to another in the corona that is particularly interesting in view of the Sun's need to get rid of all of the twist that is created in the solar dynamo [Bieber and Rust, 1995].

An example in which reconnection and transfer of twist to larger scales (inverse-cascading) seems to play a role before an eruption is shown in Figure 9 [Pevtsov et al., 1996]. In the first two frames of the figure, one sees what appear to be two separate structures shaped like fish hooks. The arrow in panel (b) shows the gap between them. Later, in panels (c) through (e), the gap seems to be bridged and a single inverse-S sigmoidal loop seems to be present. The region erupted between the times of panels (e) and (f). The conclusion of the authors is that after reconnection the total twist in the coronal part of the sigmoid exceeded the threshold for the kink instability [Priest, 1984], which led to erup-

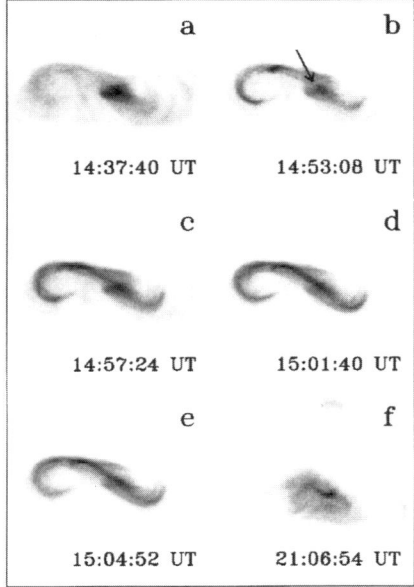

Figure 9. Sequence of Yohkoh SXT images showing the development of an inverse S loop before the flare of 1992 May 8 15:12 UT, between frames (e) and (f). By permission [Pevtsov et al., 1996].

tion. It must be noted, though, that one must not take too simplistic a view of the role of the kink mode; the Aly-Sturrock theorem [Aly, 1991; Sturrock, 1991; Antiochos, 1998] places significant requirements on the magnetic field topology. Simple kinking is not the whole story; the corona must exhibit magnetic fields that are not force-free, or not simply connected to the photosphere or infinity, or eruptions that do not completely open the appropriate flux systems. The challenge is to understand how the kink mode, so strongly implicated by observations, can operate within these constraints.

Another example of reconnection and transfer of twist to larger scales prior to eruption was observed before a famous Yohkoh flare [Canfield and Reardon, 1998]. In an MPEG movie on CDROM in the referenced publication, several preflare events can be seen in which twist is observed to propagate from an emerging flux system into a pre-existing filament, which subsequently erupts. A schematic diagram intended to explain these events in terms of the transfer of twist between flux system is shown in Figure 10, which shows two loops, a twisted one connecting A to B; and an untwisted one connecting C to D. After reconnection, a larger loop is formed from A to D and a shorter one from C to B. Just after the moment of reconnection, all of the twist in the longer A – D loop is concentrated in only part of its length, and it spreads as a twist packet until it reaches equilibrium.

What is seen in the observations is the propagation of a spinning cylindrical region from an emerging flux system, indicated by what is called an arch filament, to the filament flux system. Initially the filament flux system responds by simply undulating; after several such episodes of twist propagation, the filament erupts.

We close by noting recent observational work that links sigmoidal structures to coronal eruptions and their effects on Earth. We know from earlier work discussed above [Rust and Kumar, 1996] that transient sigmoidal structures are associated with eruptions. In particu-

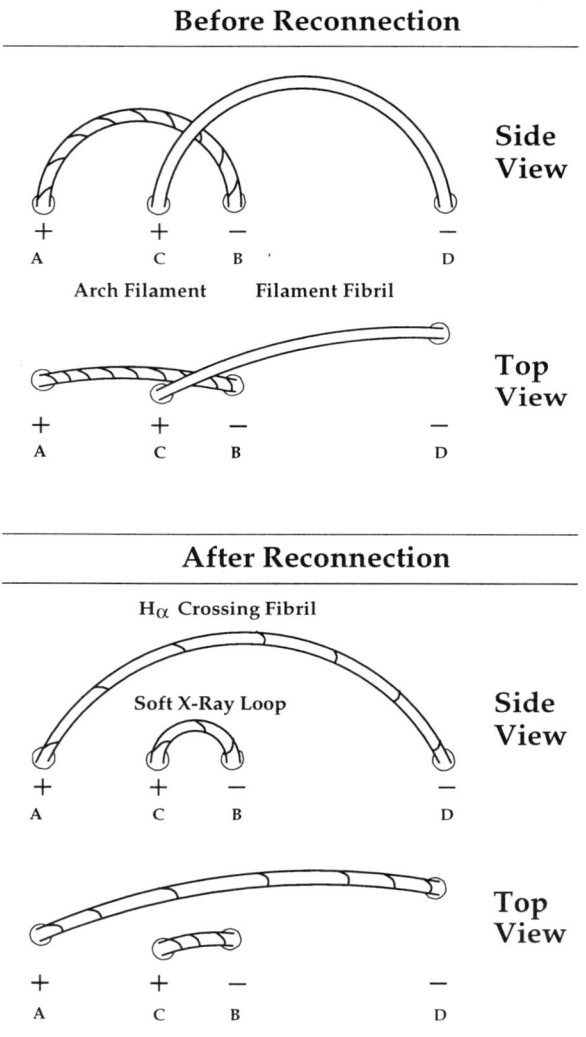

Figure 10. Schematic model of the relationship between chromospheric structures (arch filaments, filament fibrils, and crossing fibrils) and the soft X-ray loops of transient brightenings, before and after reconnection. By permission [Canfield and Reardon, 1998].

lar, sigmoidal brightenings have recently been observed prior to so-called halo CMEs, i.e., coronal mass ejections directed toward Earth [Sterling and Hudson, 1997; Hudson et al., 1998]. The implication of that finding is that improved ability to predict solar eruptions that will have an effect on space surrounding Earth will result from improved understanding of the relationship between the topology of coronal structures and their eruption.

Acknowledgments. The authors wish to thank David Rust and Kazunari Shibata for helpful comments on the manuscript. This research has been supported by NASA through SR&T grant NAG5-5043 and the Yohkoh Soft X-Ray Telescope contract NAS8-40801.

REFERENCES

Aly, J. J., How much energy can be stored in a three-dimensional force-free magnetic field?, *Astrophys. J.*, *375*, L61–L64, 1991.

Ambroz, P., About the Large-Scale Turbulent Transport of Magnetic Field, in *The Solar Cycle*, edited by Harvey, K. L., pp. 35–43, Conference Series, Vol. 27, Astronomical Society of the Pacific, 1992.

Antiochos, S. K., The Magnetic Topology of Solar Eruptions, *Astrophys. J.*, *502*, L181–L184, 1998.

Babcock, H., The Topology of the Sun's Magnetic Field and the 22-Year Cycle, *Astrophys. J.*, *304*, 542–559, 1961.

Beck, J. G., Duvall, T. L., and Scherrer, P. H., Long-Lived Giant Cells Detected at the Surface of the Sun (abstract), *Nature*, *394*, 653–655, 1998.

Bieber, J. W. and Rust, D. M., The Escape of Magnetic Flux from the Sun, *Astrophys. J.*, *453*, 911–918, 1995.

Brummell, N. H., Hurlburt, N. E., and Toomre, J., Turbulent Compressible Convection with Rotation. I. Flow Structure and Evolution, *Astrophys. J.*, *473*, 494–513, 1996.

Canfield, R. C., Pevtsov, A. A., and McClymont, A. N., Magnetic Chirality and Coronal Reconnection, in *Magnetic Reconnection in the Solar Atmosphere*, edited by Bentley, R. D. and Mariska, J. T., pp. 341–346, Proceedings of a Yohkoh Conference, Astronomical Society of the Pacific 1996.

Canfield, R. C. and Reardon, K. P., The Eruptive Flare of 15 November 1991: Preflare Phenomena, *Sol. Phys.*, *182*, 145–157, 1998.

Hale, G. E., Ellerman, F., Nicholson, S. B., and Joy, A. H., The Magnetic Polarity of Sun-Spots, *Astrophys. J.*, *49*, 153–178, 1919.

Hathaway, D. H., Bogart, R. S., and Beck, J. G., A Search for Giant Cells on the Sun, *Eos Trans. AGU*, *79*(17), S281, 1998.

Hudson, H. S., Lemen, J. R., St. Cyr, O. C., Sterling, A. C., and Webb, D. F., X-Ray Coronal Changes during Halo CMEs, *Geophys. Rev. Lett.*, *25*, 2481–2484, 1998.

Ji, H., Laboratory Studies of Magnetic Reconnection and Magnetic Helicity, in *Magnetic Helicity in Space and Laboratory Plasmas*, edited by Brown, M. R., Canfield, R. C., and Pevtsov, A. A., Geophys. Monogr. Ser., p. this volume, AGU, Washington, D.C. 1999.

Jiao, L., McClymont, A. N., and Mikic, Z., Reconstruction of the Three-Dimensional Coronal Magnetic Field, *Sol. Phys.*, *174*, 311–327, 1997.

Kosovichev, A. G., Flows in the Convection Zone, in *Magnetic Helicity in Space and Laboratory Plasmas*, edited by Brown, M. R., Canfield, R. C., and Pevtsov, A. A., Geophys. Monogr. Ser., p. this volume, AGU, Washington, D.C. 1999.

Longcope, D. W., Twist in Rising Magnetic Flux Tubes, in *Magnetic Helicity in Space and Laboratory Plasmas*, edited by Brown, M. R., Canfield, R. C., and Pevtsov, A. A., Geophys. Monogr. Ser., p. this volume, AGU, Washington, D.C. 1999.

McClymont, A. N., Jiao, L., and Mikic, Z., Problems and Progress in Computing Three-Dimensional Coronal Active Region Magnetic Fields from Boundary Data, *Sol. Phys.*, *174*, 191–218, 1997.

Melrose, D. B., A Solar Flare Model Based on Magnetic Reconnection between Current-carrying Loops, *Astrophys. J.*, *486*, 521–533, 1997.

Ogawara, Y., Takano, T., Kato, T., Kosugi, T., Tsuneta, S., Watanabe, T., Kondo, I., and Uchida, Y., The Solar-A Mission - an Overview, *Sol. Phys.*, *136*, 1–16, 1991.

Pevtsov, A. A. and Canfield, R. C., Do Photospheric Motions Cause Sheared Coronal Loops?, *Eos Trans. AGU*, *79*(17), S273, 1998.

Pevtsov, A. A. and Canfield, R. C., Helicity of the Photospheric Magnetic Field, in *Magnetic Helicity in Space and Laboratory Plasmas*, edited by Brown, M. R., Canfield, R. C., and Pevtsov, A. A., Geophys. Monogr. Ser., p. this volume, AGU, Washington, D.C. 1999.

Pevtsov, A. A., Canfield, R. C., and McClymont, A. N., On the Subphotospheric Origin of Coronal Electric Currents, *Astrophys. J.*, *481*, 973–977, 1997.

Pevtsov, A. A., Canfield, R. C., and Zirin, H., Reconnection and Helicity in a Solar Flare, *Astrophys. J.*, *473*, 533–538, 1996.

Priest, E. R., Solar magneto-hydrodynamics, in *Geophysics and Astrophysics Monographs*, Dordrecht: Reidel, *1984.*

Rust, D. M., Magnetic Helicity in Solar Filaments and Coronal Mass Ejections, in *Magnetic Helicity in Space and Laboratory Plasmas*, edited by Brown, M. R., Canfield, R. C., and Pevtsov, A. A., Geophys. Monogr. Ser., p. this volume, AGU, Washington, D.C. 1999.

Rust, D. M. and Kumar, A., Evidence for Helically Kinked Magnetic Flux Ropes in Solar Eruptions, *Astrophys. J.*, *464*, L199–L202, 1996.

Sandborgh, S. C., Canfield, R. C., and Pevtsov, A. A., Chirality of Large-Scale Flux Systems in the Solar Corona, *Eos Trans. AGU*, *79*(17), S285, 1998.

Sterling, A. C. and Hudson, H. S., Yohkoh SXT Observations of X-Ray "Dimming" Associated with a Halo Coronal Mass Ejection, *Astrophys. J.*, *491*, L55–L58, 1997.

Sturrock, P. A., Maximum energy of semi-infinite magnetic field configurations, *Astrophys. J.*, *380*, 655–659, 1991.

Tsuneta, S., Acton, L., Bruner, M., Lemen, J., Brown, W., Caravalho, R., Catura, R., Freeland, S., Jurcevich, B., and Owens, J., The soft X-ray telescope for the SOLAR-A mission, *Sol. Phys.*, *136*, 37–67, 1991.

van Ballegooijen, A. A., Photospheric Motions as a Source of Twist in the Photosphere and Corona, in *Magnetic Helicity in Space and Laboratory Plasmas*, edited by Brown, M. R., Canfield, R. C., and Pevtsov, A. A., Geophys. Monogr. Ser., p. this volume, AGU, Washington, D.C. 1999.

van Ballegooijen, A. A. and Martens, P. C. H., Magnetic fields in quiescent prominences, *Astrophys. J.*, *361*, 283–289, 1990.

Vrsnak, B., Ruzdjak, V., and Rompolt, B., Stability of prominences exposing helical-like patterns, *Sol. Phys.*, *136*, 151–167, 1991.

Richard C. Canfield and Alexei A. Pevtsov, Department of Physics, Montana State University, Bozeman, MT 59717-3840. (e-mail: canfield@physics.montana.edu)

The Role of Reconnection in the Formation of Flux Ropes in the Solar Wind

J. T. Gosling

Los Alamos National Laboratory, Los Alamos, New Mexico

New magnetic loops formed in the corona beneath coronal mass ejections, CMEs, provide strong evidence that magnetic reconnection commonly occurs within the magnetic 'legs' of the departing CMEs. Such reconnection is inherently 3-dimensional and produces a magnetic flux rope structure within a CME. Depending upon how far reconnection proceeds, the resulting flux rope can contain a mixture of doubly attached (to the Sun), singly attached (i.e., open), and disconnected magnetic field lines. The helicity associated with these flux ropes can be the result of emergence of helicity from beneath the solar surface or can be a consequence of photospheric motions, but the actual helical field lines observed in the solar wind in this scenario are a consequence of reconnection following eruption of the CMEs. A wide variety of solar and solar wind observations are consistent with this picture; however, it is also possible that some flux ropes, or portions thereof, observed in the solar wind emerge as such from beneath the solar surface or form from reconnection in the solar atmosphere prior to CME eruption.

INTRODUCTION

Coronal mass ejections, or CMEs, are among the most spectacular of all solar phenomena (see recent reviews by *Hundhausen* [1997] and *Gosling* [1997]). During a CME 10^{15} - 10^{16} g of solar material is propelled outward into the solar wind. This material arises from closed magnetic field regions where the coronal magnetic field previously was sufficiently strong to constrain the plasma from expanding outward. CMEs are commonly observed in association with eruptive filaments (prominences when observed above the solar limb) and are often followed by gradual soft x-ray flares. CMEs play a central role in the long-term evolution of the solar corona [e.g., *Low*, 1997] and are the prime link between solar activity and large solar wind and geomagnetic disturbances [e.g., *Gosling*, 1993].

Approximately 1/3 of all CMEs observed in the solar wind far from the Sun (where they often are called interplanetary CMEs, or ICMEs) appear to have the form of magnetic flux ropes [*Gosling*, 1990]. Such ropes are characterized by series of helical magnetic field lines of ever increasing pitch wrapped about central axes. The physical origin of these flux ropes is still a subject of debate. Our own contribution has been to suggest that the flux ropes are primarily a consequence of 3-dimensional reconnection occurring within the magnetic 'legs' of the departing CMEs. Our purpose here is to provide a brief overview of solar and solar wind observations that have provided insight into the magnetic structure and topology of CMEs in the solar wind and to describe how this inferred structure and topology can be explained in terms of 3-dimensional magnetic reconnection near the foot points of the CMEs.

INFERRING MAGNETIC TOPOLOGIES IN THE SOLAR WIND

Solar wind electron distributions below 1 keV contain two components: a cold and dense thermal 'core' population and a hot and tenuous suprathermal 'halo' population [e.g., *Feldman et al.*, 1975]. Beyond heliocentric distances of several solar radii the suprathermal electrons, which carry

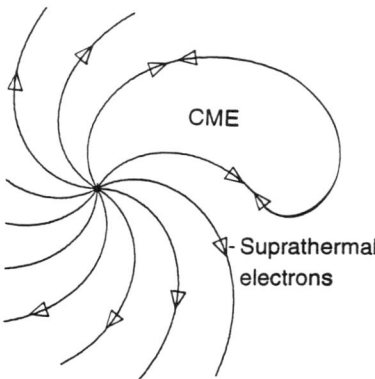

Figure 1. Sketch illustrating singly and doubly attached (to the Sun) magnetic field topologies in the solar wind and the corresponding types of suprathermal electron streaming observed. [*Gosling*, 1996].

the solar wind electron heat flux, travel almost collision-free along the interplanetary magnetic field, IMF. Because of their ubiquity, their high speeds, and their nearly collisionless nature, these electrons serve as effective tracers of magnetic field topology in the solar wind. As illustrated in Figure 1, fluxes of suprathermal electrons in the normal solar wind are unidirectional, because field lines there are effectively anchored in the corona at only one end. Field lines anchored to the hot corona at both ends can be identified by counterstreaming (along the IMF) fluxes of suprathermal electrons. Most CMEs in the solar wind can be identified by such counterstreaming fluxes, reflecting their origin in doubly anchored field regions in the corona [e.g., *Gosling*, 1990]. Finally, suprathermal electron dropouts identify IMF field lines disconnected from the hot corona [e.g., *McComas et al.*, 1989].

MAGNETIC FIELD ROTATIONS WITHIN CORONAL MASS EJECTIONS IN THE SOLAR WIND

Figure 2 shows selected plasma and magnetic field data from ISEE 3, then in orbit about the L1 point approximately 1.3 x 10^6 km upstream from Earth, for three days late in May 1979. Solid vertical lines in the figure mark the start and stop times for the interval when counterstreaming suprathermal electrons were observed, and thus indicate passage of the doubly anchored field lines that commonly identify a CME in the solar wind. Running out ahead of the CME was a shock (vertical dashed line), a result of the speed differential between the fast CME and the slower ambient wind ahead. It is of particular interest that the field orientation did not change substantially within the body of material identified as the CME. This is characteristic of many events identified as CMEs in the solar wind on the basis of counterstreaming suprathermal electrons and other signatures (see *Gosling* [1990] for a review of these other signatures). However, a subset (~1/3) of all CMEs identified in the solar wind do contain large, coherent changes in field orientation.

Figure 3 shows selected plasma and magnetic field data from Ulysses for a CME in the solar wind that exhibited a large coherent internal magnetic field rotation. Two of the solid vertical lines in the figure bracket the interval when counterstreaming electrons, which serve to identify the CME, were observed. The other two vertical lines indicate a forward-reverse shock pair produced by the expansion of the CME. Events in the solar wind exhibiting large coherent changes in field orientation such as that shown in Figure 3 are commonly known as magnetic clouds when the plasma beta is low [e.g. *Burlaga*, 1991].

MAGNETIC FLUX ROPES IN THE SOLAR WIND

Various attempts have been made to visualize the global magnetic field structure responsible for the coherent changes in field orientation observed during events such as that shown in Figure 3. The most popular interpretation is that such an event signals passage by the spacecraft of a cylindrical magnetic flux rope [*Goldstein*, 1983], much as

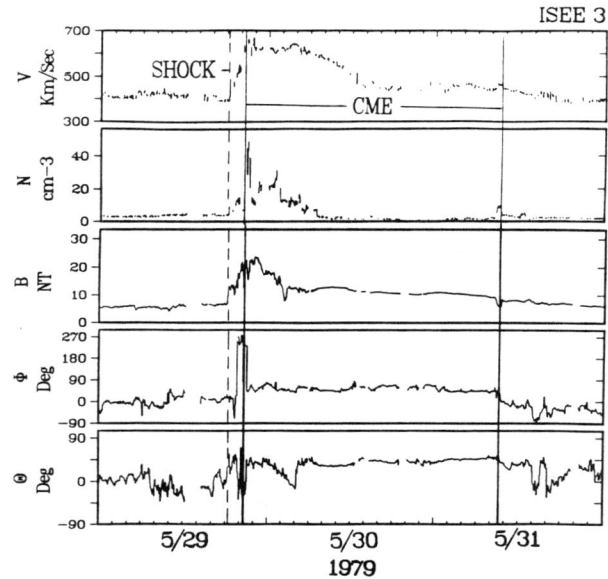

Figure 2. Solar wind plasma and magnetic field parameters during passage of a CME-driven solar wind disturbance observed by ISEE-3. From top to bottom the quantities plotted are the solar wind bulk flow speed, the proton density, the magnetic field strength, and the field azimuthal and polar angles in GSE coordinates. Adapted from *Gosling* [1990].

illustrated in the bottom portion of Figure 4. Quantitative comparisons of observed CME events exhibiting such coherent changes in field orientation with models of force-free flux ropes provide encouraging support for such an interpretation [e.g., *Marubashi*, 1986; *Burlaga*, 1988; *Lepping et al.*, 1990]. Although the flux rope interpretation is reasonably consistent with observations, it is not necessarily a unique interpretation since the flux rope field structure has been inferred from single spacecraft passages through large, 3-dimensional objects. Indeed, the possibility remains that some CMEs in the solar wind exhibiting large internal field rotations may be better explained in terms of some other type of magnetic field structure [e.g., *Vandas et al.*, 1997]. Moreover, approximately 2/3 of all CMEs identified in the solar wind, like the event shown in Figure 2, do not exhibit significant changes in field orientation. These more common events appear to consist of series of simple magnetic loops extending out from the Sun, much as illustrated in the upper portion of Figure 4.

SUGGESTED ORIGINS OF FLUX ROPE CMES

If one assumes that the cylindrical flux rope interpretation is correct for events such as that shown in Figure 3, then

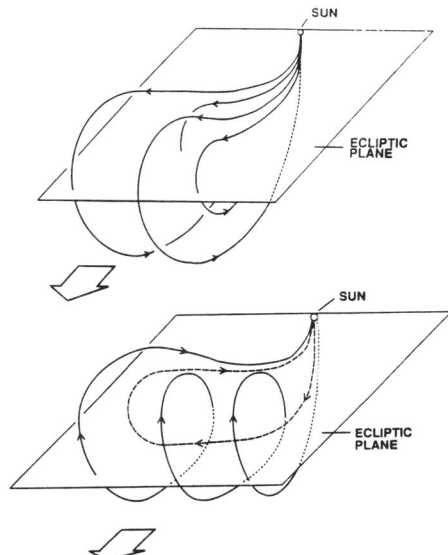

Figure 4. Sketches illustrating a simple magnetic-loop CME (above) and a magnetic flux rope CME (below). The axes of the CMEs need not lie within the ecliptic plane or be transverse to the radial direction. Adapted from *Gosling* [1990].

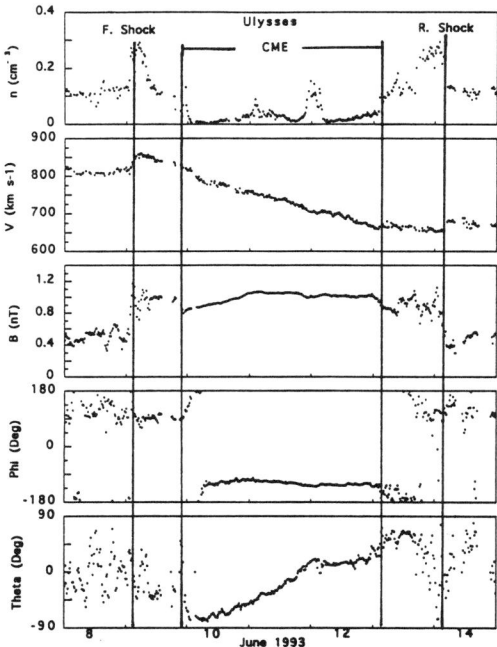

Figure 3. A CME in the solar wind observed by Ulysses at 4.6 AU and S33°. Parameters plotted from top to bottom include the proton density, the bulk flow speed, the magnetic field strength, and the field azimuthal and polar angles.

observed events of this nature can be classified according to the orientation of their axes and the sense of twist (chirality) of the helical field lines. Several studies have found a strong correlation between flux rope orientations and chiralities observed in the solar wind and that inferred for the solar filaments that presumably erupted in association with those particular CMEs [*Rust*, 1994; *Bothmer and Schwenn*, 1998]. From this it has been suggested that flux rope CMEs in the solar wind are simply heliospheric extensions of solar filaments [e.g., *Burlaga*, 1991]. It is this author's opinion that this interpretation is incorrect, since filaments typically comprise only small fractions of the volumes of CMEs when observed by coronagraphs and would be expected to be found primarily on the trailing edges of CMEs in the solar wind. On the other hand, the events identified as flux ropes in the solar wind usually are the entire CMEs, not small portions thereof.

CMEs appear to be a consequence of the evolution of the coronal magnetic field. One popular idea is that CMEs result from magnetic non equilibrium within closed field structures (see, for example, the review by *Low* [1997]). The left portion of Figure 5 shows a sketch of a meridional cut through a hypothetical coronal streamer whose core close to the Sun contains simple magnetic field loops anchored at both ends in the Sun. Although it has been argued from energy equilibria considerations that the eruption of a streamer containing simple magnetic loops

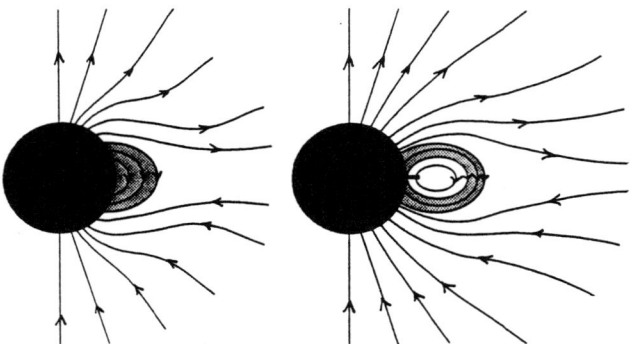

Figure 5. (Left) Magnetic lines of force of an axisymmetric corona showing anchored fields at the equator sandwiched between two hemispheres of open fields. (Right) A 2D projection of magnetic lines of force of an axisymmetric corona showing anchored bipolar fields and some completely closed field loops, all sandwiched between two hemispheres of open fields. The closed loops actually have a significant azimuthal field component so that these fields represent an azimuthal flux rope running above the equator. The thick line represents a vertical sheet of prominence material suspended at the base of the closed loop region. Adapted from *Low* [1997].

such as this can not produce a CME if the magnetic field is originally force-free [*Aly*, 1991], there is very good reason to believe that coronal streamers are, in fact, not force-free. With increasing height above the coronal base the gas pressure increasingly distorts field lines within a streamer from the force-free state, ultimately opening them up to the solar wind flow [*Pneuman and Kopp*, 1971], as illustrated in the sketch. Moreover, simulations suggest that equilibrium considerations do not really apply to a dynamical event such as a CME [*Mikic and Linker*, 1994]. We shall return to questions of field structure and topology when such coronal structures do erupt to form CMEs.

The right portion of Figure 5 shows a meridional cut through a streamer of a somewhat different nature. In this case the core of the streamer close to the Sun contains what are apparently closed loops magnetically detached from the solar surface. In a realistic 3D geometry the closed loops are actually part of a flux rope anchored at both ends in the solar surface. Overlying the flux rope are simple magnetic loops anchored at both ends in the Sun. The flux rope can be a result of emergence of helicity from beneath the solar surface or it can result from surface motions and magnetic reconnection prior to any eruption as in the work of *Ballegooijen and Martens* [1989]. The sketch includes a hypothetical prominence nested at the base of the corona within the flux rope. Low and Ballegooijen and Martens relate the overall flux rope to the density-depleted cavity that commonly overlies a prominence. Low argues that the extra magnetic energy associated with the flux rope can be used to help propel the structure, including the prominence, out into the solar wind (see also *Chen* [1997]). Moreover he suggests that CMEs would be more likely to originate from such streamers than from the type of streamer shown at the top of the figure. Note that, in contrast to the idea that flux ropes in the solar wind are simply interplanetary extensions of prominences, here the prominence is but a small part of a much larger flux rope that exists in the corona prior to an eruption.

There is some observational support for the above picture of how flux ropes in the solar wind might originate. Figure 6 shows selected solar wind plasma and magnetic field data from WIND during passage of a CME-driven solar wind disturbance in January 1997. The CME, identified as a magnetic cloud/flux rope on the basis of the smooth rotation in the field polar angle, was traveling faster than the ambient wind ahead, and was preceded by a shock. Prominence material was identified in the trailing portion of the structure on the basis of the unusually high density there and the fact that this density enhancement contained ionizationally cold material (He$^+$, not shown). *Burlaga et al.* [1998] note that the appearance of this flux rope CME event in the solar wind is almost exactly as Low and Ballegooijen and Martens would predict (compare with the right portion of Figure 5). On the other hand, we note that it is extremely rare to observe density enhancements of this nature within the trailing portions of flux rope CMEs, and significant enhancements in He$^+$ in the solar wind have been observed on only 6 occasions since 1971. Moreover, the

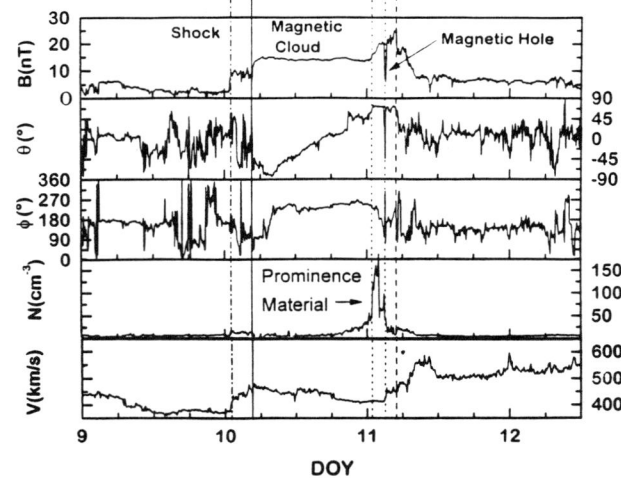

Figure 6. Plasma and magnetic field data from WIND surrounding passage of a flux rope CME that contained prominence material within its trailing portion. Adapted from *Burlaga et al.* [1998].

field rotation observed within the prominence material in the January 1997 event does not appear to be a smooth continuation of the preceding flux rope.

FLUX ROPES ORIGINATING FROM 3-DIMENSIONAL RECONNECTION

Most CMEs are observed in conjunction with long-duration soft solar x-ray events that commence close to the time that CMEs lift off from the Sun and that persist for many hours thereafter [*Webb et al.*, 1976; *Kahler*, 1977]. These long-lived events, sometimes known as gradual flares, typically are associated with a restructuring of the solar corona beneath the departing CMEs [*e.g., Sheeley et al.* 1983; *Hiei et al.*, 1993]. Figure 7 shows images of the corona obtained in soft x-rays by Yohkoh and at 10870 Å (neutral helium) by Kitt Peak taken before and some time after the eruption of a large CME from the southern solar hemisphere facing Earth. In this case the newly formed

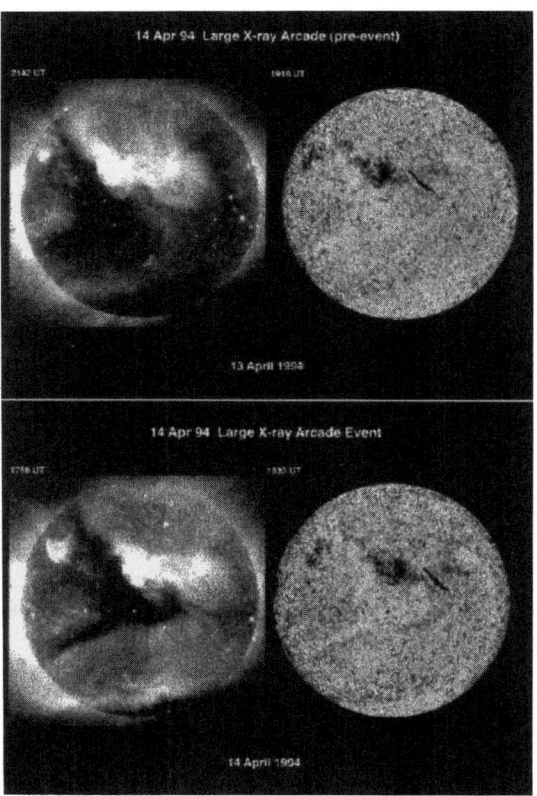

Figure 7. Images of the corona obtained prior to (top) and after (bottom) the eruption of a CME on April 14, 1994 by the soft x-ray telescope on Yohkoh (left) and in the HeI 10870 Å line by Kitt Peak (right). Courtesy of Karen Harvey.

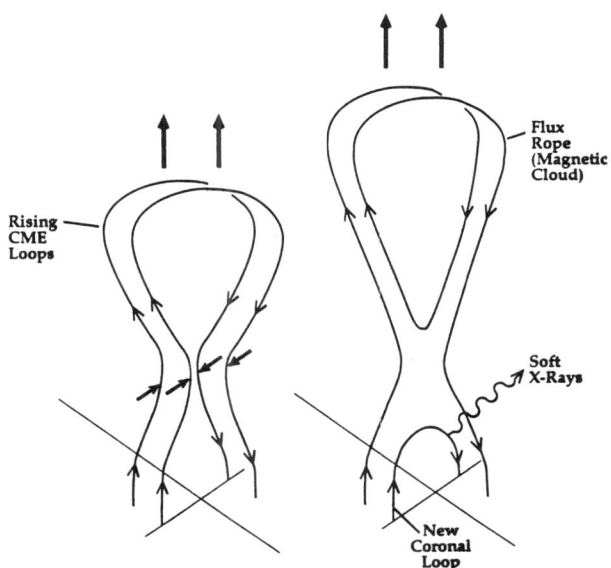

Figure 8. Sketches illustrating how reconnection of simple magnetic loops in a rising CME whose magnetic legs are sheared relative to one another produces helical field lines threading the CME and new magnetic loops in the corona below. Adapted from *Gosling* [1993].

coronal loops in the southern hemisphere extended from the east limb to the west limb and partially filled in a preexisting coronal hole lying to the north. It is commonly believed that the newly formed coronal loops result from magnetic reconnection within the magnetic legs of the rising CME [*e.g., Kopp and Pneuman*, 1976], although if reconnection is limited to the legs we would not expect to observe new coronal loops penetrating into a previously existing coronal hole. Observations reveal that loop formation often persists for hours following eruption of a CME and can fill a sizable fraction of the solar atmosphere.

When visualized in two dimensions, reconnection occurs between the opposite legs of the same magnetic loops to produce closed magnetic loops threading the CME that are completely detached from the Sun as well as new magnetic loops statically bound to the corona. However, because of the high degree of symmetry required for this to occur, it is quite unlikely that reconnection ever actually proceeds in this fashion. Any skewing or shearing of the field results in reconnection between the legs of neighboring CME loops in much the same manner as visualized by *Ballegooijen and Martens* [1989] in their prominence formation model. As illustrated in Figure 8, reconnection in the presence of shear produces helical field lines threading the CME as well as new magnetic loops in the corona below [*Gosling*, 1990]. In contrast to the 2-dimensional case, reconnection in 3 dimensions does not initially result

210 RECONNECTION AND THE FORMATION OF FLUX ROPES IN THE SOLAR WIND

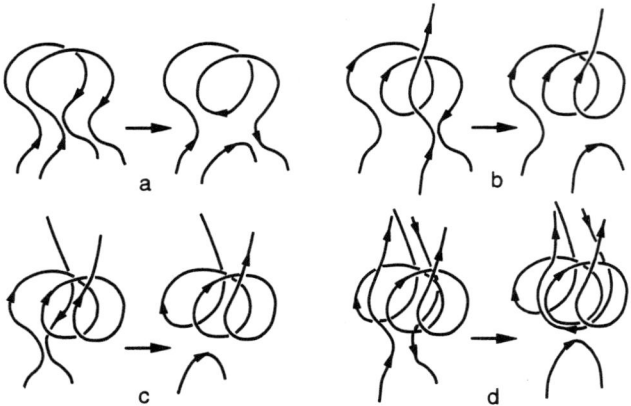

Figure 9. Sketches of successive steps in 3-dimensional reconnection in the corona beneath a departing CME. The sketches are not to scale and are intended only to illustrate successive changes in magnetic topologies resulting from reconnection. Adapted from *Gosling et al.* [1995].

in magnetic disconnection from the Sun. Further, as we shall see, when disconnection does occur the field lines are connected to the outer heliosphere at both ends.

The chirality of the helical field lines threading the CME is related to the preexisting skew of the field lines and the underlying magnetic polarities. The helicity associated with a flux rope CME can be a consequence of the emergence of helicity from beneath the solar surface or it can be a result of photospheric motions that drag field lines around, but the actual flux ropes observed in the solar wind in this scenario are a consequence of reconnection following eruption of the CME. When the plasma beta is low the helical field lines within the CME relax to the nearly force-free structure characteristic of magnetic clouds as the CME propagates out into the heliosphere. If beta is not low, the CME is still properly regarded as a flux rope even though the magnetic field does not attain the force-free state.

OPEN AND DISCONNECTED FIELD LINES WITHIN FLUX ROPES

Despite the utility of the counterstreaming suprathermal electron signature for identifying CMEs in the solar wind, it has long been apparent that not all portions of CMEs and perhaps not even all CMEs can be identified in solar wind data by that signature [e.g., *Zwickl et al.*, 1983; *Gosling*, 1990]. For example, counterstreaming suprathermal electrons characteristic of doubly anchored field lines were observed during only a small portion of a CME/flux rope event observed by WIND in October 1995 [*Larson et al.*, 1997]. The remainder of the event contained interspersed intervals of unidirectional streaming and suprathermal electron dropouts. This indicates that some field lines within CMEs in the solar wind can be open or even entirely disconnected from the Sun. Surveys indicate that open and (less frequently) disconnected field lines can be distinguished in ~20% of all CME events identified in the solar wind near 1 AU [*Gosling*, 1996]. Such field line topologies are a natural consequence of sustained magnetic reconnection within the legs of the CMEs [*Gosling et al.*, 1995].

FIELD TOPOLOGIES RESULTING FROM SUSTAINED MAGNETIC RECONNECTION

Figure 9 illustrates successive stages in 3-dimensional reconnection in the corona behind a CME. These sketches are based upon results of numerical simulations of reconnection in the qualitatively similar geometry in the Earth's magnetic tail in the presence of a cross-tail field component [*Birn and Hesse*, 1991; *Hesse and Birn*, 1991]. Panel a illustrates the initial result of reconnection as shown in Figure 8. In panel b reconnection occurs between a doubly anchored helical field line and an open field line of the normal solar wind. The result is an open helical field line and a newly formed magnetic loop in the corona below. In panel c reconnection occurs between an open helical field line and an open field line of the normal solar wind to produce a helical field line that is connected to the outer heliosphere at both ends, as well as a new magnetic loop below. Finally in panel d reconnection between two open field lines of the normal solar wind produces a simple U-shaped field line disconnected from the Sun as well as a new coronal loop below the CME.

Helical field lines threading the central portion of a CME are exposed to open solar wind field lines of opposite polarity at the base of the CME, as can be ascertained from careful examination of Figures 8 and 9. In the geomagnetic tail case [*Birn and Hesse*, 1991; *Hesse and Birn*, 1991], these tend to be the field lines that are opened up first. As a result, a variety of field line topologies can simultaneously be present within a CME flux rope. The sketch in Figure 10 illustrates an example where a simple doubly anchored magnetic loop not yet reconnected coexists with a helical field line anchored in the Sun at both ends and an open helical field line threading the interior of the flux rope. The crucial point is that reconnection along the flanks of a sheared loop system can create open or disconnected field lines within the heart of a flux rope even when the CME originates from simple doubly anchored magnetic loops.

Simulations of the geomagnetic tail case indicate that the flux rope produced by 3D reconnection typically consists entirely of open and disconnected field lines before the leading edge of the flux rope reaches 200 Earth radii down tail. On the other hand, solar wind observations indicate that reconnection in the corona seldom proceeds to the point where many field lines threading a CME are disconnected from the Sun since suprathermal electron dropouts are relatively rare anywhere in the solar wind. Indeed, as already noted, there is good evidence that most field lines threading CMEs in the solar wind are anchored to the Sun at both ends, although spectacular exceptions do occur, such as the event described by *Larson et al.* [1997].

CONCLUDING COMMENTS

Long-duration soft solar x-ray events reveal that magnetic loops commonly form in the corona beneath departing CMEs. These new loops indicate that reconnection is a common occurrence close to the Sun within the legs of the CMEs. Such reconnection is inherently 3-dimensional and produces flux rope CMEs that can be mixtures of doubly anchored, open, and disconnected field lines, as observed in the solar wind far from the Sun.

CMEs represent a process by which magnetic flux is opened up to the solar wind. The amount of new magnetic flux carried into the solar wind by a CME is reduced by the process of reconnection, thus helping to avoid a long-term buildup of magnetic flux within the heliosphere. We note that when reconnection proceeds to the point illustrated in panel b of Figure 9 that portion of the CME adds no net open magnetic flux to the solar wind. When it proceeds to the point illustrated in panel c, that portion of the CME produces a net decrease in the amount of open magnetic flux in the solar wind. In addition, the coronal loops formed by reconnection provide new interconnections between different regions on the Sun. When reconnection occurs between doubly anchored field lines and open field lines extending into the normal solar wind the boundaries of coronal holes are restructured, as in the example shown in Figure 7. In that case the newly formed coronal loops extend well into what was previously the open field region of a coronal hole.

Finally, although it is clear that reconnection commonly occurs behind departing CMEs and that such reconnection produces helical field lines within the CMEs, not all helical field lines within CMEs in the solar wind need necessarily originate in this manner. It is possible, for example, that CMEs preferentially arise from geometries such as illustrated in the right portion of Figure 5, where the inner portion of a coronal streamer already contains a flux rope prior to eruption. In that case new helical field lines formed after eruption by reconnection of the anchored magnetic loops overlying the flux rope would simply provide added helical layers to the preexisting flux rope.

Figure 10. An example of several different magnetic topologies possible in a CME that has undergone 3-dimensional reconnection within its magnetic legs. The sketch is not to scale and not all possible topologies are represented. Adapted from *Gosling et al.* [1995].

Acknowledgements. I thank Karen Harvey for providing Figure 7. This work was performed under the auspices of the U. S. Department of Energy with support from an internal Los Alamos National Laboratory research grant.

REFERENCES

Aly, J. J., How much energy can be stored in a three-dimensional force-free magnetic field? *Astrophys. J. Lett., 375*, L61, 1991.

Birn, J., and M. Hesse, MHD simulations of magnetic reconnection in a skewed three-dimensional tail configuration, *J. Geophys. Res., 96*, 23, 1991.

Bothmer, V., and R. Schwenn, The structure and origin of magnetic clouds in the solar wind, *Annales Geophysicae, 16*, 1, 1998.

Burlaga, L. F., Magnetic clouds and force-free fields with constant alpha, *J. Geophys. Res., 93*, 7217, 1988.

Burlaga, L. F., Magnetic clouds, in *Physics of of the Inner Heliosphere II*, edited by R. Schwenn and E. Marsch, Springer-Verlag, Berlin, pp. 1-22, 1991.

Burlaga, L. F., et al., A magnetic cloud containing prominence material: January 1997, *J. Geophys. Res., 103*, 277, 1998.

Chen, J., Coronal mass ejections: Causes and consequences a theoretical view, in *Coronal Mass Ejections*, Geophysical Monograph 99, edited by N. U. Crooker, J. A. Joselyn, and J. Feynman, American Geophysical Union, pp 65-81, 1997.

Feldman, W. C., J. R. Asbridge, S. J. Bame, M. D. Montgomery, and S. P. Gary, Solar wind electrons, *J. Geophys. Res., 80*, 4181, 1975.

Goldstein, H., On the field configuration in magnetic clouds, in *Solar Wind Five*, NASA Conference Publ. 2280, edited by M. Neugebauer, pp. 731-733, 1983.

Gosling, J. T., Coronal mass ejections and magnetic flux ropes in interplanetary space, in *Physics of Magnetic Flux Ropes*, edited by C. T. Russell, E. R. Priest, and L. C. Lee, Geophys. Monogr. 58, Amer. Geophys. Union, pp. 343-364, 1990.

Gosling, J. T., The solar flare myth, *J. Geophys. Res., 98*, 18,937, 1993.

Gosling, J. T., Magnetic topologies of coronal mass ejection events: Effects of 3-dimensional reconnection, in *Solar Wind Eight*, edited by D. Winterhalter, J. T. Gosling, S. R. Habbal, W. S. Kurth, and M. Neugebauer, American Institute of Physics, Conference Proceedings 383, New York, pp. 438-441, 1996.

Gosling, J. T., Coronal mass ejections: An overview, in *Coronal Mass Ejections*, Geophysical Monograph 99, edited by N. U. Crooker, J. A. Joselyn, and J. Feynman, American Geophysical Union, pp 9-16, 1997.

Gosling, J. T., J. Birn, and M. Hesse, Three-dimensional magnetic reconnection and the magnetic topology of coronal mass ejection events, *Geophys. Res. Lett., 22,* 869, 1995.

Hesse, M, and J. Birn, Plasmoid evolution in an extended magnetotail, *J. Geophys. Res., 96,* 5683, 1991.

Hiei, E., A. J. Hundhausen, and D. G. Sime, Reformation of a coronal helmet streamer by magnetic reconnection after a coronal mass ejection, *Geophys. Res. Lett., 20,* 2785, 1993.

Hundhausen, A. J., Coronal mass ejections, in *Cosmic Winds and the Heliosphere*, edited by J. R. Jokipii, C. P. Sonett, and M. S. Giampapa, Univ. of Arizona Press, Tucson, pp. 259-296, 1997.

Kahler, S. W., The morphological and statistical properties of solar x-ray events with long decay times, *Astrophys. J., 214,* 891, 1977.

Kopp, R. A., and G. W. Pneuman, Magnetic reconnection in the corona and loop prominence phenomenon, *Sol. Phys., 50,* 85, 1976.

Larson, D. E., et al., Tracing the topology of the October 18-20 1995, magnetic cloud with ~0.1 - 10^2 keV electrons, *Geophys. Res. Lett., 24,* 1911, 1997.

Lepping, R. P., J. A. Jones, and L. F. Burlaga, Magnetic field structure of interplanetary magnetic clouds at 1 AU, *J. Geophys. Res., 95,* 11957, 1990.

Low, B. C., The role of coronal mass ejections in solar activity, in *Coronal Mass Ejections*, Geophysical Monograph 99, edited by N. U. Crooker, J. A. Joselyn, and J. Feynman, American Geophysical Union, pp 39-47, 1997.

Marubashi, K., Structure of the interplanetary magnetic clouds and their solar origins, *Adv. Space Res., 6,* 1, 1986.

McComas, D. J., J. T. Gosling, J. L. Phillips, S. J. Bame, J. G. Luhmann, and E. J. Smith, Electron heat flux dropouts in the solar wind: Evidence for interplanetary magnetic field reconnection?, *J. Geophys. Res., 94,* 6907, 1989.

Mikic, Z, and J. A. Linker, Disruption of coronal magnetic field arcades, *Astrophys. J., 430,* 898, 1994.

Pneuman, G. W., and R. A. Kopp, Gas-magnetic field interactions in the solar corona, *Solar Phys., 18,* 258, 1971.

Rust, D. W., Spawning and shedding helical magnetic fields in the solar atmosphere, *Geophys. Res. Lett., 21,* 241, 1994.

Sheeley, N. R., R. A. Howard, M. J. Koomen, and D. J. Michels, Associations between coronal mass ejection events and soft x-ray events, *Astrophys. J., 272,* 349, 1983.

van Ballegooijen, A. A., and P. C. H. Martens, Formation and eruption of solar prominences, *Astrophys. J., 343,* 971, 1989.

Vandas, M., S. Fischer, D. Odstrcil, M. Dryer, Z. Smith, and T. Detman, Flux ropes and spheromaks: A numerical study, in *Coronal Mass Ejections*, Geophysical Monograph 99, edited by N. U. Crooker, J. A. Joselyn, and J. Feynman, American Geophysical Union, pp 169-176, 1997.

Webb, D. F., A. S. Krieger, and D. M. Rust, Coronal x-ray enhancements associated with Hα filament disappearances, *Solar Phys., 48,* 159, 1976.

Zwickl, R. D., J. R. Asbridge, S. J. Bame, W. C. Feldman, J. T. Gosling, and E. J. Smith, Plasma properties of driver gas following interplanetary shocks observed by ISEE 3, in *Solar Wind Five*, edited by M. Neugebauer, NASA Conf. Publ., CP-2280, pp 711-717, 1983.

J. T. Gosling, Los Alamos National Laboratory, MS D466, Los Alamos, NM 87545

Photospheric Motions as a Source of Twist in Coronal Magnetic Fields

A. A. van Ballegooijen

Harvard-Smithsonian Center for Astrophysics, Cambridge, Massachusetts

The interaction of magnetic fields with granulation and supergranulation flows causes the photospheric magnetic flux of an active region to be dispersed over the solar surface on a timescale of days to months. This photospheric "diffusion" process leads to magnetic flux cancellation at the polarity inversion line separating the leading and following polarity parts of the region. I present a magnetohydrodynamic (MHD) model which takes into account the effects of the solar differential rotation, photospheric diffusion, and flux cancellation on the three-dimensional structure of the active-region magnetic field. The model assumes that the coronal magnetic field evolves through a series of force free equilibrium states. It is shown that magnetic reconnection associated with flux cancellation causes the formation of a helical flux rope overlying the polarity inversion line. For initially twisted bipoles, the diffusion of flux to the polarity inversion line produces an S-shaped right-helical flux rope or an inverse-S-shaped left-helical flux rope, depending on the sign of the initial twist. The shapes of the simulated flux ropes are similar to the observed coronal X-ray structures. For an initially untwisted bipole, the model predicts the formation of a left-helical flux rope in the North or a right-helical flux rope in the South, in agreement with the observed weak correlation between active-region helicity and latitude. This suggests that the observed correlation may be due to the effects of the solar differential rotation acting on active-region magnetic fields after they emerge through the photosphere.

1. INTRODUCTION

Observations of solar active regions with the Soft X-ray telescope on *Yohkoh* (Tsuneta et al 1991) often show S-shaped or inverse S-shaped structures which indicate that the coronal magnetic field in these regions is strongly sheared and/or twisted (Acton et al 1992; Rust & Kumar 1996). Evidence for such large-scale shear is also seen in chromospheric Hα structure (e.g., Zirin 1972) and in photospheric vector magnetograms (e.g., Gary et al 1987; Pevtsov, Canfield & Metcalf 1995). These observations indicate that the magnetic field in the photosphere and chromosphere deviates significantly from a potential field: there exist large-scale electric currents which flow through the photosphere and up into the corona (Pevtsov, Canfield & McClymont 1997). The magnetic shear and vertical currents are often localized in areas close to the polarity inversion line (Schmieder et al 1996).

The plasma pressure in the active corona is generally much smaller than the magnetic pressure ($p \ll B^2/8\pi$). Hence, the balance of various forces on the plasma requires that the electric current density $\mathbf{j}(\mathbf{r},t)$ in the corona is nearly parallel or anti-parallel to the local magnetic field $\mathbf{B}(\mathbf{r},t)$, where $\mathbf{r}$ is a position in the corona and t is the time. The resulting *force free* magnetic field satisfies $\nabla \times \mathbf{B} = \lambda \mathbf{B}$, where $\lambda(\mathbf{r},t)$ is a helicity parameter which is constant along field lines. In general λ has different values on different field lines. Solving this force-free equation subject to the condition that $\mathbf{B}$ matches the observed photospheric vector field represents a difficult nonlinear problem (Sakurai 1993; McClymont, Jiao & Mikić 1997). Many authors use the so-called linear force-free approximation in which λ is assumed to be the same for all field lines. Measurements of λ for many active regions have shown that there is a mixture of positive and negative helicity in each hemisphere. However, there is a small excess of negative helicity regions in the northern hemisphere and positive helicity regions in the southern hemisphere (Seehafer 1990; Pevtsov, Canfield and Metcalf 1995; Canfield & Pevtsov 1998). Hemispheric patterns of helicity have also been found for quiescent filaments (Martin, Bilimoria & Tracadas 1994; Rust & Kumar 1995; Zirker et al 1997).

The role of strongly sheared magnetic fields in solar flares has been studied by many authors (e.g., Leka et al 1993; de La Beaujardiére et al 1993; Manoharan et al 1996; Hudson, Acton & Freeland 1996; Pevtsov, Canfield & Zirin 1996; Schmieder et al 1997). These studies show that the sheared core fields running under and out of the flare arcade are often S-shaped, with curved "elbows" that loop into the neighboring low corona (Moore et al 1997). These core fields may also be important for non-flare heating. Falconer et al (1997) show that strongly sheared core fields are often associated with areas of persistent enhanced coronal heating. They suggest that this heating may be due to magnetic reconnection accompanying flux cancellation at the inversion line. On the other hand, Metcalf et al (1994) find no compelling spatial or temporal correlation between the sites of vertical current and the bright X-ray structures.

What is the origin of these sheared magnetic fields in active regions? Early models assumed that active regions emerge more or less unsheared, and that the magnetic shear is produced by surface flows *after* the magnetic field has emerged through the photosphere (e.g., Tanaka & Nakagawa 1973; Krall et al 1982; Van Hoven, Mok & Mikić 1995). More recently it has been recognized that the magnetic flux bundles which make up an active region are sometimes already twisted when they first emerge in the photosphere (Kurokawa 1987; Sturrock 1987; Tanaka 1991). Leka et al (1996) study the motions of sunspots in a young active region and show that the Hα and X-ray structures associated with newly emerged bipoles within this region do not agree with potential-field extrapolations of magnetograms. The increase of the electric currents, as new flux emerges, is not consistent with their generation by photospheric motions. Van Driel-Gesztelyi et al (1997) study the development of magnetic shear and electric currents in a large sunspot group. They find that currents are induced by sunspot motions and further increased by non-potential flux emergence, leading to a series of energetic flares at the location where the newly emerged bipole interacts with the main bipole of the region. Such observations suggest that the emerging flux bundles that make up these bipoles are twisted before they emerge at the photosphere (also see McClymont, Jiao & Mikić 1997). The flux bundles are believed to originate in the deep layers of the convection zone (Parker 1955), therefore, observations of twist in emerging bipoles imply that either the deep-seated flux tubes are twisted (e.g., Rust & Kumar 1995), or they become twisted during their rise through the convection zone.

While some of the examples of highly sheared magnetic structures in active regions are clearly associated with newly emerged twisted bipoles (Leka et al 1996; van Driel-Gesztelyi et al 1997), there are other cases in which the development of strong magnetic shear and associated filament formation seem to take place well after the magnetic flux first emerges at the photosphere. This suggests that the emergence of helical magnetic fields from below the photosphere is not the only mechanism by which strongly sheared or twisted structures are formed. Pneuman (1983) proposed that helical fields can be formed by magnetic reconnection occurring in a sheared coronal arcade. Van Ballegooijen and Martens (1989) further proposed that such reconnection may be associated with *flux cancellation*, i.e., the mutual cancellation and disappearance of opposite polarity fields in the photosphere. Flux cancellation is known to take place at polarity inversion lines where opposite polarity radial fields intermix and cancel each other. Van Ballegooijen and Martens argued that flux cancellation involves the submergence of small, strongly curved magnetic loops in the photosphere. In a sheared magnetic field, such small-scale loops can be formed only by magnetic reconnection; the photospheric magnetic field must be decoupled from the axial field in the corona (i.e., the component parallel to the inversion

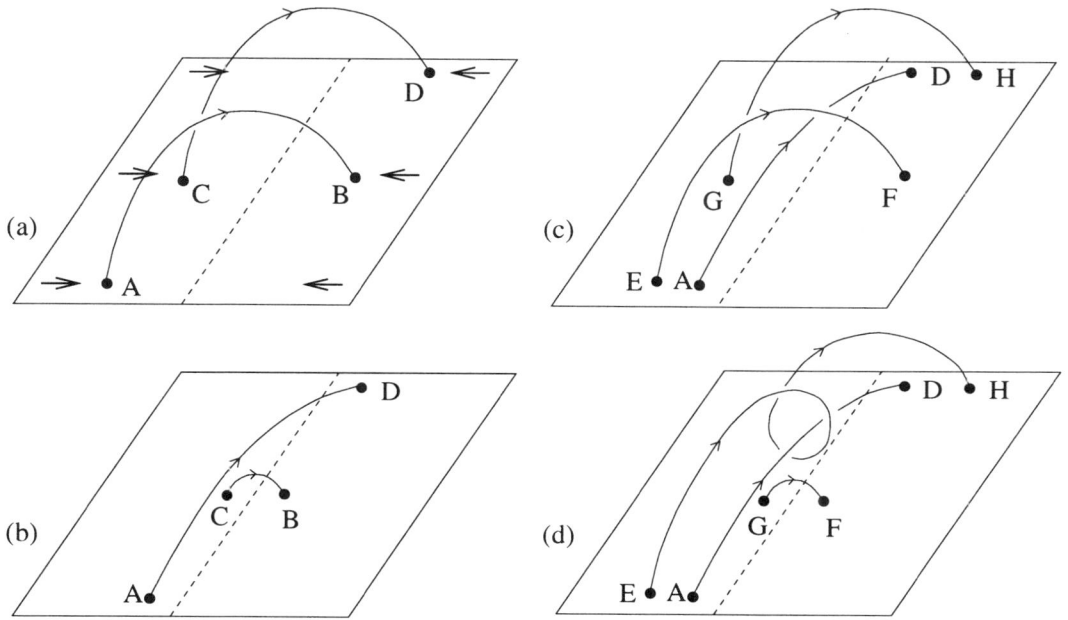

Figure 1. Flux cancellation in a sheared coronal arcade. The rectangle represents the solar photosphere, and the dashed line is the polarity inversion line. (a) Initial sheared field subject to converging flows. (b) Reconnection produces a long loop AD and a shorter loop CB, which subsequently disappears below the photosphere. (c) Overlying loops EF and GH are pushed to the inversion line. (d) Reconnection produces the helical loop EH and a short loop GF, which again submerges.

line). This reconnection process presumably takes place in the photosphere or chromosphere, and leads to the formation of a helical flux rope in the corona above the inversion line (see Figure 1).

Van Ballegooijen and Martens (1989) developed a two-dimensional, nonlinear force-free model of such a flux rope for application to solar prominences. In this model the helical flux rope is detached from the photosphere and is surrounded by a sheared coronal arcade. In the present paper I extend this model to three dimensions, and apply it to the development of strongly twisted fields in active regions, where 3-D effects are quite important. Magnetic models incorporating 3-D effects have been considered previously (see Antiochos, Dahlburg & Klimchuk 1994, and references therein), however, except for recent work by DeVore (1998), these models did not include the effects of flux cancellation.

According to the present model, magnetic bipoles emerge at the solar surface with some amount of internal twist, but usually not enough to produce strongly sheared coronal structures. These initial bipoles are then acted upon by surface flows, which cause further twisting and shearing of the coronal magnetic field (the field is assumed to be nearly "frozen" into the coronal plasma). The footpoint motions consist of *systematic flows* (e.g., solar differential rotation) and *random flows* (convective flows associated with the solar granulation and supergranulation). Random motions have the effect of transporting photospheric magnetic flux from regions of high flux density to regions of low flux density, a process that can be crudely described as "diffusion" of the radial field (Leighton 1964). Models of photospheric flux transport based on Leighton's diffusion equation are known to yield results in qualitative agreement with solar observations (e.g., Wang, Nash & Sheeley 1989). In the next section I present an MHD model which includes the effects of surface diffusion and shear flows on the three-dimensional magnetic structure of an active region. The results are further discussed in §3.

2. MAGNETIC FIELD MODEL

The coronal magnetic field of an active region is continually subjected to random footpoint motions, which cause small-scale twisting and braiding of the coronal field lines. The free magnetic energy associated with these twists can build up over time, but must eventually be released via magnetic reconnection, which leads to heating of the coronal plasma (e.g., Parker 1972). Here I consider another aspect of these random footpoint motions, namely, their effect on the large-scale structure of the coronal magnetic field. A three-dimensional model of an active region is developed. The model is based on the following assumptions: (1) the coronal plasma has high electrical conductivity, so that the magnetic field is nearly "frozen" into the plasma; (2) the coronal plasma pressure is much smaller than the magnetic pressure, as is usually the case in solar active regions; (3) the timescale of the footpoint motions is long compared to the time for an Alfvén wave to travel through the active corona, implying that the coronal field $\mathbf{B}(\mathbf{r},t)$ is approximately force free; (4) the plasma motions at the photosphere are purely horizontal, i.e., there are no vertical flows that could transport horizontal magnetic fields into or out of the corona through the base. As we will see, this last assumption is important for obtaining highly twisted magnetic fields because it prevents the submergence of axial fields that have built up in the corona.

Random footpoint motions cause magnetic diffusion both in the corona and in the photosphere. Unfortunately, a proper treatment of coronal magnetic diffusion resulting from random footpoint motions is quite difficult, and is beyond the scope of the present work. Therefore, I adopt a simpler model which includes only the photospheric diffusion and ignores the diffusion in the corona, i.e., the corona is assumed to evolve according to *ideal* MHD. Furthermore, I assume that the mean velocity of the plasma at the base of the corona is equal to the imposed mean velocity in the photosphere. These assumptions greatly simplify the problem of constructing numerical models of active regions in three dimensions. In the following I present preliminary results from such simulations.

A cartesian coordinate system (x,y,z) is used, with z the height above the photosphere. The computational domain is a rectangular box: $0 \leq x \leq L_x$, $0 \leq y \leq L_y$, and $0 \leq z \leq L_z$. I use a uniform grid, and the unit of length equals the grid spacing. The lower boundary of the box ($z=0$) is located at the photosphere. The magnetic field enters and leaves the box only through this lower boundary. The magnetic induction equation for the *interior* of the box is given by:

$$\frac{\partial \mathbf{A}}{\partial t} = \mathbf{v} \times \mathbf{B}, \quad (1)$$

where $\mathbf{v}(\mathbf{r},t)$ is the plasma velocity and $\mathbf{A}(\mathbf{r},t)$ is the vector potential ($\mathbf{B} \equiv \nabla \times \mathbf{A}$). For the photosphere ($z=0$) I use:

$$\frac{\partial A_x}{\partial t} = +v_{0,y} B_z - D \frac{\partial B_z}{\partial y}, \quad (2)$$

$$\frac{\partial A_y}{\partial t} = -v_{0,x} B_z + D \frac{\partial B_z}{\partial x}, \quad (3)$$

where $[v_{0,x}, v_{0,y}]$ is the imposed systematic flow at the photosphere, and D is the diffusion constant describing the effect of the random footpoint motions (Leighton 1964). It follows that $B_z(x,y,0,t)$ at $z=0$ satisfies the following diffusion equation:

$$\frac{\partial B_z}{\partial t} + \frac{\partial}{\partial x}(v_{0,x} B_z) + \frac{\partial}{\partial y}(v_{0,y} B_z) = D \nabla_\perp^2 B_z, \quad (4)$$

where $\nabla_\perp^2 \equiv \partial^2/\partial x^2 + \partial^2/\partial y^2$. The imposed velocity mimics the solar differential rotation on the northern hemisphere:

$$v_{0,x}(x,y) = v_0 \cos(\pi y/L_y), \quad v_{0,y}(x,y) = 0, \quad (5)$$

where v_0 is the velocity amplitude of the differential rotation. Near the left and right boundaries of the computational domain ($x \approx 0$ and $x \approx L_x$), the photospheric velocity field is modified to prevent magnetic flux from entering or leaving the domain through these boundaries. The calculations are performed with $L_x = 64$ and $L_y = L_z = 32$.

According to the present model, the corona evolves through a series of force-free equilibrium states with the current density $\mathbf{j}(\mathbf{r},t)$ parallel to the magnetic field $\mathbf{B}(\mathbf{r},t)$. To approximate such nonlinear force-free fields, I use the magneto-frictional method (see Yang, Sturrock & Antiochos 1986; Sturrock 1994, p. 214). This method assumes that the plasma experiences a frictional force as it moves with respect to the reference frame, and the inertia of the plasma is neglected. Then the plasma velocity is proportional to the Lorentz force:

$$\mathbf{v}(\mathbf{r},t) = \nu^{-1} (\nabla \times \mathbf{B}) \times \mathbf{B}, \quad (6)$$

where $\nu(\mathbf{r},t)$ is the coefficient of friction. In order to speed up the rate of convergence in the weak-field regions, I take ν to be proportional to the square of the magnetic field strength; the unit of time is chosen such that $\nu = B^2$. The imposed photospheric velocity and diffusion constant must be chosen sufficiently small that the coronal field remains close to a force-free state at all times ($v_0 = 0.01$ and $D = 0.01$ in dimensionless units). To monitor the relaxation, I compute the angle θ between the vectors **j** and **B** for each grid point, and I plot histograms of this angle sampled over all interior grid points with significant current density. In most cases $\theta < 0.2$ at almost all such grid points.

Model A simulates the evolution of an initially untwisted bipole (potential field) subject to photospheric diffusion and differential rotation. The magnetic field is followed until time $t = 2000$. The results are shown in Figure 2 (the box indicates the size of the computational domain). Note that the diffusion causes the photospheric magnetic flux to disperse over time, increasing the size of the active region and forcing magnetic flux to cancel at the polarity inversion line. This process involves reconnection of field lines at the inversion line, and produces a left-helical flux rope ($\lambda < 0$). The two ends of the flux rope are anchored in the photosphere and form an inverse S-shaped structure when viewed from above. At the end of the simulation the helical field lines make several complete turns around each other (see Fig. 2c). The axis of the helix reaches a height $z \approx 5$. The magnetic field appears to be stable.

Next I consider two models in which the initial bipole has some internal twist, as suggested by the observations (e.g., Leka et al 1996). The twist is modeled by applying vortical motions (no diffusion) to the footpoints of the potential field shown in Fig. 2a. These motions are applied for a short period of time, and are such that they do not alter the photospheric flux distribution. The velocity is given by $v_{0,x} = \partial\phi/\partial y$ and $v_{0,y} = -\partial\phi/\partial x$, where $\phi(x,y) \propto B_z^2$, i.e., the motions are *along* the contours of $B_z(x,y,0)$. In model B the footpoint motions are counterclockwise, producing a left-helical magnetic field in the corona, and in the model C the footpoint motions are clockwise, producing a right-helical field. The coronal field is allowed to relax to a force free state. The amplitude of the twist is such that the magnetic field crosses the inversion line at an angle of $\pm 45°$.

The resulting force-free configurations are used as the initial states for two time-dependent simulations that include diffusion and solar differential rotation. Figure 3 shows the resulting field at time $t = 2000$ for model B (left-helical initial bipole). In this case the diffusion and differential rotation further enhance the initial twist, producing an inverse S-shaped, left-helical flux rope which is even thicker and higher than in Fig. 2c. Figure 4 shows the result for model C (right-helical initial bipole). In this case the differential rotation counteracts the initial twist, but not enough to reverse its sign; the S-shaped flux rope has right-helical twist. Note that in both cases the central part of the flux rope is nearly

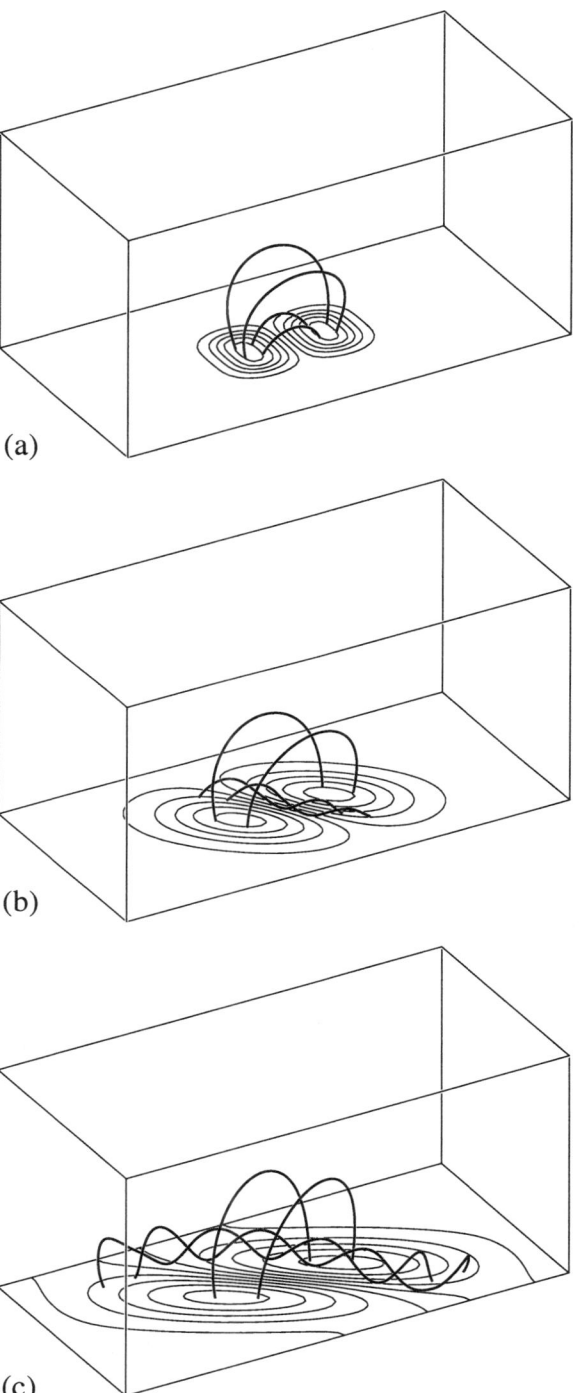

Figure 2. Formation of a left-helical flux rope in a bipolar active region in the Northern hemisphere (Model A). Magnetic field lines (*thick curves*) and contours of B_z (*thin curves*) are shown for (a) the initial potential field; (b) $t = 1000$; (c) $t = 2000$. Positive polarity is on the left, negative polarity on the right.

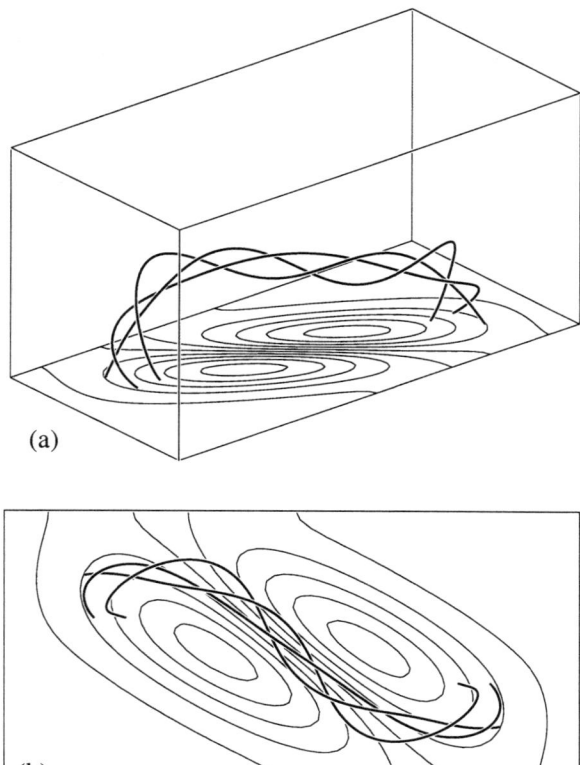

Figure 3. Model of an inverse S-shaped, left-helical flux rope in a bipolar region which initially had a small left-helical magnetic twist (model B). Magnetic field lines (*thick curves*) and contours of B_z (*thin curves*) are shown for time $t = 2000$. (a) side view; (b) top view.

parallel to the inversion line. This is a consequence of flux cancellation, which tends to concentrate the axial field at the polarity inversion line.

3. SUMMARY

It was shown that photospheric flux cancellation generally involves magnetic reconnection and changes in the topology of the coronal magnetic field. A simple model including these effects was presented. According to this model the photospheric diffusion is purely horizontal, i.e., there is no vertical transport of horizontal fields through the photosphere. This prevents axial magnetic fields from submerging below the photosphere and allows magnetic shear or twist to build up in the region just above the polarity inversion line. We presented numerical simulations of active-region magnetic fields, including the effects of photospheric diffusion and solar differential rotation; both twisted and untwisted initial fields were considered. These simulations show the formation of S-shaped and inverse S-shaped flux ropes similar to the observed coronal X-ray structures.

Observations show that active regions in each hemisphere have a mixture of both positive and negative helicity, with only a weak preference for negative helicity in the North and positive helicity in the South (Pevtsov, Canfield and Metcalf 1995). As discussed in the Introduction, most of this helicity probably originates below the photosphere. However, the results of §2 show that for active regions which emerge untwisted, a negative (positive) helicity may be generated in the northern (southern) hemisphere due to the combined effects of surface diffusion and differential rotation. This opens the possibility that some of the observed weak correlation between helicity and hemisphere may be due to the effects of surface differential rotation acting on initially untwisted bipoles. In view of this uncertainty, I suggest it is premature to conclude that the observations imply a preferred sense of twist of the subsurface magnetic flux tubes in each hemisphere.

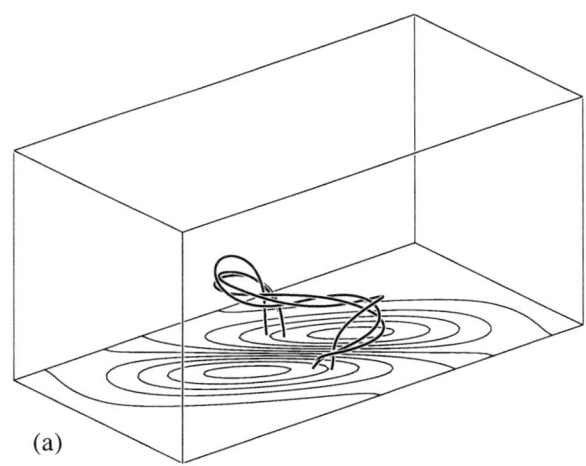

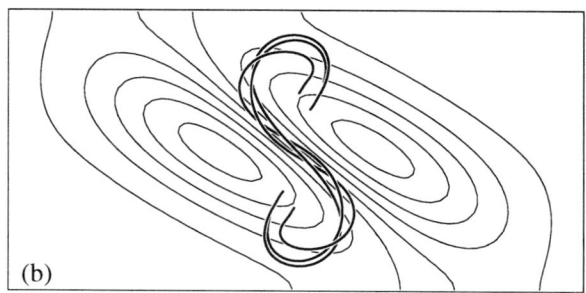

Figure 4. Similar to Fig. 3, except that the initial bipole had right-helical magnetic twist (Model C).

Acknowledgment. The author would like to thank C. R. DeVore for stimulating discussions concerning the effects of flux cancellation on coronal magnetic fields. Travel support was provided in part by NSF grant ATM-9696256 to Helio Research.

REFERENCES

Acton, L. W., et al, The Yohkoh mission for high-energy solar physics, *Science, 258*, 618-625, 1992.

Antiochos, S. K., Dahlburg, R. B., and Klimchuk, J. A., The magnetic field of solar prominences, *Astrophys. J., 420*, L41-L44, 1994.

Canfield, R. C., and Pevtsov, A. A., Helicity of solar active-region magnetic fields, in *Synoptic Solar Physics – 18th NSO Sacramento Peak Summer Workshop*, edited by K. S. Balasubramaniam, J. W. Harvey, and D. M. Rabin, ASP Conf. Series *140*, 131-143, 1998.

de La Beaujardiére, J.-F., Canfield, R. C., and Leka, K. D., The morphology of flare phenomena, magnetic fields, and electric currents in active regions. III. NOAA active regions 6233 (1990 August), *Astrophys. J., 411*, 378-382, 1993.

DeVore, C. R., Simulations of bipolar magnetic regions subjected to differential rotation and supergranular diffusion, *Eos Trans. AGU, 79*, Spring Meet. Suppl., S281, 1998.

Falconer, D. A., Moore, R. L., Porter, J. G., Gary, G. A., Shimizu, T., Neutral-line magnetic shear and enhanced coronal heating in solar active regions, *Astrophys. J., 482*, 519-534, 1997.

Gary, G. A., Moore, R. L., Hagyard, M. J., and Haisch, B. M., Non-potential features observed in the magnetic field of an active region, *Astrophys. J., 314*, 782-794, 1987.

Hudson, H. S., Acton, L. W., and Freeland, S. L., A long-duration solar flare with mass ejection and global consequences, *Astrophys. J., 470*, 629-635, 1996.

Krall, K., Smith, J., Jr., Hagyard, M., West, E., and Cummings, N., Vector magnetic field evolution, energy storage, and associated photospheric velocity shear within a flare-productive active region, *Sol. Phys., 79*, 59-75, 1982.

Kurokawa, H., Two distinct morphological types of magnetic shear development and their relation to solar flares, *Sol. Phys., 113*, 259-263, 1987.

Leighton, R. B., Transport of magnetic fields on the Sun, *Astrophys. J., 140*, 1547-1562, 1964.

Leka, K. D., Canfield, R. C., McClymont, A. N., de La Beaujardiére, J.-F., Fan, Y., and Tang, F., The morphology of flare phenomena, magnetic fields, and electric currents in active regions. II. NOAA active regions 5747 (1989 October), *Astrophys. J., 411*, 370-377, 1993.

Leka, K. D., Canfield, R. C., McClymont, A. N., and van Driel-Gesztelyi, L., Evidence for current-carrying emerging flux, *Astrophys. J., 462*, 547, 1996.

Manoharan, P. K., van Driel-Gesztelyi, L., Pick, M., and Démoulin, P., Evidence for large-scale solar magnetic reconnection from radio and X-ray measurements, *Astrophys. J., 486*, L73-L76, 1996.

Martin, S. F., Bilimoria, R., and Tracadas, P. W., Magnetic field configurations basic to filament channels and filaments, in *Solar Surface Magnetism*, edited by R. J. Rutten and C. J. Schrijver, pp. 303-338, Kluwer Acad. Publ., Dordrecht, The Netherlands, 1994.

McClymont, A. N., Jiao, L., and Mikić, Z., Problems and progress in computing three-dimensional coronal active region magnetic fields from boundary data, *Sol. Phys., 174*, 191-218, 1997.

Metcalf, T. R., Canfield, R. C., Hudson, H. S., Mickey, D. L., Wülser, J.-P., Martens, P. C. H., Tsuneta, S., Electric currents and coronal heating in NOAA active region 6952, *Astrophys. J., 428*, 860-866, 1994.

Moore, R.L., Schmieder, B., Hathaway, D.H., and Tarbell, T.D., 3-D magnetic field configuration late in a large two-ribbon flare, *Sol. Phys., 176*, 153-169, 1997.

Parker, E.N., The formation of sunspots from the solar toroidal field, *Astrophys. J., 121*, 491-507, 1955.

Parker, E.N., Topological dissipation and the small-scale fields in turbulent gases, *Astrophys. J., 174*, 499-510, 1972.

Pevtsov, A. A., Canfield, R. C., and McClymont, A. N., On the subphotospheric origin of coronal electric currents, *Astrophys. J., 481*, 973-977, 1997.

Pevtsov, A. A., Canfield, R. C., and Metcalf, T. R., Latitudinal variation of helicity of photospheric magnetic fields, *Astrophys. J., 440*, L109-112, 1995.

Pevtsov, A. A., Canfield, R. C., and Zirin, H., Reconnection and helicity in a solar flare, *Astrophys. J., 473*, 533-538, 1996.

Pneuman, G. W., The formation of solar prominences by magnetic reconnection and condensation, *Sol. Phys., 88*, 219-239, 1983.

Rust, D. M., and Kumar, A., Helical magnetic fields in filaments, *Sol. Phys., 155*, 69-97, 1995.

Rust, D. M., and Kumar, A., Evidence for helically kinked magnetic flux ropes in solar eruptions, *Astrophys. J., 464*, L199-L202, 1996.

Sakurai, T., Computational modeling of solar magnetic fields, in *Proc. IAU Coll. 141, The magnetic and velocity fields of solar active regions*, edited by H. Zirin, G. Ai, and H. Wang, ASP Conf. Series *46*, 91, 1993.

Schmieder, B., Démoulin, P., Aulanier, G., and Golub, L., Differential magnetic field shear in an active region, *Astrophys. J., 476*, 881-886, 1996.

Schmieder, B., Aulanier, G., Démoulin, P., van Driel-Gesztelyi, L., Roudier, T., Nitta, N., and Gauzzi, G., Magnetic reconnection driven by emergence of sheared magnetic field, *Astron. Astrophys., 325*, 1213-1225, 1997.

Seehafer, N., Electric current helicity in the solar atmosphere, *Sol. Phys., 125*, 219-232, 1990.

Sturrock, P. A., Solar flares and magnetic topology, *Sol. Phys., 113*, 13-30, 1987.

Sturrock, P. A., *Plasma Physics*, 335 pp., Cambridge Univ. Press, 1994.

Tanaka, K., Studies on a very flare-active δ group: peculiar δ spot evolution and inferred subsurface magnetic rope structure, *Sol. Phys., 136*, 133-149, 1991.

Tanaka, K., and Nakagawa, Y., Force-free magnetic fields and flares of August 1972, *Sol. Phys., 33*, 187-204, 1973.

Tsuneta, S., et al., The Soft X-ray Telescope for the Solar-A mission, *Sol. Phys., 136*, 37-67, 1991.

van Ballegooijen, A. A., and Martens, P. C. H., Formation

and eruption of solar prominences, *Astrophys. J., 343*, 971-984, 1989.

van Driel-Gesztelyi, L., Csepura, G., Schmieder, B., Malherbe, J.-M., Metcalf, T., Evolution of a delta group in the photosphere and corona, *Sol. Phys., 172*, 151-160, 1997.

Van Hoven, G., Mok, Y., and Mikić, Z., Coronal loop formation resulting from photospheric convection, *Astrophys. J., 440*, L105-L108, 1995.

Wang, Y.-M., Nash, A. G., and Sheeley, N. R., Jr., Magnetic flux transport on the Sun, *Science 245*, 712-717, 1989.

Yang, W. H., Sturrock, P. A., and Antiochos, S. K., Force-free magnetic fields: the magneto-frictional method, *Astrophys. J., 309*, 383-391, 1986.

Zirin, H., Fine structure of solar magnetic fields, *Sol. Phys., 22*, 34-48, 1972.

Zirker, J. B., Martin, S. F., Harvey, K., and Gaizauskas, V., Global magnetic patterns of chirality, *Sol. Phys., 175*, 27-44, 1997.

A. A. van Ballegooijen, Harvard-Smithsonian Center for Astrophysics, 60 Garden Street, MS 15, Cambridge, MA 02138. (e-mail: vanballe@cfa.harvard.edu)

Magnetic Helicity in Solar Filaments and Coronal Mass Ejections

D. M. Rust

Applied Physics Laboratory, The Johns Hopkins University, Laurel, Maryland

Erupting solar filaments are often coiled and the knots of plasma in them rotate as though constrained to follow helical magnetic field lines. Starting with this evidence of magnetic helicity in filaments, this article reviews observations and recent models of solar filaments with an emphasis on how to infer their magnetic helicity. Results from telescopic observations are often controversial. They are being supplemented by *in situ* measurements of the ejected magnetic fields and plasmas that pass by interplanetary spacecraft. Correlations of solar events with interplanetary magnetic cloud properties yield insights into the nature of magnetic helicity on the Sun. Examples include the segregation of magnetic helicity, with negative/positive helicity dominating in the north/south, and an association of filament eruptions with helical kink instabilities.

1. INTRODUCTION

Solar filaments are thread-like clouds of magnetized plasma in the Sun's atmosphere. Their density is 100 times greater than that of the surrounding corona but their temperature is 100 times lower, so they are in pressure balance with the *corona*, which is the atmosphere around them. They are suspended by magnetic fields at heights up to 100,000 km above the solar surface, or *photosphere*. Their high density (10^{10} cm^{-3}) and relatively low temperature (10,000 K) make them easy to photograph. They glow brightly because of emission in the hydrogen Balmer lines and various helium lines. For at least half a century photographs of helical filaments, such as the one in Figure 1, have been among the most popular images in astronomy textbooks. Although such images should have long ago triggered serious study of magnetic helicity on the Sun, they failed to ignite serious analysis until *House and Berger* [1987] discussed the transfer of helicity along the body of an erupting filament. Even their paper did not have much impact. Until very recently one could still get into an argument over whether such filaments truly require a helical magnetic field. In fact, there are no direct measurements of the twist of fields in filaments, but the circumstantial evidence for magnetic helicity there is convincing nevertheless. I review this evidence in Section 2. Many other issues are definitely not settled. When do the helical fields develop? Where they come from? What they have to do with filament eruptions and mass ejections from the Sun? It will be my task in Section 3 to review the pros and cons of these questions. The answers may be of fundamental importance to understanding solar activity and the internal dynamo that generates the magnetic fields.

2. EVIDENCE FOR MAGNETIC HELICITY IN FILAMENTS

The filament in Figure 1 is unusual in that it shows a clearly twisted structure even though it is relatively stable. It did erupt from the Sun a few hours after this image was made. Almost all filaments erupt eventually but they spend most of their lives suspended quietly over the borders between patches of upward- and downward-pointed magnetic fields rooted in the photosphere. That the fields in erupting filaments are twisted is fairly easily established

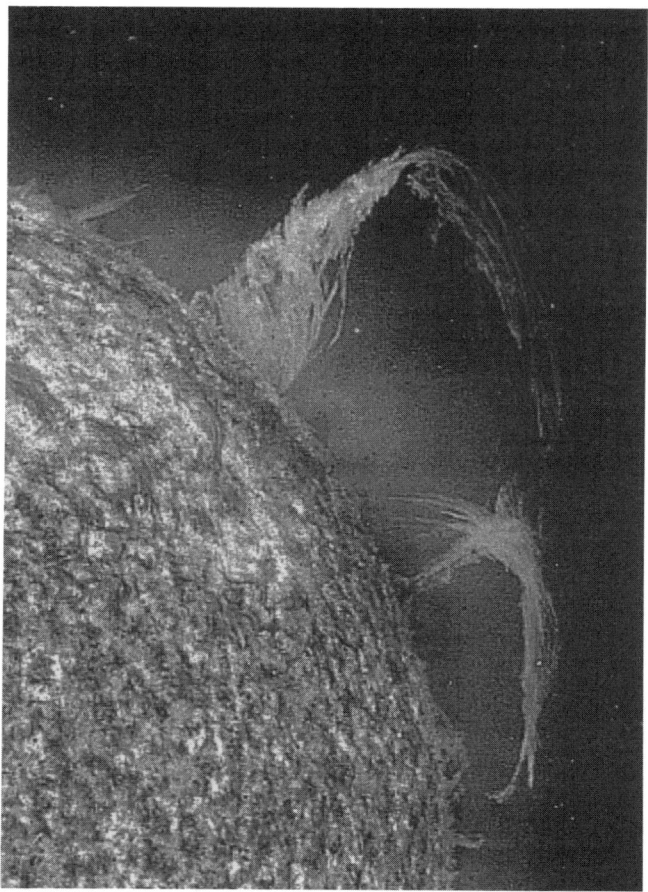

Figure 1. Photograph of a helical solar filament at 1934 UT on January 10, 1998. The filament is an ionized plasma constrained by the magnetic fields. It is seen here in the ultraviolet light emitted by ionized helium. (From the EIT experiment on the ESA/NASA SOHO spacecraft)

when spectra and a cinematic sequence of high-resolution images are available. From measured Doppler shifts and transverse motions, one can show that the trajectories of the plasma knots are indeed helical [e.g., *Kucera et al.*, 1998].

Gigolashvili [1978] used movies and spectra to determine the *chirality* (handedness) of the motions in quiescent filaments. He found that the sense of rotation was predominately right-handed in the southern hemispher and left-handed in the north, but he published no details of the measurements. Considering the controversies in recent years over just when and how the fields in filaments become helical, it is very important that his results be checked with new observations.

Liggett and Zirin [1984] reported that 5 out of 51 quiescent *prominences* they studied showed rotational motion. A prominence is a filament seen from the side, with the sky as background, as in Figure 1. The terms filament and prominence are used interchangeably because there is no physical distinction between them.

Although *Liggett and Zirin* identified rotation in prominences, they did not report on the chirality of the motion. Thus important questions remain: are the fields in quiescent filaments usually helical, and if they are, what is the global distribution of positive and negative magnetic helicity?

A systematic study of helical motions in quiescent filaments would be a difficult task. A task only slightly less difficult was undertaken by *Martin et al.* [1994] who classified hundreds of filaments according to their appearance and orientation with respect to the surrounding magnetic fields. They found that filaments whose *fibril structures* resemble the threads of a right-handed screw predominate in the northern hemisphere while filaments resembling left-handed screws predominate in the southern hemisphere. They called these filaments *dextral* and *sinistral*, respectively. The fibrils of dextral filaments would appear to an observer on either side to turn out of the filament from left to right. Similarly the fibrils of a sinistral filament would appear to flow from right to left. Upon comparing the patterns of the fields near their filaments with measurements of the axial fields, *Martin et al.* established that, for an observer in the adjacent positive-field region, the axial fields in dextral and sinistral filaments always point to the right and left, respectively.

Figure 2 shows examples of dextral and sinistral filaments and their relationship with the positive, negative and axial fields. Martin's hemispherical rule applies with the fewest exceptions to quiescent filaments and to filaments at the borders of sunspot regions. Filaments inside sunspot regions do not seem to obey the hemispheric rule so faithfully.

Certainly the global segregation of dextral and sinistral filaments suggests a global segregation according to magnetic helicity. However, lacking definitive vector magnetic field measurements or reliable 3-D trajectory data on filament knots, we cannot say for sure what is implied by filament morphology. There are many lines of evidence, however, that *the fields in dextral filaments form left-handed coils or 'flux ropes' having negative helicity* while those in sinistral filaments form right-handed coils and have positive helicity.

Very suggestive evidence appears in the records of the interplanetary disturbances called magnetic clouds (MCs) [*Klein and Burlaga*, 1982]. MCs are studied with magnetometers aboard near-Earth spacecraft. Magnetometers allow precise and almost unambiguous description of the fields in MCs, whose origins are in solar eruptions. Most MC magnetic fields are helical flux ropes and they are associated statistically with solar filament eruptions and coronal mass ejections (CMEs) [*Bothmer and Schwenn*, 1994; *Gosling*, 1990; *Marubashi*, 1986; *Rust*, 1994; *Wilson and Hildner*, 1984].

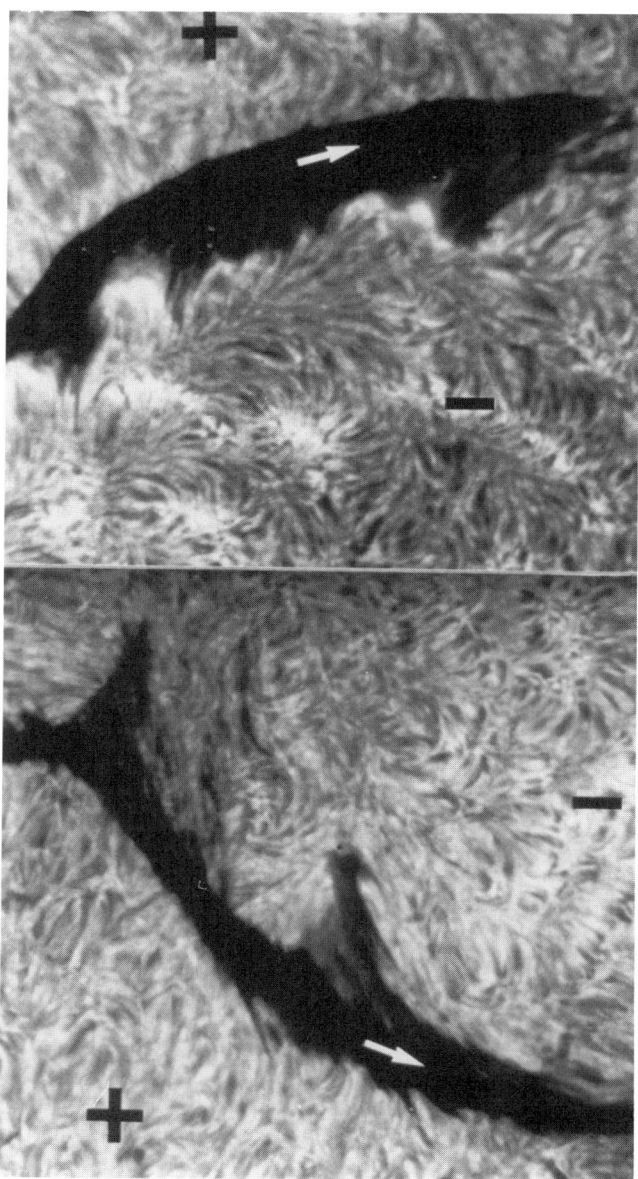

Figure 2. Two quiescent filaments. They appear dark against the solar background because they are high above the Sun's surface and scatter the light from below in all directions. They appear bright when viewed from the side, as in Figure 1. The upper/lower filament is termed 'sinistral/dextral' by *Martin et al.* [1994] because the fine threads (fibrils) give it the appearance of a left-/right-handed screw. The + and − signs indicate upward and downward magnetic fields. The arrows show the direction of the axial fields in the filaments. (Images from *Zirker* [1997b])

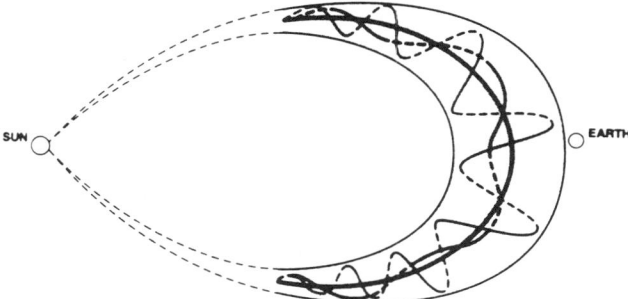

Figure 3. A twisted flux rope in a magnetic cloud extending from the Sun to Earth. Magnetic clouds pass Earth 3 - 4 days after a filament erupts.

The fields in MCs are among the most regular and precisely measured magnetic fields in all of astrophysics. They usually match the *Lundquist* [1950] solution to the force-free field equation [*Lepping et al.*, 1990] very nicely, i.e., an MC closely resembles a twisted flux rope with purely azimuthal field at the outer boundaries of a cylinder and purely axial field at the center (Figure 3). The handedness of most clouds' helicity can be determined quickly and unambiguously from the magnetometer measurements. *Lepping et al.* [1997] have determined the chirality and calculated the magnetic flux a large sample of MCs.

Rust [1994] and *Bothmer and Schwenn* [1994] independently discovered that the flux rope fields in MCs associated with filaments that erupted from the northern/southern hemisphere are predominantly left-/right-handed. From the principle of conservation of magnetic helicity [*Taylor*, 1986], we expect that the fields in filaments will retain their magnetic helicity, even as they erupt. So one can infer that filaments' chiralities are the same as those of the associated MCs. *Rust and Kumar* [1994] suggested that the MC field is simply the filament field expanded and transported through space by the eruption. Other evidence indicates that only a small part of an MC's volume is filled with filament material. This would probably not change the inferences about the chirality of filament fields. However, one might imagine (a) some transformation occurs during filament eruptions, so filament chirality might not be derived easily from MC chirality, or (b) some there is misidentification, vis., MC fields might not evolve from filament fields, but rather from the fields of the surrounding corona. The expulsion of fields and plasma from the corona is called a coronal mass ejection. CMEs accompany all filament eruptions and generally involve much larger volumes and they may become MCs in interplanetary space.

Gosling [1990; 1999] suggested that a fundamental transformation takes place during eruptions and that the helical fields eventually charted in MCs are acquired only during the eruptive process. According to Gosling, the

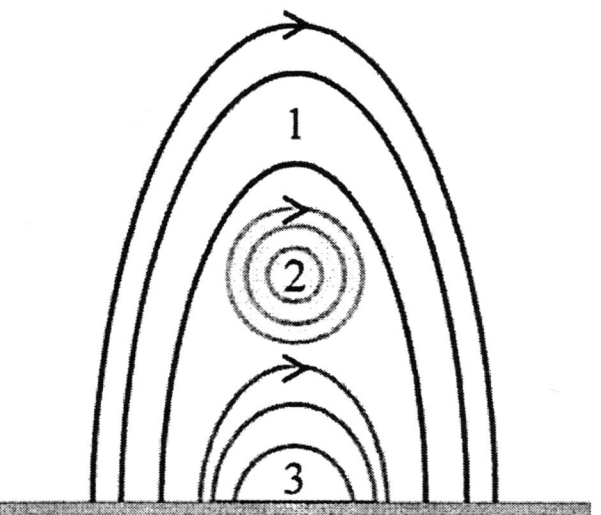

Figure 4. End-on view of the magnetic fields in a flux-rope model of a filament. Region 1 is an arcade of overlying fields. Region 2 is the flux rope in which the filament plasma is distributed. Region 3 is a low-lying arcade created in the course of a series of magnetic reconnections that also produce the flux rope [see *Berger*, 1998]

flux-rope fields of MCs could be created by field-line reconnections in the corona as an eruption occurs. Then, there would be no need for preexisting magnetic helicity in the form of a helical flux rope in the filament. All the helicity might reside in sheared coronal fields until they become unstable, reconnect and erupt from the Sun.

Berger [1998] provides a simple mathematical model of the possible effect of reconnections on sheared coronal fields. When an arcade of fields of width $2a$ is sheared a step s along the border between positive and negative photospheric fields, then the acquired magnetic helicity per unit length H is $2B_0^2 s a^2/15$, where B_0 is the field strength at a height a above the photosphere. Supposing that only the magnetic flux within a distance εa of the polarity boundary reconnects, Berger calculates that three separate magnetic regions could be formed as shown in Figure 4.

Region 1 is the coronal arcade, which has lost some of its shear to provide twist to the flux rope of region 2. By assumption, the reconnection leaves region 3, an underlying arcade, with zero helicity. Berger finds that the helicity per unit length in the flux rope is $\varepsilon^5 H$ and that if, for example $\varepsilon = 3/4$, then the helicity in the arcade (still sheared, but less so than before reconnection) is $0.3\ H$. The linking helicity, because the flux rope threads the arcade, would be $0.46\ H$, and finally, the twist helicity in the flux rope would be $0.24\ H$. Neither Gosling nor Berger goes into how the fields in the arcade could reconnect up and down the arcade to form a flux rope, but the analysis brings out two interesting points. First, the flux rope will have the same chirality as the sheared field,

and second, any shear remaining in the corona after reconnection has the same chirality as the flux rope.

The typical filament occupies part of a large arcade beneath a helmet-shaped region in the corona. According to a theoretical model by *Low and Hundhausen* [1995], the filament's support against gravity requires its fields to have the same chirality as the surrounding coronal fields. In their model, the interplanetary flux rope of an MC would be identified not only with the filament but with the arcade and filament together.

Even if the fields in quiescent filaments are not helical flux ropes, as assumed by many modelers [*Lites and Low*, 1997; *Low and Hundhausen*, 1995; *Priest et al.*, 1996; *Rust and Kumar*, 1994], and instead are sheared arcades, as advocated by *Gosling* [1999], *Antiochos* [1998] and *Martin and Echols* [1994], we may conclude that all the fields associated with a filament eruption and CME should start from the Sun with the same chirality. The inferred chirality should be the same as in the MC because of helicity conservation.

We return now to the question of whether an MC evolves from the eruptive filament or from the filament and the surrounding fields. *Burlaga et al.* [1998] have published the most persuasive evidence so far that filament plasma is a distinct feature in MCs. A small and very cold plasma 'plug' of exceptional density (150 cm^{-3}) at Earth's orbit was embedded in the MC of January 11, 1997. The ratio of density in the plug to that in the MC was about 30, which is about the ratio of density in solar prominences to that in the overlying corona. Also the plug was about 10% of the size of the cloud, i.e., about the same ratio as filament size (~50,000 km) to coronal arcade size (500,000 km). The cold material also had an unusually high $^4\text{He}^{++}/\text{He}^+$ ratio, indicating a source low in the solar atmosphere.

Although the January 11 MC was plausibly linked with a reported filament eruption, *Burlaga et al.* did not study the solar event in detail. One might conclude, however, that since the plug occupied only a small part of the MC's flux rope structure, the filament was only a small part of the flux rope that left the Sun. In this case as in virtually all other MC studies, the best available solar data have not yet been integrated with the best interplanetary data. More interplanetary spacecraft and the STEREO (Solar-Terrestrial Relations Observatory) mission will rectify this situation in the next few years. STEREO telescopes and magnetometers will provide far more comprehensive coverage of solar ejecta and MCs than ever before. It is an important goal because inferences of magnetic field properties from remote sensing will always be entangled in some ambiguity.

The magnetic helicity in a filament may become so great that an MHD instability eventuates. By examining X-ray emissions in erupting filaments, *Rust and Kumar* [1996] inferred that filaments are threaded by unstable helical flux ropes. Sigmoid X-ray emission signatures, such as the one shown in Figure 5, are usually S-shaped in the southern hemisphere and Z-shaped in the north. As shown by *Rust and Kumar* these are the signatures of right-handed and left-handed flux ropes undergoing $m = +1$ and $m = -1$ helical kink instabilities, respectively.

3. ORIGIN OF MAGNETIC HELICITY IN FILAMENTS

If filament fields are predominantly left-handed/right-handed in the north/south, then their pattern is in agreement with the global pattern of magnetic helicity as determined from observations of sunspots, coronal arcades, and interplanetary fields [*Rust*, 1994; *Zirker et al.*, 1997b]. What is the origin of this helicity?

Shear introduces helicity in fields, but the only persistent flow in the photosphere that could shear coronal arcades is the *differential rotation*. The photospheric gases at the equator rotate faster than those at the poles. Suppose that the positive magnetic footpoints of an arcade are closer to the equator than the negative footpoints. They would then rotate faster, and the sheared arcade fields would point eastward. The problem with this picture is that such shear would produce sinistral filaments in the north and dextral filaments in the south. For example, in the northern hemisphere one would see filaments with axial fields pointing to the left as viewed from the adjacent positive magnetic field region. This is the opposite of what the observations show [*Leroy et al.*, 1983; *Rust*, 1967; *Zirker et al.*, 1997a].

A computer simulation of the action of differential rotation on filaments at various orientations [*Van Ballegooijen et al.*, 1998] supports the conclusion that differential rotation produces quiescent filaments whose axial fields are opposite the observed ones. Also, such shear produces positive/negative magnetic helicity in the north/south. Helicity in MCs is the opposite.

Since the magnetic helicity of filaments cannot be attributed to the effects of shear, *Rust* [1994] suggested that the helicity is developed inside the Sun by the action of the solar dynamo. This scenario is consistent with the Babcock-Leighton [*Babcock*, 1961; *Leighton*, 1969] model of the solar cycle. It would generate positive helicity in the south and negative helicity in the north. The problem with this approach is that solar dynamo models, including the Babcock-Leighton model, are only sketches with many free parameters. They are not yet physical models with predictive powers.

Mackay et al. [1998] took a more empirical approach to the origin of magnetic helicity in filaments. They focused on the filaments that form between sunspot regions. These filaments and high latitude quiescent filaments almost always obey the hemispheric segregation rules [*Zirker et al.*, 1997a]. Mackay et al. showed how sunspot regions with negative helicity, for example, could result in

Figure 5. X-ray signature of an erupting filament. The figure shows thermal X-rays from a 10^6 K gas created probably by heating of the 10^4 K filament plasma at the onset of the eruption. The aspect ratio (height to width) of such sigmoid features is in good agreement with the predictions of MHD theory of the helical kink instability [*Rust and Kumar*, 1996].

quiescent filaments with negative helicity. *Kuijpers* [1997] offered a similar scenario, as shown in Figure 6.

The *Kuijpers* and *Mackay et al.* scenarios are based on the observation [*Martin*, 1986] that magnetic fields appear to converge on filament channels and undergo reconnection and/or submergence there [see e.g. *Wang et al.*, 1996]. When fields have negative helicity, as shown by the Z-shaped field lines in Figure 6, a component of the positive fields swept into the neutral line would point rightward along the neutral line. A filament formed there would thus have a 'dextral' axial component. It is not clear what will happen if two regions of opposite helicity converge, but coronal observations [*Canfield et al.*, 1996] suggest that no reconnection will occur, i.e., no filament will form.

Nakagawa et al. [1971] showed that the chirality of the fields in sunspots can be inferred easily and accurately because sunspots that appear to be rotating clockwise/counterclockwise must have negative/positive helical fields. What about filaments inside sunspot regions, the ones that link the two principal magnetic poles or sunspots? *Rust and Martin* [1994] studied ten sunspot regions that had both filaments and sunspots with clearly twisted fields. Whenever the helicity of the twisted fields in sunspots could be classified as positive/negative, *Rust and Martin* found that the associated filament was sinistral/dextral. In *Martin and McAllister's* [1997] filament model, the chirality of the magnetic fields is opposite to that of the

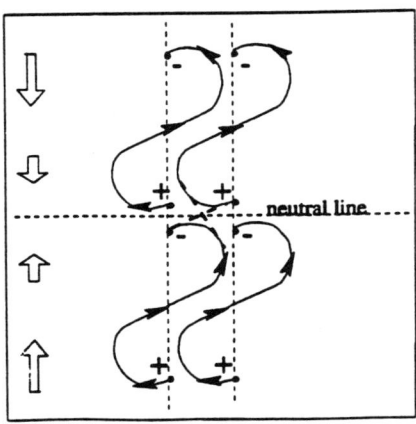 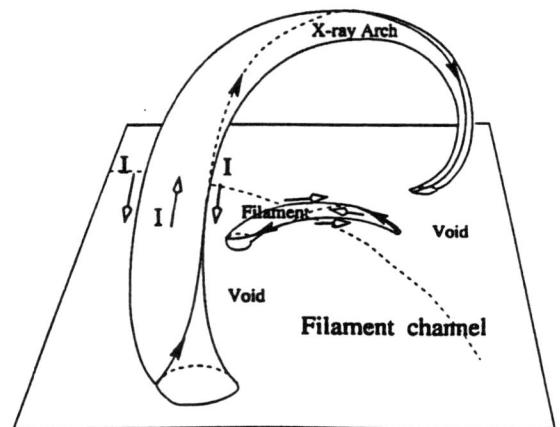

Figure 6. Method sketched by *Kuijpers* [1997] for forming the magnetic fields in filaments along the borders between sunspot regions. The regions sketched in the left panel have fields with negative magnetic helicity since the fieldlines connecting the + and - poles are shaped like Zs and not like Ss. If the fields reconnect at the neutral line, they will produce filaments whose helicity is also negative. The sketch on the right shows how the fields in a filament and in an overlying coronal arch might look after reconnection.

surrounding fields. However, the authors suppose that filaments take on the same chirality as the overlying corona when they erupt.

Martin [1998] supposes that filament fields have the opposite helicity of the surrounding fields. This is based on her observations that the fibrils extending from a filament are associated with pockets of field whose polarity is opposite that of the other bordering fields. But *Aulanier et al.* [1998a; 1998b] also studied the effect of pockets of opposite polarity field on filament morphology, and they found from a 2-1/2-D force-free helical field model that the features of sinistral filaments, for example, can be formed by minor distortions in flux ropes with positive helicity.

4. CONCLUSIONS

Although our science has yet to positively identify the individual plasmas leaving the Sun with those passing Earth, we may still reasonably conclude that there is magnetic helicity in filaments and CMEs and that it is probably some reasonable fraction of the measured helicity in MCs and that the chirality of filament fields is revealed by the chirality of the MC fields. As for the true helicity of sinistral and dextral filaments, we see that for the *Mackay et al.* Filament-formation scenario to work and for other reasons, sinistral filaments must have positive helicity and dextral filaments negative helicity. Whether the helicity is in the form of a twisted flux rope or in linked fields or in sheared fields is not decided. Eventually, some exquisite telescopic measurements at very high resolution will map the fields in filaments. Before then, however, careful study of the plasma knots in filaments, which are constrained to move along the fields, should allow field line mapping and, at least, a direct determination of filaments' chirality.

Filaments are carriers of magnetic helicity. Because of their association with the sigmoid signatures of the helical kink instability we infer that filaments and their surrounding coronal magnetic fields accumulate magnetic helicity until the conditions for instability are met. However, it would be best if we had observations of particular solar filament eruptions and their interplanetary consequences on a one-to-one basis.

Acknowledgments. This work was supported by NASA grants NAG5-4955 and NAG5-7235 and by National Science Foundation grant OPP-9615073.

REFERENCES

Antiochos, S.K., The magnetic topology of solar eruptions, *Astrophys. J. Lett.*, *502*, L181, 1998.

Aulanier, G., and P. Demoulin, 3-D magnetic configurations supporting prominences. I. The natural presence of lateral feet, *Astron. Astrophys.*, *329*, 1125-1137, 1998a.

Aulanier, G., P. Demoulin, L. Van Driel-Gesztelyi, P. Mein, and C. Deforest, 3-D magnetic configurations supporting prominences. II. The lateral feet as a perturbation of a twisted flux-tube, *Astron. Astrophys.*, *335*, 309-322, 1998b.

Babcock, H.W., The topology of the Sun's magnetic field and the 22-year cycle, *Astrophys. J.*, *133*, 572, 1961.

Berger, M., Magnetic helicity and filaments, in *New Perspectives on Solar Prominences*, ed. by D.F. Webb, D.M. Rust, and B. Schmieder, pp. 102-110, Astro. Soc. Pacific Conf. Ser., 150, 1998.

Bothmer, V., and R. Schwenn, Eruptive prominences as sources of magnetic clouds in the solar wind, *Space Sci. Rev.*, *70*, 215, 1994.

Burlaga, L., R. Fitzenreiter, R. Lepping., K. Ogilvie, A. Szabo, A. Lazarus, J. Steinberg, G. Gloeckler, R. Howard, D. Michels, C. Farrugia, R.P. Lin, and D.E. Larson, A magnetic cloud containing prominence material: January 1998, *J. Geophys. Res.*, *103*, 277-286, 1998.

Canfield, R.C., A.A. Pevtsov, and A.N. McClymont, Magnetic chirality and coronal reconnection, in *Magnetic Reconnection in the Solar Atmosphere*, ed. by R.D. Bentley, and J.T. Mariska, pp. 341 - 346, Astron. Soc. Pacific Conf. Ser., 111, 1996.

Gigolashvili, M.S., An investigation of macroscopic motions using the Ca^+ lines in the prominence of 15 October 1969, *Solar Phys.*, 60, 293, 1978.

Gosling, J.T., Coronal mass ejections and magnetic flux ropes in interplanetary space, in *Physics of Magnetic Flux Ropes*, ed. by C.T. Russell, E.R. Priest, and L.C. Lee, pp. 12-75, Amer. Geophys. Union, Geophys. Monograph 58, 1990.

Gosling, J.T., The role of reconnection in the formation of flux ropes in the solar wind, in *Magnetic Helicity in Space and Laboratory Plasmas*, ed. by A.A. Pevtsov, Amer. Geophys. Union, Geophys. Monograph (in press), 1999.

House, L.L., and M.A. Berger, The ejection of helical field structures through the outer corona, *Astrophys. J.*, 323, 406, 1987.

Klein, L.W., and L.F. Burlaga, Interplanetary magnetic clouds at 1 AU, *J. Geophys. Res.*, 87, 613-624., 1982.

Kucera, T.A., A.I. Poland, J.E. Wiik, B. Schmieder, and G. Simnett, Helical structure in an eruptive prominence related to a CME (SUMER,CDS,LASCO), in *New Perspectives on Solar Prominences*, ed. by D.F. Webb, D.M. Rust, and B. Schmieder, pp. 318-321, Astro. Soc. Pacific Conf. Ser., 150, 1998.

Kuijpers, J., A solar prominence model, *Astrophys. J. Lett.*, 489, L201, 1997.

Leighton, R.B., A magneto-kinematic model of the solar cycle, *Astrophys. J.*, 156, 1, 1969.

Lepping, R.P., J.A. Jones, and L.F. Burlaga, Magnetic field structure of interplanetary magnetic clouds at 1 AU, *J. Geophys. Res.*, 95, 11957, 1990.

Lepping, R.P., A. Szabo, C.E. DeForest, and B.J. Thompson, Magnetic flux in modeled magnetic clouds at 1 AU and some specific comparisons to associated photospheric flux, in *Correlated Phenomena at the Sun, in the Heliosphere, and in Geospace*, pp. 163-170, ESLAB, SP-415, 1997.

Leroy, J.L., V. Bommier, and S. Sahal-Brechot, The magnetic field in the prominences of the polar crown, *Solar Phys.*, 83, 135-142., 1983.

Liggett, M., and H. Zirin, Rotation in prominences, *Solar Physics*, 91, 259-267., 1984.

Lites, B.W., and B.C. Low, Flux emergence and prominences: a new scenario for 3-dimensional field, *Solar Phys.*, 174, 91-98., 1997.

Low, B.C., and J.R. Hundhausen, Magnetostatic structures of the solar corona. 2: The magnetic topology of quiescent prominences, *Astrophys. J.*, 443, 818-836, 1995.

Lundquist, S., Magnetohydrostatic fields, *Ark Fys*, 2, 361, 1950.

Mackay, D.H., E.R. Priest, V. Gaizauskas, and A.A. Van Ballegooijen, Role of helicity in the formation of intermediate filaments, *Solar Phys.*, 180, 299-312., 1998.

Martin, S.F., Recent observations of the formation of filaments, in *Coronal and Prominence Plasmas*, ed. by A.I. Poland, pp. 73-80, NASA Conf. Pub., 2442, 1986.

Martin, S.F., Filament chirality: a link between fine-scale and global patterns, in *New Perspectives in Solar Prominences*, ed. by D.F. Webb, D.M. Rust, and B. Schmieder, pp. 419-429, Astron. Soc. Pacific Conf. Ser., 150, 1998.

Martin, S.F., R. Bilimoria, and P.W. Tracadas, Magnetic field configurations basis to filament channels, in *Solar Surface Magnetism*, ed. by R.J. Rutten, and C.J. Schrijver, pp. 303-338, Springer-Verlag, 1994.

Martin, S.F., and C.R. Echols, Solar filament model, in *Solar Surface Magnetism*, ed. by R.J. Rutten, and C.J. Schrijver, pp. 339, Kluwer, 1994.

Martin, S.F., and A.H. McAllister, Predicting the sign of magnetic helicity in erupting filaments and coronal mass ejections, in *Coronal Mass Ejections*, ed. by N. Crooker, J.A. Joselyn, and J. Feynman, pp. 127-138, Amer. Geophys. Union Geophy. Monograph, 99, 1997.

Marubashi, K., Structure of the interplanetary magnetic clouds and their solar origins, *Adv. Space Res.*, 6, 335-338., 1986.

Nakagawa, Y., M.A. Raadu, D.E. Billings, and D. McNamara, On the topology of filaments and chromospheric fibrils near sunspots, *Solar Phys*, 19, 72, 1971.

Priest, E.R., A.A. Van Ballegooijen, and D.H. Mackay, A model for dextral and sinistral prominences, *Astrophys. J.*, 460, 530, 1996.

Rust, D.M., Magnetic fields in quiescent solar prominences I. Observations, *Astrophys. J.*, 150, 313, 1967.

Rust, D.M., Spawning and shedding helical magnetic fields in the solar atmosphere, *Geophys. Res. Lett.*, 21, 241-244, 1994.

Rust, D.M., Magnetic helicity, MHD kink instabilities and reconnection in the corona, in *Magnetic Reconnection in the Solar Atmosphere*, ed. by R.D. Bentley, and J.T. Mariska, pp. 353-358, Astron. Soc. Pacific Conf. Ser., 111, , 1996.

Rust, D.M., and A. Kumar, Helical magnetic fields in filaments, *Solar Phys.*, 155, 69-97, 1994.

Rust, D.M., and A. Kumar, Evidence for helically kinked magnetic flux ropes in solar eruptions, *Astrophys. J.*, 464, L199-L202, 1996.

Rust, D.M., and S.F. Martin, A correlation between sunspot whirls and filament type, in *Solar Active Region Evolution*, ed. by K.S. Balasubramaniam, and G.W. Simon, pp. 337 - 338, Astron. Soc. Pacific Conf. Ser., 68, 1994.

Taylor, J.B., Relaxation and magnetic reconnection in plasmas, *Rev. Mod. Phys.*, 58 (3), 741, 1986.

Van Ballegooijen, A.A., N.P. Cartledge, and E.R. Priest, Magnetic flux transport and the formation of filament channels on the sun, *Astrophys. J.*, 501, 866, 1998.

Wang, J., Z. Shi, and S.F. Martin, Filament disturbance and associated magnetic changes in the filament, *Astron. Astrophys.*, 316, 201-214, 1996.

Wilson, R.M., and E. Hildner, Are interplanetary magnetic clouds manifestations of coronal transients at 1 AU?, *Solar Phys.*, 91, 169, 1984.

Zirker, J.B., J.-L. Leroy, and V. Gaizauskas, The sinistral-dextral regularity: an independent test, *Solar Phys.*, 176, 279-283, 1997a.

Zirker, J.B., S.F. Martin, K. Harvey, and V. Gaizauskas, Global magnetic patterns of chirality, *Solar Phys.*, 175, 27-44, 1997b.

D. M. Rust, Applied Physics Laboratory, The Johns Hopkins University, 11100 Johns Hopkins Road, Laurel, MD 20723. (email: david.rust@jhuapl.edu)

Solar Flares, Jets, and Helicity

Kazunari Shibata

National Astronomical Observatory, Mitaka, Tokyo 181, Japan

Soft and hard X-ray telescopes aboard Yohkoh have revealed a variety of evidence of magnetic reconnection not only in large scale flares but also in small scale flares and jets. In particular, there is increasing evidence that plasmoid ejections are ubiquitous in flares and play a key role in producing flares. Plasmoids are confined in magnetic islands which are helically twisted tubes or flux ropes in three dimensional space. Thus they are a signature of magnetic helicity. There is also evidence that even small scale jets are consequences of reconnection between untwisted tubes and twisted tubes (plasmoids). Hence, a unified view has emerged on various flares, mass ejections, and jets (ranging from coronal mass ejections to X-ray jets and surges). We propose a unified model in which ejection or expulsion of magnetic helicity (plasmoids) from the system plays a key role to induce fast reconnection and results in violent energy release. In this scenario, what is important is not dissipation of helicity but ejection or expulsion of helicity from the system. It is also proposed that magnetic helicity is first transported by emerging flux (as a buoyant twisted flux tube) to the solar atmosphere and then redistributed to various parts of active regions through reconnection.

1. INTRODUCTION

Solar observers have long thought that there are two types of flares, i.e., long duration event (LDE) flares vs. impulsive flares. LDE flares typically last more than 1 hour, while impulsive flares are short lived, less than 1 hour. The latter is characterized by the impulsive hard X-ray emission whereas the former shows softer X-ray spectrum. It was also often argued that these two types might be a result of different physical origin.

After the launch of Yohkoh in 1991, it has gradually become clear that there are many common features between these two types of flares. In particular, there is increasing evidence of magnetic reconnection and associated mass ejections, such as X-ray plasmoid ejections and X-ray jets in both types of flares. It is now impossible to classify flares into two types, and we need to develop a unified model of flares.

In this article, we first review these new Yohkoh observations of flares, with emphasis on the evidence of reconnection and mass ejections, and then discuss the role of plasmoids in producing flares. It should be noted here that the *plasmoid* is plasma confined by a *magnetic island* in a two dimensional space, and is a *helically twisted flux rope* in a three dimensional space. Hence it is a signature of *magnetic helicity*. On the basis of Yohkoh observations and numerical simulations, we will discuss the *plasmoid-induced-reconnection model*, in which the ejection of plasmoids (helicity) plays a key role to induce flares (fast reconnection). We then review observations of jets associated with small scale flares, such as X-ray jets discovered by Yohkoh and spinning motion of Hα surges, and discuss how these observations can be understood by the reconnection model. Finally, a unified model is presented to account for apparently different plasmoids (from larges scale flares) and jets (from small scale flares) with a unified scheme based on the plasmoid-induced-reconnection model.

2. FLARES

2.1. LDE Flares vs. Impulsive Flares

The Yohkoh soft X-ray telescope (SXT) has revealed that many LDE flares show *cusp-shaped loop or arcade* structures (Tsuneta et al. 1992, 1996, Forbes and Acton 1996), which are quite similar to the magnetic field configuration predicted by the classical magnetic reconnection model (Carmichael 1964, Sturrock 1966, Hirayama 1974, Kopp and Pneuman 1976). This model, which is hereafter called CSHKP model, predicts that magnetic fields are first opened up by global MHD instability associated with filament (or plasmoid) eruption to form vertical current sheet, and then magnetic field lines in the current sheet succesively reconnect to form apparently growing flare loops with a temperature distribution that outer loops are hotter. There are a number of evidences of magnetic reconnection in these LDE flares, all of which have been predicted by the above reconnection model; (1) The temperature is systematically higher in outer loops. (2) The cusp-shaped loops apparently grow with time, i.e., the height of loops and the separation of two footpoints of loops increase with time. (3) The plasmoid ejections are often associated with the rise phase of LDE flares (Fig. 1).

The SXT images of impulsive flares, however, show only *simple loop* structures, as already known from Skylab era. Hence it was first thought that these impulsive flares might be created by a different mechanism from that of LDE flares and the magnetic reconnection model was questioned. Masuda (1994a) changed this situation dramatically. He carefully coaligned the SXT and the HXT (hard X-ray telescope) images of some impulsive compact loop flares observed at the limb, and showed that there is an impulsive HXR (hard X-ray) source *above* the SXR (soft X-ray) loop in addition to footpoint double HXR impulsive sources (Masuda et al. 1994b, 1995). Since the impulsive HXR sources are produced by high energy electrons and hence are thought to be closely related to the main energy release mechanism, this means that *the main energy release occurred above*

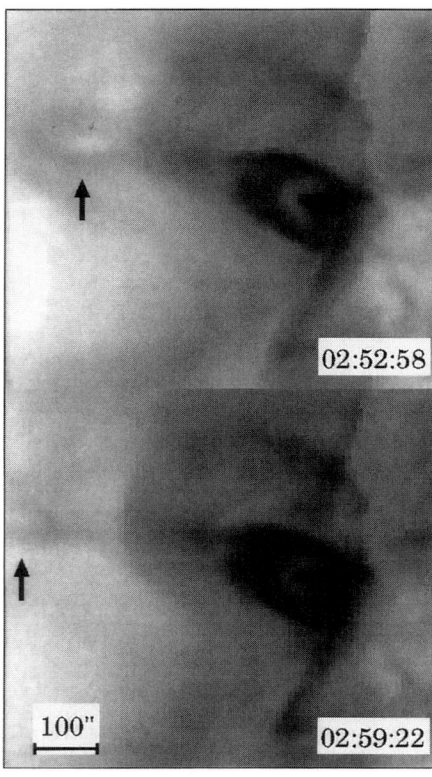

Figure 1. X-ray plasmoid ejections (indicated by an arrow) observed during the rise phase of the LDE flare on 21 Feb. 1992 (Hudson 1994).

(outside) the SXR loop. It means also that the flare models invoking the energy release mechanism inside the SXR loops (e.g., Alfven and Carqvist 1967, Spicer 1977, Uchida and Shibata 1988) must be discarded at least for these impulsive compact loop flares. Masuda et al. (1994b, 1995) postulated that the basic magnetic field configuration is similar to that of LDE flares and that the high speed jet produced by the reconnection collides with the top of the reconnected loop to produce a very hot region as well as high energy electrons.

2.2. X-ray Plasmoid Ejections from Impulsive Flares

If the impulsive compact loop flares occur as a result of reconnection in a geometry similar to that for LDE flares, plasmoid ejections would be observed high above the loop top HXR source. Shibata et al. (1995) searched for such plasmoid ejections using SXT images in 8 impulsive compact loop flares observed at the limb, which were selected by Masuda (1994a) in an unbiased manner, and indeed found that *all these flares were associated with X-ray plasma (or plasmoid) ejections*. The apparent velocities of these ejections are 50 – 400 km/s, and their height ranges are $4 - 10 \times 10^4$ km. Interestingly, flares with HXR source well above the loop top show systematically higher velocity. The temperature of plasmoids is $\sim 6 - 13$ MK, slightly less than that of flare loops, and the overall temperature distribution is consistent with that predicted by the reconnection model (Ohyama and Shibata 1997, 1998, Tsuneta 1997). The shape of these plasma ejections is usually loop-like or blob-like, which is somewhat similar to the shape of CMEs (e.g., Burkepile and St. Cyr 1993).

In many cases, strong acceleration of plasmoids occurs during the impulsive phase (Ohyama and Shibata 1997, 1998; Fig. 2), and the temporal relation between the plasmoid height and the HXR intensity is very similar to that between the CME height and the SXR intensity of an associated flare. Ohyama and Shibata (1997, 1998) also showed that the kinetic energy of plasmoids is much smaller than that of the total flare energy. This means that the kinetic energy of the plasmoid ejection cannot be the source of flare energy. Instead, the plasmoid ejection can play a role to trigger the main energy release in impulsive phase, since in some events it is found that the plasmoid starts to be ejected (at 10 km/s) well before the impulsive phase (Fig. 2).

2.3. Plasmoid-Induced-Reconnection Model

On the basis of above observations, Shibata (1996, 1998) proposed the *plasmoid-induced-reconnection mo-*

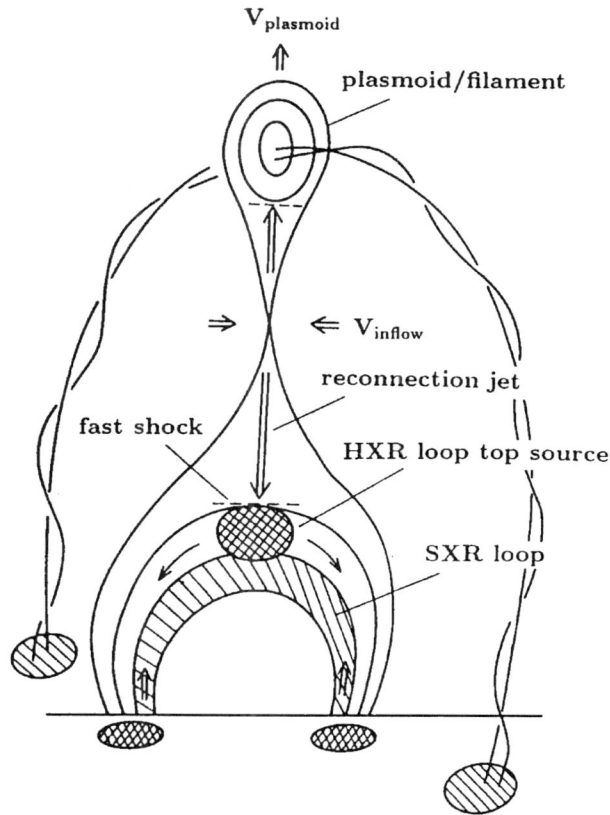

Figure 3. A unified model (*plasmoid-induced-reconnection model*) (Shibata et al. 1995).

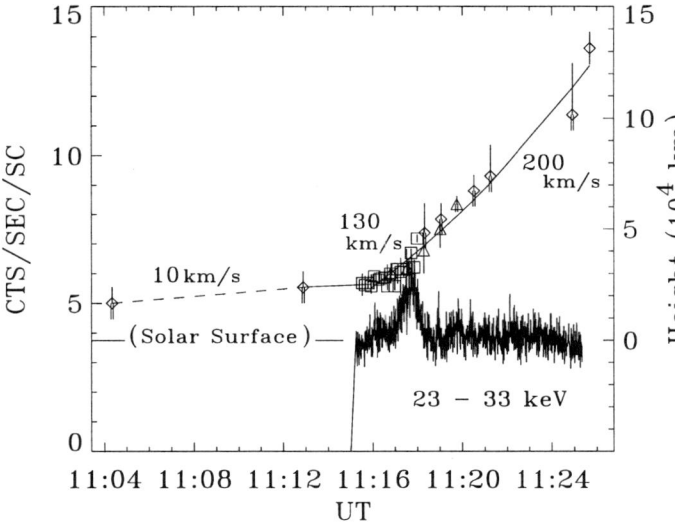

Figure 2. Temporal variations of the height of an X-ray plasmoid and the hard X-ray intensity in an impulsive flare on 11 Nov. 1993 observed by Yohkoh SXT and HXT (Ohyama and Shibata 1997).

del, by extending the classical CSHKP model. In this model, the plasmoid ejection plays a key role to trigger fast reconnection (see Fig. 3).

Let us consider the situation where a plasmoid suddenly rises at velocity $V_{plasmoid}$. (In this model, on the basis of observations, we *assume* that the plasmoid is already created before the flare, and is suddenly accelerated by some mechanism. Magnetic reconnection could also play a role in such a preflare phase as noted by Ohyama and Shibata (1997) even for accelerating plasmoids.) Since the plasma density does not change much during the eruption process, the inflow $V_{inflow} \sim V_{plasmoid} L_{plasmoid}/L_{inflow}$ must develop toward the X-point to compensate the mass ejected by the plasmoid, where $L_{plasmoid}$ and $L_{inflow} (> L_{plasmoid})$ are the typical sizes of the plasmoid and the inflow. We consider that the impulsive phase corresponds to the phase when $L_{inflow} \sim L_{plasmoid}$, i.e., $V_{inflow} \sim V_{plasmoid} \sim 50-400$ km/s. Since the reconnection rate is determined by the inflow speed, the ultimate origin of fast reconnection in this model is the fast ejection of the plasmoid. After the impulsive phase, we expect that L_{inflow} becomes larger than $L_{plasmoid}$ because the distance between the plasmoid and the X-point increases, and hence the inflow speed V_{inflow} would decrease greatly, leading to slow reconnection which corresponds to the decay phase.

The magnetic reconnection theory predicts two oppositely directed high speed jets from the reconnection point at Alfven speed, $V_{jet} \sim V_A \simeq 2000(B/100G)(n_e/10^{10} \text{cm}^{-3})^{-1/2}$ km/s, where B is the magnetic flux density and n_e is the electron density. The downward jet collides with the top of the SXR loop, producing an MHD fast shock, superhot plasmas and/or high energy electrons at the loop top, as observed in the HXR images. The temperature just behind the fast shock becomes $T_{loop-top} \sim m_i V_{jet}^2/(6k) \sim 2 \times 10^8 \ (B/100G)^2 (n_e/10^{10} \text{cm}^{-3})^{-1}$ K, where m_i is the hydrogen ion mass and k is the Boltzmann constant. This explains the observationally estimated temperature of the loop top HXR source (Masuda 1994a). The magnetic energy release rate at the current sheet (with the length of $L_{inflow} \sim L_{plasmoid} \simeq 2 \times 10^4$ km) is estimated to be

$$dW/dt = 2 \times L_{plasmoid}^2 B^2 V_{inflow}/4\pi$$

$$\sim 4 \times 10^{28} \left(\frac{V_{inflow}}{100 \text{ km/s}}\right) \left(\frac{B}{100 \text{ G}}\right)^2 \left(\frac{L_{plasmoid}}{2 \times 10^9 \text{ cm}}\right)^2 \text{ erg/s}.$$

This is comparable with the observed energy release rate during the impulsive phase, $4 - 100 \times 10^{27}$ erg/s, estimated from the HXR data, assuming the lower cut-off energy as 20 keV (Masuda 1994a).

2.4. Numerical Simulations of Plasmoid Dynamics

Magara, Shibata, & Yokoyama (1997) studied the dynamics of fast reconnection induced by plasmoid ejection during the impulsive phase of flares, peforming 2.5D MHD numerical simulations. They initially assumed a linear force free arcade in a uniform plasma without gravity and added resistive localized perturbation for a finite time. As a result of finite resistivity, a global resistive MHD instability (like the tearing instability) is excited and a thin current sheet is developed in the nonlinear stage of this instability. After some time, the current density increases enormously, so that the anomalous resistivity sets in, leading to fast reconnection (Ugai 1986). They modeled the anomalous resistivity with the formula $\eta = \eta_0(|v_d| - v_c)/v_c$ (if $v_d > v_c$), where $v_d = j/\rho$ is a relative ion-electron drift velocity and v_c is a threshold parameter. The magnetic island (plasmoid) is created as a result of reconnection and is ejected upward after the onset of anomalous resistivity. Figure 4 shows the temporal relation between the height of plasmoid and the reconnection rate ($= E_y = \eta j$), revealing that the plasmoid is accelerated during the phase of large reconnection rate. It is interesting to note that the start of plasmoid acceleration is before the peak of the reconnection rate. If the hard X-ray intensity is a measure of reconnection rate, this figure is very similar to the observed relation between the plasmoid height and the hard X-ray intensity. (In fact, the impulsive hard X-ray intensity is a measure of the total energy release rate (Masuda 1994a). Since the reconnection rate ($\propto$ inflow speed $= v_i$) is also in proportion to the total energy release rate ($dE/dt \propto v_i B^2/4\pi$), this similarity is not a mere chance coincidence.)

What is the mechanism to cause this kind of temporal relation between the plasmoid height and the reconection rate ? Magara et al. (1997) found that the reconnection rate is very small when there is a finite perpendicular magnetic field (B_y) in the current sheet. Only after the ejection of B_y (with a plasmoid), the current sheet can become very thin, so that the anomalous resistivity can set in, leading to large reconnection rate. Hence the perpendicular field (B_y) (or equivalently magnetic helicity) plays a role to inhibit current sheet thinning or collapse, and hence to inhibit fast reconnection.

If the injection of magnetic helicity continues during the preflare slow reconnection phase (i.e., when the fast reconnection is inhibited by the perpendicular field), the excess magnetic energy can be stored around the current sheet. Hence we can say that the plasmoid (helicity) plays a role of energy storage. After the ejection of plasmoid (helicity), the system settles to a lower energy state, and the same process repeats again. Eventually this leads to intermittent plasmoid ejection and associated impulsive energy release. A very nice numerical simulation of this process has been done by Kusano (1998). This explains observed recurrent and homologous behavior of solar flares.

3. JETS

3.1. X-ray Jets

Yohkoh SXT has revealed that mass ejections are also common in tiny flares, called *microflares*, whose total energy is around 10^{-6} (micro-) that of large scale flares. In these microflares, mass ejections often take the form of jets (collimated mass flow), called *X-ray jets* (Shibata et al. 1992b, 1994, 1996, Strong et al. 1992, Shimojo et al. 1996), which are quite different from blob or loop-like plasmoid ejections from large scale flares.

The occurrence frequency of microflares (and also X-ray jets) increases with decreasing their energies, and the distribution function is a power-law (Shimizu 1995), which is basically the same as that of normal flares. From a statistical point of view, it is not possible to classify *microflares* and *flares*. Nevertheless, it seems that X-ray jets tend to be ejected from microflares, whereas blob or loop like plasmoids tend to be ejected from normal flares. Why does this kind of different morphology appear in both type of mass ejections ? What is the difference between jets and plasmoids ?

We will postpone answering these questions until section 4, and now we briefly summarize basic properties of X-ray jets. The length of X-ray jets ranges from 1000 to 4×10^5 km. Their apparent velocity is 10 – 1000 km/s. The temperature of X-ray jets is about 4 – 6 MK, which is comparable to those of the footpoint microflares. The kinetic energy was estimated to be $10^{25} - 10^{29}$ erg. Figure 5 shows a typical example of X-ray jets.

Many jets show a constant or converging shape, implying the magnetic field configuration with a neutral point near the footpoint of a jet. In some jets (27 percent), a gap is seen between the footpoints of jets and the brightest part of the footpoint flares. This is also nicely explained by the reconnection model (Shibata, Yokoyama, & Shimojo 1996), since the reconnection creates two hot reconnected field lines (a loop and a jet) with a gap between them. Shimojo, Shibata, & Harvey (1998) have examined the magnetic field properties

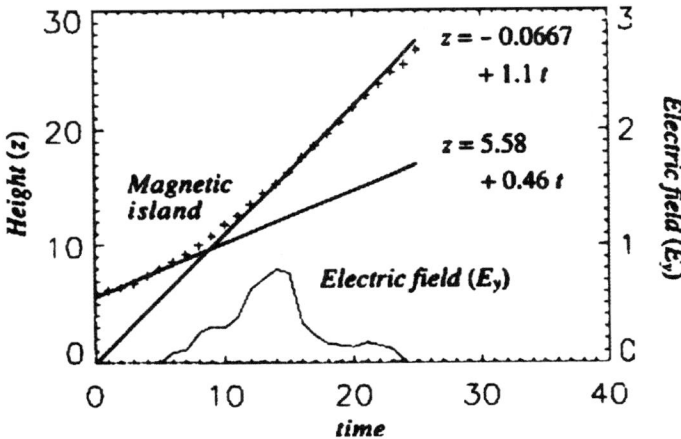

Figure 4. Temporal variations of both the y-component of electric field (E_y) at the neutral point (X-point) and the height of the magnetic island (Magara et al. 1997).

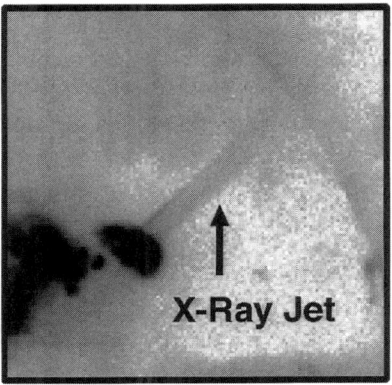

**Yohkoh SXT Image
12-Nov-91 11:30UT**

Figure 5. An X-ray jet observed with the Yohkoh/SXT on 12 Nov. 1992 (Shibata et al. 1992b).

of the footpoint of jets, and found that the footpoints usually correspond to mixed polarities or satellite spots. This is a direct evidence of existence of neutral points or a current sheet at the footpoint of jets.

3.2. Spinning Motion of Hα Surges

Canfield et al. (1996) studied the relation between Hα surges and X-ray jets and found that all Hα surges (9 events) in their observations are associated with X-ray jets. They found new evidence of reconnection in surges, such as a *moving blue shift* (as evidence of cool plasma accelerated by sling-shot effect) and *converging footpoint motion* (as evidence of a *reconnecting* loop). (Note that diverging footpoint motion is observed in *reconnected* loops.)

Moreover, they found that all Hα surges are spinning at a few 10 km/s with direction consistent with that of unwinding motion of helically twisted flux tubes observed in the same active region 7260 (Leka et al. 1996). Figure 6 shows schematic cartoons of reconnection associated with X-ray jets and Hα surges observed by Canfield et al. (1996), showing how a spinning Hα surge is created as a result of reconnection between a twisted tube and an untwisted tube. Similar observations of spinning jets have been reported by Kurokawa et al. (1987), Schmieder et al. (1995), and Mason et al. (1998).

3.3. Emerging Flux Model

Yokoyama and Shibata (1995, 1996) developed a magnetic reconnection model of X-ray jets using 2.5D MHD

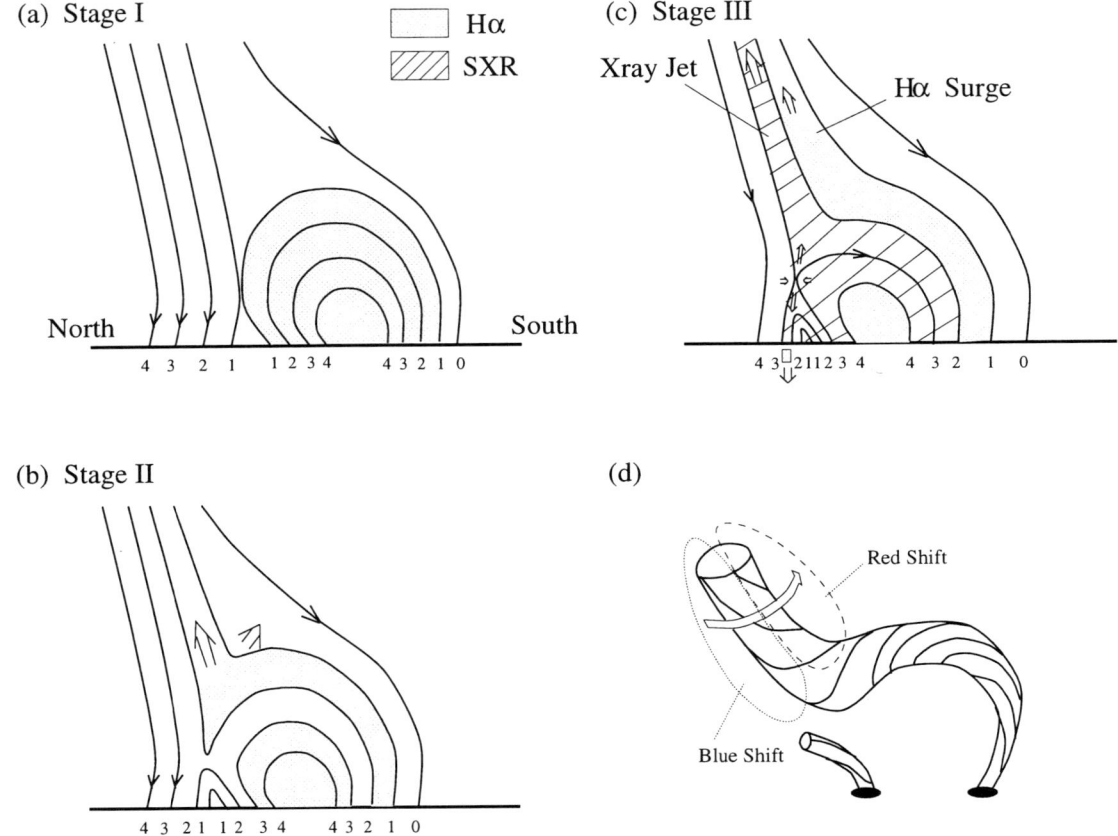

Figure 6. A magnetic twist - reconnection model for Hα surges (Canfield et al. 1996).

numerical simulations. In their model, magnetic reconnection occurs in the current sheet between emerging flux and the overlying coronal field as in the classical emerging flux model (Heyvaerts et al. 1977, Forbes and Priest 1984, Shibata et al. 1992a). They found several interesting features in their simulation results.

The reconnection starts with the formation of magnetic islands (i.e., plasmoids). These islands coalesce with each other and are finally ejected out of the current sheet. After the ejection of the biggest island, the largest energy release occurs. Not only hot jets ($T > 10^6$ K) but also cool jets ($T \sim 10^4 - 10^5$ K) are accelerated by the $\mathbf{J} \times \mathbf{B}$ force in association with reconnection. The cool jets might correspond to Hα surges associated with X-ray jets (Shibata et al. 1992b, Canfield et al. 1996). These cool jets start to be accelerated just before hot jets are formed, and are ejected originally as plasmoids and form an elongated structure after the plasmoids collides with ambient fields. The initial phase of the ejection of both cool and hot jets are seen as *whip-like motion*. In the main phase, the cool jets are situated just outside of the hot jets with nearly the same orientation. These features are indeed observed in several Hα surges associated with X-ray jets (Canfield et al. 1996).

3.4. Generation of Alfven Waves and Helical Jets by Magnetic Reconnection

Okubo et al. (1996) extended Yokoyama and Shibata (1996)'s simulations to the case in which twisted or sheared magnetic flux emerges to reconnect with the overlying field (see also Karpen et al. 1998). They found that shear Alfven waves are generated and propagate along the reconnected flux tube, as a result of reconnection between twisted (sheared) field and untwisted field. Since these Alfven waves have large amplitude, they excite large transversal motion (or spinning motion) of jets and exert nonlinear magnetic pressure force on cool/hot jets to cause further acceleration of these jets, as originally suggested by Shibata and

Uchida (1986). More recently, Yokoyama (1998) analyzed this simulation model in detail, and found that (1) the Alfven wave energy flux emitted from the reconnection region is about 3 percent of the total energy released by the reconnection, (2) the frequency spectrum of the Alfven wave is a continuum and includes high frequency modes. These Alfven waves and jets may play an important role in the acceleration of high speed solar wind.

4. UNIFIED MODEL

4.1. Unification of Emerging Flux Model and CSHKP Model

As we have seen above, Yohkoh SXT/HXT observations have revealed evidence of magnetic reconnection, especially in the common occurrence of X-ray mass ejections (plasmoids and/or jets), in LDE flares, impulsive flares, and microflares. These are summarized in Table 1. On the basis of this unified view, Shibata (1996, 1998) proposed a unified model, *plasmoid-induced-reconnection model*, to explain not only LDE and impulsive flares but also microflares and X-ray jets.

One may argue, however, that the shape of X-ray jets and Hα surges (i.e., collimated jet-like structure) is very different from that of plasmoids. How can we relate these jets with plasmoids whose shapes are blob-like (or loop-like in three dimensional space) ? The answer to this question is already given by numerical simula-

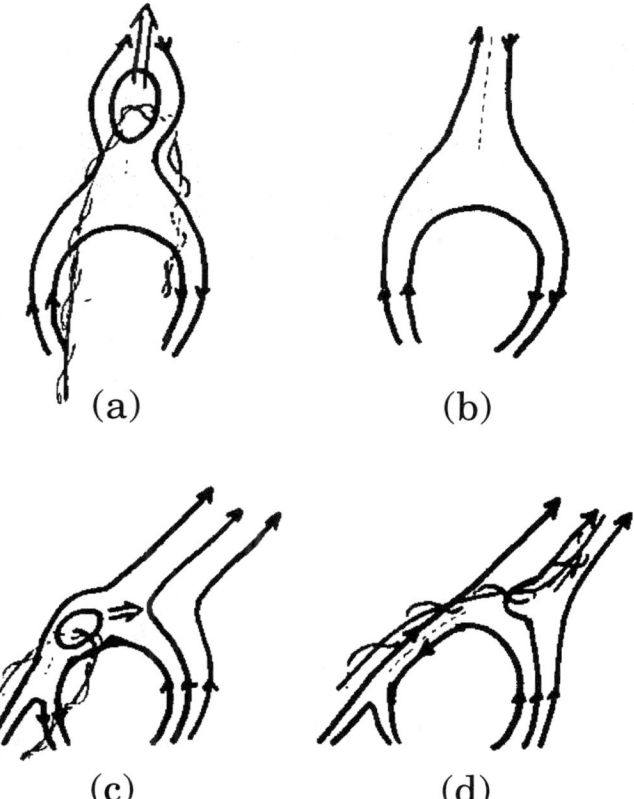

Figure 7. Unification of CSHKP model and emerging flux model by the *plasmoid-induced-reconnection* model (Shibata 1996, 1998). (a), (b): The case of a large scale flare induced by the ejection of a plasmoid. In this case, a cusp-shaped loop is remained after the ejection of the plasmoid. (c), (d): The case of a small scale flare associated with a jet. In this case, a plasmoid (a magnetic island or a helically twisted flux rope) collides and reconnects with the ambient magnetic field to disappear in a short time scale (10 – 100 sec), generating a jet with unwinding twist (torsional Alfven waves) along global field lines.

Table I Unified View of Various "Flares"

"flares"	mass ejections (cool)	mass ejections (hot)
giant arcades	Hα filament eruptions	CMEs
LDE flares	Hα filament eruptions	X-ray plasmoid ejections/CMEs
impulsive flares	Hα sprays	X-ray plasmoid ejections
transient brightenings (microflares)	Hα surges	X-ray jets

tions of Yokoyama and Shibata (1995, 1996); a blob-like plasmoid ejected from the current sheet soon collides with the ambient fields, and finally disappears (Fig. 7). The mass contained in the plasmoid is transferred into the reconnected open flux tube and forms a collimated jet along the tube. Through this reconnection, magnetic twist (helicity) is injected into the untwisted loop, resulting in the unwinding motion of the jet (Shibata and Uchida 1986), which may correspond to the spinning motion observed in some Hα surges (Canfield et al. 1996, Schmieder et al. 1995). This also explains why we usually do not observe plasmoid-like (or loop-like) mass ejections in smaller flares (e.g., microflares).

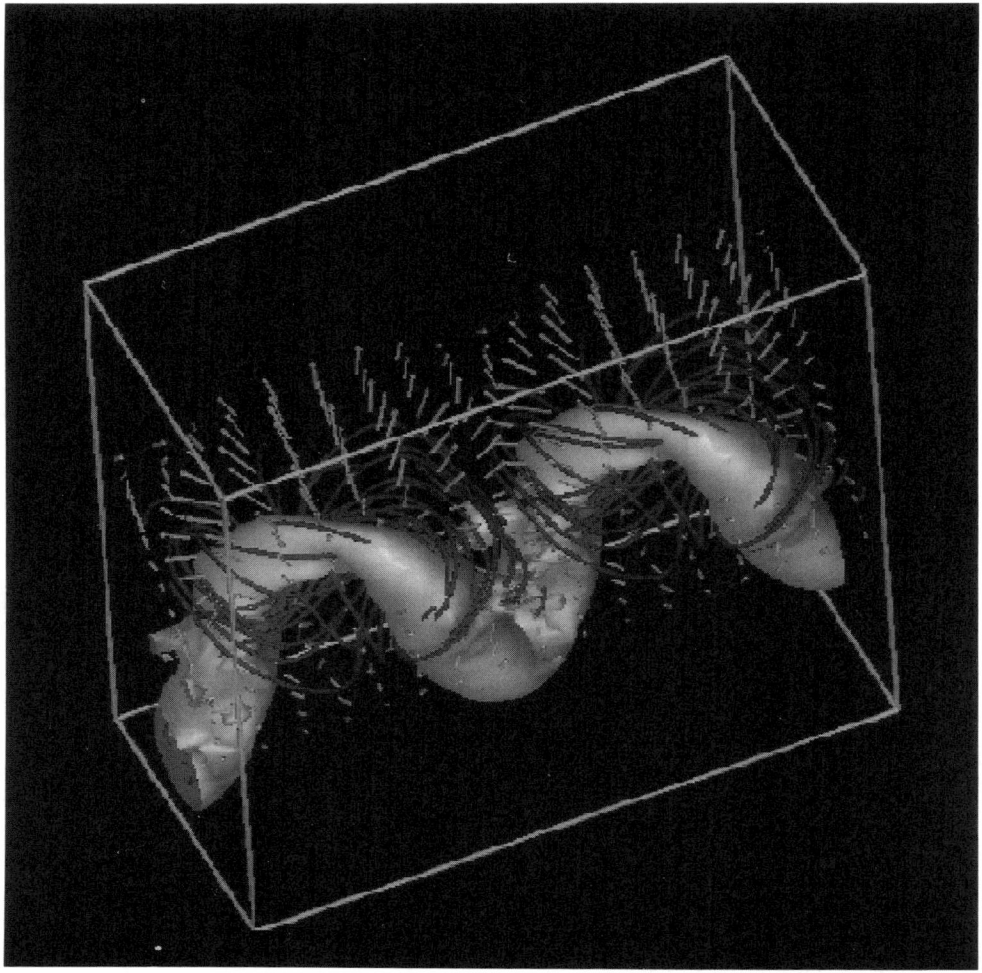

Figure 8. Results of 3D MHD simulation of emergence of twisted flux tubes (Matsumoto et al. 1998). The grey surface shows the isosurface of magnetic field strength, and arrows are velocity vectors. Grey curves show magnetic field lines.

In smaller flares, the current sheet is short, so that a plasmoid soon collides with an ambient field to reconnect with it and disappear. Hence the lifetime of the plasmoid (or loop-like) ejection is very short, of order of $t \sim L/V_{plasmoid} \sim 10 - 100$ sec. It would be interesting to test this scenario using high spatial and temporal resolution observations with Doppler shift measurement in future solar mission such as Solar-B.

4.2. Emergence of Twisted Flux Tube

Ground based observations suggest that emerging flux plays an important role in driving flares (e.g., Kurokawa 1987). For example, a famous X-class impulsive flare, the 15 Nov 1992 flare (e.g., Sakao et al. 1992), was driven by a moving satellite spot (or emerging flux). Even the 21 Feb 1992 LDE flare (e.g., Tsuneta 1996), and one homologous to it on 24 Feb 1992 (Morita et al. 1998) seem to be driven by growing flux (or emerging flux) (Zhang et al. 1998). Nevertheless, these flares clearly show filament or plasmoid ejections as well as the morphology predicted by the CSHKP model. Thus there is a need to unify the CSHKP and the emerging flux models. Such a unification is indeed possible in our plasmoid-induced-reconnection model if we consider successive emergence of twisted flux tubes and associated reconnection.

Matsumoto et al. (1998) peformed a 3D MHD numerical simulation of emergence of a twisted flux tube, and found that the resulting magnetic flux tube shows the double helix pattern (Fig. 8) and that their results explain various observations such as the peculiar motion of sunspots and the apparent sheared S structure of coronal loops found by Yohkoh. It is well known that such a sheared S structure is an actively flare producing site. From this, there is no doubt that the generation and emergence of twisted flux tubes (i.e., magnetic helicity) and resulting 3D reconnection are one of the central keys to understanding the origin of flares.

Acknowledgments. The author would like to thank R. Matsumoto, M. Ohyama, T. Yokoyama, and M. Shimojo for preparation of figures in this article.

REFERENCES

Alfven, H., and P. Carlqvist, Currents in the solar atmosphere and a theory of solar flares, *Sol. Phys.*, 1, 220, 1967.

Burkepile, J.T., and O. C. St. Cyr, A revised and expanded catalogue of mass ejections observed by the Solar Maximum Mission coronagraph, NCAR/TN-369+STR, Boulder, 1993.

Canfield, R. C., K. P. Reardon, K. D. Leka, K. Shibata, T. Yokoyama, and M. Shimojo, H α surges and X-ray jets in AR7260, *Ap. J.*, 464, 1016, 1996.

Carmichael, H., A process for flares, in *Proc. of AAS-NASA Symp. on the Physics of Solar Flares*, W. N. Hess (ed.), NASA-SP 50, p. 451, 1964.

Forbes, T. G., and L. W. Acton, Reconnection and field line shrinkage in solar flares, *Ap. J.*, 459, 330, 1996.

Forbes, T. G. and E. R. Priest, Numerical simulation of reconnection in an emerging magnetic flux region, *Sol. Phys.*, 94, 315, 1984.

Heyvaerts, J., E. R. Priest, and D. M. Rust, An emerging flux model for the solar flare phenomenon, *Ap. J.*, 216, 123, 1977.

Hirayama, T., Theoretical model of flares and prominences, I: Evaporating flare model, *Sol. Phys.*, 34, 323, 1974.

Hudson, H. S., Thermal plasmas in the solar corona: the Yohkoh soft X-ray observations, in *Proc. Kofu meeting*, eds. S. Enome, and T. Hirayama, Nobeyama Radio Observatory, p. 1, 1994.

Karpen, J. T., S. K. Antiochos, and C. R. DeVore, Dynamic responses to magnetic reconnection in solar arcades, *Ap. J.*, 495, 491, 1998.

Kopp, R. A., and G. W. Pneuman, Magnetic reconnection in the corona and the loop prominence phenomenon, *Sol. Phys.*, 50, 85, 1976.

Kurokawa, H., Y. Hanaoka, K. Shibata, and Y. Uchida, Rotating eruption of an untwisting filament triggered by the 3B flare of 25 April 1984, *Sol. Phys.*, 108, 251, 1987.

Kurokawa, H., High-resolution observations of H-alpha flare regions, *Space Sci. Rev.*, 51, 49, 1987.

Kusano, K., Magnetic helicity and stability in solar corona, in these proceedings, 1998.

Leka, K. D., R. C. Canfield, A. N. McClymont, L. van Driel-Gesztelyi, Evidence for current-carrying emerging flux, *Ap. J.*, 462, 547, 1996.

Magara, T., K. Shibata, and T. Yokoyama, Evolution of eruptive flares, I. Plasmoid dynamics in eruptive flares, *Ap. J.*, 487, 437, 1997.

Mason, H. E., C. D. Pike, and P. R. Young, Dynamic features observed with SOHO-CDS, in *Proc. Solar Jets and Coronal Plumes* (ESA SP 421), p. 95, 1998.

Masuda, S., Hard X-ray sources and the primary energy release site in solar flares, Ph. D. Thesis, U. Tokyo, 1994a.

Masuda, S., T. Kosugi, H. Hara, S. Tsuneta, and Y. Ogawara, A loop-top hard X-ray source in a copact solar flare as evidence for magnetic reconnection, *Nature*, 371, 495, 1994b.

Masuda, S., T. Kosugi, H. Hara, T. Sakao, K. Shibata, and S. Tsuneta, Hard X-ray sources and the primary energy-release site in solar flares, *PASJ*, 47, 677, 1995.

Matsumoto, R., T. Tajima, W. Chou, A. Okubo, and K. Shibata, Formation of a kinked alignment of solar active regions, *Ap. J.*, 493, L43, 1998.

Morita, S., Y. Uchida, K. Fujisaki, and S. Hirose, Homologous flare series of February 1992, in *Proc. Observational Plasma Astrophysics: Five Years of Yohkoh and Beyond*, Watanabe, T., et al. (eds), Kluwer Adademic Publ, p. 327, 1998.

Ohyama, M., and K. Shibata, Preflare heating and mass motion in a solar flare associated with a hot plasma ejection – 1993 November 11 C9.7 flare –, *PASJ*, 49, 249, 1997.

Ohyama, M., and K. Shibata, X-ray plasma ejection associated with an impulsive flare on 1992 October 5 – physical conditions of X-ray plasma ejections –, *Ap. J.*, 499, 934, 1998.

Okubo, A., R. Matsumoto, S. Miyaji, M. Akioka, K. Shibata, and T. Yokoyama, Observations and numerical studies of coronal X-ray jets and Hα surges associated with emerging magnetic fields, *Proc. Observations of Magnetic Reconnection in the Solar Atmosphere*, Bentley, R. and Mariska, J. T. (eds.), ASP conf. ser., vol. 111, p. 39, 1996.

Sakao, T., T. Kosugi, S. Masuda, K. Makishima, et al., Hard X-ray imaging observations by Yohkoh of the 1991 November 15 solar flare, *PASJ*, 44, L83, 1992.

Schmieder, B., K. Shibata, L. van Driel-Gesztelyi, and S. Freeland, Hα surges and associated soft X-ray loops, *Sol. Phys.*, 156, 245, 1995.

Shibata, K., and Y. Uchida, Sweeping-magnetic-twist mechanism for the acceleration of jets in the solar atmosphere, *Sol. Phys.*, 103, 299, 1986.

Shibata, K., S. Nozawa, and R. Matsumoto, Magnetic reconnection associated with emerging magnetic flux, *PASJ*, 44, 265, 1992a.

Shibata, K., Y. Ishido, L. W. Acton, K. T. Strong, et al., Observations of X-ray jets with the Yohkoh soft X-ray telescope, *PASJ*, 44, L173, 1992b.

Shibata, K., N. Nitta, K. T. Strong, R. Matsumoto, et al., A gigantic coronal jet ejected from a compact active region in a coronal hole, *Ap. J.*, 431, L51, 1964.

Shibata, K., S. Masuda, M. Shimojo, H. Hara, et al., Hot plasma ejections associated with compact-loop solar flares, *Ap. J.*, 451, L83, 1995.

Shibata, K., T. Yokoyama, and M. Shimojo, Coronal X-ray

jets observed with the Yohkoh soft X-ray telescope, *J. Geomag. Geoelectr., 48*, 19, 1996.

Shibata, K., New observational facts about solar flares from Yohkoh studies - evidence of magnetic reconnection and a unified model of flares -, *Adv. Space Res., 17*, (4/5)9, 1996.

Shibata, K., A unified model of flares, in *Proc. Observational Plasma Astrophysics: Five Years of Yohkoh and Beyond*, Watanabe, T., et al. (eds), Kluwer Adademic Publ, p. 187, 1998.

Shimizu, T., Energetics and occurrence rate of active-region transient brightenings and implications for the heating of the active-region corona, *PASJ, 47*, 251, 1995.

Shimojo, M., S. Hashimoto, K. Shibata, T. Hirayama, et al., Statistical study of solar X-ray jets observed with Yohkoh soft X-ray telescope, *PASJ, 48*, 123, 1996.

Shimojo, M., K. Shibata, and K. L. Harvey, Magnetic field properties of solar X-ray jets, *Sol. Phys., 178*, 379, 1998.

Spicer, D. S., An unstable arch model of a solar flare, *Sol. Phys., 53*, 305, 1977.

Strong, K. T., K. Harvey, T. Hirayama, N. Nitta, et al., Observations of the variability of coronal bright points by the soft X-ray telescope on Yohkoh, *PASJ, 44*, L161, 1992.

Sturrock, P. A., Model of the high-energy phase of solar flares, *Nature, 211*, 695, 1966.

Tsuneta, S., H. Hara, T. Shimizu, L. W. Acton, et al., Observation of a solar flare at the limb with the Yohkoh soft X-ray telescope, *PASJ, 44*, L63, 1992.

Tsuneta, S., Structure and dynamics of magnetic reconnection in a solar flare, *Ap. J., 456*, 840, 1996.

Tsuneta, S., Moving plasmoid and formation of the neutral sheet in a solar flare, *Ap. J., 483*, 507, 1997.

Uchida, Y., and K. Shibata, A magnetodynamic mechanism for the heating of emerging magnetic flux tubes and loop flares, *Sol. Phys., 116*, 291, 1988.

Ugai, M., Global dynamics and rapid collapse of an isolated current-sheet system enclosed by free boundaries, *Phys. Fluids, 29*, 3659, 1986.

Yokoyama, T., and K. Shibata, Magnetic reconnection as the origin of X-ray jets and Hα surges on the sun, *Nature, 375*, 42, 1995.

Yokoyama, T., and K. Shibata, Numerical simulation of solar coronal X-ray jet based on the magnetic reconnection model, *PASJ, 48*, 353, 1996.

Yokoyama, T., Magnetic reconnection model of coronal X-ray jets, in *Proc. Solar Jets and Coronal Plumes*, (ESA SP-421), p. 215, 1998.

Zhang, H. Q., T. Sakurai, K. Shibata, M. Shimojo, et al., Magnetic reconnection in the active region inferred by homologous soft X-ray flares in February 1992, in *Proc. Observational Plasma Astrophysics: Five Years of Yokhsooh and Beyond*, Watanabe, T., et al. (eds), Kluwer Adademic Publ, p. 391, 1998.

K. Shibata, National Astronomical Observatory, Mitaka, Tokyo 181, Japan (e-mail: shibata@solar.mtk.nao.ac.jp)

Solar-Cycle, Radial and Latitudinal Variations of Magnetic Helicity: IMF Observations

Charles W. Smith

Bartol Research Institute, University of Delaware, Newark

Solar wind dynamics span an extensive range of spatial scales from greater than an AU to less than the gyroradius of a thermal proton. Throughout this range there exist dynamics that lead to the creation, annihiliation and transport of magnetic helicity. This paper examines some of these processes with measurements of the interplanetary magnetic helicity. We find both kinetic plasma processes as well as MHD and solar source dynamics that lead to magnetic helicity in the solar wind. We attempt to characterize the interplanetary magnetic helicity with a combination of review and new material.

1. INTRODUCTION

In situ measurements of the interplanetary magnetic field (IMF) now span more than 35 years of spacecraft observations since the launch of IMP-1 in 1963. Coverage of the near-Earth IMF has been sufficient to permit long-term studies of the solar-cycle variability of the field [*King*, 1976; *Smith and Bieber*, 1991, 1993; *King and Papitashvili*, 1994]. In the 1970s the interplanetary missions of Pioneer 10 & 11 [*Hall*, 1974] and Voyagers 1 & 2 [*Behannon et al.*, 1977] created the opportunity to study the more distant outer heliosphere. Most recently, Ulysses [*Balogh et al.*, 1992] has provided the first in situ observations of the IMF and helicity over the solar poles [*Goldstein et al.*, 1995]. This paper attempts to summarize the basic observations of magnetic helicity of the IMF. A complete review is impossible within the confines of this article, so we will instead attempt to capture the fundamental attributes of the IMF magnetic helicity.

Magnetic Helicity in Space and Laboratory Plasmas
Geophysical Monograph 111
Copyright 1999 by the American Geophysical Union

To begin, we must first acknowledge that the *Parker* [1963] winding of the IMF spiral contains helicity [*Bieber et al.*, 1987]. Helicity at this scale results from nothing more than the freezing of the IMF into the radially expanding flow coupled with the rotation of the solar corona. This helicity exhibits a persistent and steady south–north asymmetry > 0 wherein the two heliospheric hemispheres display helicity of different signs and equal magnitude.

The range of spatial and temporal scales for IMF variability in the solar wind are as extensive as the measurements themselves. For spatial scales greater than the measurement's heliocentric distance (spacecraft-frame frequencies less than 4×10^{-6} Hz for measurements recorded at 1 AU) the spectrum is entirely and undisputably a manifestation of the variability of the solar source. At higher frequencies, extending to time scales of a few hours (10^{-4} Hz) the interplanetary power spectrum at 1 AU possesses an f^{-1} form. This range, which we will call the source range, is thought to result from the influence of many, uncorrelated solar sources [*Matthaeus and Goldstein*, 1986]. Figure 1 shows the high-frequency end of this range. At still higher frequencies, from 3×10^{-5} Hz (time scales on the order of 1 hour) to ~ 0.5 Hz (not shown) is the inertial range

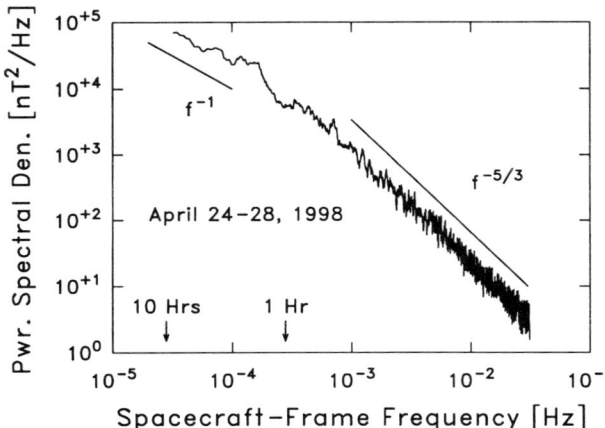

Figure 1. A fairly typical magnetic power spectrum for quiet solar wind conditions as observed by the ACE spacecraft. Only the N component of the field was used. The $f^{-5/3}$ power-law of the inertial range is clearly evident as is the high-frequency end of the f^{-1} range. The spacecraft frequencies for signals spanning 1 hour and 10 hours in spacecraft data are shown at the bottom.

which possesses an $f^{-5/3}$ power law form. In most cases the inertial range extends to frequencies more comparable to 1×10^{-5} Hz and may as well in this instance, but with poor resolution. Drawing on a simple analogy with hydrodynamics, the inertial range is thought to be an energy-conserving conduit for energy cascading from larger spatial scales down to the smallest scales for dissipation. At still higher frequencies the spectrum steepens and dissipation of magnetic energy occurs.

Following a description of how the magnetic helicity is measured in the solar wind, we present the results of recent studies of the IMF dissipation range. Then, building through the inertial range to larger scales, we will close this paper by showing the persistent south-north asymmetry of magnetic helicity at scales that are smaller than the Parker winding scale but larger than the inertial range fluctuations.

2. METHOD

The method for measuring the magnetic helicity in homogeneous, turbulent magnetofluids [*Matthaeus and Smith*, 1981; *Matthaeus et al.*, 1982; and *Oughton et al.*, 1997] is derived from the two-point autocorrelation function

$$R_{ij} \equiv \langle b_i(\mathbf{x}) b_j(\mathbf{x}+\mathbf{r}) \rangle \quad (1)$$

where $b_i(\mathbf{x}) \equiv B_i(\mathbf{x}) - \langle B_i(\mathbf{x}) \rangle$ is the ith component of the fluctuating magnetic field resulting from the subtraction of the mean field from the measurement. The ensemble average $\langle \ldots \rangle$ is typically computed from a spatial or temporal average. It is then possible to define a function

$$2\Phi(\mathbf{r}) = \int_\infty^\mathbf{r} dr \left[R_{ij}(\mathbf{r}) - R_{ji}(\mathbf{r}) \right] \quad (2)$$

where the direction given by $\mathbf{r}$ is arbitrary, but the components i and j represent directions normal to $\mathbf{r}$ such that $\hat{\mathbf{i}} \times \hat{\mathbf{j}} = \hat{\mathbf{r}}$. It is customary to define IMF measurements in heliocentric (R, T, N) components where $\hat{\mathbf{R}}$ is directed radially outward from the sun, $\hat{\mathbf{T}}$ is coplanar to the sun's rotational equator and directed in the sense of rotation, and $\hat{\mathbf{N}} = \hat{\mathbf{R}} \times \hat{\mathbf{T}}$. Since the solar wind flow is both supersonic and super-Alfvénic in the $\hat{\mathbf{R}}$ direction, it is customary to assume that the magnetic fluctuations are frozen into the flow and rewrite equation 2 in terms of temporal lags along the solar wind flow:

$$2\Phi(t) = \int_\infty^t d\tau \, (V_{SW}) \left[R_{TN}(\tau) - R_{NT}(\tau) \right]. \quad (3)$$

From either equations 2 or 3 we can obtain the net magnetic helicity in the Coulomb gauge:

$$H_M = 2\Phi(0). \quad (4)$$

We can obtain the reduced wavenumber spectrum of magnetic helicity from the spectral decomposition of the autocorrelation function according to

$$S_{ij}(k_r) \equiv (2\pi)^{-1} \int d\tau e^{-ik_r r} R_{ij}(r) \quad (5)$$

and from this the helicity spectrum is obtained to be

$$H_M(k_r) = (-i) \left[S_{TN}(k_r) - S_{NT}(k_r) \right] (k_r)^{-1} \quad (6)$$

In the above, k_r is the reduced wavenumber and decomposition is performed only along the flow direction. The magnetic helicity spectrum can be constrained according to the minimum energy required to support the helicity so that we can define

$$\sigma_M(k_r) = k_r \frac{H_M(k_r)}{E_B(k_r)} \quad (7)$$

where $E_B(k_r) \equiv \Sigma_i [S_{ii}(k_r)]$ and $-1 \leq \sigma_M(k_r) \leq +1$.

The definition of the magnetic helicity is fundamentally a spatial concept, so the above definitions have been written in terms of a reduced wavenumber. To make contact with solar wind observations, we will convert the reduced wavenumber to frequency according to

$$f = k_r V_{SW}/2\pi \qquad (8)$$

and write

$$\sigma_M(f) = \frac{2\pi f}{V_{SW}} \frac{H_M(f)}{E_B(f)} \qquad (9)$$

for the normalized magnetic helicity spectrum.

3. DISSIPATION RANGE

Recent analyses of the magnetic helicity in the dissipation range [*Goldstein et al.*, 1994; *Leamon et al.*, 1998a] reveal significant magnetic helicity signatures. A typical observation of a magnetic helicity spectrum associated with IMF dissipation is shown in Figure 2 where a 30% bias of $\sigma_M(f)$ is observed within the dissipation range. Nonzero magnetic helicity is one signature of cyclotron resonance. *Leamon et al.* [1998b, c] show that other processes, most notably ion and electron Landau damping, are also present. Electron Landau damping accounts for $\sim 1/2$ of the total dissipation, thereby providing heating of the thermal electrons as well as the thermal ions heated through the cyclotron and ion-Landau resonances. These processes and the magnetic helicity feature shown in Figure 2 are ubiquitous within the solar wind, although further study is needed to resolve the relative importance of these processes in the outer heliosphere.

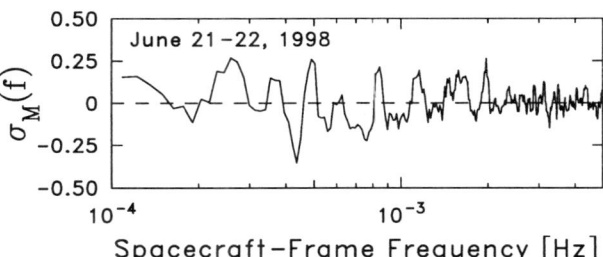

Figure 3. Normalized magnetic helicity spectrum $\sigma_M(f)$ for a frequency interval within the inertial range as recorded by ACE using 10 degrees of freedom. More aggressive smoothing to achieve higher degrees of freedom will eventually drive the spectrum to zero.

Under similar circumstances where wave-particle interactions are key, magnetic helicity signatures can easily be generated. Examples include upstream dynamics at shocks [*Smith et al.*, 1983; 1985] where energetic beams excite low-frequency fluctuations within the inertial range [*Lee*, 1984] and the waves due to pickup ions of cometary [*Tsurutani*, 1991; *Yoon and Wu*, 1991] or interstellar [*Lee and Ip*, 1987] origin. Unlike these cases where wave energy is excited, the helicity signatures present in the dissipation range result from the kinetic damping of one polarization over another in a wave field where a single propagation direction is preferred [*Leamon et al.*, 1998b]. Magnetic helicity is easily generated at energetically significant levels when non-MHD processes are involved.

4. INERTIAL RANGE

The most curious aspect of the inertial range is that any given frequency is likely to have an energetically significant amount of magnetic helicity [*Matthaeus and Goldstein*, 1982], but no net helicity exists within this range when integrated over the full frequency range or a significant fraction of the full range. An example of an inertial range spectrum is shown in Figure 3.

Matthaeus and Goldstein [1982] demonstrate that this behavior is observed out to 5 AU in the Voyager dataset and subsequent analyses [e.g., *Goldstein et al.*, 1994] continue to confirm this result wherever the inertial range helicity has been measured. The only exception to this, as noted above, is when kinetic effects excite waves of a single helicity over a limited range of the spectrum for limited times.

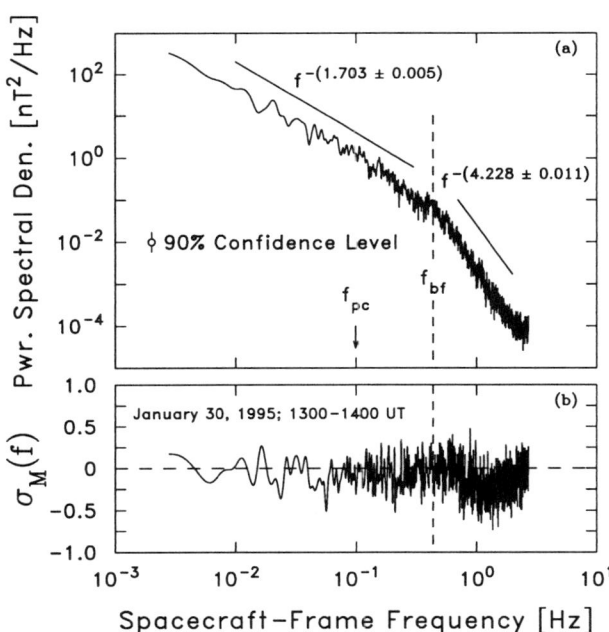

Figure 2. Power spectrum (top) and magnetic helicity spectrum (bottom) for a typical interval of WIND observations of the undisturbed solar wind at 1 AU near zero heliographic latitude. Figure reprinted from *Leamon et al.* [1998a].

5. SOURCES RANGE

In this section we show figures based both on results from equation 4 and the integrand $[R_{TN}(\tau) - R_{NT}(\tau)]$ in equation 3. It should be noted that symmetry considerations, homogeneity and integrability dictate that

$$[R_{TN}(\tau) - R_{NT}(\tau)] \stackrel{\tau \to 0}{\Longrightarrow} 0 \text{ (symmetry)} \quad (10)$$
$$\stackrel{\tau \to \infty}{\Longrightarrow} 0 \text{ (integrability)} \quad (11)$$

There are potentially two additional complications to the study proposed here. First, there exist large-scale reversals of the IMF in the near-ecliptic called magnetic sectors which are associated with crossings of the heliospheric current sheet (HCS) [*Ness and Wilcox*, 1965]. Analyzing the data without regard for sector structure represents both a potential large-amplitude "noise" signal which may contaminate the analysis of low-frequency fluctuations and the admixture of potentially unrelated observations. To this end, observations of magnetic field reversals represent the passage of the spacecraft between field lines connected to different solar hemispheres. Both removal of this signal and separation of the two measurement types are desirable. We therefore separate the IMF measurements into "toward" and "away" sector types using the expected *Parker* [1963] spiral direction [*Smith and Bieber*, 1991, 1993; *Smith and Phillips*, 1997] and separately analyze the two populations.

Second, IMF intensity variations associated either with solar-cycle, radial or latitudinal variations of the interplanetary medium can complicate comparisons of magnetic helicity results for different times or different locations. While we have developed a means of addressing this problem, it is too convoluted for presentation here. For this reason, we will simply cite the results for analyses of outer heliosphere and high-latitude datasets.

An intermediate timescale, which we take to be one solar rotation, is used to compute statistically independent estimates of the helicity. These intermediate estimates are then averaged over longer time intervals, which we generally take to be 1 year or longer, in order to obtain statistical uncertainties and refined estimates of the mean helicity asymmetry.

We impose one further constraint on the data: We demand that each estimate of the correlation function (the integrand in equation 3) obtained for a given intermediate interval be the result of at least 20 individual products of the measurements at every lag. This insures adequate coverage and reduces the likelihood that a poorly-sampled correlation function will adversely affect the resulting average. Any estimate of the correlation function derived for an individual intermediate time interval that fails to meet this criterion is discarded.

5.1. Net Helicity

We can employ the methods discussed in section 2 to compute the temporal behavior of the low-frequency magnetic helicity at 1 AU and near-ecliptic latitudes using the National Space Science Data Center (NSSDC) Omnitape dataset. The temporal resolution of Omnitape data is 1 hour. The practical concern of making a measurement requires that we limit the maximum lag in equations 2 and 3 to finite values. We will use a 36-hour maximum lag in the estimation of Φ and justify this assumption in section 5.2.

We subtract the computed helicities for northern heliospheric hemisphere measurements from those of the south making use of the association between sector polarity and the solar magnetic dipole state. Because the solar magnetic dipole is ill-defined from 1969–71 and in 1980 and 1990, we must discard these years from the analysis. The results are shown in Figure 4. We add to the Omnitape results measurements recorded by the Advanced Composition Explorer (ACE) spacecraft, which was launched in August 1997. ACE measurements, represented as a triangle in Figure 4, from shortly after launch until the end of June 1998 were used.

The striking conclusion of Figure 4 is that there exists a persistent south–north > 0 asymmetry in the net magnetic helicity of the IMF that spans 34 years of observations (neglecting the 5 years when the solar polarity is ill-determined) with only 4 exceptions (and all 4 exceptions are equivalent to zero to within 1 estimated error of the mean). The average south–north asymmetry for the magnetic helicity as computed by this method using the Omnitape dataset from 1965 through 1997 is $(4.51 \pm 1.09) \times 10^{10}$ nT^2 m. *Bieber* [this volume] discusses the implications this result holds for the propagation of cosmic rays in the interplanetary medium.

Although not shown here, the Pioneer-Venus Orbiter observations from 0.7 AU and 1978–89 as well as Pioneer 10 observations from 1973–75 and 1–6 AU, Pioneer 11 observations from 1974–79 and 1–8 AU, Voyager 1 & 2 observations from 1977–84 and 1–18 AU and Ulysses observations from 1991–92 and 2–5 AU all support this basic conclusion. The IMF fluctuation intensity decreases with increasing heliocentric distance, but in all of the above cases the relative amount of helicity scales roughly with the magnetic energy to within a factor of ~2. There is only minimal evidence that some degree of helicity injection or accumulation is active in the interplanetary medium at these scales and adding to the helicity seen at 1 AU. This evidence comes from the years 1973-75 when no magnetic helicity asymmetry is observed at 1 AU, but a finite amount of helicity with low statistical significance can be seen in the Pioneer 10 dataset. In this case, the asymmetry observed agrees with the overall results of Figure 4.

5.2. Correlation Functions

In this section we accomplish 3 things: First, we justify the assumption that a 36-hour maximum lag is sufficient to assess the net magnetic helicity at 1 AU. Second, we show that the average helicity computed across the HCS is zero. Third, we uncover a minor helicity signal associated with longer lags that is interesting, but excluded from the above results.

The first indication that the maximum lag is sufficient comes from recomputing the average asymmetry for the Omnitape dataset using a 72 hour maximum lag. The result is $(3.08 \pm 1.13) \times 10^{10}$ nT^2 m. This slightly reduced result suggests a possible feature at lags longer than 36 hours, but does not undo the earlier conclusions.

We can extract the integrand from equation 3, and neglecting the factor of V_{SW} we plot $[R_{TN}(t) - R_{NT}(t)]$. Figure 5 shows computed values of this term separately for northern (squares) and southern (triangles) hemispheric measurements. The significant feature for 1-30

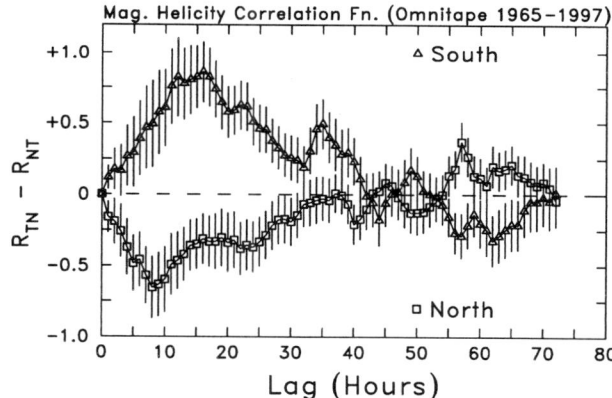

Figure 5. Integrand of $\Phi(t)$ as computed from the Omnitape dataset for the years 1965-1997, excluding the years of solar magnetic dipole reversal. The correlation function for the northern (squares) and southern (triangles) hemispheric measurements are determined separately. The statistical uncertainty is obtained from the intermediate-timescale analysis.

hour lags is evident in approximately equal and oppositely signed values of the correlation functions. This is the aspect of the correlation function that leads to the computed and persistent south-north asymmetry plotted in Figure 4. Figure 5 also demonstrates that the average H_M computed across the current sheet is approximately zero due to the approximate cancellation of these two features when summed.

The next interesting aspect of Figure 5 is the apparent reversal of the correlation functions at about 50-hour lags that results in an oppositely-signed helicity asymmetry at 50-70 hour lags. This is the feature that caused the above decrease in the computed helicity asymmetry for maximum lags of 72 hours.

The convection time for the solar wind to reach from the sun to the Earth is about 97 hours at 427 km/s (the mean wind speed for the Omnitape dataset). Therefore, any feature at lags greater than this must be a solar source effect and we choose to neglect these larger lags in this discussion. *Matthaeus* [this volume] argues that any structure greater than ~ 0.1 AU (10-hour lags) cannot arise from collective processes such as inverse cascade and must be a solar source effect. Therefore, it seems likely that the helicity asymmetry shown here and in Figure 4 is the result of solar sources and is not the result of processes such as inverse cascade working in the interplanetary medium.

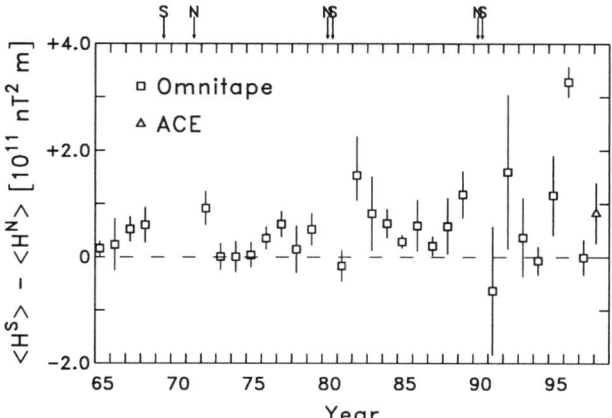

Figure 4. Net helicity asymmetry (south-north) as recorded on the Omnitape (squares) and ACE (triangle) datasets. Error estimates are determined from intermediate timescale (individual solar rotation) estimates of the asymmetry. The times of the solar magnetic reversals of the north (N) and south (S) solar poles are marked at the top of the figure.

We complete the connection with Figure 4 by subtracting the northern correlation functions from the southern forms. This is shown in Figure 6. Here, again,

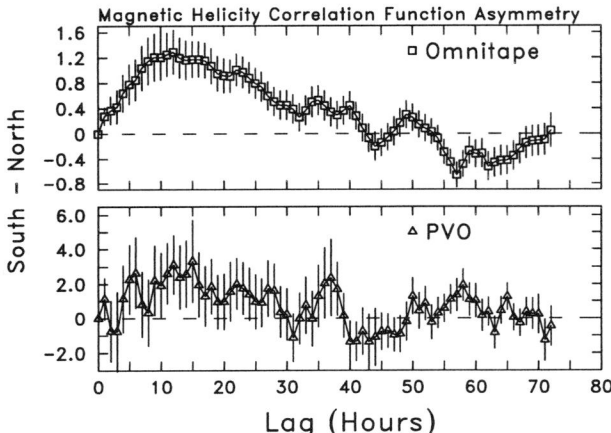

Figure 6. (top) Difference between southern and northern hemispheric correlation functions as shown in Fig. 5. The same quantity is computed from PVO observations (bottom) for the years 1978–89. Note change in scale.

the helicity feature associated with lags < 30 hours is clearly evident at high statistical significance. The same quantity is computed from PVO observations for the years 1978–89 to demonstrate the reproducibility of this result. The higher levels of the correlation function for the PVO results are associated with greater IMF intensity at 0.7 AU while the greater uncertainties are the result of using fewer years in the average.

5.3. CMEs

Coronal mass ejections (CMEs), which are generally diagnosed in interplanetary datasets through the observation of counterstreaming suprathermal electrons [*Gosling et al.*, 1987], are often observed in association with magnetic clouds [*Klein and Burlaga*, 1982; *Burlaga et al.*, 1990; *Gosling*, 1990]. Clouds are twisted flux ropes [*Burlaga et al.*, 1981; *Farrugia et al.*, 1995] that have relaxed to nearly force-free states [*Goldstein*, 1983; *Marubashi*, 1986; *Farrugia et al.*, 1992]. As such, they carry magnetic helicity at large spatial scales.

Bieber and Rust [1995] suggest that the helicity asymmetry at the large scales is the result of solar helicity injection through CMEs. *Rust* [1994] observes that magnetic clouds possess the same sign of magnetic helicity as do the filaments with which they are associated, so it is possible that there exist other sources of low-frequency magnetic helicity in the low-speed solar wind that originates from the streamer belt region [*Rust and Kumar*, 1996].

A preliminary examination of ISEE-3 observations (not shown here) using a catalog of CMEs for this spacecraft [J. T. Gosling, private communication, 1996] has revealed that half of the magnetic helicity shown in Figures 4 – 6 can be eliminated by the removal of CMEs from the dataset. Since not all magnetic clouds are coincident with CME observations, and since it is likely that the observed magnetic helicity resides within the force-free magnetic cloud configurations, it is possible that a further listing of magnetic cloud events will refine the source identification. Lastly, examination of Ulysses observations from over the solar poles (not shown) reveals very little magnetic helicity at the multi-hour scales, which is consistent with the further reduction of the signal by removal of low-speed wind originating in the streamer belt region.

6. SUMMARY

The magnetic helicity within the solar wind is observed over a very broad range of spatial scales from *Parker* [1963] spiral that defines the magnetic structure of interplanetary space, through fluctuations on the scale of 1 AU, to the smallest scales of the dissipation range. There is minimal evidence for solar sources ejecting magnetic helicity into the wind. With nonzero helicity sources at the largest and smallest scales, the intermediate scale of the inertial range appears to act as an energy- and helicity-conserving conduit for spectral transfer without buildup of helicity within its range. While processes such as inverse cascade and relaxation to force-free states are seen to be active in the evolution of CMEs, any global transfer of helicity within the undisturbed solar wind remains only speculative.

Acknowledgments. The Bartol Research Institute is the lead institution for the magnetic field experiments on the Voyagers and ACE. This work was supported by NASA through Ulysses Guest Investigator grant NAG5-6570, Cal-Tech subcontract PC251439 (ACE), and Jet Propulsion Laboratory contract 959167 (Voyager) to the Bartol Research Institute.

REFERENCES

Balogh, A., T. J. Beek, R. J. Forsyth, P. C. Hedgecock, R. J. Marquedant, E. J. Smith, D. J. Southwood, and B. T. Tsurutani, The magnetic field investigation on the Ulysses spacecraft: Instrumentation and preliminary scientific results, *Astron. Astrophys. Supp.*, *92*, 221–236, 1992.

Behannon, K. W., M. H. Acuña, L. F. Burlaga, R. P. Lepping, N. F. Ness, and F. M. Neubauer, Magnetic field experiment for Voyagers 1 and 2, *Space Science Reviews*, *21*, 235–257, 1977.

Bieber, J. W., Role of magnetic helicity in cosmic ray scattering, *this volume*, 1998.

Bieber, J. W., and D. M. Rust, The escape of magnetic flux from the Sun, *Astrophys. J.*, *453*, 911–918, 1995.

Bieber, J. W., P. Evenson, and W. H. Matthaeus, Magnetic helicity of the Parker spiral, *Astrophys. J.*, *315*, 700–705, 1987.

Burlaga, L. F., E. Sittler, F. Mariani and R. Schwenn, Magnetic loop behind an interplanetary shock: Voyager, Helios and IMP-8 observations, *J. Geophys. Res.*, *86*, 6673–6684, 1981.

Burlaga, L. F., R. P. Lepping and J. A. Jones, Global configuration of a magnetic cloud, in *Physics of Magnetic Flux Ropes*, Geophys. Monogr. Ser., vol. 58, edited by C. T. Russell, E. R. Priest, and L. C. Lee, pp. 373–377, AGU, Washington, D.C., 1990.

Farrugia, C. J., L. F. Burlaga, P. Freeman, R. P. Lepping and V. Osherovich, A comparative study of expanding force-free constant alpha magnetic configurations with application to magnetic clouds, in Solar Wind Seven, ed. by E. Marsch and R. Schwenn, pp. 611–614, Pergamon Press, New York, 1992.

Farrugia, C. J., V. A. Osherovich and L. F. Burlaga, The magnetic flux rope versus the spheromak as models for interplanetary magnetic clouds, *J. Geophys. Res.*, *100*, 12,293–12,306, 1995.

Goldstein, H., On the field configuration in magnetic clouds, *Solar Wind 5*, NASA Conference Publ. 2280, edited by M. Neugebauer, pp. 731–733, 1983.

Goldstein, B. E., E. J. Smith, A. Balogh, T. S. Horbury, M. L. Goldstein, and D. A. Roberts, Properties of magnetohydrodynamic turbulence in the solar wind as observed by Ulysses at high heliographic latitude, *Geophys. Res. Lett.*, *22*, 3393–3396, 1995.

Goldstein, M. L., D. A. Roberts, and C. A. Fitch, Properties of the fluctuating magnetic helicity in the inertial and dissipation ranges of solar wind turbulence, *J. Geophys. Res.*, *99*, 11,519–11,538, 1994.

Gosling, J. T., Coronal mass ejections and magnetic flux ropes in interplanetary space, in *Physics of Magnetic Flux Ropes*, Geophys. Monogr. Ser., vol. 58, edited by C. T. Russell, E. R. Priest, and L. C. Lee, pp. 343-364, AGU, Washington, D.C., 1990.

Gosling, J. T., D. N. Baker, S. J. Bame, W. C. Feldman, and R. D. Zwickl, Bidirectional solar wind electron heat flux events, *J. Geophys. Res.*, *92*, 8519-8535, 1987.

Hall, C. F., Pioneer 10, *Science*, *183*, 301–302, 1974.

King, J. H., A survey of long-term interplanetary magnetic field variations, *J. Geophys. Res.*, *81*, 653–660, 1976.

King, J. H., and N. E. Papitashvili, Interplanetary Medium Data Book — Supplement 5, 1988–1993, (Rep. NSSDC/WDC–A–R&S 94–08, NASA, Greenbelt, Md) 1994.

Klein, L. W., and L. F. Burlaga, Interplanetary magnetic clouds at 1 AU, *J. Geophys. Res.*, *87*, 613–624, 1982.

Leamon, R. J., C. W. Smith, N. F. Ness, W. H. Matthaeus and H. K. Wong, Observational constraints on the dynamics of the interplanetary magnetic field dissipation range, *J. Geophys. Res.*, *103*, 4775–4787, 1998a.

Leamon, R. J., W. H. Matthaeus, C. W. Smith and H. K. Wong, Contribution of cyclotron-resonant damping to kinetic dissipation of interplanetary turbulence, *Astophys. J. Lett.*, submitted, 1998b.

Leamon, R. J., C. W. Smith, N. F. Ness and H. K. Wong, Dissipation Range dynamics: Kinetic Alfvén waves, *J. Geophys. Res.*, submitted, 1998c.

Lee, M. A., Particle acceleration and MHD wave excitation upstream of interplanetary shocks, *Adv. Space Res.*, *4*, 295–304, 1984.

Lee, M. A., and W.-H. Ip, Hydromagnetic wave excitation by ionized interstellar hydrogen and helium in the solar wind, *J. Geophys. Res.*, *92*, 11,041–11,052, 1987.

Marubashi, K., Structure of the interplanetary magnetic clouds and their solar origins, *Adv. Space Sci.*, *6*, 335, 1986.

Matthaeus, W. H., Helicity in the interplanetary B field, *this volume*, 1998.

Matthaeus, W. H. and M. L. Goldstein, Measurement of the rugged invariants of magnetohydrodynamic turbulence in the solar wind, *J. Geophys. Res.*, *87*, 6011–6028, 1982.

Matthaeus, W. H., and M. L. Goldstein, Low-frequency $1/f$ noise in the interplanetary magnetic field, *Phys. Rev. Lett.*, *57*, 495–498, 1986.

Matthaeus, W. H., and C. W. Smith, Structure of correlation tensors in homogeneous anisotropic turbulence, *Phys. Rev.*, *A24*, 2135–2144, 1981.

Ness, N. F., and J. M. Wilcox, Sector structure of the quiet interplanetary magnetic field, *Science*, *148*, 1592–1594, 1965.

Oughton, S., K.-H. Rädler, and W. H. Matthaeus, General second-rank correlation tensors for homogeneous magnetohydrodynamic turbulence, *Phys. Rev. E*, *56*, 2875–2888, 1997.

Parker, E. N., *Interplanetary Dynamical Processes*, Wiley-Interscience, New York, 1963.

Russell, C. T., and M. M. Hoppe, Upstream waves and particles, *Space Science Reviews*, *34*, 155–172, 1983.

Rust, D. M., Spawning and shedding helical magnetic fields in the solar atmosphere, *Geophys. Res. Lett.*, *21*, 241, 1994.

Rust, D. M. and A. Kumar, Evidence for helically kinked magnetic flux ropes in solar eruptions, *Astrophys. J. Lett.*, *464*, L199, 1996.

Smith, C. W., and J. W. Bieber, Solar cycle variation of the interplanetary magnetic field spiral, *Astrophys. J.*, *370*, 435–441, 1991.

Smith, C. W., and J. W. Bieber, Multiple spacecraft survey of the north-south asymmetry of the interplanetary magnetic field, *J. Geophys. Res.*, *98*, 9401–9415, 1993.

Smith, C. W., and J. L. Phillips, The role of coronal mass ejections and interplanetary shocks in interplanetary magnetic field statistics and solar magnetic flux ejection, *J. Geophys. Res.*, *102*, 249–261, 1997.

Smith, C. W., M. L. Goldstein, and W. H. Matthaeus, Turbulence analysis of the Jovian upstream 'wave' phenomenon, *J. Geophys. Res.*, *88*, 5581-5593, 1983. (Correction, *J. Geophys. Res.*, *89*, 9159–9160, 1984.)

Smith, C. W., M. L. Goldstein, S. P. Gary, and C. T. Russell, Beam driven ion cyclotron harmonic resonances in the terrestrial foreshock, *J. Geophys. Res.*, *90*, 1429–1434, 1985.

Tsurutani, B. T., Comets: A laboratory for plasma waves and instabilities, in *Cometary Plasma Processes*, Geophys. Monogr. Ser., vol. 61, edited by A. D. Johnstone, pp. 189–209, AGU, Washington, D.C., 1991.

Yoon, P. H., and C. S. Wu, Ion pickup by the solar wind via wave-particle interactions, in *Cometary Plasma Processes*, Geophys. Monogr. Ser., vol. 61, edited by A. D. Johnstone, pp. 241–258, AGU, Washington, D.C., 1991.

C.W. Smith, Bartol Research Institute, University of Delaware, Newark, DE 19716.
(e-mail: chuck@bartol.udel.edu)

Magnetic Helicity and Homogeneous Turbulence Models

William H. Matthaeus

Bartol Research Institute, University of Delaware, Newark, Delaware

The role of magnetic helicity in homogeneous turbulence is discussed, both as a mathematically idealized model and as an approximation to scale-separated, weakly inhomogeneous fluctuations such as those observed in the solar wind. In this context the interpretation of helicity in terms of eigenstates of mirror reflection is useful, precise and gauge invariant. The discussion stresses physical interpretation of numerical and analytical results obtained using the periodic approximation. The expectation that magnetic helicity tends towards dynamical transfer to large scale emerges from absolute equilibrium calculations, inverse cascade and selective decay models. Transport of magnetic helicity due to weak inhomogeneities remains to be fully understood.

1. INTRODUCTION

The relevance of magnetic helicity to magnetohydrodynamic (MHD) plasma and fluid processes was recognized decades ago by *Woltjer* [1958] and others, but the special role played by helicity in homogeneous turbulence was later pointed out by [*Frisch et al.*, 1975]. An equally important but historically distinct involvement of helicity traces to the development of dynamo theory [*Krause and Rädler*, 1980; *Moffatt*, 1978], and in particular the "alpha effect," in which topological properties of the velocity and magnetic fields enter in essential ways. A third thread of investigation involving magnetic helicity is found in studies of relaxation phenomena observed in laboratory confinement devices. Relaxation to a "Taylor state" in reversed field pinches [*Taylor*, 1974] is frequently viewed as a standard dynamical scenario for a plasma bounded by a perfectly conducting surface.

Magnetic helicity is also relevant in space physics, especially in the solar wind where it is routinely observed by spacecraft [*Matthaeus et al.* 1982; *Bieber et al.*, 1987a; *Smith*, this volume] and in cosmic ray physics, where helicity can have important effects on charged particle transport [*Bieber et al.*, 1987a; *Bieber*, this volume] There are similarities to the role played by magnetic helicity in these diverse circumstances, but there are also important differences.

In laboratory plasmas enclosed by metallic walls of sufficiently high conductivity, the magnetic field can be modeled as closed, and magnetic helicity is essentially unambiguous. If the region of interest is not bounded by a flux surface, and one is interested in the topological interpretation of helicity, concerns about gauge invariance enter, but one can invoke remedies such as "relative helicity" to restore desirable properties of helicity [*Berger and Field*, 1984].

A distinct situation, the one of central interest here, is the statistically homogeneous turbulent plasma. Here magnetic helicity of the fluctuations is well defined, and even gauge invariant, provided that the usual requirements of a topological interpretation are discarded at the onset. Nevertheless, magnetic helicity remains a very useful concept, and one that has important applications, including theories of homogeneous MHD turbulence, and of charged particle scattering in random magnetic fields.

2. HELICITY AND SPATIALLY HOMOGENEOUS FLUCTUATIONS

Let us first consider a turbulent magnetic field that satisfies exactly the conditions of weak spatial homogeneity [*Panchev*, 1971]. The spatially varying total magnetic field $\mathbf{B}(\mathbf{x})$ decomposes into mean and fluctuating parts according to $\mathbf{B} = \mathbf{B}_0 + \mathbf{b}$. The ensemble averaging operator $\langle \ldots \rangle$ is defined so that it separates out the mean field, $\langle \mathbf{B} \rangle \equiv \mathbf{B}_0$. The magnetic "energy" in the fluctuations is defined by $E_b = \langle |\mathbf{b}|^2 \rangle/2$. This corresponds to energy per unit mass if $\mathbf{b}$ is in Alfvén speed units. It is convenient to define the mean magnetic helicity density as

$$H_m = \langle \mathbf{a} \cdot \mathbf{b} \rangle = \int d^3k H_m(\mathbf{k}) \qquad (1)$$

where $\mathbf{a}$ is the vector potential associated with the magnetic fluctuations, $\mathbf{b} = \nabla \times \mathbf{a}$, the Coulomb gauge $\nabla \cdot \mathbf{a} = 0$ is assumed, and $H_m(\mathbf{k})$ is the spectrum of magnetic helicity.

For weak homogeneity the first two moments of the probability distribution must be translation independent. Thus we assume for now that the mean field is spatially independent, and that the correlation function

$$R_{ij}(\mathbf{r}) = \langle \mathbf{b_i}((\mathbf{x})\mathbf{b_j}(\mathbf{x}+\mathbf{r}) \rangle \qquad (2)$$

depends only upon the separation vector or "lag" $\mathbf{r}$. For homogeneous turbulence [*Matthaeus et al.*, 1982; *Matthaeus and Goldstein*, 1982] magnetic helicity enters directly into the antisymmetric part of R_{ij}, as do other pseudoscalar quantities that describe departures from mirror-symmetry. Without loss of generality (i.e., for any rotational symmetry) the antisymmetric part is $R_{ij}^A = (R_{ij} - R_{ji})/2 = \epsilon_{ijl}\nabla_l\Phi(\mathbf{r})$, which is fully determined by a single even scalar function of spatial lag $\mathbf{r}$. The spectral tensor is defined as

$$S_{ij}(\mathbf{k}) = \frac{1}{(2\pi)^3}\int d^3r R_{ij}(\mathbf{r})e^{-i\mathbf{k}\cdot\mathbf{r}}. \qquad (3)$$

One can show that its antisymmetric part is $S_{jl}^A = (S_{jl} - S_{lj})/2 = \frac{i}{2}\epsilon_{jlm}k_m H_m(\mathbf{k})$, where $H_m(\mathbf{k})$ is the wavenumber spectrum of magnetic helicity defined in Eq.(1). Note that under the Fourier transformation there is a correspondence $\Phi(\mathbf{r}) \Rightarrow H_m(\mathbf{k})/2$. It is the fact that the antisymmetric part of the correlation tensor in homogeneous turbulence has such a simple form – involving only upon a single scalar function – that provides the basis for observational determination of helicity in the solar wind [*Oughton et al*, 1997; *Smith*, this volume].

It is possible to understand the nature of magnetic helicity in very simple terms in the present context. Let us represent homogeneous fluctuations as a Fourier series in a large periodic box of side L. The idealized infinite homogeneous limit is achieved in the usual way by holding physical scales (e.g., the correlation length) fixed, while $L \to \infty$. Prior to this limit the discussion can take place in the simpler context of Fourier series, in which we can examine the structure of a complex Fourier component $\mathbf{b}(\mathbf{k})$. Since $\nabla \cdot \mathbf{b} = 0$ implies that $\mathbf{k} \cdot \mathbf{b}(\mathbf{k}) = 0$, we may define

$$\mathbf{b}(\mathbf{k}) = b_1(\mathbf{k})\hat{e}_1(\mathbf{k}) + b_2(\mathbf{k})\hat{e}_2(\mathbf{k}) \qquad (4)$$

where the two orthogonal units vectors, $\hat{e}_1 = \mathbf{k} \times \hat{z}/|\mathbf{k} \times \hat{z}|$ and $\hat{e}_2 = \mathbf{k} \times (\mathbf{k} \times \hat{z})/|\mathbf{k} \times (\mathbf{k} \times \hat{z})|$, depend on wave vector $\mathbf{k}$ and a reference direction $\hat{z}$, typically (but not necessarily) taken to be the mean field direction. Now $\mathbf{b}(\mathbf{k})$ is represented by the two complex numbers b_1 and b_2.

Using a Fourier series representation for both $\mathbf{a}$ and $\mathbf{b}$, one easily sees that $H_m = \sum_{\mathbf{k}} Re\{\mathbf{a}^*(\mathbf{k}) \cdot \mathbf{b}(\mathbf{k})\}$ and, using the above basis, that

$$H_m = 2\sum_{\mathbf{k}} Im \frac{b_1^*(\mathbf{k})b_2(\mathbf{k})}{|\mathbf{k}|} \qquad (5)$$

It is natural to define a normalized helicity σ_m such that $\sigma_m(\mathbf{k}) = 2b_1^*b_2/(|b_1|^2 + |b_2|^2)$ so that $-1 \leq \sigma_m \leq +1$. Eq.(5) implies that magnetic helicity is essentially the 90° phase shifted correlation between b_1 and b_2.

Still another convenient way to look at the helicity is using a polarization basis $\hat{e}_\pm = 2^{-1/2}(\hat{e}_1 \pm \hat{e}_2)$, in terms of which $H_m(\mathbf{k}) = (|b_+|^2 - |b_-|^2)/|\mathbf{k}|$. These are the eigenstates of mirror symmetry of the fluctuations, in the sense that the fluctuation of the type $b_+\hat{e}_+$ corresponds to a left-handed spiral or left handed screw, while $b_-\hat{e}_-$ fluctuation is a right handed structure. Noting this, for any set of wavevectors, one can sum the energetic contributions due to left and right-handed polarizations. The corresponding normalized magnetic helicity can be expressed as

$$\sigma_m = \frac{E_L - E_R}{E_L + E_R}, \qquad (6)$$

in terms of E_L, the magnetic energy in left-handed (positive helicity) spatial structures, and E_R, the magnetic energy in right-handed (negative helicity) spatial structures. Note that $E_b = E_L + E_R$. For a single wave vector one readily sees that

$$|b_\pm|^2 = \frac{|b|^2}{2}(1 \pm \sigma_m) \quad (7)$$

where $|b|^2 = |b_+|^2 + |b_-|^2$. Reverting to the homogeneous (infinite domain) case, we can explicitly compute the energy density in left handed fluctuations E_L^b and the energy density in right handed fluctuations E_R^b according to

$$E_L^b = \frac{1}{2}\left(2E_b + \int d^3k |\mathbf{k}| H_m(\mathbf{k})\right) \quad (8)$$

and

$$E_R^b = \frac{1}{2}\left(2E_b - \int d^3k |\mathbf{k}| H_m(\mathbf{k})\right). \quad (9)$$

To summarize, the magnetic helicity H_m of homogeneous fluctuations is a well defined physical quantity, representing the decomposition of the energy into left handed and right handed structures. This is a helicity that relates to mirror symmetry but not to topology. Linkage of flux tubes does not enter into the definition nor necessarily into the interpretation of H_m, although it is possible to interpret the helicity of homogeneous fluctuations as the "self-helicity" or self-linkage of each Fourier contribution with itself [e.g., *Moffatt*, 1978].

3. WEAKLY INHOMOGENEOUS MODEL

One would sometimes like to apply ideas taken from homogeneous turbulence theory [*Batchelor*, 1970] to real systems that are not homogeneous in a global or precise way. Systems may exist in which nonuniformity is in some sense weak, so that homogeneity may be a good first approximation.

3.1. Transport and Scale Expansions

The solar wind is an example of a weakly inhomogeneous system, and it is one in which homogeneous turbulence theory appears to have some relevance [*Matthaeus and Goldstein*, 1982]. The fluctuations of interest are much smaller than the scales associated with variation of the background fields. A reasonable measure of the outer dynamical scale is the observed magnetic fluctuation correlation scale λ_c, about 1/50 AU near earth orbit at 1 AU ($= 1.5 \times 10^{13}$ cm). The local heliocentric distance, r, is the characteristic scale for variations of large scale density, plasma flow and the large scale magnetic field. So there are about two orders of "scale separation" between the fluctuation scales and the scale height associated with inhomogeneities.

The simplest type of weakly inhomogeneous treatment of solar wind fluctuations is so-called WKB theory [e.g., *Hollweg*, 1974] in which noninteracting waves propagate while also experiencing advection and expansion effects. Related but more complex theoretical frameworks have been developed to include local strong turbulence effects as well as linear interaction or "mixing" of inwards and outward propagating fluctuation types [*Tu et al.*, 1984; *Tu and Marsch*, 1989; *Zhou and Matthaeus*, 1990]. The essence of these models is that local turbulence or propagation effects are treated as they would be in a locally homogeneous model, including representation of the fluctuation properties using the theory of homogeneous tensors, sometimes called "isotropic" tensors [*Oughton et al.*, 1997]. Superposed are the effects of weak inhomogeneity.

All such theories rely on some form of scale separation, and can be developed using methods of multiple scale analysis [*Matthaeus et al.*, 1994]. The underpinning of all these approaches is that the local phenomena are viewed as homogeneous and operating on "fast" space and time scales, while inhomogeneous effects (or, "transport" effects) enter at long wavelengths and/or low frequencies. The small parameter is typically $\eta = \lambda_c/r$. It is useful to define the averaging operator $\langle ... \rangle$ so that it eliminates fast spatial variation. The ensemble averaging operator is thus directly tied to the two-scale procedure. This property facilitates development and interpretation of transport theory.

Theories of this type have been quite successful in describing many observed features of solar wind turbulence [e.g., *Zank et al*, 1996; *Tu and Marsch*, 1995]. Mean field dynamo theory [*Moffatt*, 1978; *Krause and Rädler*, 1980] can also be developed from a similar kind of multiple scale analysis, as can "K-epsilon" models used in turbulence applications [*Bradshaw*, 1981].

While in principle the length scale separated transport equations [*Zhou and Matthaeus*, 1990; *Marsch and Tu*, 1989] contain information about transport of magnetic helicity, most applications have so far only examined evolution of energy-like quantities. Helicity transport remains an essentially unexplored corner of these theories. At present we can only refer the interested reader to the bibliography, and the possibility of development of a much more detailed picture of how magnetic helicity of fluctuations behaves in a weakly inhomogeneous medium.

Development of this type of picture is feasible only if magnetic helicity of fluctuations remains a well defined concept for a weakly inhomogeneous MHD system. Recently questions have been raised about the consistency of this approach, [*Montgomery and Bates*, this volume], so it appears to be useful to examine briefly some conceptual issues regarding magnetic helicity and MHD in the locally homogeneous and periodic approximations.

4. PERIODIC MODEL

Suppose that turbulence is locally homogeneous, and correlation functions vanish sufficiently rapidly with increasing spatial lag. Then one can choose a length scale L large enough that a sample of turbulence of this size (L^3) includes subsections that are uncorrelated with one another. Such a sample is "typical" in that its volume averages should be representative of ensemble averages. Mathematically one can readily see that the mean square departure of an average based upon finite sample size (L) from exact ensemble averages [*Panchev*, 1971] is expected to vanish like some power of λ_c/L.

If we are interested in examining samples of strictly homogeneous turbulence with sample size $L >> \lambda_c$, then it would seem to make little difference whether our selected sample retains its real neighbors, or if we replace the neighbors with something convenient. Replacement of the neighbors with an exact copy of the original sample is perhaps the most natural choice, since it will not change estimates of average properties. In this way one views the spatially periodic model, with triply period boundary conditions in space, as a useful model for locally homogeneous fluctuations.

Periodic models have such a long history that even a summary cannot be attempted here. Periodic models of hydrodynamic turbulence have long been employed because of their close relationship to homogeneous turbulence, and because lend themselves to the use of powerful and accurate Fourier spectral methods for numerical simulation [*Orszag*, 1971]. Periodic spectral methods have also been a mainstay of research in MHD turbulence [e.g., *Meneguzzi et al.*, 1981; *Hossain et al.*, 1995].

The magnetic helicity in periodic geometry has some simple but useful properties. The mean magnetic helicity density for a particular realization of the periodic cube can be written as

$$\langle \mathbf{a} \cdot \mathbf{b} \rangle = \frac{1}{L^3} \int d^3 x \, \mathbf{a} \cdot \mathbf{b}. \qquad (10)$$

Additional ensemble averaging can be included if one wishes to average over a set of appropriately equivalent periodic representations (sometimes done, e.g., in simulation studies).

4.1. Conservation

Using the ideal ($\mu = 0$) MHD induction equation (see Eq.(16) below) it is straightforward to show that

$$\frac{dH_m}{dt} = \frac{d\langle \mathbf{a} \cdot \mathbf{b} \rangle}{dt} = 0 \qquad (11)$$

for periodic boundary conditions when the mean field $\mathbf{B}_0$ vanishes. In the case when $\mathbf{B}_0 \neq 0$ this conservation law is replaced [*Stribling et al.*, 1994] by the relation

$$\frac{dH_m}{dt} = -2 \int_0^t dt' \langle \mathbf{v} \times \mathbf{b} \rangle \qquad (12)$$

which can be written in the form of a conservation law by defining the time dependence for the otherwise dynamically irrelevant volume averaged vector potential $\mathbf{A}_0$, such that $d\mathbf{A}_0/dt = \langle \mathbf{v} \times \mathbf{b} \rangle$. Then in terms of the "total" or generalized helicity $\hat{H}_m$,

$$\frac{d\hat{H}_m}{dt} = \frac{d}{dt}(H_m + 2\mathbf{B}_0 \cdot \mathbf{A}_0) = 0 \qquad (13)$$

4.2. Gauge Issues

There is ample reason, as discussed above, for favoring the Coulomb gauge in periodic geometry, and in application to locally homogeneous fluctuations. In Coulomb (transverse) gauge the vector potential components are given explicitly by $\mathbf{a}(\mathbf{k}) = i\mathbf{k} \times \mathbf{b}(\mathbf{k})/k^2$, and it is clear that any further gauge transformation will involve either addition of a longitudinal contribution along $\mathbf{k}$ or a spatially uniform ($\mathbf{k} = 0$) constant. It is reasonable to suppose that $\mathbf{b}$ is derived from a vector potential that satisfies $\langle \mathbf{a} \rangle = L^{-3} \int \mathbf{a} d^3 x = 0$, since in that case $\mathbf{a}$ satisfies the same condition $\langle \mathbf{b} \rangle = 0$ that the fluctuations do. The mean vector potential should not enter into the definition of the self linkage, or self helicity of the fluctuations.

Formally the periodic mean magnetic helicity density (self helicity) is gauge invariant, since under $\mathbf{a} \to \mathbf{a} + \nabla\psi$, one readily sees that $H_m = \langle \mathbf{a} \cdot \mathbf{b} \rangle \to \langle \mathbf{a} \cdot \mathbf{b} \rangle + \langle \mathbf{b} \cdot \nabla\psi \rangle$. But $\langle \mathbf{b} \cdot \nabla\psi \rangle = \langle \psi \nabla \cdot \mathbf{b} \rangle = 0$ for periodic geometry, so the transformation does not change H_m. There is an ambiguity regarding gauge in establishing the value of $\hat{H}_m$ under the addition of a constant vector to $\mathbf{A}_0$ within the periodic cube. This cannot be resolved without looking into the system at scales longer than the local homogeneity scale, and issue we presently address.

4.3. Pre-MHD Two Scale Expansion

Physical fluctuations are never truly periodic or homogeneous. But when inhomogeneity is weak, one might employ periodic models to study local phenomena, while utilizing scale separated transport to study inhomogeneous effects. The efficacy of this approach ultimately requires empirical testing. However, at the onset one would like to understand if this is consistent with basic physical principles [see *Montgomery and Bates*, this volume]. To allay such concerns with regard to judicious application of periodic models to locally homogeneous MHD, it is useful to briefly consider a "pre-MHD" expansion in which the potentially troublesome terms from Maxwell's equations are retained. Here, in contrast to the strictly homogeneous case discussed in the previous section, we do not envision taking the limit as the box size $L \to \infty$ since that would sample inhomogeneous structure. This limit can only be taken locally, requiring definition of two spatial scales.

We introduce a two-scale expansion that differs from the one employed in transport theories (Sec. 3), in particularly with regard to the small parameter of interest. Since we are interested in dynamical behavior of the magnetic field, let us consider the relevant pair of Maxwell equations, Ampere's law,

$$\nabla \times \mathbf{B} = \mathbf{J} + \epsilon^2 \frac{\partial \mathbf{E}}{\partial t} \qquad (14)$$

and the Faraday induction equation

$$\frac{\partial \mathbf{B}}{\partial t} = -\nabla \times \mathbf{E}. \qquad (15)$$

where $\mathbf{J}$ is the electric current density. The parameter $\epsilon = V_a/c$ is discussed below. Let us in addition assume an Ohm's Law of the form $\mathbf{E} = -\mathbf{v} \times \mathbf{B} + \mu \mathbf{J}$ where $\mathbf{v}$ is the plasma fluid velocity and μ is the resistivity.

The choice of units is important. The dimensional scales are B', V', L', T' for the magnetic field, velocity field, length and time, respectively. The unit of electric current density is $J' = cB'/4\pi L'$ where c is the speed of light. The electric field is cast in units of $E' = V'B'/c$. For MHD phenomena the speed unit is the Alfvén speed $V' = V_a = B'/\sqrt{4\pi\rho}$, where the typical mass density is ρ. The length and time scales are related by $L'/T' = V'$. Eqs. (14-15) are the exact form of the two Maxwell equations in these Alfvén speed units. The parameter $\epsilon = V_a/c$ is typically very small for plasmas of interest such as the solar wind (in which $\epsilon \approx 10^{-4}$).

While no approximation has been made as yet, we now wish to exploit the smallness of ϵ. In heuristic derivations of MHD, time variations having the frequency of electromagnetic waves are dropped in favor of fluid scale motions, amounting to neglect of the displacement current ($\epsilon \equiv 0$ in Eq.(14).) Assuming high frequencies are absent, we can employ a two length scale expansion to examine whether low frequency features are properly behaved in a periodic model of locally homogeneous fluctuations.

Assume the total magnetic field is given by $\mathbf{B} = \mathbf{B}^{(0)} + \epsilon \mathbf{B}^{(1)} + \epsilon^2 \mathbf{B}^{(2)} + \ldots$, and that the other relevant variables such as current density $\mathbf{J}$ have similar expansions. Each term of the expansion depends upon both "slow" and "fast" length scale variables, where the slow scales corresponds to scales very much longer than the correlation scales, and the fast spatial coordinate is that of the local turbulence. We express the order-one "fast" scale as $\mathbf{x}_f = \mathbf{x}$, and the slowly varying length coordinate as $\mathbf{x}_s = \epsilon \mathbf{x}$. Note that $\mathbf{x}$ and t are the dimensionless position and time, respectively, in the Alfvén speed units described above. There is no need for either a very slow time scale $t_s = \epsilon t$, for a very fast time scale $t_{fast} = \epsilon^{-1} t$ in the present problem. The latter would be useful if we were including high frequency electromagnetic radiation. The time coordinate t here will remain unexpanded, in Alfvén units.

Derivatives are expanded according to $\nabla = \nabla_f + \epsilon \nabla_s$. As usual, the averaging operator eliminates fast spatial scale variations. This formal property corresponds to averaging over at least a few correlation scales, so that, for example $\langle \mathbf{B} \rangle = \mathbf{B}_0$, and $\langle \mathbf{b} \rangle = 0$.

At leading order we find that $\nabla_f \times \mathbf{B}^{(0)} = \mathbf{J}^{(0)}$, along with $\partial \mathbf{B}^{(0)}/\partial t = - = \nabla_f \times \mathbf{E}^{(0)}$, and $\mathbf{E}^{(0)} = -\mathbf{v}^{(0)} \times \mathbf{B}^{(0)} + \mu \mathbf{J}^{(0)}$. Taken together these imply that the leading order fluctuation obeys the expected MHD induction equation

$$\frac{\partial \mathbf{B}^{(0)}}{\partial t} = \nabla_f \times (\mathbf{v}^{(0)} \times \mathbf{B}^{(0)}) + \mu \nabla_f^2 \mathbf{B}^{(0)} \qquad (16)$$

while the leading order current density is determined by the curl of the leading order magnetic field, as expected in MHD. Averaging shows that $\langle \mathbf{J}^{(0)} \rangle = 0$, the leading order current density averages to zero. In addition, averaging the induction equation, we find that $\partial \langle \mathbf{B}^{(0)} \rangle / \partial t = 0$.

Up to this point, we have not invoked correspondence with periodic boundary conditions in any way. However, under the association $\langle \mathbf{B}^{(0)} \rangle \Rightarrow \mathbf{B}_0$ and $\mathbf{B}^{(0)} - \langle \mathbf{B}^{(0)} \rangle \Rightarrow \mathbf{b}$, one can examine whether periodic boundary conditions might be a consistent model for the leading order MHD-like behavior of the fluctuations. So far it seems optimistic – periodic MHD with a mean field satisfies Eq.(16), along with the condition of zero net current, $\langle \mathbf{j} \rangle = \langle \nabla \times \mathbf{b} \rangle = 0$, and constancy of the mean field, $d\mathbf{B}_0/dt = 0$. All of these are consistent with the leading order expansion. However, to fully understand consistency, including behavior of the magnetic helicity, we need to look further.

To see how the magnetic helicity behaves, we examine expansion of the vector potential. We require that

$$\mathbf{B}^{(0)} = \langle \mathbf{B}^{(0)} \rangle + \mathbf{b}^{(0)} = \nabla_s \times \langle \mathbf{A}^{(-1)} \rangle + \nabla_f \times \mathbf{A}^{(0)}. \quad (17)$$

The local mean field $\langle \mathbf{B}^{(0)} \rangle$ requires a misordered (asymptotically singular) contribution to the vector potential in these units, as one would expect. Usually, in the periodic case, there is no need to use a vector potential for the mean field. Note that the leading order magnetic field fluctuation is determined completely by $\mathbf{A}^{(0)} - \langle \mathbf{A}^{(0)} \rangle$, so that the mean value $\langle \mathbf{A}^{(0)} \rangle$ has no dynamical significance for the local leading order fluctuations. To make contact with the periodic model, we can associate $\mathbf{A}^{(0)} - \langle \mathbf{A}^{(0)} \rangle \Rightarrow \mathbf{a}$, the latter being the periodic, zero mean vector potential, which is associated with the self-linkage helicity discussed in Sec. 2.

Next order relationships clarify the role of the local mean electric field and the magnetic helicity. The divergence condition on the magnetic field implies that

$$\langle \mathbf{B}^{(1)} \rangle = \nabla_s \times \langle \mathbf{A}^{(0)} \rangle, \quad (18)$$

so that the locally volume averaged leading order vector potential, which had no dynamical role in leading order, plays the physically reasonable role of representation of corrections to the local mean magnetic field. The dynamical character of these corrections appears in the $O(\epsilon)$ induction equation, which after averaging can be formally integrated to give

$$\langle \mathbf{B}^{(1)} \rangle = -\nabla_s \times \int^t dt' \langle \mathbf{E}^{(0)} \rangle. \quad (19)$$

But according to Ohm's law the integrand is determined by the mean local turbulent electric field ("EMF"), $\langle \mathbf{v} \times \mathbf{b} \rangle$. Corrections to the mean field are induced by the turbulent electric field, as is expected in mean field dynamo theory [*Krause and Rädler*, 1980; *Moffatt*, 1978]. In addition, let us ask whether the association $\langle \mathbf{A}^{(0)} \rangle \Rightarrow \mathbf{A}_0$ is reasonable. Indeed, we see that the periodic model is consistent with the leading order of the present expansion, in consideration of Eqs. (12), (18), and (19). Extension beyond the periodic model is needed to handle the dynamical modification to the local mean field seen at higher order.

Interestingly, the potentially troublesome displacement current term in Eq.(14) has not affected any terms discussed so far. It enters for the first time at $O(\epsilon^2)$, where one cannot avoid relations like

$$\nabla_s \times \langle \mathbf{B}^{(1)} \rangle = \langle \mathbf{J}^{(2)} \rangle + \frac{\partial}{\partial t} \langle \mathbf{E}^{(2)} \rangle \quad (20)$$

Such terms appear to correspond to extremely low frequency electromagnetic fluctuations, with frequencies of order $\omega \sim kc \sim \epsilon c/\lambda_c \sim V_a/\lambda_c$, i.e., comparable to low frequency Alfvén waves. The amplitude of these corrections is small, $O(\epsilon^2)$ and, for example, would be down by a factor of about 10^{-8} for solar wind parameters. Investigation of such corrections might be worthwhile for certain parameter regimes, but it appears that non-MHD corrections to the periodic model for locally homogeneous turbulence are typically small.

5. MAGNETIC HELICITY AND TURBULENCE

There are a number of ways in which magnetic helicity enters in understanding MHD turbulence in homogeneous or periodic models. What is learned about helicity in these idealized models may be applied in appropriate circumstances to weakly inhomogeneous systems, and may also, with caution, guide understanding helicity and turbulence in other circumstances.

5.1. Absolute Equilibrium and Magnetic Helicity

Gibbs statistical mechanics techniques can be applied to a finite dimensional phase space representation of ideal MHD in periodic boundaries. This approach is useful to investigate the nature of spectral transfer of magnetic helicity and other quantities. (see review by *Kraichnan and Montgomery*, 1980.

In a landmark paper, *Frisch et al.* (1975) investigated the absolute equilibrium properties of energy and magnetic helicity, for three dimensional ideal incompressible MHD turbulence (with $\mathbf{B}_0 = 0$). Being an invariant, H_m is found to conservatively transfer around in k space, on average preferentially seeking the longest wavelengths allowed by the geometry. This preference is strong enough that in a modified thermodynamic limit, the magnetic helicity condenses entirely to the longest wavelength mode, while energy, except the part of it tied up with the helicity, distributes itself evenly across all available scales. Similar results were found when cross helicity is included in the periodic model [*Stribling and Matthaeus*, 1990], and in bounded domains [*Montgomery et al.*, 1978]. Time and again, various studies have reverberated this conclusion, that H_m dynamically seeks the large scales, while energy has a much greater affinity for the small scales.

5.2. Inverse Cascade

With care, conclusions from absolute equilibrium theory can be applied to the driven, steady state turbulent

MHD system. Propensity for H_m to seek the longest scales provides motivation for postulating a steady inverse cascade [*Frisch et al.*, 1975], in which steady "back-transfer" of H_m is expected, producing an inverse cascade powerlaw. Numerical evidence supports these conclusions, both in closures [*Pouquet et al.*, 1976] and in direct numerical simulations [*Pouquet and Patterson*, 1978; *Meneguzzi et al.*, 1981].

5.3. Selective Decay

The nonsteady counterpart of inverse cascade, the selective decay model [*Matthaeus and Montgomery*, 1980; *Riyopoulos et al.*, 1982] makes use of the inherent disparity between the decay rate expected for a quantity that is back transferred to longer scales, versus the decay rate for a quantity that is direct cascaded to smaller scale. Thus, in view of the fact that standard MHD dissipation mechanisms are much more effective at small scales, magnetic helicity is expect to decay much more slowly than energy for high magnetic Reynolds number. This conclusion is operative for cases in which H_m is ideally conserved, such as periodic geometry with $B_0 = 0$, or in suitable bounded geometry [*Montgomery et al.*, 1978].

Conservation of H_m is lost in favor of conservation of $\hat{H}_m$ in periodic geometry with $B_0 \neq 0$. What actually happens in simulations [*Stribling et al*, 1994] is essentially this: A pulse of mean induced electric field is generated on the time scale of a few nonlinear times, and at the same time an initial net H_m decays, and is "replaced" by an appropriate amount of $2\mathbf{B}_0\mathbf{A}_0$ in accordance with Eq.(13). Above we suggested, based upon the multiple scale treatment, that $\mathbf{A}_0$ is associated with MHD activity at scales much larger than the locally homogeneous sample to which the periodic model is applied. Therefore the vanishing of H_m in the periodic domain signifies, once again, that magnetic helicity is being transferred to very long scales. This phenomenon can also be interpreted as an alpha dynamo effect [*Stribling et al*, 1994], looked at from the perspective of the small scale homogeneous fluctuations.

5.4. Relaxation and the Taylor State

One of the immediate conclusions of the selective decay hypothesis [*Matthaeus and Montgomery*, 1980] is that it provides a dynamical basis for relaxation through turbulence to specific states long lived states. The basic procedure is that the direct transferred quantity is minimized subject to constancy of the back transferred quantity, with possibly other constraints included. In the absence of other influences or constraints, this leads to the so called Taylor state [*Taylor*, 1974], which is characterized by minimum energy and constant magnetic helicity.

However probably more often than not relaxation seems to lead to states that are more complicated than the simplest Taylor state, that has, for example, no kinetic energy in the relaxed configuration. In the present volume *Montgomery and Bates* give a review of ideas beyond simple selective decay that may predict more realistic relaxed states, especially in bounded laboratory geometries. Even in periodic models, numerically computed turbulence frequently singles out long lived states that are not the simplest selective decay, Taylor-like configuration. In particular the competition between selective decay and so called dynamic alignment processes has been studies in both two dimensional [*Ting et al*, 1986] and three dimensional MHD turbulence [*Stribling and Matthaeus*, 1991]. While selective decay (in three dimensions) tries to minimize E/H_m, or $\langle|\mathbf{v}|^2 + |\mathbf{b}|^2\rangle/H_m$, dynamic alignment tries to maximize $|\langle\mathbf{v}\cdot\mathbf{b}\rangle|/\langle|\mathbf{v}|^2 + |\mathbf{b}|^2\rangle$. These tendencies cannot be satisfied fully, since the target states are exclusive. However, a generalized framework for relaxation can be described [*Ting et al*, 1986; *Stribling and Matthaeus*, 1991], in which finite flow kinetic energy is predicted for the relaxed state. For these cases the target state is often one for which the "Taylor state" is not a good approximation.

Other possibilities exist as well for further generalizations of the idea of turbulent relaxation to special long lived states, including maximum entropy principles, and others (see *Montgomery and Bates*, this volume). A maximum entropy approach [*Montgomery et al.*, 1992] has been particularly well established in two dimensional hydrodynamics, but requires further development for MHD [*Montgomery et al*, 1979], especially with regards to a three dimensional formulation. The interested reader is referred to the bibliographies of the relevant articles for further reading with regard to the as yet unfinished subject of MHD relaxation processes. For now the bottom line seems to be that the simplest Taylor states are rarely expected to be a precise characterization of states favored by MHD turbulence, and therefore the principle of conservation of magnetic helicity with minimization of energy should not be elevated to strict orthodoxy.

5.5. Helicity in the Solar Wind

Ideas from homogeneous MHD turbulence [*Matthaeus and Goldstein*, 1982; *Tu et al,*, 1984; *Tu and Marsch*,

1995; *Zank et al*, 1996] have been applied with some success to the study of MHD scale fluctuation in the solar wind. Distinctive features of the magnetic helicity and its possible in interplanetary MHD are reviewed in the present volume by *Smith*. So far there is not any indication that the interplanetary medium favors strongly any particular "relaxed state." On the other hand the distinctive observational features of the magnetic helicity appears to be very useful in understanding dynamical activity in the energy containing range, the inertial range and in the dissipation range (see *Smith*, this volume, and references therein).

6. CONCLUSION

Study of magnetic helicity in homogeneous, nearly homogeneous and spatially periodic models represents a somewhat idealized approach that nevertheless appears to be useful in both solar wind studies and cosmic ray scattering theory. In addition there are certain essential features of the behavior of magnetic helicity in turbulence that may be common to these simplified models and other more complex applications. Provided one is interested in magnetic helicity as it relates to mirror symmetry, in contrast to more global considerations of topology, investigation of these simpler circumstances captures certain aspects of the basic physics of magnetic helicity.

Acknowledgments. I would like to thank my colleagues J. Bieber, M. Goldstein, D. Montgomery, S. Oughton, K.-H. Radler, and C. Smith, with whom I have collaborated, and had many illuminating discussions, on the subject of magnetic helicity. This work supported by NSF grant ATM-9713595 at Bartol.

REFERENCES

Batchelor, G. K., *The Theory of Homogeneous Turbulence*, Cambridge University Press, Cambridge, UK, 1970.

Bieber, J.W., P. Evenson and W. Matthaeus, Magnetic Helicity of the IMF and the Solar Modulation of Cosmic Rays, *Geophys. Res. Lett. 14*, 864, 1987a.

Bieber, J.W., P. Evenson and W. H. Matthaeus, Magnetic Helicity of the Parker Field, *Astrophys. J. 315*, 700, 1987b.

Berger, M., and G. Field, The topological properties of magnetic helicity, *J. Fluid. Mech., 147*, 133, 1984.

Berger, M., Magnetic Helicity in a periodic domain, *J. Geophys. Res., , 102*, 2637, 1997.

Bradshaw, P., T. Cebeci, and J. H. Whitelaw, *Engineering Calculation Methods for Turbulent Flow*, Academic, San Diego, Calif., 1981.

Frisch, U., A. Pouquet, J. Léorat and A. Mazure, Possibility of an Inverse Cascade of Magnetic Helicity in Magnetohydrodynamic Turbulence, *J. Fluid Mech. 68*, 769, 1975.

Hollweg, J. V., Transverse Alfvén waves, *J. Geophys. Res., 79*, 1539, 1974.

Hossain, M., P.Gray, D.Pontius Jr, and W. Matthaeus, Phenomenology for the decay of energy-containing eddies in homogeneous MHD turbulence, *Phys. Fluids 7*, 2886, 1995.

Kraichnan, R. H. and Montgomery, D. C., Two-dimensional Turbulence, *Rept. Prog. Phys., 43*, 547, 1980.

Krause, F. and Rädler, K.-H., *Mean-Field Magnetohydrodynamics and Dynamo Theory*, Akademie-Verlag, Berlin, 1980.

Marsch, E. and Tu, C-Y, Dynamics of Correlation Functions with Elsässer Variables for Inhomogeneous MHD Turbulence, *J. Plasma Phys., 41*, 479, 1989.

Matthaeus, W., Y. Zhou, G. Zank and S. Oughton, Transport Theory and the WKB Approximation for Interplanetary MHD Fluctuations, *J. Geophys. Res., 99*, 23421, 1994,

Matthaeus, W., and D. Montgomery, *Selective Decay Hypothesis at High Mechanical and Magnetic Reynolds Number*, Ann. N. Y. Acad. of Sciences *357*, 203, 1980.

Matthaeus, W., M. Goldstein and C. Smith, *Evaluation of Magnetic Helicity in Homogeneous Turbulence*, Phys. Rev. Lett., *48*, 1256, 1982.

Matthaeus, W., and M. Goldstein, *Measurement of the Rugged Invariants of Magnetohydrodynamic Turbulence in the Solar Wind*, J. Geophys. Res. *87*, 6011, 1982.

Meneguzzi, M., U. Frisch and A. Pouquet, *Helical and Nonhelical Turbulent Dynamos*, Phys. Rev. Lett. *47*, 1069, 1981.

Moffatt, H.K. *Magnetic Field Generation in Electrically Conducting Fluids*, Cambridge University Press, Cambridge, 1978.

Montgomery, D., L. Turner and G. Vahala, *Three-Dimensional Magnetohydrodynamic Turbulence in Cylindrical Geometry*, Phys. Fluids *21*, 757, 1978.

Montgomery, D., L. Turner, G. Vahala, Most Probable States in Magnetohydrodynamics, *J. Plasma Phys., 21*, 239, 1979.

Montgomery, D., Matthaeus, W., Stribling, W. T., Martínez, D. and Oughton, S., Relaxation in Two Dimensions and the "Sinh-Poisson" Equation, *Phys. Fluids A, 4*, 3, 1992.

Orszag, S. A., Numerical Simulation of Incompressible Flows within simple Boundaries: I. Galerkin (Spectral) Representations, *Stud. Appl. Math., 50*, 293, 1971.

Oughton, S., W. Matthaeus and K.-H. Rädler, Correlation tensors for general homogeneous turbulence *Phys. Rev. E., 56*, 2875, 1997.

Panchev, S., *Random Functions in Turbulence* , Pergamon Press (New York), 1971.

Pouquet, A., U. Frisch and J. Léorat, *Strong MHD Helical Turbulence and the Nonlinear Dynamo Effect*, J. Fluid Mech. 77, 321, 1976.

Pouquet, A., and G. S. Patterson, *Numerical Simulation of Helical Magnetohydrodynamic Turbulence*, J. Fluid Mech. *85*, 305, 1978.

Riyopoulos, S., A. Bondeson and D. Montgomery, *Relaxation toward States of Minimum Energy in a Compact Torus*, Phys. Fluids *25*, 107, 1982.

Stribling, T., and W. H. Matthaeus, *Statistical Properties of Ideal Three-Dimensional Magnetohydrodynamics*,

Phys. Fluids B *2*, 1979, 1990.

Stribling, T., and W. H. Matthaeus, *Relaxation Processes in a Low-Order Three-Dimensional Magnetohydrodynamics Model*, Phys. Fluids B *8*, 1848, 1991.

Stribling, T., W. Matthaeus and S. Ghosh, Nonlinear Decay of Magnetic Helicity in Magnetohydrodynamic Turbulence with a Mean Magnetic Field, *J. Geophys. Res.*, *99*, 2567, 1994.

Taylor, J.B., *Relaxation of Toroidal Plasma and Generation of Reverse Magnetic Fields*, Phys. Rev. Lett. *33*, 1139, 1974.

Ting, A.C., W. Matthaeus and D.Montgomery Turbulent Relaxation Processes in Magnetohydrodynamics, *Phys. Fluids*, *29*, 3261, 1986.

Tu, C., Z. Pu, and F. Wei, The power spectrum of interplanetary Alfvénic fluctuations: Derivation of the governing equation and its solution, *J. Geophys. Res.*, *89*, 9695, 1984.

Tu, C.Y., and E. Marsch, MHD Structures, Waves and Turbulence in the Solar Wind, Kluwer (Dordrecht, Netherlands), 1995.

Woltjer, L., On hydromagnetic equilibrium, *Proc. Natl. Acad. Sci. U. S. A.*, *44*, 833, 1958.

Zank, G.P., W.H. Matthaeus and C.W. Smith, Evolution of turbulent magnetic fluctuation power with heliospheric distance, *J. Geophys.Res.*, *101*, 17093, 1996.

Zhou, Y., and W. H. Matthaeus, Transport and turbulence modeling of solar wind fluctuations, *J. Geophys. Res.*, *95*, 10,291, 1990.

W. Matthaeus, Bartol Research Institute, University of Delaware, Newark DE 19716 U.S.A. (e-mail: yswhm@bartol.udel.edu)

Role of Magnetic Helicity in Cosmic Ray Scattering

John W. Bieber

Bartol Research Institute, University of Delaware, Newark, Delaware

Magnetic helicity has a strong influence on particle scattering by turbulence, because a resonant interaction occurs only for magnetic fluctuations whose handedness matches the handedness of the helical particle orbit. Coupled with the observed north-south segregation of magnetic helicity in the solar wind, helicity produces particle transport effects that are sensitive both to the Sun's magnetic polarity and to the particle charge sign. Evidence for the role of helicity in cosmic ray transport includes an observed polarity dependence of the ratio of cosmic helium to cosmic electrons and observed 22-year cyclic variations of the cosmic ray streaming anisotropy.

1. INTRODUCTION

Only two mechanisms are known that can cause an explicit charge sign dependence of cosmic ray transport in the interplanetary magnetic field: drifts and magnetic helicity. Inhomogeneities of the large scale magnetic field produce a net motion of particles perpendicular to the field. These motions, called "gradient and curvature drifts" [Rossi and Olbert, 1970; Jokipii, Levy, and Hubbard, 1977], are in opposite directions for positive and negative charges. The drift motions also reverse direction throughout the heliosphere when the Sun's magnetic polarity reverses.

The second mechanism for producing transport effects sensitive to particle charge sign and magnetic polarity is resonant scattering by magnetic turbulence that possesses nonzero magnetic helicity. The unperturbed orbit of a cosmic ray particle is a helix, which of course has a definite left or right handedness determined by the particle charge sign and its direction of motion along the magnetic field. The scattering fluctuations likewise have a definite handedness, or polarization, quantified by the magnetic helicity. Resonant scattering requires that the handedness of the particle orbit match the handedness of the scattering fluctuation, as illustrated in Figure 1. If there is an imbalance between fluctuations with left circular polarization (positive helicity) and those with right circular polarization (negative helicity), then positive and negative charges will scatter at different rates.

In the following two sections I present the theory of scattering by helicity–containing turbulence, first at the level of pitch angle scattering, and then at the level of bulk transport parameters. The next section describes expected effects of helicity on the solar modulation of cosmic rays, in light of observed properties of magnetic helicity in the solar wind. I conclude with a discussion of observational evidence for magnetic helicity effects in solar modulation.

2. PITCH ANGLE SCATTERING

Resonant scattering of cosmic rays by turbulence is quantified by the Fokker–Planck coefficient for pitch angle scattering, $\Phi(\mu) \equiv \langle \Delta \mu^2 \rangle / \Delta t$, where μ is cosine of the particle pitch angle, and Δt represents an increment of time. This coefficient describes a random walk

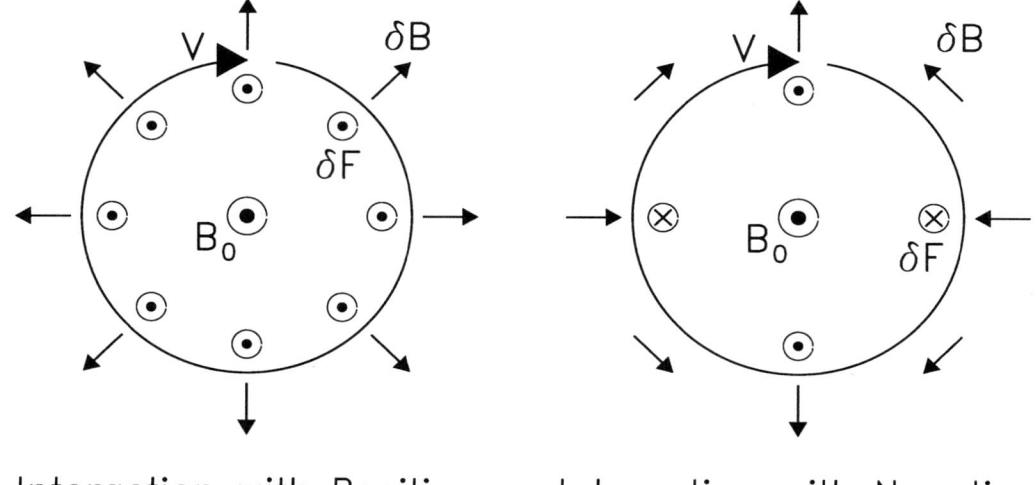

Figure 1. Head–on view of two protons spiralling towards the viewer. The mean magnetic field (B_0) points out of the page, and the protons circulate clockwise about this field. Their orbits form left–hand helices. The proton on the left interacts with a left circularly polarized (positive helicity) fluctuation. As seen in the frame of the particle, the fluctuating field (δB) rotates in the same direction as the particle, and the perturbing force ($\delta F \sim V \times \delta B$ with V the proton velocity) is maximized when averaged over the particle orbit. The proton on the right interacts with a right circularly polarized (negative helicity) fluctuation, and the perturbing force averages to zero over a full orbit. (The figure follows the usual convention that a circled dot denotes a vector pointing out of the page, and a circled "X" denotes a vector pointing into the page.)

of particle pitch angle resulting from interactions with random magnetic fluctuations. Cosmic rays in the solar wind have speeds much larger than the Alfvén speed, so it is usually adequate to treat the magnetic fluctuations as static. The particle then resonates with fluctuations whose wavelength equals the distance the particle travels along the mean field during one complete gyration, $\lambda_{res} = 2\pi\mu R_L$. The corresponding resonant wavenumber is $k_{res} = (\mu R_L)^{-1}$.

In this section, I will consider only scattering by fluctuations with "slab" geometry, for which the fluctuation wavevectors are aligned with the mean magnetic field. In this circumstance, the Fokker–Planck coefficient, $\Phi(\mu)$, can be written [Hasselmann and Wibberenz, 1968; Goldstein and Matthaeus, 1981],

$$\Phi(\mu) = \frac{\pi V}{R_L^2 B_0^2} \frac{(1-\mu^2)}{|\mu|}$$
$$\times\ E(k_{res})\left[1\ +\ \sigma(k_{res})\mathrm{sgn}(eB_0\mu)\right]\ ,\quad (1)$$

where V is particle speed, R_L Larmor radius, B_0 a signed quantity (see below) whose magnitude equals that of the mean magnetic field, $E(k_{res})$ the energy spectrum of fluctuations evaluated at the particle resonant wavenumber, σ the normalized helicity spectrum (see below), and e the signed particle charge. The "sgn" function is $+1$ or -1 according to whether its argument is respectively positive or negative. It is convenient to adopt a convention that pitch angles are measured with respect to an axis aligned with the mean field but pointing away from the Sun, so that particles propagating away from the Sun are always represented by pitch angles between $0°$ and $90°$. With this convention, the mean field B_0 is a signed quantity, positive for "Away" magnetic polarity and negative for "Toward" polarity.

If the fluctuating fields are expressed in a circular polarization basis, then the helicity parameter σ can be defined,

$$\sigma(k) = \frac{E^{(+)}(k)\ -\ E^{(-)}(k)}{E^{(+)}(k)\ +\ E^{(-)}(k)}\ ,\quad (2)$$

where $E^{(+)}$ and $E^{(-)}$ denote respectively the energy in positive helicity (left hand circularly polarized) and negative helicity (right hand circularly polarized) modes. This conceptually simple definition of the normalized magnetic helicity is equivalent to one based upon vector potentials [Matthaeus and Goldstein, 1982] provided the Coulomb gauge is employed.

It is now possible to see the connection between the mathematical formulation of resonant scattering given by equation (1) and the intuitive picture shown in Figure 1. The key point is that handedness of the particle orbit is determined by the sign of $(eB_0\mu)$. This quantity is positive for left handed helical orbits and negative for right handed orbits. For instance, let $\text{sgn}(eB_0\mu) > 0$ (left handed orbit). Then, using equation (2) and the identity $E = E^{(+)} + E^{(-)}$, the second line of equation (1) reduces to $2E^{(+)}(k_{res})$ — i.e., the particle resonates only with the positive helicity (left hand polarized) fluctuations. Similarly, if $\text{sgn}(eB_0\mu) < 0$ (right handed orbit), then only the negative helicity fluctuations produce pitch angle scattering.

3. SPATIAL DIFFUSION AND CONVECTION

Global models of cosmic ray transport in the heliosphere (or "solar modulation" as it is often called) generally do not treat cosmic rays at the level of the pitch angle distribution. Rather, they adopt a fluid approximation that attempts to describe only the cosmic ray density and flow in terms of bulk parameters such as the cosmic ray diffusion tensor and the particle convection speed [e.g., Kóta and Jokipii, 1983; Potgieter and Moraal, 1985]. It therefore becomes necessary to determine if and how the helicity dependence of the Fokker–Planck coefficient translates into a corresponding dependence of the bulk transport parameters.

Theoretical understanding of this problem is currently incomplete. It is not known whether magnetic helicity plays an important role in processes of perpendicular diffusion or drift. A completely general theory for convection and parallel diffusion applicable to the Parker field is also lacking, but a theory does exist for these processes in the special case that the mean magnetic field is aligned with the plasma flow. In the heliosphere, this condition is approximately satisfied in regions where the mean field is nearly radial, i.e., in the high latitude heliosphere and in the inner heliosphere at lower latitudes.

Using Luhmann's [1976] formalism for cosmic ray transport in a geometry where the mean field is aligned with the plasma flow, Bieber and Burger [1990] derived the following expressions for the parallel diffusion coefficient, $K_\parallel$, and cosmic ray convection speed, V_c:

$$K_\parallel = -\frac{LV}{2} \int_{-1}^{+1} \mu e^{-G(\mu)} d\mu, \qquad (3)$$

$$V_c = \frac{3}{4} V_w \int_{-1}^{+1} (1-\mu^2) e^{-G(\mu)} d\mu. \qquad (4)$$

Here, V_w is the solar wind speed, and L is the magnetic focusing length, defined as

$$L = -\left\{\frac{1}{B_0}\frac{\partial B_0}{\partial z}\right\}^{-1}, \qquad (5)$$

where z is distance along the mean field. In accordance with our pitch angle convention, z increases away from the Sun, so that L is positive in the Parker field. The function $G(\mu)$ is defined to satisfy

$$\frac{\partial G}{\partial \mu} \equiv \frac{V}{L}\frac{(1-\mu^2)}{\Phi(\mu)}, \qquad (6)$$

with boundary condition $G(0) = 0$.

In the special case that σ does not vary with wavenumber, and the magnetic energy spectrum is a power law, $E(k) \propto |k|^{-q}$, $G(\mu)$ has the analytic form,

$$G(\mu) = \frac{R_L^2 B_0^2}{\pi L E(\frac{1}{R_L})} \frac{|\mu|^{2-q}}{(2-q)(1-\sigma^2)}$$
$$\times \quad [\text{sgn}(\mu) - \sigma \text{sgn}(eB_0)]. \qquad (7)$$

A crucial aspect of this equation, which is key to understanding helicity effects in solar modulation, is that the sign of helicity enters into the bulk transport parameters only in the combination $\sigma \text{sgn}(eB_0)$.

Figure 2 displays the dependence upon normalized helicity σ of the bulk transport parameters for protons and electrons of 10 GV rigidity. (Rigidity is defined as particle momentum per unit charge multiplied by the speed of light. Particles with the same rigidity have the same Larmor radius.) The parallel mean free path, $\lambda_\parallel$, is related to the parallel diffusion coefficient by $\lambda_\parallel \equiv 3K_\parallel/V$. The figure employs representative 1 AU values for solar wind quantities such as B_0, L, and the energy spectrum E. A value $q = 1$ was used, which is appropriate for the low–wavenumber regime where 10 GV particles are resonant [Bieber et al., 1993].

Both the mean free path and the convection speed exhibit an explicit dependence upon particle charge sign, with the direction of the effect determined by the sign of σB_0. When $\sigma B_0 > 0$, protons have a larger mean free path and convection speed than electrons of the same rigidity. When $\sigma B_0 < 0$, the electron bulk parameters are larger.

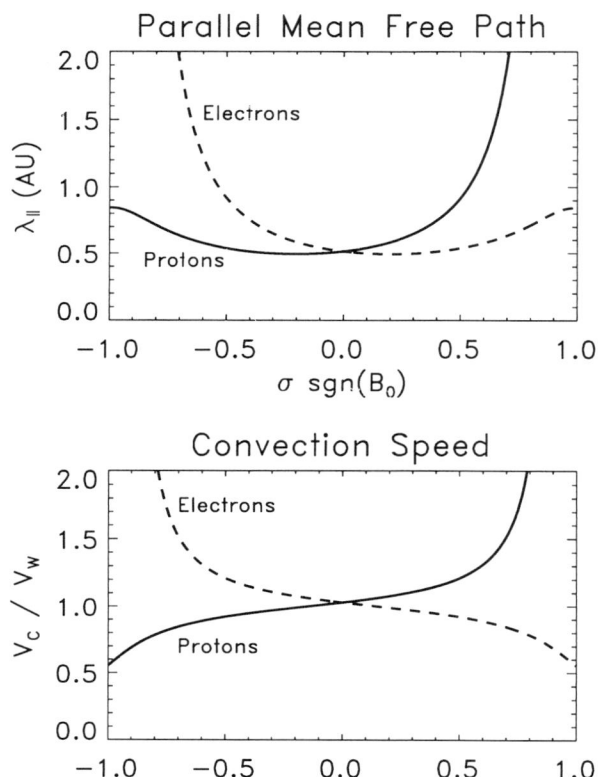

Figure 2. Dependence of parallel mean free path $\lambda_\parallel$ and convection speed V_c upon normalized magnetic helicity σ for 10 GV protons and electrons at 1 AU. The quantity $\text{sgn}(B_0)$ is $+1$ for "Away" magnetic polarity and -1 for "Toward" polarity.

4. MAGNETIC HELICITY AND SOLAR MODULATION

Direct measurements of the interplanetary magnetic field (IMF) reveal the presence of finite net (integrated over wavenumber) magnetic helicity in the solar wind [Bieber, Evenson, and Matthaeus, 1987a; Smith, 1999]. The sign of helicity is correlated with IMF polarity in a manner dependent upon the Sun's magnetic polarity. Specifically, during positive solar polarity (outward magnetic field in the Sun's northern hemisphere, e.g., the 1990's), the helicity tends to be positive in Toward IMF polarity and negative in Away IMF polarity — i.e., the product $\sigma B_0 < 0$ on average throughout the heliosphere. During negative solar polarity (inward magnetic field in the Sun's northern hemisphere, e.g., the 1980's), the reverse tendency exists, and the product $\sigma B_0 > 0$ on average throughout the heliosphere.

This observed association between IMF polarity and the sign of helicity is consistent with a pattern of helicity that is negative north of the current sheet and positive south of the sheet, regardless of the Sun's magnetic polarity. In other words, it is the same negative–north, positive–south pattern of helicity that is observed on the Sun. (See articles by A. A. Pevtsov and by D. M. Rust in this volume.) Insofar as cosmic ray transport is concerned, however, reinterpretation of the correlation between helicity and polarity as a north–south effect is not crucial. It is sufficient to know that σ and B_0 are correlated, because this combination of parameters governs the direction of helicity effects, as shown in Figure 2.

With this information it is possible to identify several key properties of helicity effects in solar modulation:

(1) For a given solar polarity and charge sign, the effect is in the same direction in both hemispheres of the heliosphere. Because both σ and B_0 reverse across the current sheet, their product has the same sign on both sides of the current sheet.

(2) During positive solar polarity, σB_0 tends to be negative, and electrons have a larger mean free path than protons of the same rigidity. See Figure 2. Other things being equal, this implies that electrons should experience less solar modulation than protons during positive polarity.

(3) During negative solar polarity, σB_0 tends to be positive. This corresponds to a relatively larger proton mean free path, and less modulation of protons than electrons.

Another factor governing helicity effects is the wavenumber spectrum, which determines the particle energies affected by magnetic helicity. At high wavenumbers, the helicity parameter σ varies with wavenumber in a random fashion [Matthaeus and Goldstein, 1982]. Its sign reverses frequently, such that the value of σ tends to zero in smoothed data [Smith, 1999].

At sufficiently small wavenumbers, however, steady helicity with a definite sign does emerge [Smith and Bieber, 1993]. Figure 3 displays the magnetic helicity difference between positions south and north of the current sheet, as computed from the National Space Science Data Center's OMNI dataset. Position relative to the current sheet was inferred from IMF polarity.

The figure shows that the helicity difference has a definite dominant sign for frequencies below 10^{-5} Hz, while the helicity above this frequency fluctuates and is generally consistent with zero. The sense of helicity difference is consistent with the familiar negative–north, positive–south pattern. Using the frozen flow

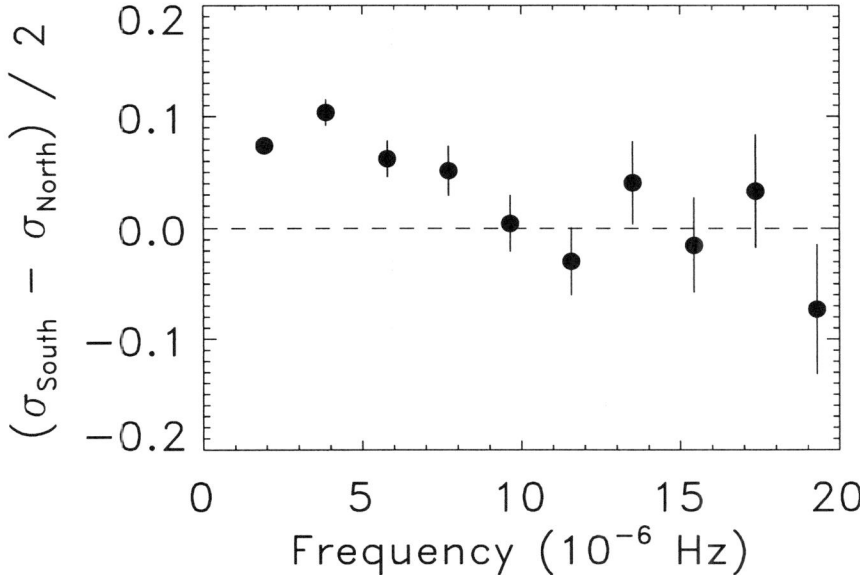

Figure 3. Normalized magnetic helicity has a definite dominant sign at very low frequencies (large scales), consistent with the same negative-north, positive-south pattern observed on the Sun. These average values were computed from more than 20 years of 1 AU data. Adapted from Smith and Bieber [1993].

approximation, $k \sim 2\pi f/V_w$, a frequency $f = 10^{-5}$ Hz corresponds to a wavenumber $\sim 10^{-10}$ m^{-1}.

For typical conditions at 1 AU, a resonant wavenumber of 10^{-10} m^{-1} corresponds to a particle energy of 10 GeV. Therefore, at 1 AU particles with energy greater than 10 GeV experience a helicity effect, while those with lesser energy do not.

The energy range of particles influenced by helicity may, however, be broader than this computation suggests. Beyond 1 AU, the Larmor radius of a particle of fixed energy increases as $\sim r$ (equatorial heliosphere) or $\sim r^2$ (polar heliosphere), which implies that its resonant wavenumber decreases at a rate between r^{-1} and r^{-2}. The question now arises: How does the wavenumber defining the transition from fluctuating to steady helicity change with heliocentric distance? If it remains fixed or decreases more slowly than r^{-1}, as is suggested by some turbulence evolutionary scenarios [Zank et al., 1998], then particle energies not resonant with the steady helicity regime at 1 AU will migrate into this regime as their paths are traced back through the heliosphere. In other words, even though cosmic rays with energy below 10 GeV typically do not experience a helicity effect at 1 AU, they may nonetheless have been influenced by helicity at an earlier time.

The strength of the helicity effect may also be greater than a naive interpretation of Figure 3 might indicate. Recent observations suggest that turbulence in the solar wind is composed of a dominant ($\sim 85\%$) two-dimensional (2D) component, and only a minor ($\sim 15\%$) slab component [Bieber, Wanner, and Matthaeus, 1996]. The 2D component is a very weak scatterer of particles, because its wavevectors are normal to the mean field which is the wrong direction to resonate with the particles. If the field fluctuations of the 2D component are also normal to the mean field, then a simple proof (which is, however, outside the context of this article) shows that this component of turbulence cannot contain magnetic helicity. Because the helicity in Figure 3 is normalized to the total energy (sum of 2D and slab), this implies that the helicity of the slab component, which is what scatters the particles, may be $|\sigma_{SLAB}| \sim 0.5$ or even larger. As a glance at Figure 2 will show, the charge sign dependence of the bulk transport parameters can be quite pronounced when σ is this large.

5. OBSERVED HELICITY EFFECTS

5.1. The Steady-State Anisotropy

The steady-state flow of cosmic rays in the heliosphere produces an anisotropy. At 1 AU, the dominant effects are inward diffusion parallel to the Parker spiral and outward convection with the solar wind flow. These

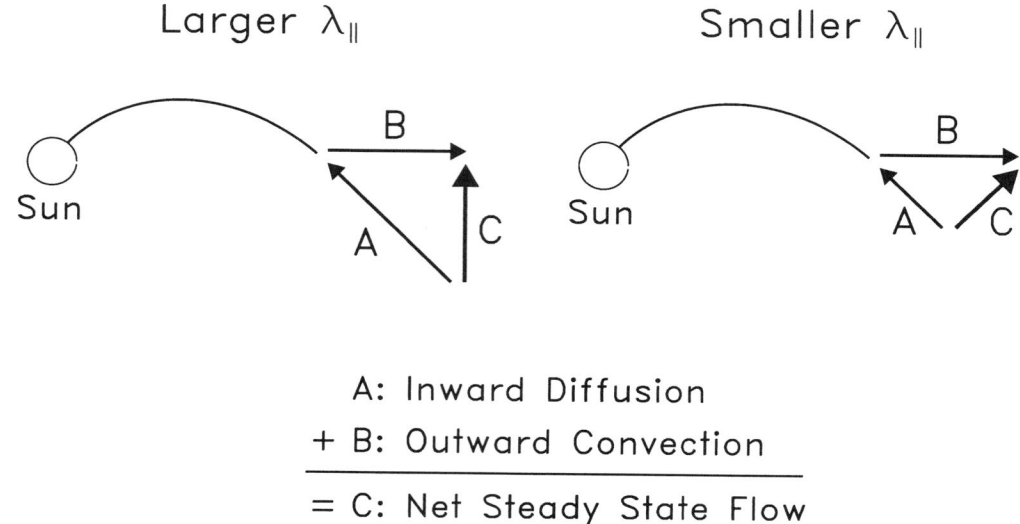

Figure 4. (Left) The steady-state anisotropy of cosmic rays results primarily from a competition between inward diffusion along the magnetic field and outward convection by the solar wind. For the case illustrated, a vertically viewing detector on Earth's surface observes the maximum flux at 1800 h Local Time, when it is looking directly into the flow. (Right) Reducing the inward diffusive anisotropy shifts the local time of maximum towards earlier hours.

effects combine to produce a net anisotropy more or less in the direction of Earth's rotation about the Sun, as depicted in Figure 4.

The steady-state anisotropy is manifested as a diurnal variation of the count rate observed by a ground-based detector. The local time of maximum of the variation occurs when the detector is looking into the oncoming particle flow, i.e., at 1800 h Local Time if the flow is exactly in the direction of Earth's rotation.

Figure 5 presents 32 years of measurements made by the Newark, Delaware neutron monitor together with sunspot number. The neutron monitor detects atmospheric neutrons generated primarily by cosmic ray protons with a typical energy of 17 GeV.

The local time of maximum of the diurnal variation displays a prominent dependence on the Sun's magnetic polarity. It is typically about 1800 h Local Time during sunspot minima with negative solar polarity (1965, 1986) and about 1500 h Local Time during sunspot minima with positive polarity (1976, 1996). This magnetic cycle variation is often called the "twenty-year wave." It has persisted since systematic cosmic ray observations began in 1936 [Forbush, 1967; Duggal, Forbush, and Pomerantz, 1970; Bieber and Chen, 1991].

Initially, the twenty-year wave was cited as evidence for drift effects in solar modulation [Levy, 1976]. The inward diffusive anisotropy is proportional to the product, $\lambda_{\parallel} G_r$, of the parallel mean free path ($\lambda_{\parallel}$) and the radial density gradient of cosmic rays (G_r). Drift models predict that the radial gradient, and hence the diffusive anisotropy, should be smaller during periods of positive solar polarity than during negative polarity [Kóta and Jokipii, 1983; Potgieter and Moraal, 1985]. Hence the local time of maximum should shift towards earlier hours during positive polarity, as observed.

The drift explanation for the twenty-year wave became untenable, however, when subsequent analyses revealed that the radial gradient at neutron monitor energies is nearly the same in the two different solar polarity states [Bieber and Pomerantz, 1986; Chen and Bieber, 1993]. It then became necessary to invoke a solar polarity dependence of the parallel mean free path in order to explain the required change in the diffusive anisotropy.

Magnetic helicity provides a ready explanation for an explicit solar polarity dependence of the mean free path. As shown in Figure 2, the proton mean free path is expected to have a larger value during negative solar polarity ($\sigma B_0 > 0$) than during positive solar polarity ($\sigma B_0 < 0$). This corresponds to a smaller inward diffusive anisotropy during positive polarity, and, as illustrated in Figure 4, the net anisotropy shifts towards earlier hours as observed. (I neglect possible changes of convection speed, because first, the dependence of V_c upon helicity is weaker than that of $\lambda_{\parallel}$, and sec-

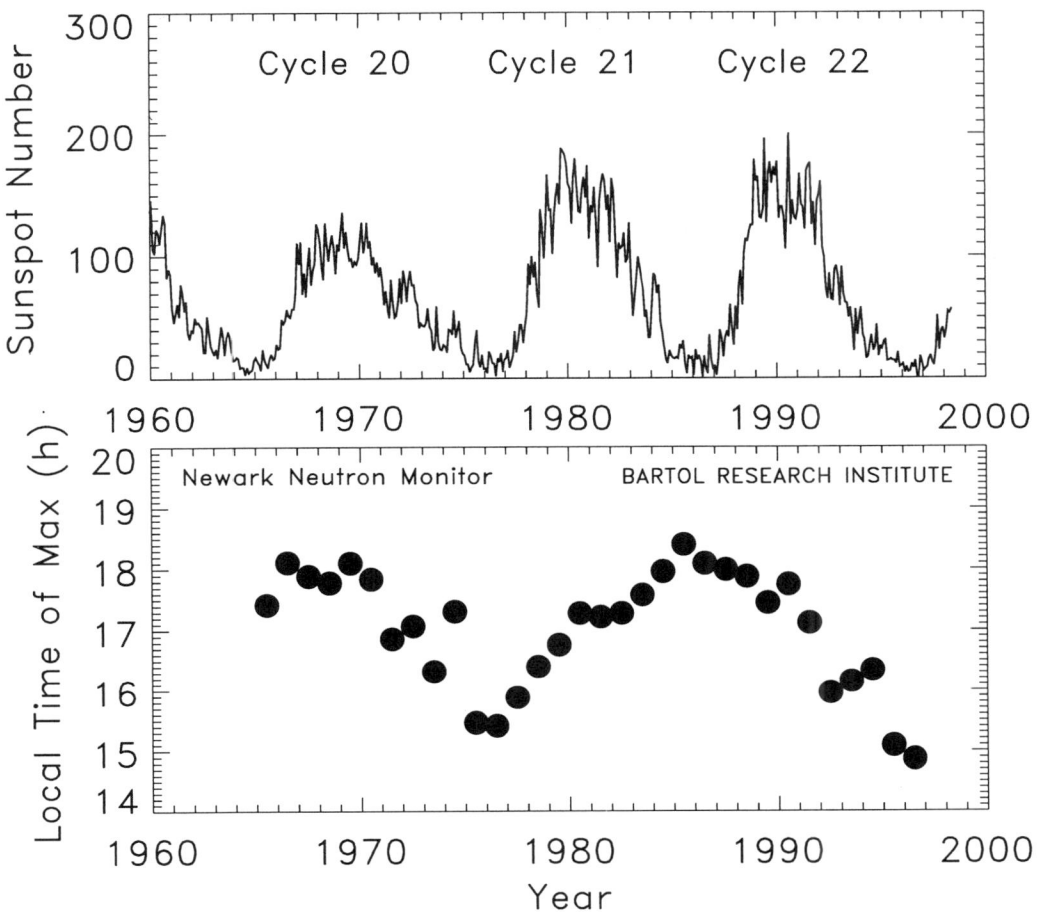

Figure 5. Sunspot number and local time of maximum observed by the Newark neutron monitor, which responds primarily to high energy (∼ 17 GeV) protons. The neutron monitor displays a prominent magnetic cycle variation called the "20 year wave." Local time of maximum has been adjusted to remove effects of the geomagnetic field.

ond, the solar wind flow is not perfectly aligned with the mean magnetic field as envisioned in Luhmann's [1976] quasilinear equation.) Chen and Bieber [1993] concluded that $|\sigma| \sim 0.4$–0.5 is required to explain the size of the observed shift, which as noted in Section 4 is consistent with the observed helicity spectrum (Figure 3) if most of the helicity resides in the minority slab component of the turbulence.

5.2. Charge Sign Dependent Modulation

The most direct probe of magnetic helicity effects in solar modulation is to measure intensities of negative and positive charge species simultaneously at the same rigidity. Figure 6 displays the ratio of cosmic electrons to cosmic helium observed over a 25 year period [Garcia–Munoz et al., 1991] together with observations of the electron to proton ratio made aboard Ulysses [Raviart et al., 1997].

The behavior of the ratios in Figure 6 may reflect a combination of drift and magnetic helicity effects. Drift models predict that the time profile of the ratio of negative to positive charges (at similar rigidity) should have the shape of a "V" centered on solar minimum during negative solar polarity (e.g., the 1980's) and a "Λ" centered on solar minimum during positive solar polarity. There is little indication of such behavior in the 1970's, but the minimum in e^-/He^{++} in 1987 and the local maximum in the Ulysses ratio near the beginning of 1997 are consistent with the predicted behavior.

The most prominent variation of the ratios, however, occurs near solar maximum as the solar magnetic field

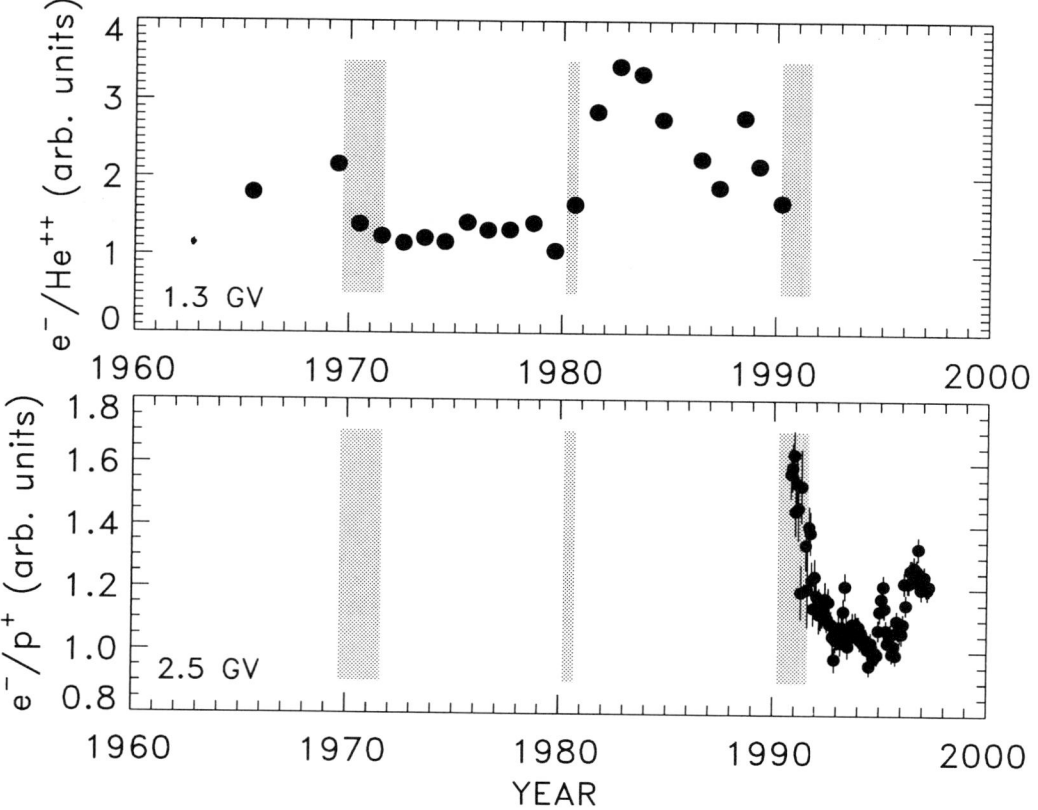

Figure 6. Ratio of (top) cosmic electrons to cosmic helium at 1.3 GV rigidity and (bottom) cosmic electrons to cosmic protons at 2.5 GV rigidity. Shaded areas delimit time periods when the Sun's poloidal field was reversing. Data are from Garcia-Munoz et al. [1991] and from Raviart et al. [1997].

reverses. There is an abrupt change from higher to lower ratios in 1970 and 1990 and a change in the opposite direction in 1980.

Whether this effect can be understood based purely upon drift theory is currently under active investigation. According to one school of thought, drift effects become globally unimportant around solar maximum when the heliospheric current sheet becomes highly inclined [Haasbroek, Potgieter, and Le Roux, 1995; Potgieter, 1998]. If this is correct, then drifts cannot be responsible for the significant charge sign dependent modulation occurring through the magnetic polarity reversal. Another recent study, however, concludes that drifts remain an important source of charge sign dependent modulation even when the current sheet is highly inclined [Burger and Hattingh, 1998]. The changes expected through the polarity reversals are consistent in direction with those shown in Figure 6.

On the other hand, rapid variations around the time of solar field reversal are also to be expected if the ratios are being strongly influenced by magnetic helicity. This is the time when the governing factor σB_0 reverses sign and produces an abrupt change in the bulk modulation parameters. See Figure 2.

Although magnetic helicity can in principle produce rapid ratio changes through the solar polarity reversal, there is a problem with the direction of the changes. Larger mean free paths correspond to less modulation and a higher expected intensity of cosmic rays. One therefore expects the electron to proton ratio to display the same qualitative behavior as the ratio of their mean free paths. According to Figure 2, however, the ratio of mean free paths (electron to proton) is expected to be larger in positive polarity ($\sigma B_0 < 0$) than in negative polarity ($\sigma B_0 > 0$), exactly opposite the behavior displayed by the intensity ratios in Figure 6.

One possible resolution of this dilemma is to invoke turbulence theory arguments suggesting that the helicity spectrum evolves in such a way that the helicity at small scales (roughly, scales smaller than a magnetic field correlation length) develops the opposite sign from the helicity at large scales [Bieber, Evenson, and Matthaeus, 1987a; 1987b]. Based upon resonance considerations, a reversal in helicity sign at the magnetic field correlation scale should result in a change in particle behavior at ~ 10 GeV. This energy does in fact lie above the energies shown in Figure 6, and below the median energy (17 GeV) of the data shown in Figure 5.

The suggestion of such a change in particle behavior is also consistent with cosmic ray energy spectra derived from neutron monitor latitude surveys. These observations consistently show a crossover at about 8 GeV between solar minimum spectra measured during positive solar polarity and those measured during negative polarity [Moraal et al., 1989; Bieber et al., 1997].

Above 8 GeV, higher intensities are observed during negative solar polarity, which is the "expected" direction for protons based upon the 1 AU magnetic helicity spectrum. Below 8 GeV, however, higher intensities are observed during positive polarity which is consistent with a helicity origin only if the helicity at the smaller scales responsible for scattering these particles has the opposite sign of the large scale helicity.

Acknowledgments. I thank Adri Burger, Paul Evenson, and Bernd Heber for useful discussions. This work was supported by NASA grant NAG5-7142 and NSF grant ATM-9616610.

REFERENCES

Bieber J. W., and R. A. Burger, Cosmic-ray streaming in the Born approximation, *Astrophys. J., 348*, 597-607, 1990.

Bieber, J. W., and J. Chen, Cosmic-ray diurnal anisotropy, 1936-1988: Implications for drift and modulation theories, *Astrophys. J., 372*, 301-313, 1991.

Bieber, J. W., and M. A. Pomerantz, Solar cycle variation of cosmic ray north-south anisotropy and radial gradient, *Astrophys. J., 303*, 843-848, 1986.

Bieber, J. W., P. Evenson, and W. H. Matthaeus, Magnetic helicity of the IMF and the solar modulation of cosmic rays, *Geophys. Res. Lett., 14*, 864-867, 1987a.

Bieber, J. W., P. Evenson, and W. H. Matthaeus, Magnetic helicity of the Parker field, *Astrophys. J., 315*, 700-705, 1987b.

Bieber, J. W., W. Wanner, and W. H. Matthaeus, Dominant two-dimensional solar wind turbulence with implications for cosmic ray transport, *J. Geophys. Res., 101*, 2511-2522, 1996.

Bieber, J. W., J. Chen, W. H. Matthaeus, C. W. Smith, and M. A. Pomerantz, Long-term variations of interplanetary magnetic field spectra with implications for cosmic ray modulation, *J. Geophys. Res., 98*, 3585-3603, 1993.

Bieber, J. W., P. Evenson, J. E. Humble, and M. L. Duldig, Cosmic ray spectra deduced from neutron monitor surveys, *Proc. 25th Internat. Cosmic Ray Conf. (Durban), 2*, 45-48, 1997.

Burger, R. A., and M. Hattingh, Toward a realistic diffusion tensor for galactic cosmic rays, *Astrophys. J., 505*, 244-251, 1998.

Chen, J., and J. W. Bieber, Cosmic-ray anisotropies and gradients in three dimensions, *Astrophys. J., 405*, 375-389, 1993.

Duggal, S. P., S. E. Forbush, and M. A. Pomerantz, The variation with a period of two solar cycles in the cosmic ray diurnal anisotropy for the nucleonic component, *J. Geophys. Res., 75*, 1150-1156, 1970.

Forbush, S. E., A variation, with a period of two solar cycles, in the cosmic-ray diurnal anisotropy, *J. Geophys. Res., 72*, 4937-4939, 1967.

Garcia-Munoz, M., P. Meyer, K. R. Pyle, J. A. Simpson, P. Evenson, J. Esposito, and E. Tuska, The dependence of solar modulation on the sign of the cosmic ray particle charge during the 22-year solar magnetic cycle, *Proc. 22nd Internat. Cosmic Ray Conf. (Dublin), 3*, 497-500, 1991.

Goldstein, M. L., and W. H. Matthaeus, The role of magnetic helicity in cosmic ray transport theory, *Proc. 17th Internat. Cosmic Ray Conf., (Paris), 3*, 294-297, 1981.

Haasbroek, L. J., M. S. Potgieter, and J. A. Le Roux, The time-dependent recovery after the large cosmic-ray decrease in 1991, *Proc. 24th Internat. Cosmic Ray Conf., (Rome), 4*, 710-713, 1995.

Hasselmann, K., and G. Wibberenz, Scattering of charged particles by random electromagnetic fields, *Zs. Geophys., 34*, 353-388, 1968.

Jokipii, J. R., E. H. Levy, and W. B. Hubbard, Effects of particle drift on cosmic-ray transport. I — General properties, application to solar modulation, *Astrophys. J., 213*, 861-868, 1977.

Kóta, J., and J. R. Jokipii, Effects of drift on the transport of cosmic rays. VI. A three-dimensional model including diffusion, *Astrophys. J., 265*, 573-581, 1983.

Levy, E. H., Theory of the solar magnetic cycle wave in the diurnal variation of energetic cosmic rays — Physical basis of the anisotropy *J. Geophys. Res., 81*, 2082-2088, 1976.

Luhmann, J. G., A quasi-linear kinetic equation for cosmic rays in the interplanetary medium, *J. Geophys. Res., 81*, 2089-2093, 1976.

Matthaeus, W. H., and M. L. Goldstein, Measurement of the rugged invariants of magnetohydrodynamic turbulence in the solar wind, *J. Geophys. Res., 87*, 6011-6028, 1982.

Moraal, H., M. S. Potgieter, P. H. Stoker, and A. J. van der Walt, Neutron monitor latitude survey of cosmic ray intensity during the 1986/1987 solar minimum, *J. Geophys. Res., 94*, 1459-1464, 1989.

Potgieter, M. S., The modulation of galactic cosmic rays in the heliosphere: Theory and models, *Space Sci. Rev., 83*, 147-158, 1998.

Potgieter, M. S., and H. Moraal, A drift model for the modulation of galactic cosmic rays, *Astrophys. J., 294*, 425-440, 1985.

Raviart, A., P. Ferrando, B. Heber, C. Paizis, V. Bothmer,

W. Dröge, H. Kunow, R. Müller–Mellin, and G. Wibberenz, Evolution of cosmic ray electron spectra above 350 MeV along the Ulysses trajectory, *Proc. 25th Internat. Cosmic Ray Conf.* (Durban), *2*, 37–40, 1997.

Rossi, B., and S. Olbert, *Introduction to the Physics of Space,* 454 pp., McGraw–Hill, New York, 1970.

Smith, C. W., Solar–cycle, radial and latitudinal variations of magnetic helicity: IMF observations, *this volume*, 1999.

Smith, C. W., and J. W. Bieber, Detection of steady magnetic helicity in low–frequency IMF turbulence, *Proc. 23rd Internat. Cosmic Ray Conf.* (Calgary), *3*, 493–496, 1993.

Zank, G. P., W. H. Matthaeus, J. W. Bieber, and H. Moraal, The radial and latitudinal dependence of the cosmic ray diffusion tensor in the heliosphere, *J. Geophys. Res., 103,* 2085–2097, 1998.

John W Bieber, Bartol Research Institute, University of Delaware, Newark, DE 19716 U.S.A. (e-mail: john@bartol.udel.edu)

The Role of Magnetic Helicity in Magnetospheric Physics

Andrew N. Wright

Mathematical and Computational Sciences Dept., St Andrews University, St Andrews, Scotland

The principal use of magnetic helicity in magnetospheric physics has been to study the topological structure of reconnected flux tubes. A change in flux tube linkage (mutual helicity) is balanced by a change in internal structure of the reconnected flux tubes (self-helicity), and this permits limits to be placed upon the amount of twist produced by different reconnection models.

1. INTRODUCTION

The Earth's magnetosphere is the environment around the Earth permeated by terrestrial field lines. *Dungey* [1961] suggested reconnection between the terrestrial field and that in the solar wind could produce an "open" magnetosphere, in which field lines from the polar caps extend out into the solar wind. The boundary between these open field lines and lower latitude "closed" field lines (having both ends in the ionosphere) is the auroral zone. *Dungey* described how reconnection on the sunward side (see Figure 1) could produce open field lines that are dragged antisunward to form an extended magnetotail (labelled 2 and 3, in the figure) where further reconnection can close the earthward portion of the field lines, allowing them to return earthward and establish a global convection pattern of flux tubes.

Today we still regard Dungey's open magnetosphere as the best way to understand magnetospheric structure. The advances to the original open magnetosphere model have taken the form of filling in details of the reconnection process. For example, the dayside/sunward reconnection often takes place in a sporadic fashion in which isolated bundles of magnetic flux are reconnected. In the extended tail, reconnection seems to occur in a quasi-steady fashion at a distant reconnection line (100-140 R_E downtail) and occasionally in an explosive fashion at a near-Earth reconnection line (20-30 R_E downtail). The latter process is referred to as a magnetic substorm. An excellent source of articles on magnetospheric reconnection may be found in the AGU monograph (30) edited by Hones [1984a].

2. TOPOLOGICAL CONSIDERATIONS

Evidently reconnection changes the topology of the magnetic field and this is now described in terms of magnetic helicity. Originally this was not the case, and the origin of twisted flux tubes (particularly on the dayside) remained an enigma for several years.

2.1. Dayside Reconnection

On the dayside of the magnetosphere reconnection can occur in a quasi-steady or, more commonly, in a sporadic fashion. [See the review by *Sonnerup*, 1984, and references therein.] Sporadic reconnection gives the clearest signatures in data, and was termed a Flux Transfer Event (FTE) by *Russell and Elphic* [1978]. Figure 2 shows a schematic of two reconnected flux tubes threading the magnetopause (the plane). Magnetometer data from satellites crossing the magnetopause may be interpreted in terms of isolated reconnected flux tubes being convected past the satellite. The surprising feature of these tubes is that they were twisted. Moreover, on some occasions the twist appeared to be an Alfvén wave [*Saunders et al.*, 1984] propagating away from the reconnection site. Other observations sug-

Magnetic Helicity in Space and Laboratory Plasmas
Geophysical Monograph 111
Copyright 1999 by the American Geophysical Union

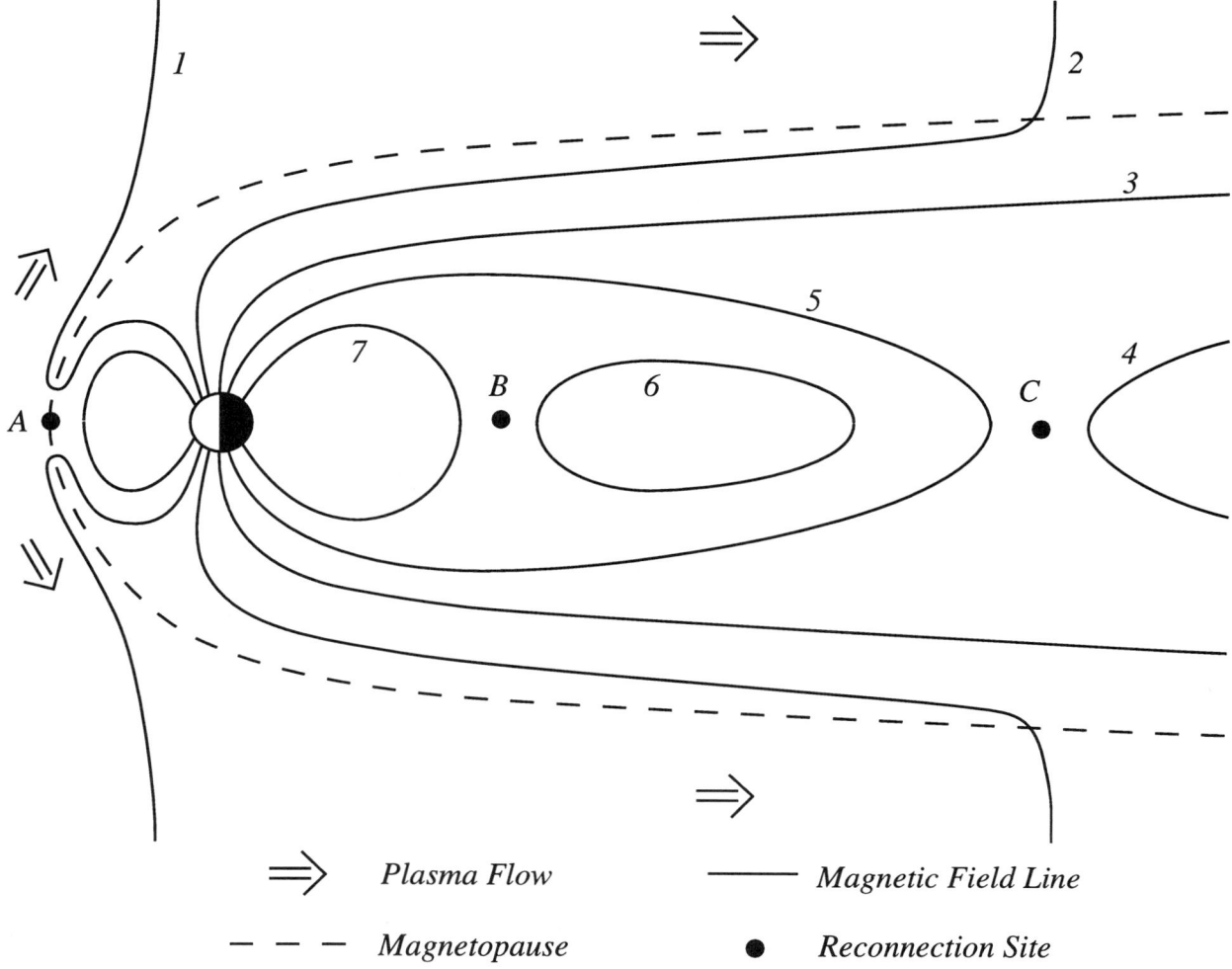

Figure 1. A sketch of magnetic field lines in the noon-midnight meridian plane. The magnetopause (dashed line) separates the magnetosphere, dominated by the terrestrial field, and the fast antisunward flowing magnetosheath (shocked solar wind). The sunward reconnection site (A) is about $10\,R_E$ away from the Earth. (Not drawn to scale).

gested that the twisted tube of plasma was in equilibirium [*Paschmann et al.*, 1982].

Lee and Fu [1985] suggested that the twist could be produced by simultaneous reconnection at multiple reconnection lines located on the dayside magnetopause. Whilst this will certainly produce twisted flux tubes, it is not obvious why reconnection should occur simultaneously, at two sites or whether the strong antisunward convection of solar wind magnetic field lines would permit the retention of field and plasma between multiple reconnection lines.

Wright [1987] considered the detailed 3-dimensional structure of a reconnected flux tube (Figure 3a). As reconnection progresses, successive sheets of field lines are added onto the outside of the existing reconnected flux. Figure 3a shows the state of an idealized reconnected flux tube following a short burst of reconnection. It is similar to a strip of paper that has been folded to have a kink. *Wright* noted that the tube actually had a half twist, and showed how the Maxwell stresses could propagate the twist away as the Alfvén waves reported by *Saunders et al.* [1984]. Such an Alfvén wave carries a field-aligned current along the centre of the tube, and a return current on the surface of the tube. There also exist perpendicular currents connecting the field aligned currents, e.g., at the leading and trailing edges of the Alfvén wave-packet.

Sonnerup [1987] also favoured the formation of recon-

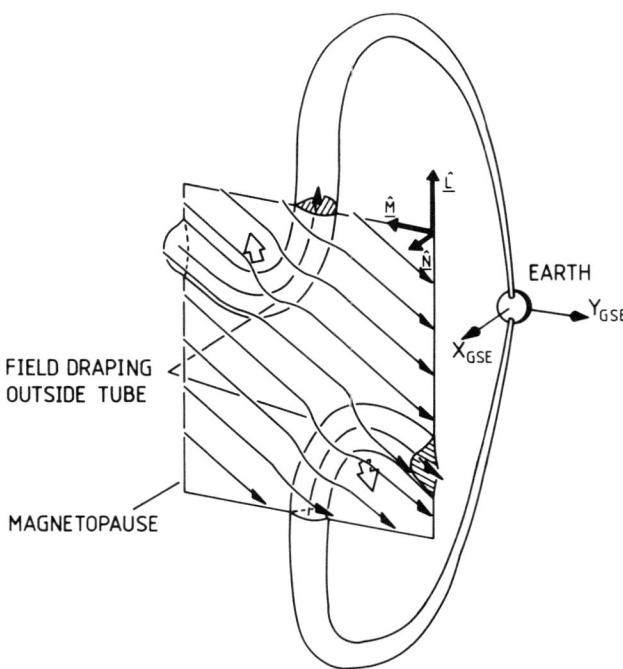

Figure 2. A section of the dayside magnetopause (the plane) and two flux tubes following reconnection at the centre of the plane.

nected flux tubes like those in Figures 2 and 3, but noted a mechanism for producing additional twist: He argued that the magnetopause was not really a discontinuity in which the magnetic field switched directions, but rather a layer of finite width through which the field smoothly switched directions. *Sonnerup* represented this as a layer containing field of intermediate direction to that on either side (Figure 4), and showed how the retreating flux tubes would gather up the weaker "magnetopause field lines". Besides discussing the stress balance of this system, *Sonnerup* argued that reconnection could unlink any excessive collection of magnetopause field lines to produce the configuration in Figure 4c, in which the original flux tube has become more twisted. It is hard to visualise the origin of the extra twist in the manner described by *Sonnerup*, but quite easy using the concepts of magnetic helicity to which we now turn.

Wright and Berger [1989] and *Song and Lysak* [1989a] both independently advocated the use of magnetic helicity to understand the origin of the twist in FTEs. *Wright and Berger* showed how magnetic helicity should be conserved (to at least 90%) during dayside reconnection. Previously, *Berger* [1984] had shown how the total helicity of a collection of flux tubes could be thought of as arising from the internal structure of each tube, and the linking of different tubes with each other.

Wright and Berger introduced the terms "self-helicity" and "mutual helicity" to describe these two contributions, in analogy with the geometrical interpretations of magnetic inductance. The total helicity (which is conserved) is simply the sum of the self- and mutual helicities, H_s and H_m, respectively. (H_s and H_m are not necessarily conserved independently.)

$$H = H_s + H_m \qquad (1)$$

Song and Lysak [1989a] considered the quantities "twist" and "kink" helicities, whose sum is the self-helicity. In ideal MHD self- and mutual helicities are independently conserved, but twist and kink helicities are not. Hence, ideal motions can transform a kinked tube to a twisted tube (Figure 3), as noted by *Wright* [1987], and *vice-versa*.

Wright and Berger [1989] and *Song and Lysak* [1989a] [see also *Berger and Wright* 1990; and, *Lysak and Song*, 1990] both showed that during reconnection, which forms flux tubes like those in Figures 2 and 3, mutual helicity is converted into self-helicity. Figure 5 is a sketch of two thin flux tubes (of flux Φ) before and after reconnection. Note that the flux tube axes (which need not be straight) turn about each other by an angle $\theta = \phi(L) - \phi(0)$. Using formulae derived by *Berger* [1988], the self-helicity and mutual helicity in Figure 5a are $H_s = 0, H_m = (\theta/\pi)\Phi^2$. The two planes in Figure 5 do not represent physical boundaries, but are there to provide a well-defined region of space within which to calculate the magnetic helicity. They may be moved off to infinity if desired. (Note that since $\mathbf{B} \cdot \mathbf{n} \neq 0$ on these boundaries H is strictly the gauge invariant relative magnetic helicity.) Following reconnection (Figure 5b) the tubes turn through an angle $\theta' = \theta - \pi$. Thus the mutual helicity has decreased by $-\Phi^2$, and so the self-helicity must increase by $+\Phi^2$. *Wright and Berger* showed that the increase in self-helicity is shared equally between the two tubes. The details of the internal structure of each tube have considerable freedom. Indeed, both *Wright and Berger*, and *Song and Lysak* [1989b] considered turbulent or random reconnection over a patch of the magnetopause, which may produce small flux filaments that reorganise to a larger scale twist. The simplest choice is a uniform twist, in which case each tube has a half twist. Thus *Wright and Berger* and *Song and Lysak* [1989a] provided a theoretical foundation for *Wright's* [1987] claim that the tube in Figure 3 must have half a twist.

If the tubes in Figure 5a are antiparallel ($\phi(L) - \phi(0) = 0$, and the sign of one of the fluxes is reversed) the original configuration has $H_s = H_m = 0$. Following reconnection H_m is still zero, and so must H_s be. Thus reconnection of antiparallel fields produces no

Figure 3. A detailed sketch of the upper flux tube in Figure 2. The tube is given a rectangular cross-section to emphasize the structure of the tube. The kink that is present in (a) relaxes to a twist (b) which propagates as a torsional Alfvén wave. (After *Wright*, 1987.)

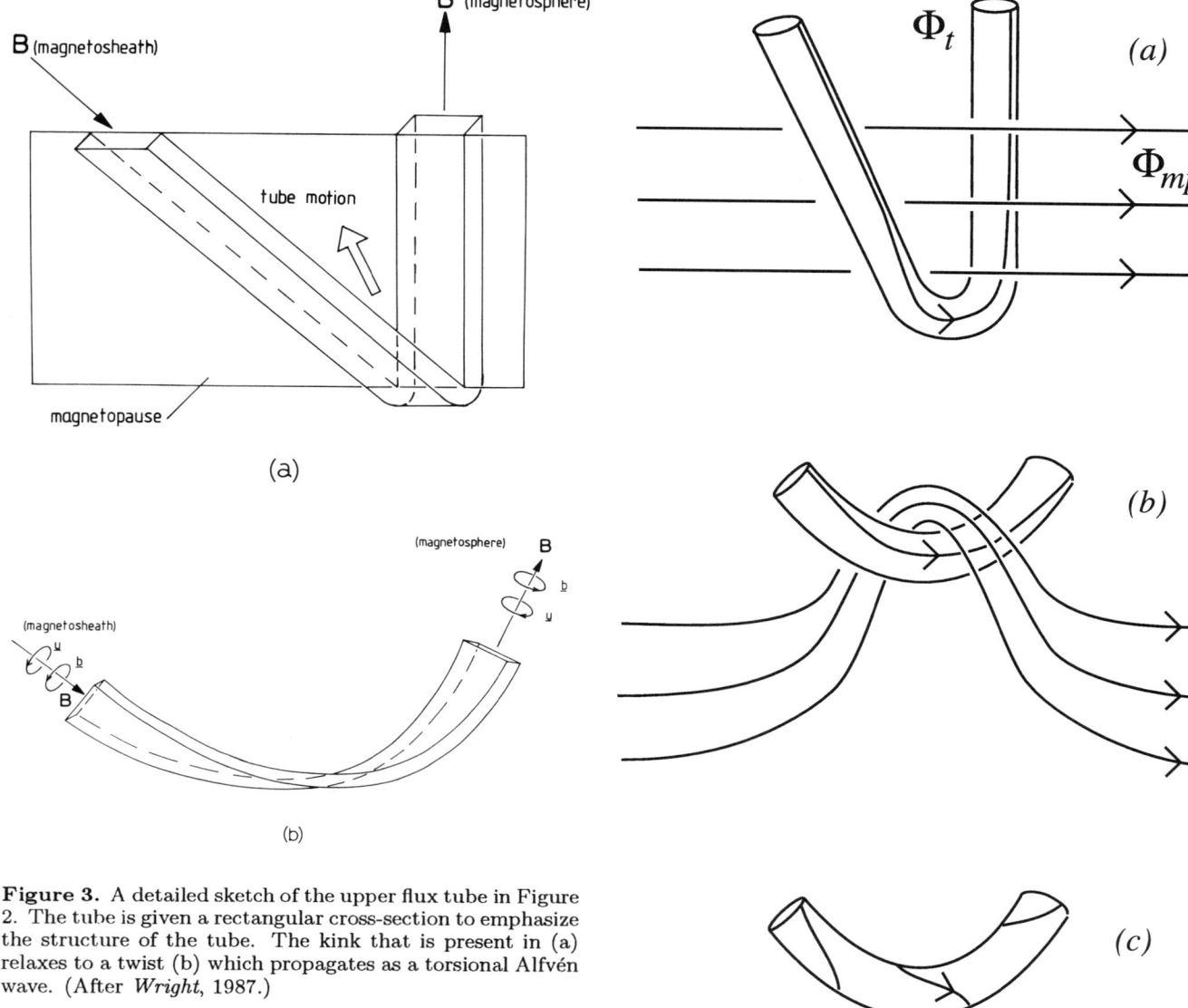

twist. However, *Wright and Berger* [1990] showed that the velocity shear across the magnetopause could inject up to a half twist. This is shown in Figure 6, which is a box containing a velocity shear. Figure 6a shows the field lines as rectangular flux tubes prior to reconnection, and Figure 6b shows how the footpoint motion can introduce a half twist into the reconnected tubes. The sense of twist depends upon the direction of the shear velocity: If $\mathbf{V}_{A1}$ and $\mathbf{V}_{A2}$ are the equilibrium Alfvén speeds on either side of the magnetopause, and $\mathbf{V}_{21}$ the jump in plasma flow velocity, the vector

$$\mathbf{U} = \mathbf{V}_{A1} + \mathbf{V}_{A2} \mp V_{21} \qquad (2)$$

Figure 4. Sonnerup's twisting mechanism relies upon the magnetopause not being a discontinuity, but a thin layer containing flux (Φ_{mp}) of direction between that in the magnetosheath and magnetosphere. (b) As the reconnected tube (Φ_t) relaxes it gathers up the magnetopause field lines. (c) Unlinking of Φ_t and Φ_{mp} by reconnection can result in the flux tube accumulating extra twist.

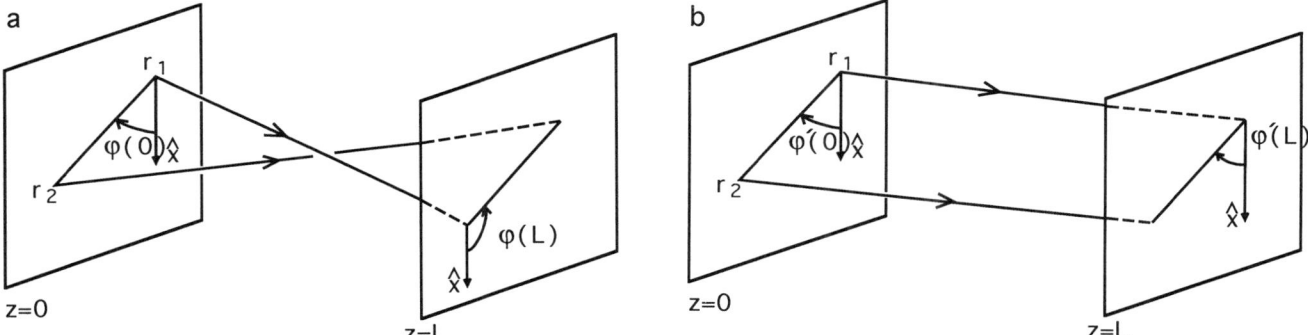

Figure 5. The line joining the centres of two thin straight flux tubes emerging from $r_1(z=0)$ and $r_2(z=0)$ make an angle (a) $\phi(z)$ to the x-axis, before reconnection, and (b) an angle $\phi'(z)$ after reconnection. The angle the two tubes turn about each other ($\theta = \phi(L) - \phi(0)$) changes by π as a result of reconnection. Note that the tube axes need not be straight. The relative helicity of the volume between the planes is conserved for ideal distortions of the flux tubes. (After *Wright and Berger*, 1989).

corresponds to the velocity at which information is communicated along the magnetopause. (Note the equilibrium fields need not be antiparallel now. The upper/lower signs in equation (2) correspond to the upper/lower tubes in Figure 2.) *Wright and Berger* [1990] show how the direction of **U** relative to $\mathbf{V}_{A1}$ and $\mathbf{V}_{A2}$ determines whether the flow affects the sense of twist in the tubes or whether the magnetic geometry is the dominant consideration.

The rate at which information of reconnection propagates along the magnetopause (**U**) and the length of the reconnection line (ℓ) defines a time $\tau \approx \ell/u$. For time t following reconnection such that

$$t > \tau \qquad (3)$$

the tube is well described as a 3-dimensional twisted and bent flux tube. However, if

$$t < \tau \qquad (4)$$

the reconnected flux may look like a quasi-2D structure near the reconnection line. Figure 7 shows a sketch of the possible structure in a plane perpendicular to the reconnection line. Indeed *Southwood et al.* [1988] and *Scholer* [1988] have suggested that long reconnection lines may produce this sort of structure at the dayside. The magnetometer signatures of this structure are also consistent with observations. The twisted/kinked internal field in the centre of the structure will not be observed as Alfvén waves, although it is possible that field lines leaving the ends of this structure may carry such waves. The curved field line region is more likely to be in local pressure balance, similar to the observations reported by *Paschmann et al.* [1982]. Whether the 2D or 3D notions are applicable probably depend upon the inequalities (3) and (4).

Song and Lysak [1989a] also showed how helicity ideas could be applied to Sonnerup's mechanism (Figure 4) to provide a qualitative estimate for the amount of twist produced while progressing from Figure 4a and 4c. With a little imagination, we can represent this with Figure 5; let the flux from r_1 represent the flux tube (Φ_t) and that from r_2 the "magnetopause flux", Φ_{mp}. (This flux is represented by a thin flux tube, but this is fine if the separation of the endpoints in the planes is much greater than the cross section of the fluxes.) Letting $\theta = \phi(L) - \phi(0)$ in Figure 5a be $\sim 2\pi$ can be accomplished by leaving the tube from r_2 straight, but allowing the tube from r_1 to be curved and spiral about once around the straight tube. This represents the configuration in Figure 4a. Allowing the reconnection to produce the topology in Figure 4c, we now have $\theta'(L) = \theta(L) - 2\pi$, and the mutual helicity has changed $\Delta H_m = -2\Phi_t \Phi_{mp}$. Evidently the self-helicity of the flux tube has increased by $2\Phi_t \Phi_{mp}$, and if distributed uniformly would have added an extra $2\Phi_{mp}/\Phi_t$ twists to the tube's field lines. *Song and Lysak* [1989a] estimate this could add about an extra half twist for typical dayside parameters, which is in addition to the half-twist present following formation of the main flux tube. (Note we can not exclude the possibility that some of the increase in self-helicity is present in the magnetopause field.)

2.2. Nightside Reconnection

Magnetic reconnection on the Earth's nightside (i.e., in the geomagnetic tail) has been described in 2-dimensional terms in the noon-midnight meridian by *Hones*

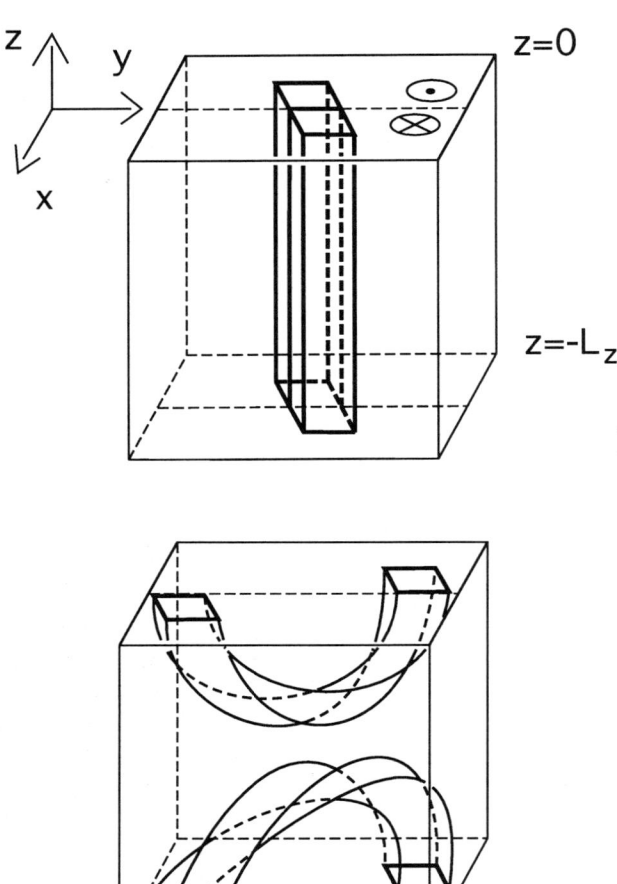

Figure 6. Reconnection of antiparallel fields $\mathbf{B} = -B\hat{z}, x > 0$; $\mathbf{B} = B\hat{z}, x < 0$) in a shear flow ($\mathbf{v} = -v\hat{y}x > 0$; $\mathbf{v} = v\mathbf{y}, x < 0$). The two columns of flux are shown in (a) prior to reconnection, and (b) following reconnection. The reconnected tubes are initially untwisted, but the shearing of the footpoints injects a half twist. (After *Wright and Berger*, 1990).

[1979, 1984b] and references therein, for the case when the magnetic field normal to this meridian plane is zero. Whilst the topology of fields between the two nightside reconnection lines is instructive in this limit, it is highly idealized and has a magnetic helicity that is identically zero.

Subsequently, *Hughes and Sibeck* [1987] considered the 3-dimensional structure of the plasmoid, or flux rope, that is formed between the reconnection lines. *Wright and Berger* [1989] also considered this question. These studies found that the flux ropes were a mixture of four classes of field lines: they can originate in either the ionosphere or the solar wind and finish either in the ionosphere or solar wind. The details of which fluxes are present are sensitive to the overlap of the ends of the two reconnection lines, and the times during which reconnection occurs at the lines. (Indeed, it is this timing consideration which causes a problem for multiple reconnection line models on the dayside.) Figure 8 shows two reconnection lines (bold lines), viewed normal to the current sheet, and the boundaries of fluxes that may be reconnected at them. For nightside reconnection, the angle β would be close to 90°, and the separation of the reconnection lines much greater than their lengths. Nonetheless, the configuration in Figure 8 accurately represents the nightside topology, and can also be used to analyse the dayside multiple reconnection line models of *Lee and Fu* [1985].

The large separation of the two nightside reconnection sites, and the fact that the distant line has quasi-steady reconnection of approximately antiparallel fields, means that it is easy to form a flux rope when recon-

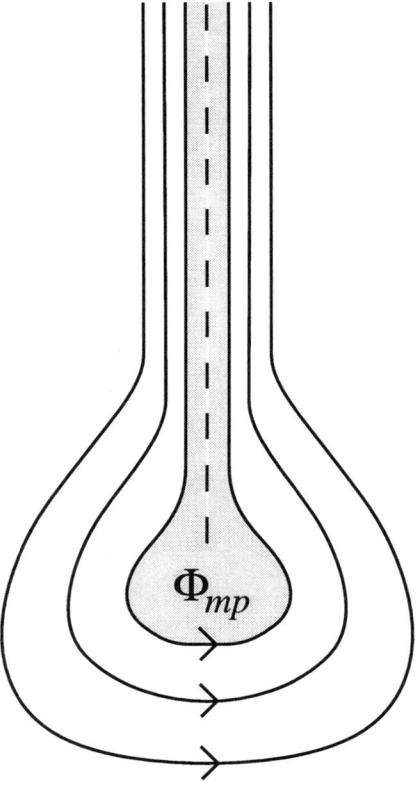

Figure 7. When the reconnection line is sufficiently long a quasi-2D structure may be produced of which the cross section is shown. The magnetopause field (Φ_{mp}) is directed predominantly out of the paper, and is gathered up by the overlying reconnected field lines to give a strong axial field. The overlying field may also have a field component out of the paper.

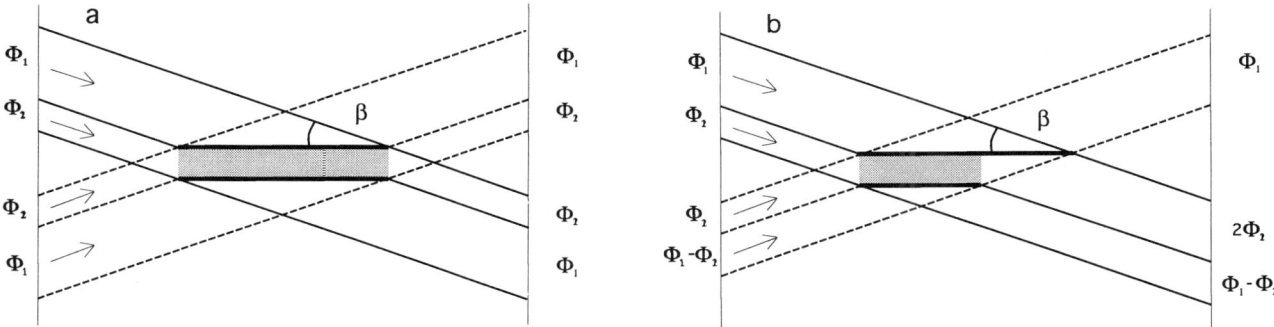

Figure 8. (a) Two parallel reconnection lines (shown in bold) and the edges of fluxes that are reconnected. (b) The overlap of the reconnection lines is important in determining the geometry of the reconnected fluxes. (After *Wright and Berger*, 1989.)

nection commences at the near-Earth line. *Wright and Berger* [1989] and *Song and Lysak* [1989a] showed how helicity conservation ideas can be used to obtain quantitative estimates for the amount of twist in the flux ropes. Indeed, this process can be very efficient at producing twist, and will certainly do so in the tail. It is not so obvious that it operates on the dayside because (a) it requires simultaneous reconnection at two lines, and dayside reconnection is often short-lived; (b) the field lines reconnected at one line must map to the other line, and so imposes geometrical constraints; and (c) multiple reconnection lines produce complex connections, for which there is no observational evidence on the dayside. The mixed nature of the flux rope (the shaded region in Figure 8a) is evident if we imagine flux from the upper half of the diagram as being connected to the ionosphere, and flux from the lower half mapping to the solar wind. The rope has equal fluxes (Φ_2) entering on the left from ionospheric and solar wind origins. The ionospheric components map between the reconnection lines several times before emerging at the right-hand end. When it does emerge, it will (in general) map in part to the ionospheric flux, and part to the solar wind flux. The exact partition depending upon β and the length and separation of the reconnection lines. The same analysis applies to the solar wind flux entering on the left, so there are four classes of field lines [*Wright and Berger*, 1989].

From magnetic stress balance the flux rope in Figure 8a is in equilibrium and would not want to move up or down. If the overlap of the lines is different, flux ropes composed of 2 classes of field line may be produced (Figure 8b), which have a net magnetic stress and would move (downward for this configuration). By extending the upper reconnection lines in Figure 8b a similar amount to the left, a flux rope of one class of field line could be produced and under these geometrical requirements the flux ropes envisaged by *Lee and Fu* [1985] may be produced. However, it is not obvious why such a geometry should be preferred, and other equally likely geometries produce flux ropes that begin and end in the solar wind, or are connected solely to the ionosphere. The dynamics of such tubes could be quite different to FTEs.

The simple systems depicted in Figure 8 assume reconnection happens at both lines simultaneously and at the same rate. This is certainly not true in the geomagnetic tail. The distant reconnection line (labelled C in Figure 1) has a relatively slow but steady reconnection rate. The near-Earth reconnection line (B) does not exist most of the time, but when it forms has a relatively high reconnection rate. In the early stages a flux rope (or plasmoid) is formed between the reconnection lines, but eventually the near-Earth line dominates the reconnected flux and field lines like those labelled "3" in Figure 1 are reconnected at B not C. At this point the system behaves almost as if there were a single reconnection line (the near-Earth one). The reconnected flux is accelerated away from this line, including the flux rope, and eventually the near-Earth reconnection line (B) migrates down the tail to become the new "distant" reconnection line (C), while the flux rope has been expelled downtail [*Hones*, 1979; 1984b]. Then the whole process can repeat itself. If the lengths, locations and reconnection rates for the reconnection lines are known, the whole process could be described in terms of magnetic helicity. (Note that both the near-Earth and deep tail reconnection sites can sometimes be composed of multiple reconnection lines.)

3. DYNAMICAL CONSIDERATIONS

The reconnection process which produces twisted flux tubes is clearly a dynamic process, as evidenced by the propagation of twist, field-aligned currents, and bulk motion of flux tubes. The dynamical evolution is naturally described in terms of waves, while generation of twist and current can be described as a dynamo.

3.1. Magnetohydrodynamic Waves

It was noted by *Wright* [1987] that the kink in a newly reconnected flux tube was not in equilibrium, and would propagate away as a torsional Alfvén wave. *Song and Lysak* [1994] have also found a wave description useful. They derived two "macro particle" solutions to the wave equations; the compressional and shear "Alfvénon", associated with the MHD fast and Alfvén modes, respectively. They also showed how the compression of a current sheet (such as the magnetopause) could be viewed as the collision of two compressional Alfvénons which interact with each other (and the current sheet) to radiate two shear Alfvénons along the reconnected field lines. The shear Alfvénon is of particular interest because it carries a net flux of helicity away from the reconnection site and into the magnetosphere. There is also a flux of energy into the magnetosphere, and *Wright* [1996] showed that the Alfvén wave is the natural agent for this task on open field lines.

Song et al. [1994] considered the distribution of twist over the dayside magnetopause as a result of the field shear and velocity shear [*Wright and Berger*, 1990]. This enabled them to calculate the helicity input as a function of longitude. The propagation of the helicity (i.e., twist) from the reconnection site, along the background magnetic field to the ionosphere was considered by *Lysak et al.* [1994]. They use a numerical solution in dipole geometry with a dissipative ionosphere. The results showed long-lived wave activity close to the magnetopause. The boundary condition for the injection of helicity also injects vorticity (which is also carried by the shear Alfvénon) and suggests there may be a dawn/dusk and north/south asymmetry.

3.2. Dynamos

The twisting of a magnetic flux tube is described in terms of adding azimuthal flux, and so may be viewed as a dynamo. The Alfvén waves which carry the azimuthal flux along the reconnected tube also transport field-aligned currents and so constitute a current dynamo, too [*Song and Lysak*, 1989b; *Lysak and Song*, 1990]. These studies noted a curious feature of reconnection in that it is locally a dissipative process ($\mathbf{j} \cdot \mathbf{E} > 0$), but it results in the dynamo generation of magnetic field and currents ($\mathbf{j} \cdot \mathbf{E} < 0$) and injection of energy to the magnetosphere.

The details of reconnection may be turbulent and cause field lines to reconnect across a patch of the current sheet [*Song and Lysak*, 1989b; *Wright and Berger*, 1989] which can result in a reconnected flux bundle with strong current filaments. *Lysak and Song* [1990] note that the energy in these filaments should dissipate much more rapidly than the helicity associated with them. As a result there is an inverse cascade to larger scales, and the current filaments may condense to give a smoother field structure such as a flux tube with a uniform twist.

4. HELICITY "DISSIPATION"

Most of the ideas described here rely upon the conservation, or approximate conservation, of magnetic helicity. The rate of change of helicity due to resistivity (η) is the volume integral of $-2\eta \mathbf{j} \cdot \mathbf{B}$. This was used by *Berger* [1984] and *Wright and Berger* [1989] to estimate the fraction of helicity that is dissipated, although we shall see that the term "dissipation" can be a little misleading.

There are several elegant geometrical interpretations for the helicity of a variety of magnetic field configurations (e.g., a twisted flux tube, braided flux tubes, linked arched flux tubes, linked closed flux tubes, etc). *Wright and Berger* [1991] derived geometrical interpretations for the change in helicity associated with the reconnection process. (This places the limit of helicity conservation on a firm theoretical basis.) For simple 2D configurations like those in Figure 9, an elegant result was found. The loop of flux in Figure 9a has an azimuthal flux (per unit length in the invariant z direction) of Φ_l, and also a flux coming out of the figure of Φ_z. The block in the centre of the loop is the reconnection site and it contains a flux out of the figure of Φ_{zr}. One can imagine that, in the vicinity of the reconnection site, reconnection occurs in a steady fashion so that all the fields and fluxes are independent of time in that region. Now consider the change in helicity as the loop shown in Figure 9a is swept through the reconnection site to achieve the configuration in Figure 9b. *Wright and Berger* [1991] showed that to leading order the dissipation of helicity is associated with the unlinking of Φ_l and Φ_{zr}; $\Delta H = -2\Phi_l \Phi_{zr}$. (This is strictly the relative helicity per unit length in z.) It is interesting to note that, for suitable driving, we may change from the topology in Figure 9b to that in Figure 9a. Now the helicity has increased, so $-2\eta \mathbf{j} \cdot \mathbf{B}$ can be a source of helicity despite the dissipative nature of reconnection. Clearly the term helicity dissipation is inappropriate here as helicity is generated. Nevertheless, for small η the generation of helicity is small, and we find helicity is approximately conserved.

5. CONCLUSION

The three-dimensional time-dependent nature of magnetospheric reconnection makes a full solution of the

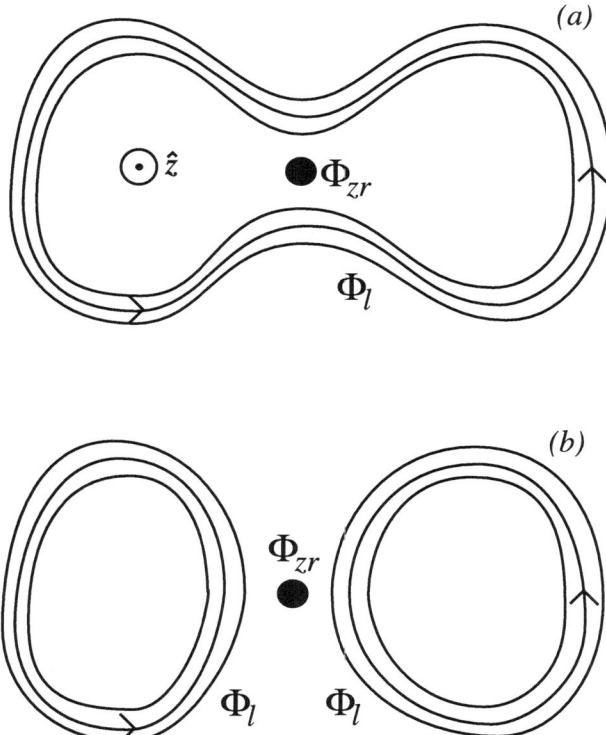

Figure 9. A two-dimensional field configuration (independent of z) consists of concentric surfaces of flux and a reconnection site (the black circle). The flux per unit length of z directed in the (x,y) directions is Φ_l, and the reconnection site contains a flux Φ_{zr} in the z direction. Geometrically, the change in helicity due to reconnection is the change in linkage of Φ_l and Φ_{zr}. Reconnection from (a) to (b) decreases the helicity, while changing from (b) to (a) increases the helcity.

equations formidable. An alternative approach is to reformulate the governing equations as invariants, one of which is the magnetic helicity. The gross features of the reconnection process coupled with helicity conservation enable strict limits to be put upon the amount of twist that is generated in the reconnected fields. This is viewed as the conversion of mutual helicity to self-helicity. When used in conjuction with observations it should be possible to rule out some of the proposed reconnection models. Dynamical considerations show there is a flux of energy and helicity across the current sheet. Alfvén waves are the natural agent for carrying these fluxes, and are radiated from the reconnection site.

Over a decade ago observations of reconnected flux tubes showed they had internal twist and carried field aligned currents. There was much debate over why the tubes had these features, but now we have one way of understanding: Reconnection does not dissipate helicity rapidly enough for them to be untwisted and current-free.

Acknowledgments. This article was written while the author was funded by a UK PPARC Advanced Fellowship. Numerous enlightening discussion with Mitch Berger and Yan Song are gratefully acknowledged. Mark Saunders kindly supplied Figure 2.

REFERENCES

Berger, M. A., Rigorous new limits on magnetic helicity dissipation in the solar corona, *Geophys. Astrophys. Fluid Dyn., 30,* 79, 1984.

Berger, M. A., Three dimensional reconnection from a global viewpoint, p83-86, *Reconnection in Space Plasma,* edited by T. D. Guyenne and J. J. Hunt, ESA SP-285, *11,* Paris, 1988.

Berger, M. A., and A. N. Wright, The generation of twisted flux ropes during magnetic reconnection, *The physics of magnetic flux ropes,* C. T. Russell, E. R. Priest, and L. C. Lee (eds.), AGU Monograph 58, p. 521, 1990.

Dungey, J. W., Interplanetary magnetic fields and the auroral zones, *Phys. Rev. Lett., 6,* 47, 1961.

Hones, E. W., Jr., Transient phenomena in the magnetotail and their relation to substorms, *Space Sci. Rev., 23,* 339, 1979.

Hones, E. W., Jr. (eds.), *Magnetic reconnection in space and laboratory plasmas,* AGU Monograph 30, 1984a.

Hones, E. W., Jr., Plasma sheet behavior during substorms, *Magnetic reconnection in space and laboratory plasmas,* Hones, E. W., Jr. (eds.), AGU Monograph 30, p. 178, 1984b.

Hughes, W. J., and D. G. Sibeck, On the 3-dimensional structure of plasmoids, *Geophys. Res. Lett., 14,* 636, 1987.

Lee., L. C., and Z. F. Fu, A theory of magnetic flux transfer at the Earth's magnetopause, *Geophys. Res. Lett., 12,* 105, 1985.

Lysak, R. L., and Y. Song, Formation of flux ropes by turbulent reconnection, *The physics of magnetic flux ropes,* C. T. Russell, E. R. Priest, and L. C. Lee (eds.), AGU Monograph 58, p. 525, 1990.

Lysak, R. L., Y. Song and D.-H. Lee, Generation of ULF waves by fluctuations in the magnetopause position, *Solar wind sources of Ultra-Low-Frequency waves,* M. J. Engebretson, K. Takahashi, and M. Scholer (eds.), AGU Monograph 81, p. 273, 1994.

Paschmann, G., G. Haerendel, I. Papamastorakis, N. Sckopke, S. J. Bame, and J. T. Gosling, Plasma and magnetic field characteristics of magnetic flux transfer events, *J. Geophys. Res., 87,* 2159, 1982.

Russell, C.T., and R. C. Elphic, Initial ISEE magnetometer results: Magnetopause observations, *Space Sci., Rev., 22,* 681, 1978.

Saunders, M. A., C. T. Russell, and N. Sckopke, A dual-satellite study of the spatial properties of FTEs, *Magnetic reconnection in space and laboratory plasmas,* Hones, E. W. Jr. (eds.), AGU Monograph 30, p. 145, 1984.

Scholer, M., Magnetic flux transfer at the magnetopause based on single X line bursty reconnection, *Geophys. Res. Lett., 15,* 291, 1988.

Song, Y., and R. L. Lysak, Evaluation of twist helicity in flux transfer event tubes, *J. Geophys. Res., 94,* 5273, 1989a.

Song, Y., and R. L. Lysak, Current dynamo effect of 3-D time-dependent reconnection in the dayside magnetopause, *Geophys. Res. Lett., 16,* 911, 1989b.

Song, Y., and R. L. Lysak, Alfvénon, driven reconnection and the direct generation of the field-aligned current, *Geophys. Res. Lett., 21,* 1755, 1994.

Song, Y., R. L. Lysak and N. Lin, Control of the generationn of field-aligned currents and transverse ULF waves by the magnetic helicity input, *Solar wind sources of Ultra-Low-Frequency waves,* M. J. Engebretson, K. Takahashi, and M. Scholer (eds.), AGU Monograph 81, p. 223, 1994.

Sonnerup, B. U. O., Magnetic field reconnection at the magnetopause: An overview, *Magnetic reconnection in space and laboratory plasmas,* Hones, E. W., Jr. (eds.), AGU Monograph 30, p. 92, 1984.

Sonnerup, B. U. O., On the stress balance in flux transfer events, *J. Geophys. Res., 92,* 8613, 1987.

Southwood, D. J., C. J. Farrugia, and M. A. Saunders, What are flux transfer events?, *Planet. Space. Sci., 37,* 503, 1988.

Wright, A. N., The evolution of an isolated reconnected flux tube, *Planet. Space Sci., 35,* 813, 1987.

Wright, A. N., and M. A. Berger, The effect of reconnection upon the linkage and interior structure of magnetic flux tubes, *J. Geophys. Res., 94,* 1295, 1989.

Wright, A. N., and M. A. Berger, The interior structure of reconnected flux tubes in a sheared plasma flow, *J. Geophys. Res., 95,* 8029, 1990.

Wright, A. N., and M. A. Berger, A physical description of magnetic helicity evolution in the presence of reconnection lines, *J. Plasma Phys., 46,* 179, 1991.

Wright, A. N., Transfer of magnetosheath momentum and energy to the ionosphere along open field lines, *J. Geophys. Res, 101,* 13,169, 1996.

A.N. Wright, Mathematical and Computational Sciences Dept., St Andrews University, St Andrews KY16 9SS, Scotland. (e-mail: andy@dcs.st-and.ac.uk)

The Role of Coarse-Grained Helicity and Self-Organized Criticality in Magnetotail Dynamics

Tom Chang

Center for Space Research, Massachusetts Institute of Technology, Cambridge, MA, and International Space Science Institute, Bern, Switzerland

In 1992, *Chang* suggested that the substorm dynamics of the Earth's magnetotail may be described by the stochastic behavior of a nonlinear dynamical system near self-organized criticality (SOC). Subsequently, *Chang* (1998a,b) demonstrated that multiscale intermittent turbulence of overlapping plasma resonances is the underlying physics that can lead to the onset and evolution of substorms. Such a description provides a convenient explanation of the localized and sporadic nature of the reconnection signatures (bursty bulk flow) that are commonly observed in the magnetotail region (*Angeloupolos et al.*, 1996; *Lui*, 1998). It also furnishes a natural explanation of the phenomenon of inverse cascade in intermittent MHD turbulence. These concepts provide a new paradigm for the understanding of the ever-elusive phenomenon of magnetic substorms.

1. MAGNETOTAIL, PLASMA SHEET AND SUBSTORMS--A BRIEF INTRODUCTION

The Earth's magnetosphere is a cavity carved out from the solar wind by the magnetic field of the Earth much as the wake region created by a blunt object in a flowing stream. The rarefied medium contained within the magnetosphere is a mixture of charged particles (plasma) of both ionospheric and solar wind origins. Most of the plasma particles in the tail region of the magnetosphere (the magnetotail) reside in a central region called the "plasma sheet". The plasma sheet is quite dynamic. It quasi-periodically expands and contracts. A popularly accepted scenario describing the dynamic process of the plasma sheet is based on the classical idea of "magnetic reconnection", Figure 1. In this picture, the dynamic process is initiated by the thinning of the midsection of the plasma sheet and the subsequent forming of a near-Earth reconnection region of the magnetic field lines. Portion of the plasma sheet then expands Earthward. During this expansion phase, charged particles are accelerated toward the pole-regions of the Earth accompanied by violent magnetic disturbances. These magnetic disturbances and the associated dynamical phenomena (including plasma waves and particle diffusion/acceleration) in the magnetosphere have been christened the magnetic "substorms", [*Akasofu*, 1964]. In the meantime, the remaining portion of the plasma sheet upon reconnection is propelled tailward, sometimes in the form of plasma blobs termed "plasmoids" [*Hones*, 1976]. The plasma sheet then recovers until another near-Earth reconnection process occurs and a new cycle of such events is initiated.

Although some of the dynamical characteristics of the magnetotail prescribed by the above scenario has been observed by in-situ satellite measurements, there are a number of notable inconsistencies. For example, the observed "reconnection" signatures are quite sporadic and localized randomly at various sites within the plasma sheet [*Angelopoulos et al.*, 1996; *Lui*, 1998]. In addition,

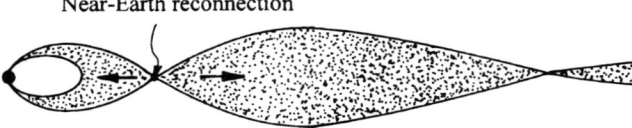

Figure 1. Classical picture of plasma sheet near-Earth reconnection. Arrows indicate general directions of plasma flow.

the typical size of the observed localized merging (reconnection) processes and the duration of these events are at least an order of magnitude smaller than those suggested by the above classical reconnection scenario. Generally, the turbulence that is prevalent in the magnetotail, before, during and after the onset of substorms, is intermittent and multiscale in nature. Recently, a model involving the generation, dispersing, and merging of multiscale coherent plasma structures and associated fluctuations based on the concept of coarse-grained helicity has been suggested [*Chang*, 1998a,b] to address the implications of such observations. In this description, the dynamics of the magnetotail during (as well as before and after) magnetic substorms is relegated to the stochastic behavior of a multiscale dynamical system near self-organized criticality (SOC) and associated global instability or instabilities..

In several recent interesting papers [*Baker et al.*, 1990; *Vassiliadis et al.*, 1990; *Klimas et al.*, 1992; *Sharma et al.*, 1993, *Klimas et al.*, 1998], it has been demonstrated that certain substorm characteristics could be modeled by deterministic chaos of low-dimensional dynamical systems (i.e., systems characterized by a small number of relevant physical parameters) with fractal characteristics. Based on the theory of the dynamic renormalization group, it was shown by *Chang* [1992; 1993] that nonlinear stochastic systems near self-organized criticality (SOC) such as the Earth's magnetotail during substorms generally are expected to exhibit such low-dimensional and fractal behavior.

Both the physical concepts and mathematical techniques associated with the above ideas are non-traditional. It is the purpose of this paper to provide an introductory background for those who are interested in understanding of some of these fundamental concepts. For the most part, only physical and topological descriptions will be provided. The readers are referred to the various original papers for further in-depth studies.

2. PLASMA RESONANCES, COARSE-GRAINED HELICITY AND COHERENT STRUCTURES

We start our discussion by assuming that the dynamics of the Earth's magnetotail is characterized primarily by the MHD equations. The relevant equations of induction and motion are:

$$\frac{d\mathbf{B}}{dt} = (\mathbf{B} \cdot \nabla)\mathbf{V} + \cdots, \text{ and} \quad (1)$$

$$\rho \frac{d\mathbf{V}}{dt} = (\mathbf{B} \cdot \nabla)\mathbf{B} + \cdots. \quad (2)$$

where the ellipsis represent the compressible and anisotropic pressure terms, and all notations are standard. Generally, of course, dissipative terms must be included on the right hand sides of Eqs. (1, 2) as well. It is clear from above that one of the wave modes allowed by these equations is the Alfvén wave. For such modes to propagate, the propagation vector $\mathbf{k}$ must contain a field-aligned component, i.e., $\mathbf{B} \cdot \nabla \to i\mathbf{k} \cdot \mathbf{B} \neq 0$. However, at sites where the parallel component of the propagation vector vanishes, $k_{\parallel} = \mathbf{k} \cdot \mathbf{B} = 0$, energies are localized and the field lines may be distorted effortlessly. We shall call these singularities (points, curves or surfaces) at which $k_{\parallel} = 0$ "Alfvén resonances". As it will be demonstrated below, the existence of these resonance sites will lead to the formation of nearly-nonpropagating and essentially closed macroscopic magnetic structures. Because of the presence of the pressure tensor term in Eq. (2), there exists also the possibility of other macroscopic as well as kinetic resonances. [To consider the merging of such magnetic structures, particularly near the "neutral sheet" region, it will be necessary to include the effects of the pressure tensor and the associated particle kinetics.] We shall address these effects later briefly in discussing the idea of multifractals in the turbulence spectra.

2.1 Coarse-Grained Helicity.

We now consider the magnetic field structures near the Alfvén resonances. Neglecting the pressure effects, it is clear from Eqs. (1, 2) that the forces arise from the fluctuations just away from these resonance sites, i.e., $\delta\mathbf{B} \cdot \nabla$, will tend to restore the field lines towards the resonance sites, thereby forming essentially closed coherent magnetic structures. In the following we shall consider the general topology of such coherent structures.

For an ideal MHD system, any physically acceptable magnetic field must satisfy $\nabla \cdot \mathbf{B} = 0$. Also, any variation of the field away from the initial value must satisfy the constraints:

$$\partial \mathbf{B}/\partial t + \nabla \times \mathbf{E} = 0, \quad (3)$$

$$\mathbf{E} + \mathbf{V} \times \mathbf{B} = 0, \quad (4)$$

Taylor [1974] demonstrated that Eqs. (3,4) may be replaced by an infinite set of integral constraints involving the *helicity K*, such that

$$K = \int_V \mathbf{A} \cdot \mathbf{B}\, dV \quad (5)$$

is an invariant for any volume V enclosed by a flux surface, where **A** is the vector potential. It can be shown that as the system relaxes to its minimum energy state satisfying the helicity conservation constraints, the magnetic structure will be in a force-free state, i.e.,

$$\mathbf{j} \times \mathbf{B} = 0. \quad (6)$$

2.2 Taylor's Conjecture.

Let us now consider our present situation at hand. We are interested in the more realistic situation that characterizes the dynamics of the magneto-tail where the plasma is slightly dissipative and in addition, there are stochastic macroscopic (as well as microscopic fluctuations) fluctuations. The dissipation and magnetic stochasticity will allow the field lines to merge, mix, and break.. It is obvious that it no longer makes sense to discuss the topology of individual field lines. Nevertheless, it was suggested by *Taylor* [1974; 1986] that when the volume integral for Eq. (5) is taken over the "stochastic region", the coarse-grain averaged helicity in a relaxed state will be essentially conserved. This indicates that when considering the stochastic do-main, the average magnetic structure will again be essentially force free, with $\mathbf{j} \times \mathbf{B} = 0$, where **j** and **B** are now to be interpreted as the mean current and magnetic field, respectively. This result can also be arrived at using the clump theory of MHD turbulence [*Tetreault*, 1992; and references contained therein].

We are, of course, interested in the magnetotail at dynamic states that are far from equilibrium. Thus, in visualizing the relaxed states from the point of view of the *Taylor's conjecture*, we shall consider timescales such that "nearly coherent" magnetic structures are formed. These structures actually move, mix and sometimes merge together while immersed in an otherwise turbulently diffusing plasma medium.

3. MERGING OF COHERENT FLUX TUBES

Let us now apply these concepts to the sheared magnetic field geometries that are generally found in the "neutral sheet" region of the magnetotail. The nearly force-free condition for the coarse-grain averaged coherent structures would then orient themselves more-or-less in the average cross-tail current direction in the form of twisted flux tubes. In general, there will be a constellation of such coherent structures immersed in the turbulent plasma medium, Figure 2.

As these coherent structures migrate toward each other, they will merge and form new coherent structures. Depending on the polarities and intensities of the currents that orient these flux tubes, the resulting coherent structures will be either larger or smaller than the original individual structures. The final states of the new coherent structures will again be essentially force-free in the coarse-grained sense. As these new structures are generated, new MHD fluctuations are produced; and thereby spontaneously set up new resonance sites. Thus, an interesting scenario of intermittent turbulent mixing, diffusing, and merging sets in. This type of intermittent turbulence is anisotropic, inhomogeneous and multiscale in the magnetotail [Figure 2]. In the following, we first discuss the individual localized merging processes.

Let us consider the most probable situation of merging, i.e., the merging of two coherent structures. Viewed in a section normal to the average direction of the cross-tail current, the topologies of the field lines during such a merging process mimic that is generally considered for a classical magnetic "reconnection" process [Figure 3]. However, we note that this localized merging process can take place without the requirement of **B** = 0 and/or the existence of a true "neutral line." In fact, as seen above, the pre-requisite for the existence of many such coherent structures as well as the sporadic merging of these structures is the existence of many "Alfvén resonance" sites with $k_\parallel = 0$. This occurs when the background magnetic field is three-dimensional and nonzero and when there are three-dimensional macroscopic MHD fluctuations.

Thus, we suggest that as a spacecraft flies through the neutral sheet region of the magnetotail, there is a finite probability for the instruments on the spacecraft to detect classical-like reconnection signatures. Such signatures can be detected nearly anywhere in the plasma sheet, but more probably in the "neutral sheet" region, particularly during substorm times. The duration of interaction of these observed localized merging processes should be the approximate time required for the new relaxed coherent structures to emerge and in general, would be rather

280 HELICITY AND SOC IN MAGNETOTAIL DYNAMICS

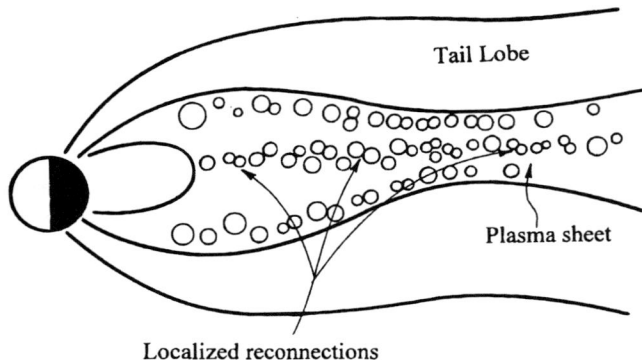

Figure 2. Multiscale intermittent turbulence in the magnetotail. Cross-sectional view of sporadically distributed flux tubes in the plasma sheet.

sporadic. We suggest that these are the origins of the observed "bursty bulk flows" [*Angelopoulos et al.*, 1996; *Lui*, 1998; *Kivelson and Kepco*, private communication]. The time scale, size, and energy contents involved in these localized merging processes will generally be much smaller then those that are considered to be relevant for the full dynamics of the magnetotail, particularly during the full duration of the magnetic substorms.

Most of the observed localized reconnection signatures to date seem to indicate that these localized merging processes take place in domain sizes comparable to that of the ion gyroradius, especially during substorm times. Thus, very probably most of these processes will be influenced by microscopic kinetic effects. During these dynamic processes, the ions can probably be assumed to be unmagnetized and the electrons fully magnetized and the plasma nearly collisionless. This, of course, would lead to electron-induced Hall currents. Depending on the underlying magnetic geometry (since these processes can occur at any arbitrary underlying magnetic field configuration), the relevant kinetic instability that can initiate the localized merging (or reconnection) can be any of the many recently suggested microscopic instabilities such as the collisionless tearing instability, cross-field two-stream instability [*Lui*, 1998], *etc.* It is very probable that the nonlinear state of merging for each of these localized reconnections again entails the phenomenon of overlapping resonances [*Galeev, Zelenyi, and Kosnetsova*, 1986]. (Now these resonances will arise from the localization of microscopic fluctuations, e.g., the whistler resonances, and multiple tearing modes.)

We therefore envision a multiscale intermittent turbulence with coherent and plasma fluctuations ranging from the MHD macroscopic scales down to microscopic kinetic sizes. In the following sections, we shall discuss the consequences of such type of anisotropic, inhomogeneous turbulence, particularly during the onset and the various evolutionary stages of substorms.

4. SELF-ORGANIZED CRITICALITY

In 1987, *Per Bak et al.* suggested that dynamically interacting systems generally organize themselves into complex states similar to the critical states under equilibrium phase transitions. For systems at or near criticality, the correlation lengths of the fluctuations are long-ranged and involve infinitely-many (numerous) spatial and time scales. It is known from the renormalization-group (RG) theory of critical phenomena for nonequilibrium systems [see, e.g., Chang et al., 1992] that although a system at or near the dynamical critical state is rather complex, the system will exhibit certain general invariant behavior because of the long-ranged correlations. This result can be understood phenomenologically as follows: Let us view the system from a distance and then gradually stepping away from it. Because of the long-ranged correlations, the system should look rather similar although it may appear smaller (self-similar scaling). Let us denote the stepping parameter as "s", then it is reasonable to recognize that the parameters P_i which characterize the dynamical system would scale with s as follows:

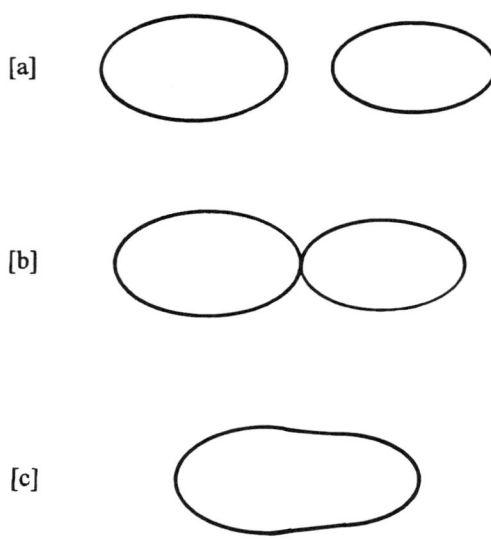

Figure 3. Cross-sectional view of coherent structures at various stages of merging. (a) Just prior to merging, (b) During the process of merging, (c) Relaxed state after merging.

$$P_i' = (s'-s)^{a_i} P_i, \quad (7)$$

where "a_i" are constants. If "a_i" is positive, then the corresponding parameter P_i will become more and more dominant as one is stepping away from the system. Such a parameter is called a "relevant" parameter. On the other hand if "a_i" is negative, the corresponding parameter becomes essentially irrelevant. Thus, for a system near criticality only the parameters with positive "a_i" are important. If we had used rigorous theoretical arguments based on the dynamical renormalization group, the above result can be demonstrated to be generally correct (with some special exceptions which we shall not consider here). In addition, it can be shown using the renormalization theory that typically there are only a set of small number of "relevant" parameters for systems at criticality.

It is then easy to verify from Eq. (7) that the ratio

$$P_i / P_j^{a_i/a_j} = \text{constant} \quad (8)$$

at any value of s. Thus, these are invariants for the system at or near criticality. In a log-log plot, the relations between the relevant parameters are therefore power laws. Because there are only a few relevant scaling parameters, we expect that the minimum number of independent invari-ants for a system at criticality is usually quite small. Since the parameters are related by the physics that characterize the dynamics of the dynamical system, there are definitive relations among the a_i's. These are called scaling laws.

Generally the power laws (8) for dynamic systems near criticality will deviate from that could be deduced by straight-forward dimensional analyses. Thus, near criticality, the stochastic system exhibits "anomalous dimensions". It has become fashionable sometimes to truncate the description of a stochastic system near criticality into a dynamic system expressed in terms of an arbitrarily small number of parameters or "dimensions" (i.e., low-dimensionality). From the above discussions, it is reasonable to believe that such a prescription is viable (provided one is reasonably sure of what are the relevant parameters to be incorporated in the truncated dynamical equations). When the truncated system exhibits chao-like behavior, it is then claimed to have a fractal dimension. This fractal dimension is, of course, intimately related to the anomalous dimensions discussed above and can be calculated using renormalization-group techniques.

All the above results seem to indicate that the magnetotail dynamics (particularly before, during and after substorms) is a stochastic system near self-organized criticality.

Under favorable conditions (e.g., with the availability of a free energy source such as the enhancement of the cross-tail current due to the change of certain global controlling parameters for the magnetotail), the state of intermittent turbulence discussed above may grow by producing more and larger coherent structures and fluctuations as well as new resonance sites [Figure 2.]. This type of instability, by definition, is genuinely "nonlinear," and usually global in nature (in the sense that the nonlinear dissipation is determined self-consistently through the induced turbulence throughout the medium subject to the global parameters that control the magnetotail dynamics).

For the onset and growth of a classical nonlinear instability, there generally exists a prescribed minimum finite amplitude of disturbance (measured, for example, by the root-mean-square of fluctuations) beyond which the fluctuations and coherent structures can grow provided that there is an available abundance of efficient convertible free energy [Figure 4]. Much attention has been paid recently to the onset of substorms associated with the phenomenon of self-organized criticality. During the onset of a substorm, the effect of the fluctuations becomes an important factor in determining the critical threshold of onset. In this situation, the nonlinear instability is no longer described by its classical threshold and the nonclassical instability should generally depend on certain overall global parameters that characterize the dynamics of the magnetotail [Figure 4]. And it is expected that the resulting fluctuation spectra will generally exhibit fractal structures [*Chang, 1997*].

5. MULTIFRACTAL SPECTRA

In previous sections, we demonstrated that the dynamics of the magnetotail, particularly during the onset and evolution of substorms, is characterized by the sporadic merging (localized reconnection) of coherent macroscopic magnetic flux structures. The resulting turbulence is multi-scale, intermittent, anisotropic and inhomogeneous. In addition, the localized merging process provides a natural explanation of the phenomenon of inverse cascade in MHD turbulence. A standard technique to characterize the behavior of such type of turbulence is through the properties of the spectra of the turbulent fluctuations.

For example, in the "neutral sheet" of the magnetotail, one of the more important spectra to consider is that of the square of the magnetic fluctuations in the cross-tail direction $\langle \delta B^2 \rangle$. We expect the spectra to generally exhibit fractal characteristics (i.e., nonclassical slopes

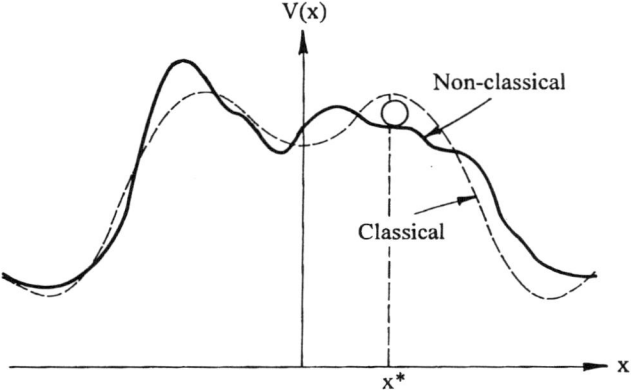

Figure 4. Marble rolling over a hill. Schematic representation of classical and nonclassical, nonlinear instability. x*: Classical threshold for nonlinear instability.

with discernible deviations from those obtainable by naive dimensional arguments). [See Figure 5.] In regions where the fluctuations and merging dimensions are much larger than that of the local ion gyroradius, the spectrum is expected to exhibit two distinguishable parts: a domain characterized by the larger scale coherent structures and a fractal domain characterized by the predominantly MHD fluctuations. On the other hand, in regions within the narrow cross-tail current sheet, we expect the spectra to exhibit at least three distinguishable parts: a domain that contains predominantly large scale coherent structures, an MHD fractal domain and a kinetic fractal regime whose fractal dimension(s) generally depends on the type(s) of microscopic fluctuations and microinstabilities that are relevant for the merging and diffusion processes. Such type of fluctuation spectra has been recently observed [*Hosino et al.*, 1994; *Milovanov et al.*, 1996; *Zelenyi et al.*, 1997; and references contained therein.]. The shapes (slopes) of these spectra in the distant-tail region have been compared with results based on theoretical scaling ideas involving fractal dimensions [*Milovanov et al.*, 1996; *Zelenyi et al.*, 1997; *Lui*, 1998]. The difference of slopes of the various domains of an individual spectrum indicates that the scaling (fractal) behavior of each domain belongs to a different "universality class". Such type of change of scaling behavior from one universality class to another is called "symmetry breaking". In addition to the scaling properties of individual discernible domains, there are also intermediate regimes whose fractal properties are much more complicated (as indicated by the circled region of Figure 5). The scaling laws for these regions are generally expected to exhibit multiple-power or other nonlinear characteristics. These results depend on the details of the symmetry breaking and overlapping be-havior among the universality classes and were discussed in detail by Chang and co-workers in a series of papers on critical phenomena [*Chang and Stanley*, 1973; *Chang et al.*, 1973a; 1973b; *Nicoll et al.*, 1974; 1975; 1976; *Chang et al.*, 1992].

6. SYNOPSIS AND THE NEW MILLENNIUM MAGNETOTAIL

In summary, we have introduced a multiscale intermittent turbulence model for the dynamics of the magnetotail. The theory is based on the overlapping resonances of plasma fluctuations. It provides a physical picture of sporadic and localized merging of coherent magnetic structures of varied sizes. Such a picture seems to depict the observational properties of "bursty bulk flow" (sporadic localized reconnections) in the magnetotail [*Angelopoulos et al.*, 1996; *Lui*, 1998]. In this picture, the onset of substorm is due a global nonclassical nonlinear instability and the dynamics of the magnetotail during the evolution of the substorm is characterized by the phenomenon of forced or self-organized criticality.

The consequence of this is the prediction of multi-fractal characteristics of the fluctuation spectra [*Hosino et al.*, 1994; *Milovanov et al.*, 1996; *Zelenyi et al.*, 1997; *Chang*, 1992, 1993; and references contained therein] and the dynamics of the magnetotail behaves essentially as a low-dimensional system. This conclusion seems to agree with the results of some of the recent nonlinear dynamics calculations [*Baker et al.*, 1990; *Klimas et al.*, 1991, 1992; 1998; *Baker*, 1998].

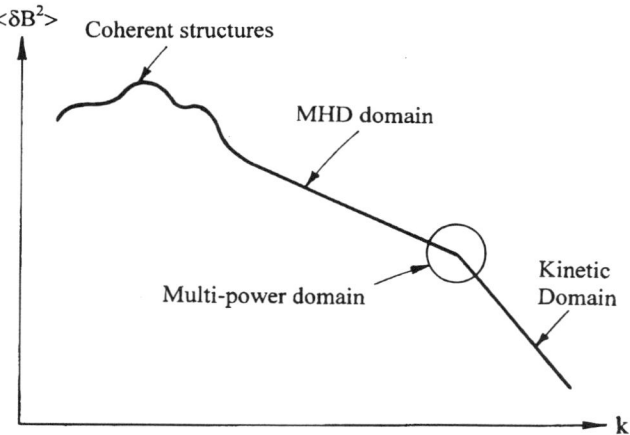

Figure 5. Multifractal spectrum near the "neutral sheet".

We suggest this as a new paradigm of magnetotail dynamics for the new millennium.

As an addendum, we note that recently *Chapman et al.* [1998] presented the simulation results of a simple avalanche model, that displayed many of the characteristic features (such as self-organized criticality and global instability, etc.) of magnetospheric activity that was advocated in this treatise.

Acknowledgment. The author wishes to acknowledge useful discussions with A.T.Y. Lui, C.C. Wu, C.F. Kennel, D. Baker, M. Yamada, H.E. Petschek, D. Tetreault, M. Kivelson, L. Kepko, L. Zelenyi, V. Angelopoulos, M. Hoshino, S. Chapman, A. Klimas, D. Vassiliadis, D. Vvedensky, J.F. Nicoll, J. Kan, N. Watkins, A.S. Sharma, R. Stenzel, and J. Büchner. a number of the conceptual ideas discussed in this paper is echoed in a recent book authored by *C. Kennel* [1995]. This research is partially supported by NASA, NSF, AFOSR and AFRL. A portion of this research was completed at the International Space Science Institute, Bern, Switzerland.

REFERENCES

Akasofu, S.I., The development of the auroral substorms, *Planet. Space Sci.*, 12, 273, 1964.

Angelopoulos, V., F.V. Coroniti, C.F. Kennel, M.G. Kivelson, R.J. Walker, C.T. Russell, R.L. McPherron, E. Sanchez, C.I. Meng, W. Baumjohann, G.D. Reeves, R.D. Belian, N. Sato, E. Fris-Christensen, P.R. Sutcliffe, K. Yumoto and T. Harris, Multi-point analysis of a BBF event on April 11, 1985, *J. Geophys. Res.*, 101, 4967, 1996.

Bak, P., C. Tang and K. Wisenfield, Self-organized criticality. An explanation of 1/f noise, *Phys. Rev. Lett.*, 59. 381, 1987.

Baker, D.N., Substorms: A global magnetospheric instability, in *Proc. 4th Intern. Conf. on Substorms*, edited by S. Kokubun and Y. Kamide, p. 231 (Terra Scientific Publishing Company, Tokyo, and Kluwer Academic Publishers, Dordrecht) 1998.

Baker, D.N., A.J. Klimas, R.L. McPherron and J. Büchner, The evolution from weak to geomagnetic activity: An interpretation in terms of deterministic chaos, *Geophys. Res. Lett.*, 17, 41, 1990.

Baker, D., A.J. Klimas, D. Vassiliadis and T.I. Pulkkinen, The magnetospheric dynamical cycle: role of microscale and mesoscale processes in the global substorm sequence, *Physics of Space Plasmas (1995)*, 14, 41, 1996.

Büchner, J., Kinetic effects controlling tail reconnection in the course of magnetospheric substorms, in *Proc. 4th Intern. Conf. on Substorms*, edited by S. Kokubun and Y. Kamide, p. 461 (Terra Scientific Publishing Company, Tokyo, and Kluwer Academic Publishers, Dordrecht) 1998.

Chang, T., Low-dimensional behavior and symmetry breaking of stochastic systems near criticality - can these effects be observed in space and in the laboratory?, *IEEE Trans. on Plasma Science*, 20, 691, 1992.

Chang, T., Path Integral Approach to Stochastic Systems near Self-Organized Criticality, in Nonlinear Space Plasma Physics, Research Trends in Physics, p. 165, Editer-in Chief, R.Z. Sagdeev, (American Institute of Physics, New York) 1993.

Chang, T., J.F. Nicoll, and J.E. Young, A closed-form differential renormalization-group generator for critical dynamics, *Phys. Lett.*, 67A, 287, 1978.

Chang, T., D.D. Vvedensky and J.F. Nicoll, Differential renormalization-group generators for static and dynamic critical phenomena, *Physics Reports*, 217, 279, 1992.

Chang, T., Multiscale intermittent turbulence in the magnetotail, in *Proc. 4th Intern. Conf. on Substorms*, edited by S. Kokubun and Y. Kamide, p. 431 (Terra Scientific Publishing Company, Tokyo, and Kluwer Academic Publishers, Dordrecht) 1998a.

Chang, T., Sporadic, localized reconnections and multiscale intermittent turbulence in the magnetotail, in *Geospace Mass and Energy Flow*, edited by Horwitz, J.L., D.L. Gallagher, and W.K. Peterson, AGU Monograph Number 104, p. 193 (American Geophysical Union, Washington, D.C.) 1998b.

Chang, T., and H.E. Stanley, Renormalization-group verification of crossover with respect to lattice anisotropy parameter, *Phys. Rev.*, B8, 1973.

Chang, T., A. Hankey, and H.E. Stanley, Double-power scaling functions near tricritical points, *Phys. Rev.*, B7, 4263, 1973a.

Chang, T., A. Hankey, and H.E. Stanley, Generalized scaling hypothesis in multicomponent systems. I. Classification of critical points by order and scaling at tricritical points, *Phys. Rev.*, B8, 346, 1973b.

Chapman, S.C., N.W. Watkins, R.O. Dendy, P. Helander and G. Rowlands, A simple avalanche model as an analogue for magnetospheric activity, *Geophys. Res. Lett.*, 25, 2397, 1998.

Galeev, A.A., M.M. Kuznetsova and L.M. Zeleny, Magnetopause stability threshold for patchy reconnection, *Space Science Reviews*, 44, 1, 1986.

Hones, E.W., The magnetotail: Its generation and dissipation, in *Physics of Solar Planetary Environments*, vol. II, edited by D.J. Williams, p. 558 (AGU, Washington, D.C.) 1976.

Hosino, M. A. Nishida and T. Yamamoto, Turbulent magnetic field in the distant magnetotail: Bottom-up process of plasmoid formation, *Geophys. Res. Lett.*, 21, 2935, 1994.

Kennel, C., *Convection and Substorms: Paradigms of Magnetospheric Phenomenology* (Oxford University Press, Oxford) 1995.

Klimas, A.J., D.N. Baker, D.A. Roberts, D.H. Fairfield and J. Büchner, A nonlinear dynamical analogue model of geomagnetic activity, *J. Geophys. Res.*, 97, 12253, 1998.

Klimas, A.J.,D. Vassiliadis, J.A. Valdivia, D.N. Baker, Analogue analysis of the solar wind Vb_z to AL electrojet index relationship, in *Proc. 4th Intern. Conf. on Substorms*, edited by S. Kokubun and Y. Kamide, p. 669 (Terra Scientific Publishing Company, Tokyo, and Kluwer Academic Publishers, Dordrecht) 1998.

Kivelson, M., and L. Kepco, private communication.

Lui, A.T.Y., Current disruptions in the Earth's magnetosphere: observations and models, *J. Geophys. Res.*, 101, 4899, 1996.

Lui, A.T.Y., Multiscale and intermittent nature of current disruption in the magnetotail, *Physics of Space Plasmas*, 15, 233, 1998.

Milovanov, A., L. Zelenyi and G. Zimbardo, Fractal structures and power law spectra in the distant Earth's magnetotail, *J. Geophys. Res.*, 101, 19903, 1996.

Nicoll, J.F., T. Chang, and H.E. Stanley, Nonlinear solutions of renormalization-group equations, *Phys. Rev. Lett.*, 1446, 1974.

Nicoll, J.F., T. Chang, and H.E. Stanley, Nonlinear crossover between critical and tricritical behavior, *Phys. Rev. Lett.*, 36, 113, 1976.

Shan, L.H., C.K. Goertz, and R.A. Smith, Chaotic appearance of the AE index, *Geophys. Res. Lett.*, 18, 1647, 1991.

Sharma, A.S., D. Vassiliadis and K. Papadopoulos, Reconstruction of low-dimensional magnetospheric dynamics by singular spectrum analysis, *Geophys. Res. Lett.*, 20, 335, 1993.

Taylor, J.B., Relaxation of toroidal plasma and generation of reverse magnetic fields, *Phys. Rev. Lett.*, 33, 1139, 1974.

Taylor, J.B., Relaxation and magnetic reconnection in plasmas, *Rev. Mod. Phys.*, 58, 741, 1986.

Tetreault, D., Turbulent relaxation of magnetic fields: 1. coarse-grained dissipation and reconnection, *J. Geophys. Res.*, 97, 8531, 1992.

Vassiliadis, D.V., A.S. Sharma, T.E. Eastman, and K. Papadopoulos, Low-dimensional chaos in magnetospheric activity from AE time series, *Geophys. Res. Lett.*, 17, 1841, 1990.

Zelenyi, L.M., A.V. Milovanov, and G. Zimbardo, Multiscale magnetic structure of the distant tail: self-consistent fractal approach, *AGU Monograph on "The Earth's Magnetotail: New Perspectives"*, American Geophysical Union, Washington, D.C., to be published in 1998.

T. Chang, Center for Space Research, Massachusetts Institute of Technology, Room 37-261, 77 Massachusetts Avenue, Cambridge, MA 02139.

Galactic and Accretion Disk Dynamos

Ethan T. Vishniac

Department of Physics and Astronomy, Johns Hopkins University, Baltimore, Maryland

Dynamos in astrophysical disks are usually explained in terms of the standard alpha-omega mean field dynamo model where the local helicity generates a radial field component from an azimuthal field. The subsequent shearing of the radial field gives rise to exponentially growing dynamo modes. There are several problems with this model. The exponentiation time for the galactic dynamo is hard to calculate, but is probably uncomfortably long. Moreover, numerical simulations of magnetic fields in shearing flows indicate that the presence of a dynamo does not depend on a non-zero average helicity. However, these difficulties can be overcome by including a fluctuating helicity driven by hydrodynamic or magnetic instabilities. Unlike traditional disk dynamo models, this 'incoherent' dynamo does not depend on the presence of systematic fluid helicity or any kind of vertical symmetry breaking. It will depend on geometry, in the sense that the dynamo growth rate becomes smaller for very thin disks, in agreement with constraints taken from the study of X-ray novae. In this picture the galactic dynamo will operate efficiently, but the resulting field will have a radial coherence length which is a fraction of the galactic radius.

1. CONTEXT

The traditional focus of astrophysical dynamo theory has been on stars, where spherical symmetry is a reasonable first approximation, and the inward pull of gravity is balanced by the radial pressure gradient. In spite of the eponymous role of stars in astrophysics, this ignores the importance of magnetic fields in disks, where gravity is balanced by centrifugal forces. This traditional bias can be explained by the fact that we can observe the magnetic fields of at least one star in some detail, whereas the magnetic field of the Galactic disk presents itself as a bewildering mixture of structure on a range of scales. However, recent years have witnessed an accumulation of data concerning the structure of the magnetic field in our galaxy, and in external galaxies. In addition, it has become clear that magnetic fields play a critical dynamical role in accretion disks of all sizes, including some of the most luminous objects in the universe. Here I will summarize recent theoretical progress in understanding disk dynamos. In an unexpected twist, we will see that the role of global helicity in magnetic field generation may be small.

We start by considering the context of disk dynamos. Astrophysical disks can be divided into two general categories, galactic disks and accretion disks. The latter category includes disks around the supermassive black holes, as in active galactic nuclei (AGN), and stellar disks surrounding protostars or members of binary star systems. Although the physical conditions in these disks span an enormous range we will restrict ourselves

Magnetic Helicity in Space and Laboratory Plasmas
Geophysical Monograph 111
Copyright 1999 by the American Geophysical Union

in only two ways. First, we consider disks that are sufficiently ionized that ohmic dissipation is negligible on the scale of the disk thickness. This may exclude parts of protostellar disks, and some regions in the interstellar medium (ISM). Second, we restrict ourselves to disks that are primarily supported by rotation, and are consequently geometrically thin. There are cases where a large fraction of the disk support comes from radial pressure gradients or from the magnetic field of the accreting object. We ignore these cases because of the complicated physics involved, not because we believe them to be unimportant.

With this in mind, we can summarize the differences between galactic and accretion disks in the following manner. Galactic disks are confined vertically by the gravity of the disk and halo acting together. The typical galactic disk has an angular velocity $\Omega \propto r^{-1}$ over a broad range in radii. The gaseous disk itself is composed of a heterogeneous interstellar medium with a complicated history. It is typically marginally stable against local gravitational collapse. The number of dynamical time scales since the formation of the galaxy is limited, and it is unclear whether or not the magnetic field in the disk can be regarded as having reached a stationary state, or whether initial conditions might be important in understanding its structure. Finally, the orbital period is the longest natural time scale in these systems, followed by the dynamical time scale of local random motions within the disk, followed by the particle collision time in the gas. This is, in turn, greater than plasma time scales, such as the inverse of the ion cyclotron frequency or the inverse of the plasma frequency. Treating the gas in a galactic disk as a fluid is clearly a dangerous approximation, both because of its complicated substructure, and because the hydrodynamic approximation is unlikely to be accurate even within a relatively homogeneous volume of the ISM. Most of the volume of the gaseous disk is occupied by gas that is sufficiently ionized that ohmic dissipation is negligible on disk scales.

In contrast, vertical confinement in accretion disks is supplied by the vertical component of the gravity of the central object. This leads to an orbital frequency $\Omega \propto r^{-3/2}$ and, through the condition of hydrostatic equilibrium a thickness $H \propto c_s/\Omega$, where c_s is the local sound speed. Accretion disks are relatively homogeneous, in the sense that the vertical sound crossing times are short and pressure equilibrium is a good approximation. In the absence of magnetic fields accretion disks are stable, although strongly unstable when they are present. Their age is greater than all other relevant time scales, so initial conditions can be ignored. The time it takes material to spiral inward to the central object is greater than the time scale for local thermal equilibrium, which is greater than an orbital period. Inverse plasma frequencies are typically much less than an orbital period, and usually much greater than the mean free collision time for a particle. Accretion disks are good fluids, although accretion disk coronae are not. Accretion disks are not always good conductors, but the exceptions are cold and difficult to observe.

2. CLUES AND CONSTRAINTS

Our knowledge of astrophysical magnetic fields is never as complete as we would like. For galactic magnetic fields we have a variety of diagnostics which tells us about the current state of the field. Direct observations of evolutionary effects are, of course, impossible. These diagnostics include the intensity and polarization of synchrotron radiation, the polarization of starlight, the polarization of infrared dust emission, Faraday rotation, and Zeeman splitting. (For a general review of galactic magnetic fields see Zweibel and Heiles 1997; Vallée 1997). It is important to note that each of these diagnostics involves other quantities, for example electron density or the physical properties of interstellar dust grains, for which we have only rough estimates. In addition, the direction of the magnetic field can be derived only from Faraday rotation, and only for the component along the line of sight.

Keeping these uncertainties in mind, we note that a rough concurrence among these methods allows us to conclude that the mean value of the magnetic field in the disk is approximately $10^{-5.5}$ Gauss, with comparable power in the large scale and 'random' (i.e. small scale) components. The large scale field is approximately aligned with the azimuthal direction, but tilted somewhat towards the direction of the local spiral arms. The number of large scale field reversals in the disk is unknown, but cannot be very large, since observations of Faraday rotation tend to give be consistent with a large scale field coherence length of at least several hundreds of parsecs.

Models of galactic magnetic field generation usually assign a rather large role to the galactic shear. We note that this is about 10^{-15}sec^{-1} at our position in the Galaxy. Given a galactic disk age of $\sim 10^{10}$ years this gives a maximum growth of roughly 300 e-foldings. There are various suggestions for modifying fundamental physics in order to obtain a large scale primordial magnetic field, but these proposals are all highly speculative. Simply positing a primordial field as an initial condition poses severe problems for the successful standard cosmological model. If we restrict ourselves to magnetic fields generated by the stresses that accompany the formation of a galactic disk, then we obtain large scale seed fields in the range 10^{-18} to 10^{-19} Gauss (Lazarian 1992; Kulsrud, Cen, Ostriker and Ryu 1997) by invoking the Biermann battery in a realistic protogalaxy (Biermann 1950; for an exposition in English see Kemp 1982). This implies about 30 e-foldings of growth up to the present day, or a galactic dynamo growth rate which is no less than ten percent of the local shear rate. Since the current epoch in the history of our galaxy is unlikely to be special, in the sense that the magnetic field is unlikely to have just reached equipartition with the gas pressure, we would prefer a dynamo growth rate comfortably above this minimum.

For accretion disks we face a major observational difficulty. The magnetic field inside an accretion disk is completely unobservable. However, there are indirect constraints on the magnetic field strength. The luminosity of an accretion disk depends on the mass transport through the disk, and indirectly on the average radial velocity of the disk material. This is related to the dimensionless 'viscosity' α by

$$V_r \approx \alpha \frac{c_s^2}{r\Omega}. \tag{1}$$

When a magnetic field is present, local instabilities in the field (see below) imply

$$\alpha \approx \frac{B_r B_\theta}{4\pi P} \propto \frac{V_A^2}{c_s^2}. \tag{2}$$

In other words, the efficiency of radial mass transport is a measure of the ratio of magnetic pressure to gas pressure in an accretion disk.

For stationary systems this does not allow us to constrain the mean magnetic field, but the evolution of time varying systems is sensitive to the actual value of α. In particular, a variety of systems, including dwarf novae and X-ray novae, undergo recurrent transitions between hot, ionized, luminous states and cold, mostly neutral quiescent states. The luminous outburst state is marked by a relatively high mass flux through the disk while the quiescent state transfers little mass through the disk. Consequently, each system undergoes a thermal limit cycle, in which material accumulated near the outer edge of the disk during quiescence is spread through the disk, and onto the central object, during an outburst (for a review see Cannizzo 1993). Typical bright outbursts are marked by a fast rise and exponential decay. The rise marks appearance and spread of the hot state, typically starting far from the central object. The decay corresponds to the reappearance of the cold state, typically near the outer edge, and the subsequent progress of a cooling front to small radii. The duration of the outburst is sensitive to the rate at which a significant fraction of the total disk mass can be deposited on the central star, and therefore is a direct measure of α_{hot}, the average value of α in the hot state. Conversely, the duration of a quiescent phase is a measure of how much mass can be accumulated without forcing the disk into outburst, and is therefore a measure of α_{cold}. Finally, the shape of the luminosity decay at the end of an outburst is a measure of how the cooling front velocity depends on radius. All of this data can be fit by taking

$$\alpha \sim 35 \left(\frac{c_s}{r\Omega}\right)^{3/2}, \tag{3}$$

which also fits the difference in time scales between black hole candidate systems, with a central mass $\sim 7 M_\odot$, and white dwarf systems (Cannizzo, Chen, and Livio 1995; Vishniac and Wheeler 1996). The ratio $c_s/(r\Omega)$ is not necessarily a sign that the orbital velocity of the disk material is directly connected to the dynamo rate. This is also the ratio of the disk height to radius and may have a purely geometric origin.

3. LOCAL MAGNETOHYDRODYNAMIC INSTABILITIES IN DISKS

In a purely hydrodynamic disk, i.e. when no magnetic field is present, there are no local instabilities aside from those induced by self-gravity or tidal effects from a companion. This encourages us to treat the evolution of a magnetic field in a smooth background. The dispersion relation is

$$\left[1 - \frac{(x^2 + 3x_A^2)}{(x^2 - x_A^2)^2(1+\kappa^2)} + \frac{9}{2}\frac{\kappa^2 x_A^2}{(1+\kappa^2)x^2(x^2 - x_A^2)}\right]u_r =$$
$$= \frac{9}{4}\frac{\kappa^2}{1+\kappa^2}\partial_x^2 u_r + \frac{9}{2}\frac{\kappa^2}{1+\kappa^2}\frac{x_A^2}{(x^2 - x_A^2)x}\partial_x u_r, \tag{4}$$

for radial scales $\ll r$ and ignoring the vertical structure of the disk (Vishniac and Diamond 1992; Matsumoto and Tajima 1995). In this equation

$$x \equiv \frac{\bar{\omega}}{\Omega} = \frac{\omega}{\Omega} + k_\theta r, \tag{5}$$

$$x_A \equiv \frac{\omega_A}{\Omega} = \frac{\vec{k}\cdot\vec{B}}{(4\pi\rho)^{1/2}\Omega}, \tag{6}$$

$$\kappa \equiv \frac{k_\theta}{k_z}, \qquad (7)$$

and u_r is the radial velocity perturbation. The frequency $\bar{\omega}$ is the frequency measured by an observer rotating with the local fluid speed and ω is the frequency measured by an external observer. Since ω is a global quantity, while the dynamics of the perturbation are determined by $\bar{\omega}$, which is a function of radius, the radial dependence cannot be generally assumed to be described by some radial wavenumber. Here we have taken advantage of the radial dependence of $\bar{\omega}$ to use x as a radial coordinate.

In the axisymmetric limit this expression gives an instability. It is less obvious when $k_\theta \neq 0$ but this instability is generally present. It was first discovered by Velikhov (1959), and independently by Chandrasekhar (1961), and first applied to accretion disks by Balbus and Hawley (1991). Physically it is related to the famous tethered satellite experiment, except that it works. If magnetic field lines in the vertical or azimuthal direction are perturbed radially, then gas at smaller radii can transfer angular momentum outward to the slower moving gas on the same field line. This works whenever Ω increases inward while specific angular momentum increases outward. In a accretion disk the large scale azimuthal field tends to dominate, so the non-axisymmetric case is the most important. One additional subtlety is that local nonaxisymmetric disturbances do not correspond to global linear modes, and only grow $\sim k_z/k_\theta$ e-foldings before dissipating, but this is sufficient to ensure local instability in any practical sense of the phrase.

Our expectation, based on this linear dispersion relation, is that the dominant modes will have growth rates comparable to Ω, and wavelengths of roughly V_A/Ω in all directions.

4. NUMERICAL SIMULATIONS

Linear theory gives us some understanding of the driving force behind the transition to turbulence, and consequently a set of dimensional estimates for the nature of the turbulent regime. However, any hope of obtaining a quantitative understanding of real systems has to rest with numerical simulations. A number of groups have attempted simulations of the growth of the Balbus-Hawley instability in accretion disks (see, for example Brandenburg, Nordlund, Stein, and Torkelsson 1996; Hawley, Gammie, and Balbus 1996; Stone, Hawley, Gammie, and Balbus 1996). While these simulations have not completely overcome the technical difficulties involved in following MHD turbulence over a broad dynamical range, they do show some common results which we can take as a guide in considering real accretion disks. Since the nature of the simulations may play a large role in the results, we need to consider their common elements. First, in order to reduce the problem to a manageable size, the disk is idealized as a fluid in a shearing flow, with a scale height which is comparable to its radial extent. Rather than simulate an entire annulus the usual procedure is to make the box periodic in the azimuthal direction, with a total length which is typically about 2π vertical scale heights. (Although there have been simulations with azimuthal lengths up to four times longer.)

What do the results look like? First, naive expectations based on linear theory appear to be correct. There is a transition to turbulence, with the scales expected from the linear analysis. The resulting eddies are moderately anisotropic with $\lambda_\theta > \lambda_r > \lambda_z$.

Second, the evolution of the magnetic field typically has two phases. At first the magnetic field strength grows exponentially, with a rate $\sim \Omega$. However, this growth involves short wavelength components of the field. When this phase saturates, a slower growth appears, in which the large scale field components acquire a substantial fraction of the total magnetic energy. This latter phase frequently includes large scale field reversals, with a frequency which is roughly comparable to the growth rate of the large scale field.

Third, at saturation the field typically shows $\langle v^2 \rangle$ a fraction of V_A^2, which is in turn a large fraction of c_s^2. We expect α to scale with $(V_A/c_s)^2$, but in practice it remains small, typically less than a percent. However, the value of α varies from one simulation to another and appears to increase with increasing numerical resolution. It is plausible to suppose that for realistic Reynolds numbers α would reach reasonable values, although this involves a considerable amount of extrapolation.

Finally, one of the more striking features of this work is that the results are not qualitatively different for simulations which include vertical stratification and those that simply confine the fluid in a box with periodic vertical boundary conditions. In other words, vertical symmetry breaking does not play an important role in the dynamo present in these simulations.

5. DYNAMO THEORY

5.1. Conventional $\alpha - \Omega$ Dynamos

What generates the large scale field in the simulations, or, for that matter, in astrophysical disks? The usual answer is to appeal to mean field dynamo theory. In the context of strongly shearing astrophysical disks, the evolution equations for the large scale field can be written in a simplified form, i.e.

$$\partial_t B_r \approx -\partial_z(\alpha_{\theta\theta} B_\theta) + \partial_z D_T \partial_z B_r, \quad (8)$$

and

$$\partial_t B_\theta \approx -\frac{3}{2}\Omega B_r + \partial_z D_T \partial_z B_\theta, \quad (9)$$

where D_T is the turbulent diffusivity and

$$\alpha_{\theta\theta} = \langle v_z \partial_\theta v_r - v_r \partial_\theta v_z \rangle \tau. \quad (10)$$

Here τ is the velocity correlation time. This formulation of mean field dynamo theory is referred to as the '$\alpha - \Omega$ dynamo, since the radial field is generated from the azimuthal field by helicity and the cycle is closed by the shearing of the radial field to create azimuthal field.

In order have a non-zero $\alpha_{\theta\theta}$ we need to have some systematic violation of symmetry with respect to the $\hat{z}$ direction. The same is also true for radial and azimuthal motions, but coriolis forces can be relied upon to generate correlations between motions and gradients in these two directions. Vertical symmetry breaking requires the presence of vertical stratification. However, as we saw in the last section, this does not play a crucial role in the simulations. Whatever dynamo is operating in them is indifferent to whether or not $\alpha_{\theta\theta} = 0$.

Notwithstanding this point, there have been several attempts to derive a dynamo theory for accretion disks using magnetic field buoyancy, or more specifically, the Parker instability (see, for example Tout and Pringle 1992). These models all face a basic theoretical problem. The growth rate for the Parker instability is of order $(V_A/c_s)\Omega$ with the fastest growing modes having azimuthal wavelengths similar to the pressure scale height, or c_s/Ω. Shearing constraints imply that the corresponding radial wavelengths are of order V_A/Ω, which is also the typical radial scale for the Balbus-Hawley instability. Consequently, rising and falling sections of the magnetic field are mixed at a rate $\sim \Omega$. Unless the magnetic field is already strong (i.e. $V_A \sim c_s$) this is much faster than the growth rate of the Parker instability. In fact, numerical simulations with vertical stratification show little sign of the Parker instability, even when V_A is large.

5.2. Incoherent and Chaotic Dynamos

What are the alternatives to the standard $\alpha - \Omega$ dynamo? One idea is that the magnetic field is sustained through a local, chaotic process, in which local field stretching amplifies the field up to equipartition with the ambient pressure. This picture was originally suggested by Batchelor (1950), although the first detailed treatment is due to Kazantsev (1967). It can be rigorously justified only in the limit of a weak magnetic field, which is never the case when the turbulence itself is driven by the field. In any case, if we accept this possibility in accretion disks then the large scale field would then be explained as the result of some sort of inverse cascade within a turbulent fluid. This model is not consistent with accretion disk phenomenology, in particular the thermal limit cycle and the decay from outburst of dwarf novae and soft X-ray transients mentioned above. It is also unclear why the very largest scales, with wavelengths equal to several eddy scales, always end up with a significant fraction of the total power.

An alternative explanation is that the large scale field is generated by an extension of the $\alpha - \Omega$ dynamo developed by Vishniac and Brandenburg (1997) called the 'incoherent dynamo'. In the simplest version of this model the vertical symmetry is assumed to be unbroken, so that $\langle \alpha_{\theta\theta} \rangle = 0$. However, at any moment a magnetic domain containing N eddies will have a helicity of

$$\alpha_{\theta\theta} \sim \frac{\alpha_{\theta\theta,E}}{N^{1/2}}, \quad (11)$$

where $\alpha_{\theta\theta,E}$ is the helicity associated with a single eddy, which is comparable to the turbulent eddy velocity, V_T. In this case equation (8) can be written as a stochastic equation. It is also helpful to rewrite it in terms of the evolution of $\langle B_r^2 \rangle$ or

$$\partial_t \langle B_r^2 \rangle = 2\langle [\partial_z(\alpha_{\theta\theta} B_\theta)]^2 \rangle \tau - 2D_T \langle (\partial_z B_r)^2 \rangle. \quad (12)$$

(Here the brackets denote only spatial averaging.) Combining equations (9) and (12) we can estimate the incoherent dynamo growth rate as

$$\gamma \approx \left(\frac{\langle \alpha_{\theta\theta,E}^2 \rangle \tau \Omega^2}{L_z^2 N}\right)^{1/3}, \quad (13)$$

where L_z is the vertical height of a magnetic domain. This growth is a combination of random walk in B_r, driven by fluctuations in B_θ, and the shearing of B_r. The fact that it gives exponential growth results from a tendency for the distribution of B_r/B_θ to be biased towards negative numbers. When this ratio becomes sufficiently positive the field undergoes a sudden reversal and B_θ switches sign. Typically non-axisymmetric domains are sheared out faster than they can grow, so the usual expression for N in isotropic turbulence will be

$$N \approx \frac{L_z L_r 2\pi r}{\lambda_T^3}. \tag{14}$$

Incoherent dynamos are intrinsically noisy. In addition to the field fluctuations on eddy scales, the large scale field will undergo spontaneous field reversals with a frequency not far below the dynamo growth rate. Furthermore, the coupling between different domain scales implies that there is constant 'crosstalk' between different Fourier modes of the large scale magnetic field. Consequently, there are no well-defined linear eigenfunctions of this dynamo. Since individual annuli switch polarity on a regular basis, there seems little chance that the disk magnetic field will become uniform on radial scales larger than a few disk scale heights. Furthermore, this will reduce the strength of any large scale poloidal field produced via magnetic buoyancy. Different disk annuli will contribute randomly to any global field.

Finally, if we compare the growth rate γ to the dissipation rate, $\sim V_T^2 \tau / L_z^2$, we see that the largest vertical scale domains will accumulate most of the energy. (This line of reasoning can't be used to argue for larger radial scales since extending magnetic domains radially will lower the growth rate while leaving the dissipation rate unchanged.)

6. THE INCOHERENT DYNAMO IN ASTROPHYSICAL DISKS

6.1. Accretion Disks

If we wish to apply the incoherent dynamo to accretion disks then the obvious source of small scale turbulence is the Balbus-Hawley instability. Aside from the point that this is the only source of turbulence which is guaranteed to accompany a successful dynamo, only very strong convection is likely to survive the turbulent mixing caused by magnetic field instabilities. In this case we can write the dynamo growth rate as

$$\gamma \sim \left[\left(\frac{V_A}{c_s} \right)^5 \frac{H}{r} G(\beta) \right]^{1/3}, \tag{15}$$

where $G(\beta)$ describes the saturation of this mechanism as the ratio of magnetic to ambient pressure (β^{-1}) approaches unity. Here I have assumed that the magnetic domain is about as thick and wide as a disk vertical scale height. The dissipation rate is proportional to $\beta^{-1}\Omega$. This implies that the growth rate of the dynamo scales as the magnetic field strength to the 5/3, while the dissipation rate scales as the magnetic field strength squared. Consequently, the saturated state will be sensitive to other aspects of the model, including numerical viscosity in the computer simulations.

We can get a sense of how this works for accretion disks by assuming

$$\alpha = \frac{1}{3} \left(\frac{V_A}{c_s} \right)^2, \tag{16}$$

and

$$G = 1 - \frac{V_A^2}{c_s^2}. \tag{17}$$

Both of these are meant to be illustrative rather than serious predictions, however they have roughly the properties we expect for the exact solution. The function G should cut off sharply as $V_A \to c_s$, since in this limit the Balbus-Hawley instability disappears. Furthermore, for $V_A \ll c_s$ we expect G to have a leading order correction term of order β^{-1}. The scaling of α is roughly consistent with the numerical simulations, but a bit on the high side, reflecting our expectation that current simulations tend to underestimate its value. Balancing dynamo growth and turbulent dissipation we find that

$$\alpha = \frac{C_0^6}{3} \left(\frac{c_s}{r\Omega} \right)^2 (1 - 3\alpha)^2. \tag{18}$$

The value of C_0 is difficult to estimate, and in any case is raised to such a high power that it has to be regarded as essentially a free parameter.

Applying equation (18) to real disks requires us to fit to phenomenological models of dwarf novae and soft X-ray transients. If we take $C_0 \sim 3$ then we can produce an acceptable fit to equation (3). In this case we find that for values of $c_s/(r\Omega)$ between 1 and 1/4 the predicted value of α drops from 0.32 to 0.29, i.e. negligibly. For values of $c_s/(r\Omega)$ more appropriate for dwarf novae systems, in the range 0.04 to 0.025, α drops to the range 0.15 to 0.1, with a slope with respect to $c_s/(r\Omega)$ of 0.75 to 1. Finally, if we take $c_s/(r\Omega)$ down to one percent, as expect for soft X-ray transients, we get $\alpha \approx 0.03$ with a slope of $\sim 5/3$. These values and slopes are consistent with models of these systems and with the results of computer simulations. The extremely weak response of α to changes in the disk height to radius ratio when that ratio is not extremely small seems a bit odd. However, it is mostly the result of taking $C_0^6/3$ large, which is required by the thin disk models. A considerably smaller contribution to this effect comes from the sharp cutoff in G as $V_A \to c_s$. Both of these effects are intrinsic to the incoherent dynamo model and would be expected in any phenomenologically acceptable version of the model.

6.2. Galactic Disks

Aside from the differences already noted between galactic and accretion disks, there is another point which is critical for any application of the incoherent dynamo to galactic disks. Since galactic magnetic fields start out weak, the scale of turbulence due to magnetic instabilities would have been small, and the incoherent dynamo would have been relatively ineffective. In order to have a strong dynamo from very early times we need to appeal to other sources of turbulent motion. In the case of a galactic disk, one plausible source would be local gravitational instabilities. Another might be violent outflows from star forming regions. In either case it is difficult to assign length scales and velocities from first principles.

Suppose we take the point of view that the kinds of motions present at early times were not very different from what we see today. If we take

$$V_T \sim 10 \text{ km/sec}, \quad (19)$$

$$L_T \sim 300 \text{ parsecs}, \quad (20)$$

and assume a magnetic field vertical scale of 1 kpc, then we get a growth rate of

$$\gamma \sim 10^{-16} \text{ sec}^{-1}, \quad (21)$$

with a slightly smaller dissipation rate. This estimate is just marginally fast enough, but ignores factors of order unity, which are bound to be important in this case. The only conclusion we can draw from this exercise is that it is *possible* that the incoherent dynamo is responsible for the growth of large scale galactic fields, but any real answer will require a firmer understanding of turbulence in the Galactic disk. On the other hand, the incoherent dynamo does make a testable prediction. Since the growth time is only slightly less than the reversal time, and since the typical magnetic domain has a radial extent comparable to the disk thickness, it follows that we should expect the large scale B_θ to reverse over radial scales slightly larger than the disk thickness. This is consistent with current observations (see references in Zweibel and Heiles 1997), but the number of galaxies with observed reversals is still very small.

7. SUMMARY

We note several points in conclusion. First, disk dynamos do not require an average fluid helicity. They may require a mean square helicity, but this is a by-product of turbulence in general.

Second, incoherent dynamo effects match phenomenological constraints on accretion systems. They are not inconsistent with numerical simulations, but are not yet clearly confirmed by such work. A clear signature of their presence would be a turn-down in the value of α in the limit of very long computational boxes.

Third, the incoherent dynamo may be relevant for the rapid growth of galactic fields. However, models are sensitive to assumptions about the properties of turbulence in galactic disks. The only clear prediction is that large scale field reversals should be common on radial scales of a kiloparsec or more.

Fourth, at odds with the general theme of this conference, it is difficult to find a major role for either fluid or magnetic helicity in simulations of disk dynamos, or, perhaps, inside real astrophysical disks. The interaction of the disk field with its environment may present a mechanism for the generation of magnetic helicity by disks (cf. R. Matsumoto's contribution to this volume).

Acknowledgments. The work presented here was supported in part by part NAG5-2773 and NSF grant AST-9318185 (ETV). I am grateful for a number of helpful discussions with A. Brandenburg and E. Zweibel as well as the hospitality of MIT and the CfA for the 1997-98 academic year.

REFERENCES

Balbus, S.A., and Hawley, J.F., A powerful local shear instability in weakly magnetized disks. I. - Linear analysis, *Astrophys. J.*, *376*, 214-233, 1991.

Batchelor, G.K., On the spontaneous magnetic field in a conducting fluid in turbulent motion, Proc. R. Soc. Lond. *A201*, 405-416, 1950.

Biermann, L., Uber den Ursprung der Magnetfelder auf Sternen und interstellaren Raum, Zs. Naturforschung A *5*, 65-71, 1950.

Brandenburg, A., A. Nordlund, R.F. Stein, and U. Torkelsson, The disk accretion rate for dynamo-generated turbulence, *Astrophys. J.*, *458*, 45-48, 1996.

Cannizzo, J.K., The Limit Cycle Instability in Dwarf Nova Accretion Disks, in *Accretion Disks in Compact Stellar Systems*, edited by J.C. Wheeler, pp. 6-40, World Scientific Press, Singapore, 1993.

Cannizzo, J.K., W. Chen, and M. Livio, The accretion disk limit cycle instability in black hole X-ray binaries, *Astrophys. J.*, *454*, 880-894, 1995.

Chandrasekhar, S. *Hydrodynamic and Magnetohydrodynamic Stability*, 652 pp., Oxford University Press, Oxford UK, 1961.

Hawley, J.F., C.F. Gammie, and S.A. Balbus, Local three dimensional simulations of an accretion disk hydromagnetic dynamo, *Astrophys. J.*, *464*, 690-703, 1996.

Kazantsev, A.P., Enhancement of a magnetic field by a conducting fluid, JETP *53*, 1806-1813, 1967.

Kemp, J.C., The Biermann mechanism and spontaneous field generation in stars, *Publ. A. S. P.*, *94*, 627-633, 1982.

Kulsrud, R.M., R. Cen, J.P. Ostriker, and D. Ryu, The protogalactic origin for cosmic magnetic fields, *Astrophys. J.*, *480*, 481-491, 1997.

Lazarian, A., Diffusion generated electromotive force and seed magnetic field problem, *Astron. & Astrophys.* *264*, 326-330, 1992.

Matsumoto, R., and T. Tajima, Magnetic viscosity by localized shear flow instabilities in magnetized accretion disks, *Astrophys. J.*, *445*, 767-779, 1995.

Stone, J.M., J.F. Hawley, C.F. Gammie, and S.A. Balbus, Three dimensional magnetohydrodynamical simulations of vertically stratified accretion disks, *Astrophys. J.*, *463*, 656-673, 1996.

Tout, C.A., and J.E. Pringle, A simple model for a magnetic dynamo, *Mon. Not. R. Astron. Soc.*, *259*, 605-612, 1992.

Vallée, J.P., Observations of the magnetic fields inside and outside the Milky Way, starting with globules (∼ 1 parsec), filaments, clouds, superbubbles, spiral arms, galaxies, superclusters, and ending with the cosmological universe's background surface (at ∼ 8 teraparsecs), *Fundamentals of Cosmic Physics* *19*, 1-89, 1997.

Velikhov, E.P., Stability of an ideally conducting liquid flowing between cylinders rotating in a magnetic field, *Sov. Physics - JETP Lett.* *36*, 1398-1404, 1959.

Vishniac, E.T., and A. Brandenburg, An incoherent $\alpha - \Omega$ dynamo in accretion disks, *Astrophys. J.*, *475*, 263-274, 1997.

Vishniac, E.T., and P.H. Diamond, Local MHD instabilities and the wave-driven dynamo in accretion disks, *Astrophys. J.*, *398*, 561-568, 1992.

Vishniac, E.T., and J.C. Wheeler, The speed of cooling fronts and the functional form of the dimensionless viscosity in accretion disks, *Astrophys. J.*, *471*, 921-929, 1996.

Zweibel, E.G., and C. Heiles, Magnetic fields in galaxies and beyond, *Nature* *385*, 131-136, 1997.

E.T. Vishniac, Department of Physics and Astronomy, Johns Hopkins University, 3400 N. Charles St., Baltimore MD 21218. (e-mail: ethan@pha.jhu.edu)

Jets from Magnetized Accretion Disks

Ryoji Matsumoto

Department of Physics, Faculty of Science, Chiba University, 1-33 Yayoi-Cho, Inage-ku, Chiba 263-8522, Japan

When an accretion disk is threaded by large scale poloidal magnetic fields, the injection of magnetic helicity from the accretion disk drives bipolar outflows. We present the results of global magnetohydrodynamic (MHD) simulations of jet formation from a torus initially threaded by vertical magnetic fields. After the torsional Alfvén waves generated by the injected magnetic twists propagate along the large-scale magnetic field lines, magnetically driven jets emanate from the surface of the torus. Due to the magnetic pinch effect, the jets are collimated along the rotation axis. Since the jet formation process extracts angular momentum from the disk, it enhances the accretion rate of the disk material. Through three-dimensional (3D) global MHD simulations, we confirmed previous 2D results that the magnetically braked surface of the disk accretes like an avalanche. Owing to the growth of non-axisymmetric perturbations, the avalanche flow breaks up into spiral channels. Helical structure also appears inside the jet. When magnetic helicity is injected into closed magnetic loops connecting the central object and the accretion disk, it drives recurrent magnetic reconnection and outflows.

1. INTRODUCTION

One of the most spectacular phenomena in cosmic plasmas is the formation of well collimated bipolar jets observed in active galactic nuclei (AGN), galactic superluminal sources (microquasars), binary stars such as SS433, and in star forming regions. The jet formation often accompanies rotating disks called accretion disks. Accretion disks are differentially rotating disks which are formed when matter with angular momentum accretes to the central object.

In active galactic nuclei and in galactic microquasars, the energy source of various activities (e.g., X-ray emission and jet formation) is believed to be the gravitational energy of matter accreting to the central black hole. In a jet-forming Seyfert galaxy NGC4258, the existence of subparsec scale Keplerian rotating disk was confirmed by using water masers [*Miyoshi et al.*, 1995]. Circumnuclear gas torus is also observed in active galactic nuclei such as NGC4261 [*Jaffe et al.*, 1993]. In protostars, the observations by Hubble space telescope revealed the existence of obscuring torus which is believed to be the protoplanetary disk. High velocity (~ 200 km/s), collimated, ionized bipolar jet emanates from the central region of the torus.

The most promising model of astrophysical jets is the magnetically driven jets from accretion disks. When an accretion disk is threaded by large-scale poloidal magnetic fields, centrifugal force and magnetic pressure can drive outflows (Figure 1a). Theory of steady, axisymmetric magnetohydrodynamic (MHD) outflows from accretion disks has been developed by many authors [e.g., *Blandford and Payne*, 1982; *Pudritz and Norman*, 1986; see *Kudoh and Shibata*, 1995, 1997 and

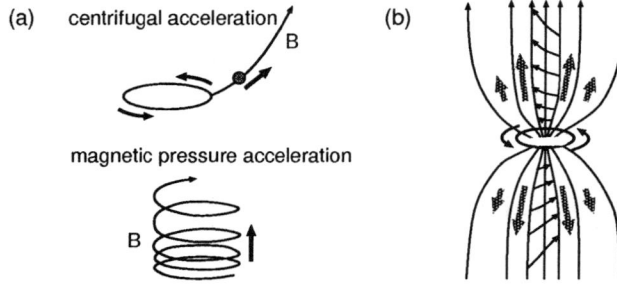

Figure 1. (a) A schematic picture of the driving mechanisms of magnetically driven jets. (b) The generation and relaxation of magnetic twists driven by the rotation of the disk. The outflows are collimated along the rotation axis due to the magnetic pinch effect.

references therein]. *Blandford and Payne* [1982] obtained a self-similar solution of the centrifugally driven wind ejected from a Keplerian disk. *Sakurai* [1987] obtained a self-consistent two-dimensional MHD jet solution ejected from an accretion disk without assuming self-similarity.

Nonlinear, time-dependent, two-dimensional (axisymmetric) MHD simulations of magnetically driven jets were first carried out by *Shibata and Uchida* [1985]. They showed that collimated jets are formed when magnetic twists accumulated in a local region of magnetized plasma relax by emitting torsional Alfvén waves which propagate along large-scale magnetic field lines. Subsequently, *Uchida and Shibata* [1985] and *Shibata and Uchida* [1986] combined this model with the magnetic twist injection from an accretion disk threaded by vertical magnetic field lines. They showed by two-dimensional (2D) MHD simulations that a bipolar jet is formed through the accumulation and relaxation of magnetic twists injected from the rotating disk (Figure 1b). The outflow is collimated along the rotation axis due to the pinch effect of the toroidal component of magnetic fields. The terminal speed of the jet generated by this mechanism was found to be the order of the Keplerian rotation speed of the disk. They called this mechanism as the "sweeping magnetic twist mechanism" and applied it to various astrophysical jets such as the bipolar flows in star forming regions [*Uchida and Shibata*, 1985; *Shibata and Uchida*, 1990] and to the Galactic center radio lobes [*Uchida, Sofue and Shibata*, 1985; *Shibata and Uchida*, 1987].

The numerical results by *Uchida and Shibata* [1985] and *Shibata and Uchida* [1986] have been confirmed by *Stone and Norman* [1994]. *Matsumoto et al.* [1996] applied this mechanism to jet formation from a geometrically thick disk in AGNs. *Kudoh, Matsumoto and Shibata* [1998] studied the dependence of the mass accretion rate and mass outflow rate on magnetic field strength.

The Uchida and Shibata's model of jet formation is intrinsically time-dependent because the disk gas accretes to the central object by losing angular momentum. In order to obtain steady state solutions through time dependent simulations, several authors have carried out MHD simulations by fixing the boundary conditions at the surface of the disk and by neglecting the effects of disk accretion due to magnetic braking. *Ustyugova et al.* [1995], *Romanova et al.* [1997] and *Ouyed and Pudritz* [1997] have carried out these two-dimensional simulations for many disk rotation periods and obtained steady-like solutions. On the other hand, *Ouyed, Pudritz and Stone* [1997] have shown that when the initial magnetic field is uniform and parallel to the rotation axis of the disk, outflows occur episodically. *Meier et al.* [1997] proposed a magnetic "switch", in which the outflow speed becomes much larger than the escape speed when the Alfvén speed exceeds the escape speed.

In these models, the surface conditions of the disk are given arbitrarily. However, the mass injection rate from the surface of the accretion disk, for example, is related to the global structure of magnetic field lines and velocity fields. The surface conditions needs to be determined self-consistently.

2. PROPERTIES OF STEADY AXISYMMETRIC MHD FLOW

In this section, we summarize the properties of axisymmetric MHD flows according to the paper by *Kudoh and Shibata* [1997]. They assume the shape of a poloidal magnetic field line, for simplicity. The equations of steady and axisymmetric MHD flow consist of five conservation equations along a stream line (i.e., a poloidal magnetic field line); (1) $P = K\rho^\gamma$, (2) $\rho v_p = \lambda B_p$, (3) $(v_\phi - \Omega r)B_p = v_p B_\phi$, (4) $r[v_\phi - B_\phi/(4\pi\lambda)] = L$, and the Bernoulli's equation

$$\frac{1}{2}v_p^2 + \frac{1}{2}v_\phi^2 + \frac{\gamma}{\gamma-1}\frac{P}{\rho} + \Psi_g - \frac{r\Omega B_\phi}{4\pi\lambda} = E,$$

where K, λ, Ω, L, and E are constants along a stream line, Ψ_g is the gravitational potential, γ the adiabatic index, and ρ, P, v_p, v_ϕ, B_p, and B_ϕ are density, pressure, and poloidal and toroidal components of the velocity and magnetic field, respectively. The gravitational potential is taken as that of the central star, i.e.,

$\Psi_g = -GM/(r^2+z^2)^{1/2}$ where M is the mass of the central star, G is the gravitational constant, and z is the height from the equatorial plane. The Alfvén radius r_A, where the poloidal velocity equals $B_p/(4\pi\rho)^{1/2}$ is related to the angular momentum as $L = \Omega r_A^2$.

Steady wind solutions can be obtained by requiring that they should pass through the slow and fast magnetosonic points at which the poloidal flow speed equals to these magnetosonic speeds. In the following, we use the non-dimensional parameters $E_{th} = (a_0/V_{K0})^2/\gamma$ and $E_{mg} = (V_{Ap0}/V_{K0})^2$ where the subscript zero denotes the value at the footpoint of the jet at $(r,z) = (r_0, 0)$, a_0 the sound speed, V_{Ap0} the poloidal Alfvén speed, and V_{K0} is the Keplerian rotation speed. When E_{mg} is small, the ratio of B_ϕ to B_p at the slow point becomes $|B_\phi/B_p|_s \propto E_{mg}^{-1/2} \gg 1$. The acceleration mainly takes place after the flow passes through the Alfvén point, whose radius is the typical scale length of the angular momentum transfer from the magnetic field to the fluid. This means that in the weak field regime, the magnetic pressure plays an important role in the acceleration of the flow. On the other hand, when the magnetic field is strong, acceleration mainly takes place between the slow point and the Alfvén point. In the strong magnetic field regime, the terminal velocity is roughly expressed as $v_\infty \sim r_A\Omega$.

Figure 2 shows the dependence of mass flux and terminal speed of the wind solutions on E_{mg}. When E_{mg} is large, the mass flux $\dot{M}$ tends toward a constant value which corresponds to $\rho a \Sigma$ at the slow point, where Σ is a cross section of a flux tube and a is the sound speed. When the magnetic energy is smaller, however, the mass flux tends toward $\dot{M} \propto E_{mg}^{0.5}$ [Kudoh and Shibata, 1995]. The dependence of the terminal velocity on E_{mg} can be obtained from the Michel's minimum energy solution $V_\infty = [\Phi^2\Omega^2/(4\pi\dot{M})]^{1/3}$ [Michel, 1969], where $\Phi = B_p\Sigma$. In non-dimensional form, $V_\infty/V_{K0} \sim (E_{mg}/\dot{m})^{1/3}$. Since the non-dimensional mass flux $\dot{m}$ is constant when E_{mg} is large, $V_\infty/V_{K0} \propto E_{mg}^{1/3}$. When E_{mg} is small, $V_\infty/V_{K0} \propto E_{mg}^{1/6}$ because $\dot{m} \propto E_{mg}^{1/2}$. These weak dependence of the terminal speed on E_{mg} explains why the terminal speed in nonsteady simulations of magnetically driven jets is the order of the Keplerian rotation speed for wide range of parameters.

3. ACCRETION AVALANCHES AND THE JET FORMATION

In this section we present typical results of two-dimensional MHD simulations of nonsteady jets [Matsumoto et al., 1996]. We assume that a rotating poly-

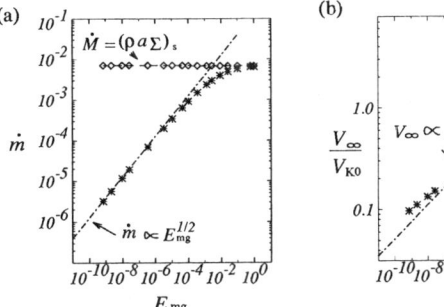

Figure 2. The dependence of (a) mass flux of the jet $\dot{m} = \dot{M}/(\rho_0 V_{K0} \Sigma_0)$, and (b) the terminal speed V_∞ on magnetic energy when $E_{th} = 1.0 \times 10^{-3}$ and $\Theta \simeq 53°$, where Θ is the minimum angle between the disk surface and the magnetic field line.

tropic torus with constant angular momentum distribution $L = L_0$ is imbedded in a spherical, non-rotating isothermal halo. The gravitational field is assumed to be given by a point mass M. In a cylindrical coordinate (r, φ, z), the dynamical equilibrium of the disk is described by $\Psi_g + L_0^2/(2r^2) + (n+1)P/\rho = const.$ where Ψ_g is the gravitational potential, and n is the polytropic index. We take the radius of the pressure maximum of the disk $[r = L_0^2/(GM)]$ as the reference radius r_0.

The initial magnetic field is assumed to be uniform and vertical. The vertical magnetic field assumed here is either a part of the large-scale poloidal magnetic field brought in from the interstellar space during the formation process of the central object and torus, or the central part of the dynamo-generated global magnetic field system. We use the normalization $r_0 = V_{K0} = \rho_0 = 1$, where V_{K0} is the Keplerian rotation speed at $r = r_0$. The halo parameters are $1/\alpha = C_{sh}^2/(\gamma V_{K0}^2)$ and ρ_h/ρ_0 where C_{sh} and ρ_h are the sound speed and density in the halo at $(r,z) = (0, r_0)$, respectively. We use $\alpha = 1.0$ and $\rho_h/\rho_0 = 10^{-3}$. We solved the ideal MHD equations in a cylindrical coordinate by using a modified Lax-Wendroff method [Rubin and Burstein, 1967] with artificial viscosity.

Figure 3 shows numerical results for a typical model (model B3 in Matsumoto et al., 1996) at $t = 2\pi r_0/V_{K0}$. The model parameters are $E_{th} = (a_0/V_{K0})^2/\gamma = 0.05$, $E_{mg} = (V_{AP0}/V_{K0})^2 = 10^{-3}$. The initial ratio of gas pressure to magnetic pressure $(\beta = P_{gas}/P_{mag})$ in the torus at $(r,z) = (r_0, 0)$ is $\beta_0 = 100$. The plasma β in the halo at $(r,z) = (0, r_0)$ is $\beta_h = 2.0$. After the torsional Alfvén wave generated by the rotation of the disk propagates into the corona, the surface layer of

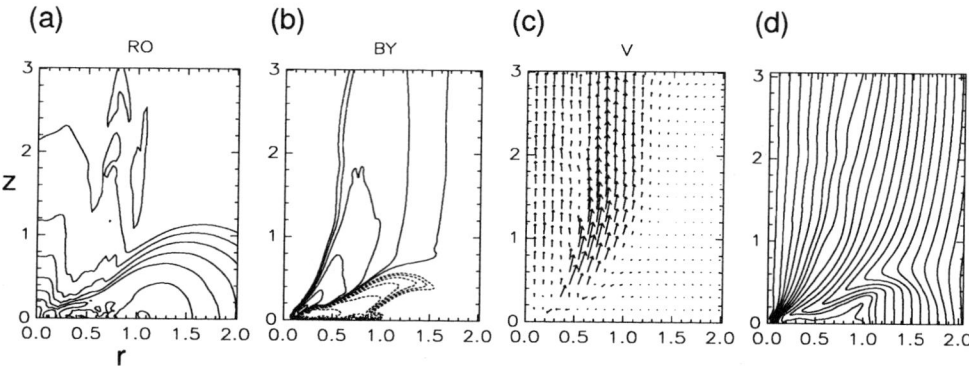

Figure 3. A result of 2.5D MHD simulation of a typical model at $t = 2\pi r_0/V_{K0}$. (a) Isocontours of density. (b) Isocontours of toroidal magnetic field component. (c) Velocity vectors. (d) Magnetic field lines.

the torus loses angular momentum and infalls like an avalanche. Subsequently, the cold material in the disk surface is accelerated and ejected as a bipolar jet. The outflow is collimated along the rotation axis due to the toroidal pinch effect. The maximum speed of the jet is $V_{max} = 1.7V_{K0}$.

The avalanching motion which appear in our simulation can be considered as the global version of the "two channel flow" which appeared in the nonlinear stage of the magneto-rotational (or Balbus and Hawley) instability [*Balbus and Hawley*, 1991; *Hawley and Balbus*, 1992]. The wiggling of magnetic field lines inside the torus (see Figure 3d) is also due to the growth of the Balbus and Hawley instability. The relation between the Balbus and Hawley instability and magnetic braking has been discussed by *Stone and Norman* [1994] and by *Matsumoto et al.* [1996].

Recently, *Kudoh et al.* [1998] carried out 2D MHD simulations of nonsteady jets from a torus by using a newly developed CIP-MOCCT code. They compared the numerical results with steady solutions and discussed the ejection mechanism of a jet in nonsteady MHD simulations. The model parameters are the same as those in model B3. The top panel of Figure 4 shows the trajectories of Lagrangian fluid elements along a magnetic field line. The test particles initially located above a open circle move out as an outflow. The bottom panels of Figure 4 show the poloidal speed and the effective potential $\Psi_{eff} = \Psi_g - \Omega_F^2 r^2/2$ where Ω_F is the angular frequency of the magnetic field line which is defined as $\Omega_F = v_\phi/r - v_{p\|}B_\phi/(rB_p)$, where $v_{p\|}$ is the poloidal velocity component parallel to the poloidal field line. The Lagrangian fluid elements which are in the region of $d\Psi_{eff}/ds < 0$ (the fluid elements right-

side of the open circle) are ejected as an outflow. The poloidal velocity exceeds the slow magnetosonic speed near the maximum of the effective potential. The fluid elements are accelerated up to the Alfvén speed. These results indicate that the mechanism proposed by *Blandford and Payne* [1982] works also in our time dependent simulation, i.e., the centrifugal force along the poloidal field line accelerate the jet within an Alfvén radius. It should be noted, however, that the disk and jet obtained by our nonsteady simulations never reach steady sate; the structure of the disk and jet change in the time scale of Keplerian orbit. Nevertheless, the ejection mechanism in our simulation is essentially the same as that in the steady models.

4. MASS OUTFLOWS DRIVEN BY TWIST INJECTION INTO CLOSED MAGNETIC LOOPS

Even if a large scale open magnetic field does not exist, magnetically driven jets can be created if closed magnetic loops thread an accretion disk (Figure 5). Figure 6 shows a result of 2.5D axisymmetric resistive MHD simulation of an accretion disk rotating in the dipole magnetic field of the central star [*Hayashi et al.*, 1996]. Similar simulations have been carried out by *Miller and Stone* [1997] and *Goodson, Winglee and Böhm* [1997]. *Hayashi et al.* [1996] assumed anomalous resistivity which sets in when J/ρ exceeds a critical value, where J is the current density. As magnetic twists are injected from the rotating disk, the magnetic loops connecting the disk and the star begin to inflate. A current sheet is formed inside the expanding loops. Magnetic reconnection taking place in the current sheet creates an outgoing magnetic island and post flare loops. The magnetic reconnection is Petchek type because we can identify slow shocks. The released magnetic energy heats up the flaring plasma. The speed of hot plasmoid ejected by the reconnection is $2-5$ times the Keplerian rotation speed. Dense, cold, magnetically driven wind emanates from the surface of the disk along the partially open magnetic field lines threading the disk.

Observations by the ASCA satellite revealed that protostars are hard X-ray sources [*Koyama et al.*, 1996]. Furthermore, some protostars show hard X-ray flaring activities. The total energy released by a protostellar flare ($10^{35}-10^{36}$ erg) is 10^{5-6} times larger than solar flares. The size of the flaring region is estimated to be

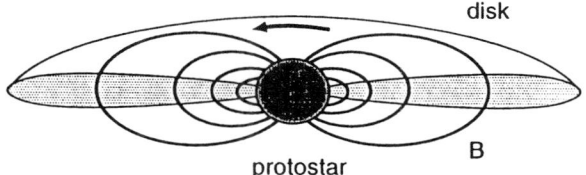

Figure 5. A schematic picture of the initial condition of 2.5D MHD simulations of the disk-star interaction. A rotating disk is threaded by the dipole magnetic field of the central star.

larger than the radius of the protostar. These characteristics of protostellar flares can be explained by our twist injection model. Figure 6(b) schematically shows numerical results applied to the star forming region. We can explain hard X-ray flares observed in protostars [*Koyama et al.*, 1996], optical jets, and high velocity neutral winds. Recently, *Tsuboi et al.* [1998] reported that in a protostar observed by ASCA, three hard X-ray flares occured recursively. Numerical simulations also indicate that the magnetic reconnection takes place intermittently because magnetic twists are continuously injected into the post flare loops connecting the central star and the disk.

Let us show the topological change of magnetic structure as a result of magnetic reconnection. Figure 7 shows three-dimensional structure of magnetic field lines after magnetic reconnection. The reconnected magnetic field lines in the magnetic island are detached from the central star and create a rotating spheromak which carry away magnetic helicity.

5. GLOBAL 3D MHD SIMULATIONS OF A TORUS THREADED BY VERTICAL MAGNETIC FIELDS

The disks and jets can subject to non-axisymmetric instabilities. *Curry and Pudritz* [1996] carried out a global linear analysis of a differentially rotating cylinder threaded by vertical magnetic fields and obtained growth rates for non-axisymmetric perturbations $\exp(im\varphi)$ where φ is the azimuthal angle. When the initial angular momentum is constant, low-m modes preferentially grow. On the other hand, in Keplerian disks, high-m modes dominate. Differentially rotating disks threaded by azimuthal magnetic fields also subject to non-axisymmetric instabilities. By local 3D MHD simulations of an accretion disk, *Hawley, Gammie and Balbus* [1995] showed that the two channel flow which appears in the nonlinear stage of the Balbus and Hawley

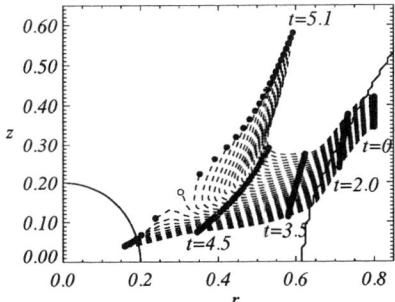
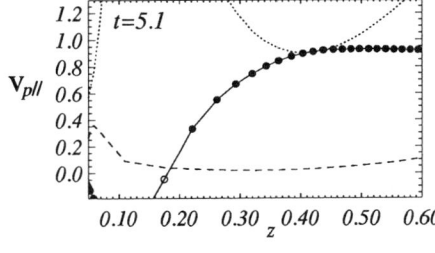
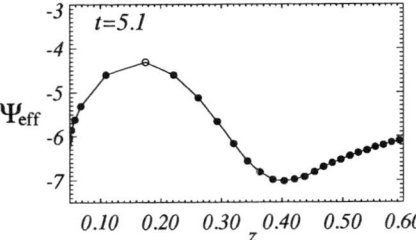

Figure 4. The top panel show the trajectories of Lagrangian test particles initially located on a magnetic field line. The bottom panel shows poloidal speed and effective potential of test particles at $t=5.1$. The dashed curve and the dotted curve show the slow magnetosonic speed and the Alfvén speed, respectively.

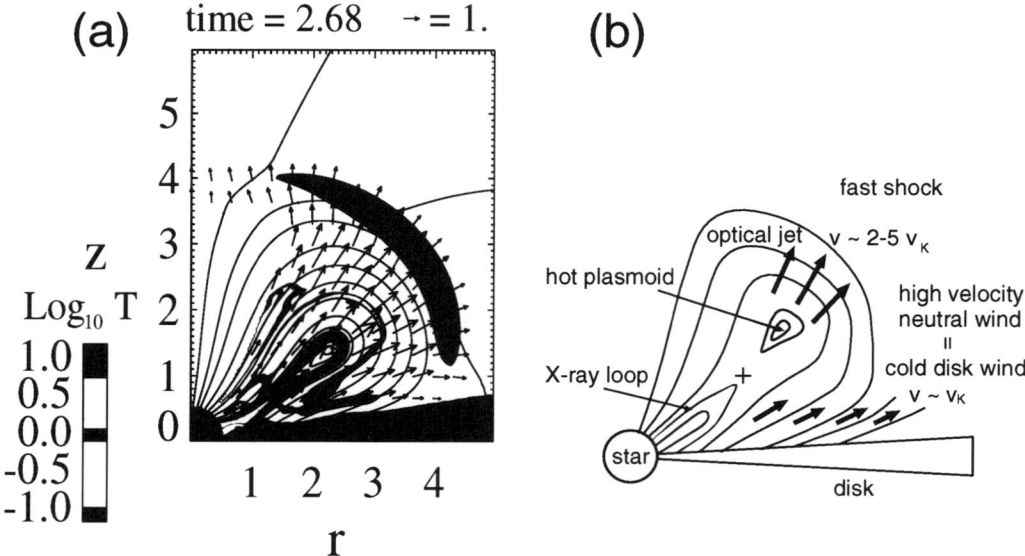

Figure 6. (a) The result of 2.5D MHD simulation of the interaction between the dipole magnetic field of the central star and a Keplerian disk. (b) A schematic picture of numerical results.

instability breaks up due to non-axisymmetric instabilities and generates turbulence in accretion disks. These results have been confirmed by *Matsumoto and Tajima* [1995] and *Brandenburg et al.* [1995]. The growth of non-axisymmetric instabilities inside the disk may also affect the structure of jets.

We extended the 2D cylindrical MHD code to 3D and carried out 3D simulations of jet formation from a torus. The model parameters are the same as those in model B3. We initiate the non-axisymmetric evolution by imposing perturbations for azimuthal velocity as $\delta v_\varphi = 0.01 v_\varphi \sin(m\varphi)$. Figure 8 shows numerical results when one armed ($m = 1$) perturbation is imposed. We confirmed the results of previous 2D axisymmetric simulations [*Matsumoto et al.*, 1996] that bipolar jet is formed and that the surface layer of the disk accretes faster than the equatorial part. The avalanche flow creates a radial component of magnetic fields which is further twisted by the differential rotation of the disk. The magnetic field lines at $t = 12.86 r_0/V_{K0}$ indicate that toroidal field components dominate inside the torus. The density isosurface shows that the dense region of the torus is deformed into a disk-shape. Due to the growth of non-axisymmetric instabilities, the magnetic field lines are bunched into helical bundles in the jet. Helical filamentary structures can also be seen in the density distribution of the jet.

Figure 9 shows the projected magnetic field lines and isocontours of β at $t = 11.4 r_0/V_{K0}$ when initially $m = 2$ perturbation is imposed. Inside the disk, accretion proceeds along spiral channels. In the innermost region of the disk where toroidal magnetic fields become dominant, spirally shaped, magnetic pressure dominated ($\beta < 1$) regions appear.

6. SUMMARY

We have shown through 2D and 3D MHD simulations that when an accretion disk is threaded by large scale poloidal magnetic fields, magnetically driven jets emanate from the surface of the disk. The outflows

Figure 7. Three dimensional structure of magnetic field lines after magnetic reconnection.

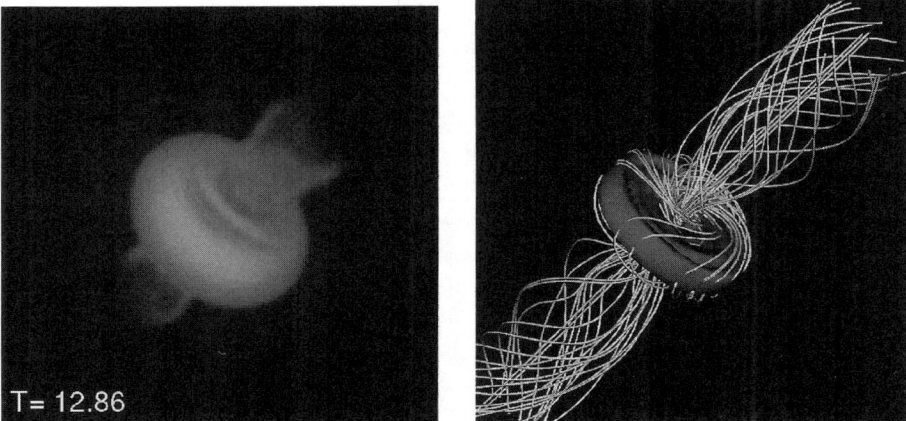

Figure 8. Results of 3D MHD simulation of a typical model with $m = 1$ perturbation. The left panel shows the volume rendered image of density distribution. The right panel shows magnetic field lines and isosurface of density.

are magnetically collimated along the rotation axis. Magnetized disks and jets can subject to global non-axisymmetric instabilities and local non-axisymmetric Balbus and Hawley instability. The 3D simulation results we presented here indicate that the avalanche flow breaks up into spiral channels due to the growth of non-axisymmetric modes. Spirally shaped, magnetic pressure dominated regions appear inside the disk. Since magnetic turbulence driven by the magnetic instabilities efficiently redistributes angular momentum inside the disk, a geometrically thick torus evolves toward a flattened, Keplerian accretion disk.

Acknowledgments. We thank Drs. K. Shibata, T. Kudoh, Y. Uchida and T. Tajima for discussion. Numerical computations were carried on Fujitsu VPP300/16R at National Astronomical Observatory, Japan. This work is supported in part by the Grant-in-Aid of the Ministry of Education, Science, Sports and Culture, Japan (07640348).

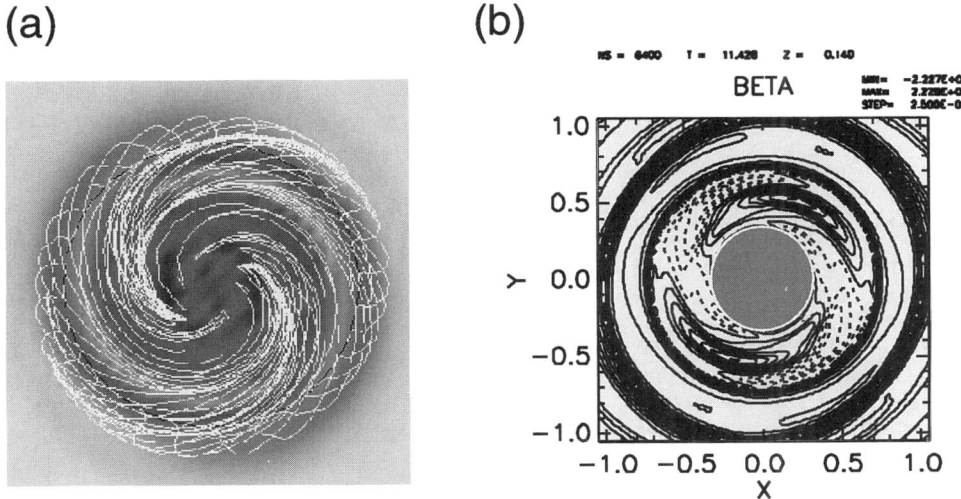

Figure 9. Results of 3D MHD simulations of a typical model with $m = 2$ perturbation. (a) Projection of magnetic field lines onto the equatorial plane at $t = 11.4 r_0/V_{K0}$. Gray scale shows density distribution. (b) Isocontours of β. Dashed curves show low-β region.

REFERENCES

Balbus, S.A., and J.F. Hawley, A powerful local shear instability in weakly magnetized disks. I - Linear analysis, *Astrophys. J.*, *376*, 214-233, 1991.

Blandford, R.D., and D.G. Payne, Hydromagnetic flows from accretion discs and the production of radio jets, *MNRAS*, *199*, 883-903, 1982.

Brandenburg, A., A. Nordlund, R. Stein, and U. Torkelsson, Dynamo-generated turbulence and large-scale magnetic fields in a Keplerian shear flow, *Astrophys. J.*, *446*, 741-754, 1995.

Curry, C. and R.E. Pudritz, On the global stability of magnetized accretion discs - III. Non-axisymmetric modes, *MNRAS*, *281*, 119-136, 1996.

Goodson, A.P., R.M. Winglee, and K.H. Böhm, Time-dependent accretion by magnetic young stellar objects as a launching mechanism for stellar jets, *Astrophys. J.*, *489*, 199-209, 1997.

Hawley, J.F., and S.A. Balbus, A powerful local shear instability in weakly magnetized disks. III - Long-term evolution in a shearing sheet, *Astrophys. J.*, *400*, 595-621, 1992.

Hawley, J.F., C.F. Gammie, and S.A. Balbus, Local three-dimensional magnetohydrodynamic simulations of accretion disks, *Astrophys. J.*, *440*, 742-763, 1995.

Hayashi, M.R., K. Shibata, and R. Matsumoto, X-ray flares and mass outflows driven by magnetic interaction between a protostar and its surrounding disk, *Astrophys. J.*, *468*, L37-L40, 1996.

Jaffe, W., H. Ford, L. Ferrarese, F. van den Bosch, and R.W. O'Connell, The nuclear disk of NGC 4261: Hubble space telescope images and ground-based spectra, *Astrophys. J.*, *460*, 214-224, 1996.

Koyama, K., S. Ueno, N. Kobayashi, and E.D. Feigelson, Discovery of hard X-rays from a cluster of protostars, *PASJ*, *48*, 87-92, 1996.

Kudoh, T., and K. Shibata, Mass flux and terminal velocities of magnetically driven jets from accretion disks, *Astrophys. J.*, *452*, L41-L44, 1995.

Kudoh, T., and K. Shibata, Magnetically driven jets from accretion disks. I. steady solutions and application to jets/winds in young stellar objects, *Astrophys. J.*, *474*, 362-377, 1997.

Kudoh, T., R. Matsumoto, and K. Shibata, Magnetically driven jets from accretion Disks. III. 2.5-dimensional non-steady simulations for thick disk case, *Astrophys. J.*, *508*, 186-199, 1998.

Matsumoto, R., and T. Tajima, Magnetic viscosity by localized shear flow instability in magnetized accretion disks, *Astrophys. J.*, *445*, 767-779, 1995.

Matsumoto, R., Y. Uchida, S. Hirose, K. Shibata, M.R. Hayashi, A. Ferrari, G. Bodo, and C. Norman, Radio jets and the formation of active galaxies: accretion avalanches on the torus by the effect of a large-scale magnetic field, *Astrophys. J.*, *461*, 115-126, 1996.

Meier, D.L., S. Edgington, P. Godon, D.G. Payne, and K.R. Lind, A magnetic switch that determines the speed of astrophysical jets, *Nature*, *388*, 350-352, 1997.

Michel, F.C., Relativistic stellar-wind torques, *Astrophys. J.*, *158*, 727-738, 1969

Miller, K.A., and J.M. Stone, Magnetohydrodynamic simulations of stellar magnetosphere-accretion disk interaction, *Astrophys. J.*, *489*, 890-902, 1997.

Miyoshi, M., J. Moran, J. Herrnstein, L. Greenhill, N. Nakai, P. Diamond, and M. Inoue, Evidence for a black-hole from high rotation velocities in a sub-parsec region of NGC4258, *Nature*, *373*, 127, 1995.

Ouyed, R., and R.E. Pudritz, Numerical simulations of astrophysical jets from Keplerian disks. I. stationary models, *Astrophys. J.*, *482*, 712-732, 1997.

Ouyed, R., R.E. Pudritz, and J.M. Stone, Episodic jets from black holes and protostars, *Nature*, *385*, 409-414, 1997.

Pudritz, R.E., and C.A. Norman, Bipolar hydromagnetic winds from disks around protostellar objects, *Astrophys. J.*, *301*, 571-586, 1986.

Romanova, M.M., G.V. Ustyugova, A.V. Koldoba, V.M. Chechetkin, and R.V.E. Lovelace, Formation of stationary magnetohydrodynamic outflows from a disk by time-dependent simulations, *Astrophys. J.*, *482*, 708-711, 1997.

Rubin, E.L., and S.Z. Burstein, Difference methods for the inviscid and viscous equations of a compressible gas, *J. Comp. Phys.*, *2*, 178-196, 1967.

Sakurai, T., Magnetically collimated winds from accretion disks, *PASJ*, *39*, 821-835, 1987.

Shibata, K., and Y. Uchida, A magnetodynamic mechanism for the formation of astrophysical jets. I - Dynamical effects of the relaxation of nonlinear magnetic twists, *PASJ*, *37*, 31-46, 1985.

Shibata, K., and Y. Uchida, A magnetodynamic mechanism for the formation of astrophysical jets. II - Dynamical processes in the accretion of magnetized mass in rotation, *PASJ*, *38*, 631-660, 1986.

Shibata, K., and Y. Uchida, A magnetodynamical model for the Galactic center lobes, *PASJ*, *39*, 559-571, 1987.

Shibata, K., and Y. Uchida, Interaction of molecular bipolar flows with interstellar condensations - Sweeping magnetic twist mechanism and the blobs in Lynds 1551 molecular flow, *PASJ*, *42*, 39-67, 1990.

Stone, J.M., and M.L. Norman, Numerical simulations of magnetic accretion disks. *Astrophys. J.*, *433*, 746-756, 1994.

Tsuboi, Y., K. Koyama, K. Kamata and S. Yamauchi, ASCA observations of class I protostars in the Rho Oph dark cloud, *The hot universe, eds. K. Koyama, S. Kitamoto, and M. Itoh, Kluwer Academic, Dordrecht*, 236, 1998.

Uchida, Y., and K. Shibata, Magnetodynamical acceleration of CO and optical bipolar flows from the region of star formation, *PASJ*, *37*, 515-535, 1985.

Uchida, Y., Y. Sofue, and K. Shibata, Origin of the galactic centre lobes, *Nature*, *317*, 699-701, 1985.

Ustyugova, G.V., A.V. Koldoba, M.M. Romanova, V.M. Chechetkin, and R.V.E. Lovelace, Magnetohydrodynamic simulations of outflows from accretion disks, *Astrophys. J.*, *439*, L39-L42, 1995.

R. Matsumoto, Department of Physics, Faculty of Science, Chiba University, 1-33 Yayoi-Cho, Inage-Ku, Chiba 263-8522, Japan. (e-mail: matumoto@c.chiba-u.ac.jp)

Magnetic Helicity in Space and Laboratory Plasmas: Editorial Summary

Michael Brown[1], Richard Canfield[2], George Field[3], Russell Kulsrud[4], Alexei Pevtsov[2], Robert Rosner[5], Norbert Seehafer[6]

A summary of discussions at the Chapman Conference on Magnetic Helicity in Space and Laboratory Plasmas is presented. Attention is focussed on the role magnetic helicity plays in four magnetofluid systems: the Sun, the solar wind, astrophysical dynamos and laboratory plasmas.

1. INTRODUCTION: UTILITY OF THE HELICITY CONCEPT

As evidenced by several discussions at the conference, the concept of helicity plays an important role in the understanding of several magnetofluid systems. The overarching theme in all of these applications is that magnetic helicity H_M appears to be more "rugged" than magnetic energy E_M in the presence of small scale turbulence and dissipation. One argument presented is that since $\dot{H}_M \sim k$ whereas $\dot{E}_M \sim k^2$ (where k is a measure of the inverse scale length), dissipation at the smallest scales is more effective on the magnetic energy. A related argument comes from the Schwartz triangle inequality $E(k) \geq kH(k)$ which states that for a given amount of helicity H(k) it takes less energy to put the helicity at small k (or at large scales, see Berger, this volume). Put another way, conservation of magnetic helicity prevents the magnetic energy from vanishing. A ramification of the ruggedness of helicity is the natural evolution of large scale helical structures. A problem with these arguments is that one might expect relaxation to occur slowly over a characteristic decay time for the system ($t_{L/R}$) whereas relaxation is observed both experimentally and computationally to occur rapidly in a few characteristic flow times of the system (t_{Alfven}).

Since helicity is closely linked to the magnetic topology of the system (helicity is related to the linking number = twist + writhe, see Berger, this volume) and is relatively well conserved, the concept of helicity is most useful in determining the final, large scale, "relaxed" state of a magnetofluid system. Helicity indirectly determines the orientation of magnetic field structures even in the presence of reconnection. While conservation of magnetic helicity is a useful guiding principle, it should not be a substitute for investigation into detailed magnetofluid dynamics.

In what follows, we summarize a consensus gauge invariant definition of helicity then summarize the role of helicity in four magnetofluid systems: the Sun, the solar wind, astrophysical dynamos and laboratory plasmas.

2. GAUGE INVARIANT DEFINITIONS OF HELICITY

In general, the helicity of a vector function is defined as the volume integral of the function dotted into its

[1] Department of Physics and Astronomy, Swarthmore College, Pennsylvania
[2] Department of Physics, Montana State University, Bozeman, Montana
[3] Harvard-Smithsonian Center for Astrophysics, Cambridge, Massachusetts
[4] Princeton Plasma Laboratory, Princeton, New Jersey
[5] Department of Astronomy and Astrophysics, University of Chicago, Illinois
[6] Institute of Physics, University of Potsdam, Germany

curl and is a measure of the extent to which the field wraps or coils upon itself. In particular, magnetic helicity is defined: $H_M = \int \mathbf{A} \cdot (\nabla \times \mathbf{A})\, d^3x = \int \mathbf{A} \cdot \mathbf{B}\, d^3x$ where $\mathbf{A}$ is the magnetic vector potential and $\mathbf{B}$ is the magnetic induction. In a closed system where no magnetic flux leaves the boundary this definition poses no problems. However, in open systems where flux enters and leaves the boundary, the definition needs to be more precise (since $\mathbf{A}$ is not gauge invariant, see Low, this volume). Like any potential function, the magnetic vector potential is not unique. For example, the gradient of any scalar function can be added to $\mathbf{A}$ with impunity since $\nabla \times \nabla f = 0$ (see Bellan, this volume). Perhaps the most natural re-definition involves using the helicity of a vacuum or potential field (with $\nabla \times \mathbf{B} = 0$) as a reference (see Berger, this volume). A gauge invariant form of the helicity can be written: $H_{rel} = \int (\mathbf{A} + \mathbf{A}_P) \cdot (\mathbf{B} - \mathbf{B}_P)\, d^3x$, where $\mathbf{B}_P$ is the vacuum magnetic field with the flux at the boundaries defined and $\mathbf{A}_P$ is the vector potential for $\mathbf{B}_P$. The potential field used as a reference is unique as long as $\mathbf{B}_P \cdot \hat{\mathbf{n}}$ is prescribed at the boundary.

The utility of the "current helicity" ($H_J = (\nabla \times \mathbf{B}) \cdot \mathbf{B} = \mathbf{J} \cdot \mathbf{B}$) was also discussed (mostly in solar contexts). Strictly speaking, H_J should be referred to as a current helicity "density" since it is not defined as an integral quantity. It happens that H_J is straightforward to measure on the solar surface and for an isolated single twisted flux tube it has the same sign as the self magnetic helicity of the flux tube. However, it should be emphasized that H_J is not a "rugged" invariant in the same sense that H_M is. In fact, H_J is related to the helicity *dissipation* rate when integrated over an appropriate volume $\dot{K} = -2 \int \eta \mathbf{J} \cdot \mathbf{B}\, d^3x$. Care should be taken in the interpretation of such quantities.

Finally, the notion of the total helicity was discussed stemming from the definition of the canonical momentum $\mathbf{p} = m\mathbf{v} + q\mathbf{A}$. The volume integral of this vector function dotted into its curl can be written: $\int (m\mathbf{v} + e\mathbf{A}) \cdot \nabla \times (m\mathbf{v} + e\mathbf{A})\, d^3x$ which generates four terms (including H_M, the kinetic helicity $H_K = \int \omega \cdot \mathbf{v}\, d^3x$ and the cross-helicity $H_C = \int \mathbf{v} \cdot \mathbf{B}\, d^3x$). Applying these additional constraints to the minimum energy problem will generate different states with larger energy than the simple "Taylor state". The kinetic helicity is found to play an important role in dynamo theories.

3. HELICITY CONSERVATION IN THE SUN

Twisted magnetic structures have been observed on the Sun for decades. There was agreement among conference attendees that magnetic helicity generation must ultimately be a consequence of the rotation of the Sun, acting through the Coriolis force. Despite the evidence presented that dynamos can exist without H_K, the consensus of the group was that the solar dynamo is a helical one. It is generally accepted that the solar dynamo produces twisted flux tubes which buoy to the surface. The sense of twist is left-handed in the northern hemisphere and right handed on the south. This handedness is observed in magnetograms as the loops emerge. Furthermore, this handedness persists in the observed twist of ejected flux in coronal mass ejections (CMEs) and even in the statistical fluctuations in the solar wind. This is a consistent picture that tends to rule out twist of pre-formed loops by shear in the photosphere (at least at the large scales we can observe with magnetograms).

Two independent data sets were presented that point to the conclusion that H_J has opposite sign in the two hemispheres (negative in the north and positive in the south) (see Longcope and Pevtsov, this volume). The data have significant scatter but the result is consistent with dynamo theory and with the net helicity in the solar wind above and below the ecliptic (see Matthaeus and Smith, this volume). In addition, the observation of sigmoidal (S-shaped) coronal structures allows qualitative discussion of H_J for coronal magnetic fields (see Canfield, this volume).

Naively, one would predict that the northern and southern hemispheres would produce mean fields of opposite helicity. Given that the magnetic helicity of an isolated system is conserved in the absence of resistivity, this would conform to the production of zero net helicity when integrated over the Sun. However, this could in principle be accomplished in different ways. One is that the two hemispheres are somehow coupled so that the production of positive helicity in the south is exactly balanced by the production of negative helicity in the north. This is analogous to twisting a stretched rubber band in the middle; a left-handed twist propagates up balanced by a right-handed twist which propagates down. Another possibility is that even without communication, each hemisphere produces exactly zero net helicity, the observed large-scale helicity being compensated by production of helicity of opposite sign at the smaller scales, where it can be destroyed by Ohmic dissipation. The majority of the group favored a model in which large-scale helicity is produced with opposite sign in the two hemispheres. This would assure that the helicity production in each hemisphere stays in phase with the solar cycle. Most everyone seemed to agree that observations of magnetic helicity on the Sun are useful in constraining the operation of the solar dynamo.

4. SOLAR WIND: STATISTICAL ASPECTS OF HELICITY

In the fully developed turbulence of the solar wind magnetofluid, a statistical description of magnetic helicity and energy is appropriate (see chapters by Matthaeus Smith, Bieber and Montgomery in this volume). In a spectral representation of a dynamic turbulent process ($H_M = \int H(k) d^3k$), magnetic helicity tends to be transported to longer scales (smaller k) by a process known as an inverse cascade. This tendency is strong enough that in a modified thermodynamic limit, the magnetic helicity condenses entirely into the longest wavelength mode, while the energy (except for the part of it tied up with the helicity) distributes itself evenly across all available scales (see Matthaeus, this volume). In the spirit of the Schwartz inequality argument noted above, magnetic helicity is the quantity that seeks the longest scale in the system.

Helicity is observed in the solar wind at all scales and with the same asymmetry as observed in the solar photosphere and corona (left-handed or negative in the north, right-handed or positive in the south). At the largest scales, the Parker spiral structure of the interplanetary magnetic field (IMF) contains helicity. The source of this large scale helicity is due to the freezing of the IMF in the radial outflow of the wind coupled with the rotation of the corona. Virtually all the flux generated by the Sun is ejected. Large scale CMEs ejected from the Sun after a violent reconnection process retain their helicity. The helicity of the ejected CME plays a role in the subsequent interaction with the earth's magnetosphere (see Rust and Wright, this volume).

At smaller scales, it is useful to define a helicity normalized to the maximum allowed by the Schwartz inequality $\sigma_K = k H_m(k)/E_B(k)$. Using magnetic data from several satellites spanning 34 years, a remarkable, persistent asymmetry can be shown in the net helicity of the IMF (see Smith, this volume). Since the transit time of the solar wind from Sun to earth is about 100 hours, structures correlated for times even as short as 10 hours are likely to be of solar origin. Because of this, measured helicity asymmetries in the solar wind are not likely due to dynamical processes such as an inverse cascade.

Another manifestation of the statistical aspect of helicity is in the scattering of cosmic rays off magnetic fluctuations in the solar wind. Magnetic helicity has a strong influence on particle scattering by turbulence, because a resonant interaction occurs only for magnetic fluctuations whose handedness matches the handedness of the helical particle orbit (see Bieber, this volume). Convincing evidence was presented for a polarity dependence of the ratio of cosmic helium to cosmic electrons due to magnetic helicity asymmetry.

5. ROLE OF HELICITY IN THE ASTROPHYSICAL DYNAMO PROBLEM

As we move into the realm of large scale astrophysical plasmas, the role of helicity is less clear. It may well be that magnetic helicity is less important in high β plasmas where the magnetic pressure is weak (like accretion disks and galactic dynamos) than in low β plasmas (like the solar corona and laboratory plasmas). In general, helicity is generated at large scales by differential rotation and at small scales by the alpha effect (generation of an electromotive force along a mean field by turbulence). Numerical evidence suggests that unlike solar/stellar dynamos, accretion disk dynamos do not require an average fluid helicity H_K nor is there a significant role for the magnetic helicity H_M. They may require a mean square helicity due to turbulence (see Vishniac, this volume). Other numerical simulations showed that when an accretion disk is threaded by large scale poloidal magnetic fields, magnetically driven jets with helical structure emanate from the surface of the disk (see Matsumoto, this volume).

6. HELICITY IN LABORATORY PLASMAS

It is in controlled laboratory plasmas that the concept of magnetic helicity has its most utility (see the chapters by Prager, Yamada, Ji, and Stenzel, this volume). There are several reasons for this. First of all, helicity is defined as an integral over an entire volume of interest ($H_M = \int \mathbf{A} \cdot \mathbf{B} \, d^3x$). It is in laboratory plasmas that (at least in principle) H_M can be measured everywhere. Secondly, the subtle issues of a gauge invariant definition of H_M are less critical in the laboratory setting. Generally, the plasma under study is entirely enclosed in a highly conducting boundary (with $\mathbf{B} \cdot \hat{\mathbf{n}} = 0$) which acts as a flux conserver and helicity barrier. If $\mathbf{B} \cdot \hat{\mathbf{n}} \neq 0$ then the normal flux is typically well known or straightforward to measure. Finally, and most importantly, laboratory plasmas are typically formed with particular amounts of initial magnetic energy, flux and helicity (flux linkage). The system is then free to evolve in relative isolation to a minimum energy state subject to the constraint of fixed helicity.

Magnetic helicity provides a constraint to the Taylor state in the following way. Turbulence, allied with small resistivity, allows the plasma rapid access (in a time short compared to the usual resistive diffusion time) to a minimum-energy force-free state. It can be shown (see Bellan, this volume) that if one minimizes the magnetic energy of the system $E_M = \int B^2 \, d^3x$ subject to the constraint that the H_M is fixed (using the technique of Lagrange multipliers), the resulting magnetic states satisfy the force-free condition $\nabla \times \mathbf{B} = \lambda \mathbf{B}$ (with $\lambda = E_M/H_M = J/B$ is the Lagrange multiplier). Given the caveat that Taylor's principle is only valid in those regions where the effects of resistivity (magnetic reconnection) occur, any initial configuration will self-organize to the relaxed state after sufficient time.

It was generally agreed that helicity conservation provides an excellent tool to determine final states in laboratory plasmas (when one isn't interested in the dynamics of evolution). RFPs and spheromaks consistently relax (at least partially) to a force-free or Taylor state (see Prager, this volume). Relaxation theory provides a simple and general method for determining the outcome of arbitrarily complicated dynamics.

M. Brown, Department of Physics and Astronomy, Swarthmore College, Swarthmore, PA 19081. (e-mail: mbrown3@swarthmore.edu)

R. Canfield, A. Pevtsov, Department of Physics, Montana State University, Bozeman, MT 59717-3840. (e-mail: canfield@physics.montana.edu; pevtsov@physics.montana.edu)

G. Field, Harvard-Smithsonian Center for Astrophysics, 60 Garden St., Cambridge, MA 02138. (e-mail: gfield@cfa.harvard.edu)

R. Kulsrud, Princeton Plasma Physics Laboratory, P.O. Box 451, Princeton, NJ 08543. (e-mail: rkulsrud@astro.princeton.edu)

R. Rosner, Department of Astronomy and Astrophysics University of Chicago. 5640 S. Ellis Ave., Chicago, IL 60637. (e-mail: rrosner@oddjob.uchicago.edu)

N. Seehafer, Institute of Physics, University of Potsdam, PF 601553, D-14415 Potsdam, Germany. (e-mail: seehafer@agnld.uni-potsdam.de)